SINKHOLES: THEIR GEOLOGY, ENGINEERING AND ENVIRONMENTAL IMPACT

PROCEEDINGS OF THE FIRST MULTIDISCIPLINARY CONFERENCE ON SINKHOLES
ORLANDO / FLORIDA / 15-17 OCTOBER 1984

SINKHOLES: THEIR GEOLOGY, ENGINEERING AND ENVIRONMENTAL IMPACT

Edited by
BARRY F. BECK
Florida Sinkhole Research Institute, University of Central Florida, Orlando

Sponsored by the
Florida Sinkhole Research Institute, College of Engineering,
University of Central Florida

A.A. BALKEMA / ROTTERDAM / BOSTON / 1984

Cover photographs: Development and repair of the catastrophic Winter Park Sinkhole, Florida, USA.
The sinkhole is estimated to have caused more than four million dollars in damages. Investigation and repair was undertaken by Jammal and Associates, Inc., Consulting Geotechnical Engineers.

*The texts of the various papers in this volume were set individually by typists
under the supervision of each of the authors concerned.*

ISBN 90 6191 570 8 cloth edition

ISBN 90 6191 571 6 paper edition

A.A.Balkema, P.O.Box 1675, 3000 BR Rotterdam, Netherlands
Distributed in USA & Canada by: A.A.Balkema Publishers, P.O.Box 230, Accord, MA 02018
Printed in the Netherlands

Table of contents

Sinkhole terminology *Barry F.Beck*	IX

1. *The geologic framework and mechanisms of sinkhole development*

Review of induced sinkhole development *J.G.Newton*	3
Florida karst – Its relationship to geologic structure and stratigraphy *Walter Schmidt & Thomas M.Scott*	11
Karst progression *Dennis J.Price*	17
Impact of ground-water chemistry on sinkhole development along a retreating scarp *Sam B.Upchurch & Fred W.Lawrence*	23
Sinkhole collapse induced by groundwater pumpage for freeze protection irrigation near Dover, Florida, January 1977 *Susan J.Metcalfe & Larry E.Hall*	29
Influence of a karstified limestone surface on an open-marine, marsh-dominated coastline: West Central Florida *Joan G.Hutton, Albert C.Hine, Mark W.Evans, Eric B.Osking & Daniel F.Belknap*	35
Seismic-reflection studies of sinkholes and limestone dissolution features on the northeastern Florida shelf *Peter Popenoe, F.A.Kohout & F.T.Manheim*	43
Structural and hydrogeologic applications of remote sensing data, eastern Yucatan Peninsula, Mexico *C.Scott Southworth*	59
A comparison of sinkhole depth frequency distributions in temperate and tropic karst regions *J.W.Troester, Elizabeth L.White & William B.White*	65
Sinkhole distribution in the central and northern Valley and Ridge province, Virginia *David A.Hubbard, Jr.*	75
Sinkhole distribution in Winona County, Minnesota *Janet Dalgleish & E.Calvin Alexander*	79
Tectonics and geology in karst development of Northern Lower Michigan *Tyrone J.Black*	87
Pattern and antiquity of sinkholes along an alluviated karstified valley: Friars Hole, West Virginia *S.R.H.Worthington & D.C.Ford*	93
Karst and subsidence in China *Zhang Shouyue*	97
A contour map, volume estimate, and description of Teague's Sinkhole *James J.Hollingshead*	105
Development, occurrence, and triggering mechanisms of sinkholes in the carbonate rocks of the Lehigh Valley, eastern Pennsylvania *Paul B.Myers, Jr. & Michael Perlow, Jr.*	111
Submarine 'sinkholes': A review *Marco Taviani*	117
A brief review of the South African sinkhole problem *Anthony B.A.Brink*	123

2. *Site studies and evaluation of sinkhole-susceptibility*

Catastrophic subsidence: Shelby County, Alabama *Philip E.LaMoreaux*	131
Sinkhole development in North-Central Puerto Rico *Mikolaj Wegrzyn, Alejandro E.Soto & Juan A.Pérez*	137

Collapse sinkholes in the blanket sands of the Puerto Rico karst belt 143
Alejandro E.Soto & Wanda Morales

Predicting the location of surface collapse within karst depressions: A Jamaican example 147
Michael Day

Investigation techniques on dolomites in South Africa 153
Peter W.Day & Fritz von M.Wagener

New Jersey sinkholes: Distribution, formation, effects, geotechnical engineering 159
Joseph A.Fischer & Richard W.Greene

Sinkhole risk analysis for a selected area in Warren County, New Jersey 167
D.Raghu & Charles Tiedeman

Use of percussion probes for the design and construction of foundations in and on carbonate formations 171
D.Raghu, J.J.Lifrieri & F.C.Rhyner

Methods for describing and predicting the occurrence of sinkholes 177
Albert E.Ogden

Factors affecting the collapse of cavities 183
Thomas F.Beggs & Byron E.Ruth

Relationship of modern sinkhole development to large scale-photolinear features 189
J.R.Littlefield, M.A.Culbreth, S.B.Upchurch & M.T.Stewart

Application of double Fourier series analysis to ground subsidence susceptibility mapping in covered karst terrain 197
Marcus J.W.Thorp & George A.Brook

Evaluation of subsidence or collapse potential due to subsurface cavities 201
Richard C.Benson & Lester J.La Fountain

Examination of sinkholes by seismic reflection 217
Don W.Steeples, Ralph W.Knapp & Richard D.Miller

Geophysical characteristics of fracture traces in the carbonate Floridan Aquifer 225
Mark Stewart & John Wood

Sinkhole prediction – Review of electrical resistivity methods 231
Eberhard Werner

3. *Sinkhole-like features (subsidence pits)*

Hydrocompaction sinkholes in the San Joaquin Valley, California 237
Nikola P.Prokopovich

Sinkholes in southeastern North Carolina – A geologic phenomenon and related engineering problems 243
Henning F.Koch

Soil cavities formed by piping 249
Ralph J.Hodek, Allan M.Johnson & Dean B.Sandri

Sinkholes at Tarbela Dam project 255
Izharul Haq & Altaf-ur-Rehman

Self-healing sinkholes in an earth dam foundation 261
Neil H.Wade & Lloyd R.Courage

Sinkhole problem related to dam engineering 267
Robert C.Lo

4. *Environmental / societal impact of sinkholes*

The influence of urbanization on sinkhole development in central Pennsylvania 275
Elizabeth L.White, Gert Aron & William B.White

Sinkhole flooding associated with urban development upon karst terrain: Bowling Green, Kentucky 283
Nicholas C.Crawford

Litigious problems associated with sinkholes, emphasizing recent Kentucky cases alleging liability when sinkholes were flooded 293
James F.Quinlan

Toxic and explosive fumes rising from carbonate aquifers: A hazard for residents of sinkhole plains 297
Nicholas C.Crawford

Characterization of the shallow groundwater system in an area with thin soils and sinkholes (Door Co., WI) 305
James H.Wiersma, Ronald D.Stieglitz, DeWayne L.Cecil & Glenn M.Metzler

Altura Minnesota lagoon collapses 311
E.Calvin Alexander & Paul R.Book

Part 1: The applicability of the Florida Mandatory Endorsement for Sinkhole Collapse Coverage – Legal aspects 319
William G.Salomone

Part 2: The applicability of the Florida Mandatory Endorsement for Sinkhole Collapse Coverage – Case history, foundation settlement of a residential structure – Was it a sinkhole? 329
William G.Salomone

5. Case histories: Remedial engineering of sinkholes

Remedial measures associated with sinkhole-related foundation distress 335
John E.Garlanger

A geological survey's cooperative approach to analyzing and remedying a sinkhole related disaster in an urban environment 343
Robert Canace & Richard Dalton

Sinkhole activity in the vicinity of the Sunny Point, N.C., Military Ocean Terminal 349
Earl F.Titcomb, Jr. & Jack M.Keeton

Evaluation, repair and stabilization of the Boling Sinkhole FM 442, Wharton County, Texas 353
Boyd V.Dreyer & Clyde E.Schulz

Engineering problems associated with sinkholes in the Valley of Virginia 359
H.G.Larew & E.O.Gooch

Maturation of the Winter Park sinkhole 363
S.E.Jammal

6. Engineering in sinkhole-prone areas

Correction and protection in limestone terrane 373
George F.Sowers

Sinkhole and subsidence damage and protective measures 379
Nath S.Parate

Geotechnical considerations in the location, design, and construction of highways in karst terrain – 'The Pellissippi Parkway extension', Knox-Blount Counties, Tennessee 385
Harry L.Moore

A model study of a proposed concrete road pavement over a potential sinkhole area 391
J.Marius Louw, Paul H.Goedhart & Frederick J.van Zyl

Foundation problems on karstic limestone formations in Western Thailand – A case of Khao Laem Dam 397
Dennes T.Bergado, Chanin Areepitak & Friedrich Prinzl

Construction on dolomite in South Africa 403
Fritz von M.Wagener & Peter W.Day

High-volume grouting to control sinkhole subsidence 413
Christopher R.Ryan

Collapse and compaction of sinkholes by dynamic compaction 419
Christian A.Guyot

Late paper to Part 3

Sinkhole development in reclaimed smectitic spoil 425
Maurice B.Dusseault, J.Don Scott & Steve Moran

Sinkhole terminology

BARRY F. BECK *University of Central Florida, Orlando, USA*

The term sinkhole suffers a wide variety of uses and misuses. It, of course, generally refers to an area of localized land surface subsidence, or collapse, due to karst processes, which results in a closed hollow of moderate dimensions (Monroe, 1970; Sweeting, 1973). In geologic research the synonymous term doline, or dolina, has precedence and has become standard in the literature (Sweeting, 1973). However, sinkhole has maintained acceptance not only in the U.S. but also internationally as is demonstrated by its use in innumeralbe professional publications (e.g., Wolters, 1973). Among practicing engineers the term sinkhole is widely used and broadly applied, as can be seen from the papers in this volume or from engineering reports such as Jammal and Assoc. (1982). Brink (this volume) even uses the terms sinkhole and doline to contrast similar forms arising from different processes.

Within the framework of localized subsidence or collapse phenomena which ultimately can be attributed to karst processes, there are several distinctly different mechanisms by which rock dissolution may lead to surface lowering. There are also numerous other geomorphic processes which may lead to similar results--localized land surface subsidence or collapse (for a few examples refer to the papers under "Sinkhole-like Features" in this volume). Terminology for these various occurrences should be unambiguous.

The terms sinkhole and doline are synonymous and both should be restricted to enclosed depressions caused ultimately by dissolution of the underlying rocks, i.e. karst processes. The term sinkhole should not be used for mine collapse or piping features, or other depressions which are not due to bedrock dissolution. The term "subsidence pit", as used by Hodek (this volume) would appear to be a concise, descriptive term which the author proposes be used for all localized subsidence or collapse features of non-karstic origin. Within this broad category, specific subtypes may be defined: e.g., a mine collapse subsidence pit or a soil piping subsidence pit.

The terminology of mining-related subsidence features becomes even more confusing when it is realized that mining may cause both sinkholes and subsidence pits. In the case of mine roof collapse, the resulting depression is obviously a subsidence pit. The cause is not karstic. Similarly, the piping of unconsolidated surficial sediments into abandoned mine adits by downward flowing groundwater might also be termed a subsidence pit. However, where the dewatering of mines or quarries in limestone may accelerate the downwashing of sediment into karstic voids, this may trigger the development of true sinkholes rather than subsidence pits.

The basic cause of karst topography is, of course, the dominance of bedrock solution over mechanical abrasion. One type of karstic surface depression is that caused purely by bedrock solution: a solution doline, or solution sinkhole. "These are due primarily to pronounced surface solution of the karst bedrock around some favorable point such as a joint intersection. The solutes and some insoluble residues are removed down solution-widened planes of weakness, though once the latter are enlarged to shaft dimensions there will be sliding and falling of residues and rock fragments brought to their apertures. As soon as a focus of downward percolation is established by solution, it will gather drainage to itself and the embryonic doline will further its own development. In fairly uniform rock the interaction of solution, mechanical slope wash and mass movements of material with the angle of the doline sides can result in conditions of dynamic equilibrium on uniform slopes. A conical shape is therefore characteristic of dolines of this kind. However, residues may accumulate at the bottom too rapidly for removal down widened joint planes, so that they form flat and perhaps swampy floors or even cause pools." (Jennings, 1971, p. 121-122). While this is the classic karst sinkhole, it is a slow, gradual phenomenon and certainly not a collapse problem. In fact, solution sinkholes should not present the engineering foundation problems which other more rapid and unexpected sinkholes do, although they may be related to foundation settling and cracking. However, solution sinkholes do serve as discrete sources of groundwater recharge and, as such, they may be related to contamination problems.

Another major cause of karstic land surface depressions is the collapse of the roof of a bedrock cavern. Such cave collapse produces a steep-sided, bedrock-walled hole possibly widening into interconnected cave passages at depth. This type of hole is generally termed a collapse sinkhole. If the underlying cave system is water filled, a cenote (Yucatan, Mexico) is the result. Sweeting (1973, p. 68) states that "Collapse of limestones is a fairly frequent phenomenon, despite the fact that it is not often actually recorded." However, it is the present author's opinion that examples of collapse sinkholes are common because their rock-walled character gives them geomorphic persistence. When incidents of recent collapse are analyzed, subsidence dolines (see below) appear much more common and actual cave roof collapse appears rare (Sowers, 1975; Williams and Vineyard, 1976; Newton, 1976). Of the 650 sinkholes recorded in Florida in recent years (Beck, 1984) only one might possibly be a cave roof collapse, and this is not certain.

The vast majority of damaging sinkholes which collapse today fall in the category of subsidence dolines or subsidence sinkholes (Jennings, 1971).* These are more often termed ravelling sinks by engineers (Sowers, 1975; Jammal and Assoc., 1982). Jennings describes subsidence doline thus: "Where superficial deposits or thick residual soils overlie karst rocks, dolines can develop through spasmodic subsidence and more continuous piping of these materials into widened joints and solution pipes in the bedrock beneath. They vary much in size and shape. A quick movement of subsidence may temporarily produce a cylindrical hole which rapidly weathers into a gentler conical or bowl-shaped depression." (1971, p. 126). Jammal and Assoc. describe the concept of ravelling similarly: "In the ravelling process, unconsolidated materials (clay, silts, and/or sands) filter downward into voids in the underlying limestone. This movement is promoted by water infiltrating downward.... The usual expression is initially a bowl-shaped sinkhole. Where overburden thickness is greater, the ravelling process may continue until the soil shear strength can no longer support the arch of overburden; collapse then occurs." (1982, p. 8-3). The terms and processes are obviously synonymous.

It would be inappropriate for either geologists or engineers to abandon their terminology; however, the definition should be applied precisely and confined to occurrences related to underlying karstic voids. Most sinkhole collapse problems arise from this mechanism. Man's activities, such as groundwater withdrawal, may trigger the development of subsidence, or ravelling, sinkholes. These are generally referred to as induced sinkholes; see Newton (this volume).

In summary, the term sinkhole (or doline) should refer only to localized land surface depressions arising from karst processes. Solution sinkholes form from the slow dissolution of bedrock and are not generally an engineering problem although they may be an avenue for groundwater pollution. Collapse sinkholes arise when the roof of a bedrock cavern collapses; such incidents are rare. Subsidence sinkholes (geology) or ravelling sinks (engineering) form by the piping of unconsolidated overburden into karstic openings in the underlying soluble bedrock, usually limestone. Localized land surface subsidence, rapid or slow, may also arise from numerous non-karstic causes, particularly mining and soil piping. It is suggested that these features be referred to collectively as subsidence pits to distinguish them from true sinkholes.

References

Beck, Barry F., 1984, A computer based inventory of recorded, recent sinkholes in Florida: Florida Sinkhole Research Inst. (U. of Central Fl., Orlando, Fl.) Report 84-1, 12 p.

Jammal and Assoc., 1982, The Winter Park Sinkhole: Winter Park, Florida, Jammal and Assoc., Inc., 256 p.

Jennings, J.N., 1971, Karst: Cambridge, Mass., M.I.T. Press, 252 p.

Monroe, W.H., 1970, A glossary of karst terminology: U.S. Geol. Surv., Water Sup. Paper 1899K.

Newton, J.G., 1976, Induced sinkholes--a continuing problem along Alabama highways: Internat. Assoc. of Hydro. Sci. Proceedings, Anaheim Symp., 1976, no. 21, p. 453-463.

Sowers, G.F., 1975, Failures in limestones in humid subtropics: J. Geotech. Eng. Div., Proc. ASCE, GT8, p. 771-787.

Sweeting, M.M., 1973, Karst landforms: New York, N.Y., Columbia U. Press, 362 p.

Williams, J.H. and Vineyard, J.D., 1976, Geologic indicators of catastrophic collapse in karst terrain in Missouri: Transportation Research Record 612 (NTIS PB-272844) p. 31-37.

Wolters, R., (ed.), 1973, Proceedings--Symposium on sinkholes and subsidence: engineering - geological problems related to soluble rocks: Hanover, Germany, Int. Assoc. of Eng. Geol., 320 p.

---

*Sweeting (1973, p. 59) refers to these as "alluvial dolines." This term is easily confused with stream sinks (ponors or swallets) and is not descriptive of the mechanism of formation. Subsidence doline appears preferable.

1. The geologic framework and mechanisms of sinkhole development

Review of induced sinkhole development

J.G.NEWTON *Consultant, Tuscaloosa, Alabama, USA*

ABSTRACT

Induced sinkholes are those caused or accelerated by man's activities. They are divided into two types: those resulting from a decline of water level due to pumpage and those resulting from construction. Almost all occur where cavities develop in unconsolidated deposits overlying openings in carbonate rocks. Triggering mechanisms resulting from water level declines are (1) loss of buoyant support, (2) increase in water velocity, (3) water-level fluctuations, and (4) induced recharge. Those resulting from construction include piping, saturation, and loading.

Sinkholes resulting from man's water activities are predictable in some instances but only in the context that they will occur within the area impacted. This is limited largely to those resulting from groundwater withdrawals.

Introduction

The sudden formation of large sinkholes in recent years has focused attention on a little understood geologic hazard. Few realize that thousands of similar but smaller sinkholes have formed in the United States since 1950. Costly damage has resulted from their sudden collapse beneath highways, railroads, buildings, dams, reservoirs, pipelines, vehicles, and people.

Sinkholes can be separated into two categories, even though most factors involved in their occurrence are the same. These categories are defined as "induced" and "natural." Induced sinkholes are those caused or accelerated by man's activities whereas natural ones are not. Recognition of induced sinkholes, the subject of this paper, and limited investigations of them are confined to this century. Almost all investigations dealing with triggering mechanisms or processes have been made since 1950.

The purpose of this paper is to present a review of geologic and hydrologic mechanisms or forces involved in the development of induced sinkholes and any predictive capabilities relating to their occurrence. Because of limited work and findings in recent years, this paper consists largely of excerpts from similar reports and papers by the author (1976a, 1976b, 1984a, and 1984b).

Geologic and Hydrologic Setting

The terrane used to illustrate sinkhole development, hereafter referred to as the "selected terrane", is a youthful basin underlain by carbonate rocks (fig. 1). It contains a perennial or near-perennial stream. Water is stored in underlying rocks and moves through interconnected openings along bedding planes, joints, fractures, and faults that often are enlarged by solutioning. It moves in response to gravity, generally toward the stream channel where it discharges and becomes streamflow.

Water in carbonate rocks occurs under water-table and artesian conditions; however, this study is concerned primarily with water-table conditions. The configuration of the water table conforms to that of the topography but is influenced by geologic structure, water withdrawal, and precipitation. The lowest water level occurs where the water intersects the stream channel. Bedrock openings underlying lower parts of the basin are water filled and those underlying highland areas are air filled.

A mantle of unconsolidated deposits resulting from the solution of the underlying rocks, consists chiefly of residual clay (residuum). This clay, commonly containing chert debris, covers most bedrock. Alluvial or other unconsolidated deposits often overlie the clay. Other unconsolidated deposits commonly fill openings in bedrock. The contact between residuum and underlying bedrock, because of differential solution, can be highly irregular.

Induced Sinkholes

Induced sinkholes were first classified by separating those caused by lowering the water level from those caused by raising the water level (Aley and others, 1972). This classification was modified slightly by Newton (1976a) by separating those caused by declines in water levels due to ground-water withdrawals from those caused by construction. Collapses resulting from construction included those caused by the erection of a structure, the impounding or diverting of surface water, and any modification of the land surface.

The sudden development of induced sinkholes results from (1) collapse of the roof of a cavity or cavern in rock due to its progressive enlargement by solution or (2) the downward migration of soil and other unconsolidated deposits into openings in the top of bedrock. Collapse of bedrock roofs, in comparison to the migration of unconsolidated deposits into underlying openings, is rare. According to Williams and Vineyard (1976) in a study involving 97 collapses in Missouri, "Although cavern roof collapse in bedrock has been cited as a cause of catastrophic sinkhole formation, no contemporary event that could be attributed directly to this cause has been observed." Similar observations have been made in Alabama (Newton and Hyde, 1971) and elsewhere.

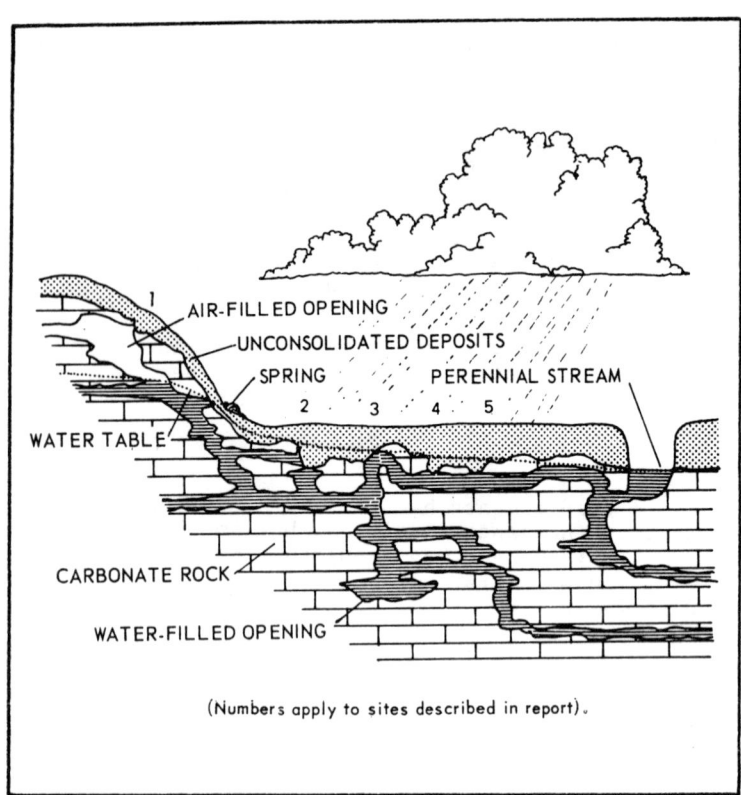

Figure 1: Schematic cross-sectional diagram of basin showing geologic and hydrologic conditions. (From Newton, 1976a.)

Most collapses forming sinkholes result from roof failures of cavities in unconsolidated deposits overlying carbonate rocks. Their occurrence, growth, and collapse has been described by Donaldson (1963), Jennings and others (1965), and many others. These cavities are created when the unconsolidated deposits migrate or are eroded downward into openings in the top of bedrock. When this occurs, a void is created. The typical cavity in unconsolidated deposits is circular with the configuration of the top resembling a dome or arch. The sides at the bottom generally coincide with pinnacles or irregularities in the top of bedrock and the walls are usually vertical as the opening grows toward the land surface (fig. 2A). This configuration however, can be modified by the shape of the underlying opening in bedrock and by variations in the cohesion or competence of overlying beds. It can change when its upward growth reaches a more competent bed. The roof flattens, the growth continues laterally, and the walls taper toward the opening in bedrock (fig. 2B). Other cavity configurations such as horizontal tunneling has also been observed (Newton, 1976b).

Decline of Water Level
    Foose (1953), in the first investigation of this type sinkhole activity, associated the occurrence of sinkholes with pumping and a subsequent decline in the water table. He determined that their formation was confined to areas where a drastic lowering of the water table had occurred, that their occurrence ceased when the water table recovered, and that the shape of collapses indicated a lowering of the water table and withdrawal of support. Robinson and others (1953) attributed sinkhole occurrence in a cone of depression to the increased velocity of groundwater movement.

Jennings and others (1965) associated development of sinkholes with pumpage and creation of cones of depression and determined that sinkhole and subsidence problems increased where the water table was lowered.

Spigner (1978) attributed intense sinkhole development near Jamestown, South Carolina to a water level decline resulting from pumpage, provided descriptions indicating loss of support and attributed some downward movement of unconsolidated deposits to piping. Sinclair

(1982) attributed similar activity in Florida to loss of support and water-level fluctuations.

Subsidence rather than collapse sometimes results in the formation of sinkholes due to declines in water level. Movement of unconsolidated materials into bedrock where the strength of overlying material is not sufficient to maintain a cavity roof, will result in subsidence at the surface (Donaldson, 1963).

Cited reports have described only indirectly or in part the hydrologic mechanisms resulting from a decline that cause the downward migration of unconsolidated deposits. These mechanisms, based on studies in Alabama (Newton, 1976a) are (1) loss of buoyant support to roofs of cavities or caverns in bedrock and to residual clay or other unconsolidated deposits overlying openings in the top of bedrock, (2) increase in the velocity of movement of ground water, (3) increase in the amplitude of water-level fluctuations, and (4) movement of water from the land surface to openings in bedrock where recharge had previously been largely rejected.

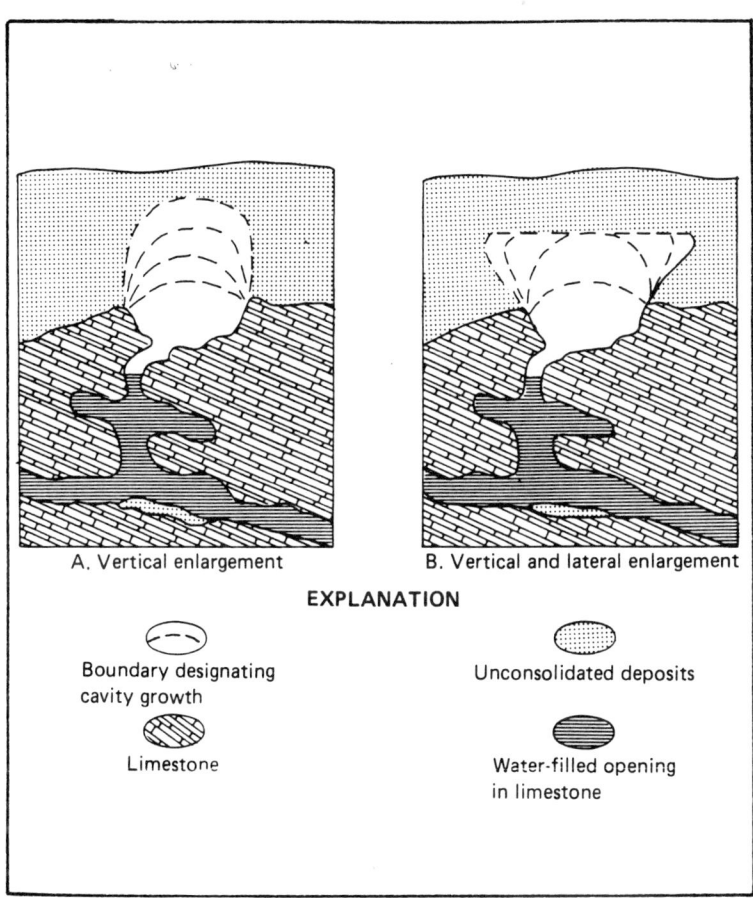

Figure 2: Development of cavities in unconsolidated deposits. (Modified from Newton, 1976a.)

A cone of depression resulting from pumpage from a quarry is superimposed on the selected terrane to illustrate the downward migration of unconsolidated deposits, creation of cavities in them, and sinkhole development (fig. 3). A solutionally enlarged opening in the stream has been sealed, a common mine dewatering practice, to prevent flooding in the cone of depression.

The loss of buoyant support following the water-level decline can result in an immediate collapse of roofs of openings in bedrock and unconsolidated deposits or can cause a downward migration of unconsolidated deposits spanning openings in the top of bedrock. The support exerted by water on a solid and, hypothetically, unsaturated clay overlying an opening in bedrock, for instance, would be equal to about 40 percent of its weight.

A collapse triggered by a loss of buoyant support is illustrated at site 5 in figure 3. Loss of support also triggered the downward movement of residual clay and the creation of a cavity in unconsolidated deposits at site 3. Openings in the top of bedrock at both sites were overlain by unconsolidated deposits prior to the decline of the water table (fig. 1).

The creation of a cone of depression in an area of water withdrawal results in an increased hydraulic gradient toward the point of discharge (fig. 3) and a corresponding increase in the velocity of ground-water movement. Erosion caused by this movement through unobstructed openings and against joints, fractures, faults, or other openings filled with clay or other unconsolidated sediments results in the creation of cavities that enlarge and eventually collapse.

Pumpage results in water-level fluctuations greater in magnitude than those occurring under natural conditions. The repeated movement of water through openings in bedrock against overlying unconsolidated deposits causes repeated addition and subtraction of buoyant support to them and repeated saturation and drying. Both result in the downward migration of the deposits that creates or enlarges cavities in them. All collapses and cavities in unconsolidated deposits illustrated in figure 3 could have resulted from this mechanism.

A drastic decline of the water table in the selected terrane (fig. 3) results in induced recharge of surface water. This recharge was partly rejected prior to the decline because the openings were water filled. The inducement of recharge through openings in unconsolidated deposits interconnected with openings in bedrock results in the creation of cavities in the deposits. The material immediately overlying the bedrock openings is eroded to lower elevations. The water table, previously located above the top of bedrock (fig. 1), is no longer in a position to dissipate the mechanical energy of downward moving recharge. Repeated rains result in the progressive enlargement of this type cavity. A corresponding thinning of the cavity roof due to its enlargement toward the surface eventually results in collapse. The position of the water table below unconsolidated deposits and openings in the top of bedrock favorable to induced recharge is illustrated at sites 2 and 4 in figure 3. Cavities in unconsolidated deposits at these sites were formed primarily or in part by induced recharge. The creation and eventual collapse of cavities in the deposits by induced recharge is described by many authors as "piping."

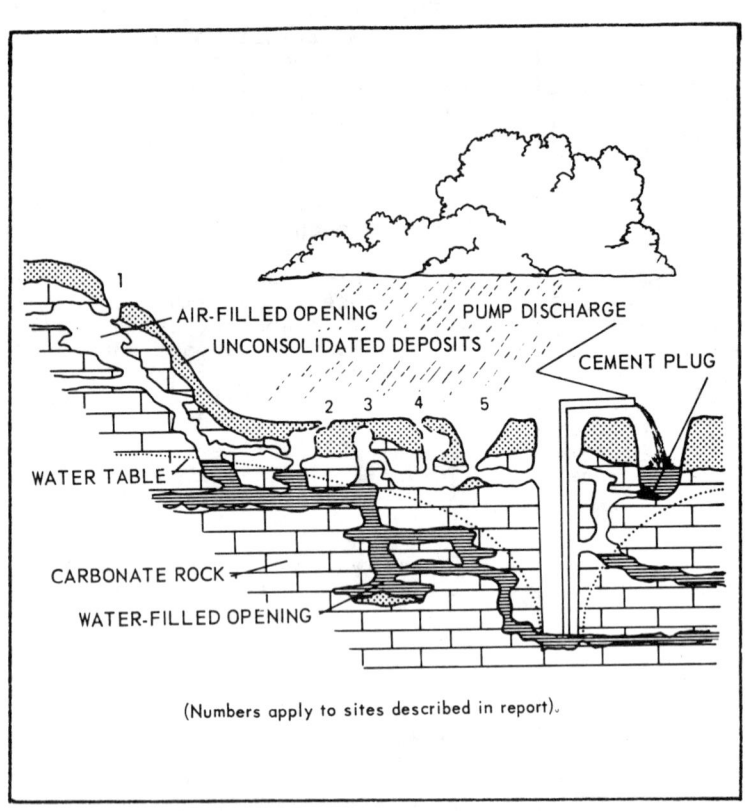

Figure 3: Schematic cross-sectional diagram showing changes in geologic and hydrologic conditions resulting from water withdrawal. (From Newton, 1976a.)

Where the cone of depression is maintained by constant pumpage (fig. 3), all mechanisms described are active even though one may be responsible for the development of a specific collapse. In contrast, a cavity resulting from a loss of support (site 3) can be enlarged and collapsed by induced recharge if it has intersected openings interconnected with the surface. Similarly, in an area near the outer margin of the cone (site 2), the creation of a cavity and its collapse can result from all mechanisms. It can originate from a loss of support, can be enlarged by water-level fluctuations, can be enlarged by increased velocity of water movement against sediment that originally filled the openings (fig. 1), and can be enlarged and collapsed by induced recharge.

Occurrence and Size

Sinkhole activity due to water-level declines is confined to small areas that generally vary in size from a square meter to about 26 km^2. Water withdrawals from sinkhole active areas has generally ranged from 1,900 to 76,000 m^3 per day with the resulting water level declines generally varying from 6 to 107 m. Activity within the areas is often intense and prolonged. Near Tampa, Florida, 64 collapses reportedly occurred within a 1.6 km radius of a well field in 1964 (Sinclair, 1982). Five other sites experiencing one to thirty collapses in the same general area were also reported. Near Jamestown, South Carolina, 42 collapses occurred within a cone of depression (Spigner, 1978). In Alabama, an estimated 1,700 collapses or related features have occurred in five areas examined (Newton, 1976a). In Pennsylvania, about 100 collapses have occurred in a cone of depression near Hershey (Foose, 1953). Near Friedensville, records indicated that 128 sinkholes formed in an area around a point of withdrawal from 1953-57 and 25 new sinkholes were recorded during a four month period ending January 1971 (Pennsylvania Department of Transportation, 1971). Sites of similar intense development, in addition to those described, were identified in Alabama, Georgia, Maryland, North Carolina, Pennsylvania, South Carolina, and Tennessee (Newton, 1984a).

Data available relating to the size of sinkholes resulting from ground water withdrawals are limited. Most are described as being "small." Of 42 collapses described in South Carolina (Spigner, 1978), the largest was about 6 m in diameter, and the greatest depth exceeded 3 m. The largest of 64 collapses near Tampa, Florida, was reported to have the same dimensions (Sinclair, 1982). Collapses in Alabama generally range from 1 to 100 m in diameter and from 1 to 30 m in depth. An inventory of 243 collapses in two areas (Newton and Hyde, 1972; Newton and others, 1973) showed that the average sinkhole in the first area was 3.7 m long, 3 m wide, and 2.4 m deep. The average sinkhole in the second area was about 6.1 m long, 4 m wide, and 2.1 m deep. In Shelby County, Alabama, six collapses observed had diameters approaching or exceeding 30 m. Collapses with diameters generally ranging from 7.6 to 15.2 m were not uncommon. Some near Sylacauga, Alabama, had surface dimensions of 9 to 30 m. In Hershey Valley, Pennsylvania, 100 new sinkholes were reported to be 0.3 to 6.1 m in diameter and 0.6 to 3 m deep (Foose, 1953).

Construction

The term "construction" applies to the erection of a structure, to any modification of the land surface, and to the diversion and impoundment of water. Diversion of drainage also includes any activity that results in changes in the downward movement of recharge. These activities include removal of timber and drilling, coring, and augering where pumpage is not involved. Also included is leakage from sewers, pipes, and similar facilities.

Construction practices often "set the stage" for sinkhole occurrence. Grading results, in cuts, in the thinning of unconsolidated deposits. Emplacement of weight on thinned roofs of existing cavities in residual clay or on those of shallow bedrock cavities can cause their failure. The occasional collapse beneath heavy equipment during construction is probably attributable to this cause. Rainfall and saturation of roofs of underlying cavities in residual clay after grading can also result in their failure. Differential compaction caused by the weight of a structure on unconsolidated deposits overlying the irregular surface of the top of bedrock also results in subsidence and foundation problems.

Shocks or vibrations resulting from blasting can also cause or contribute to the failure of roofs of cavities in bedrock and unconsolidated deposits. About four percent of collapses identified in Missouri have been attributed to this cause (Williams and Vineyard, 1976).

Concentration of water by drainage diversion may increase recharge to underlying bedrock openings. It also can cause saturation and weakening of roofs of existing cavities in unconsolidated deposits. Collapses due to this occurrence are less common than those caused by the creation and enlargement of openings in the deposits that result from the movement of water to and through existing openings in the top of bedrock. The subsurface erosion of unconsolidated deposits and the creation or enlargement of resulting cavities has been described by Newton (1976a) and Moore (1980). The former described it as being the same as the "piping" process resulting from induced recharge caused by a decline in the water table. The latter identified the process as being responsible for collapse failures in Tennessee. Collapses resulting from the piping process would be most common where the water table is located below the top of bedrock (site 1 on fig. 1). The erosive energy of downward moving water dissipates when it encounters a water table located above an opening in bedrock.

Collapses resulting from leakage from underground pipes are well documented in the literature. Resulting collapse mechanisms are the piping process and saturation. Saturation causes loss of cohesion of residual clays and also causes loading due to the addition of the weight of water.

Collapse also can result where surface water gains access to uncased or unsealed holes created by drilling, augering, or coring. The piping process is generally responsible. The same process may be responsible for collapses that occur at drainage wells.

Collapses caused by the impounding of drainage occur, in part, in the same manner as those resulting from diversions of drainage. The impounding of water results in the saturation and loss of cohesiveness of unconsolidated deposits overlying bedrock openings. This, accompanied by loading resulting from the weight of impounded water, can result in the collapse of the deposits into the bedrock opening and a draining of the impoundment (Aley and others, 1972). If the impoundment is located where the water table is below the top of bedrock and openings at the surface are interconnected with those in bedrock, a collapse can result from the piping process. Collapses resulting from saturation and piping have been described by Warren (1974). The piping process can also result in sinkholes when water is impounded on unconsolidated deposits where the water table was originally located above the top of bedrock. On the floor of the impoundment, water moving under increased head through openings in the deposits into openings in underlying rocks can both form and cause collapse of cavities in the deposits. This would occur where there is considerable pressure exerted by the impounded water and where openings in underlying carbonate rocks have a discharge point outside of the impoundment at a lower altitude.

Occurrence and Size

Sinkholes resulting from diversions of drainage have occurred in most States in which active subsidence has been reported. Numerous sites attributable to this cause have been inventoried in Alabama and Georgia. Collapses associated with highway construction in Missouri are due to changes in the water regimen (Williams and Vineyard, 1976). Collapses attributable to this cause along highways in Tennessee have been described and illustrated (Moore, 1980). Similarly, collapses beneath drains in Pennsylvania and Kentucky are attributed to this cause. About 20 collapses resulting from concentration of drainage occurred at one site in Pennsylvania (Knight, 1970).

Sinkholes resulting from the impounding of water have occurred in numerous States. Impoundments affected extend from Florida in the south to Minnesota in the north. Some of the larger impoundments are located in Alabama, Arkansas, Georgia, Kentucky, Missouri, and North Carolina.

Most collapses due to construction are relatively small. Most in Alabama are less than 5 m in diameter. The largest surface area involved was about 18 m in diameter and the greatest depth about 8 m. These dimensions are similar to the largest reported along roadways in Florida (Florida Department of Transportation, written commun., 1981). The largest reported in Missouri was about 23 m long, 11 m wide, and 8 to 9 m deep (Aley and others, 1972).

Prediction

Induced sinkholes resulting from man's water activities are predictable in some instances but only in the context that they will occur within the area impacted. This capability is also restricted to certain terranes. It would not, for instance, relate to possible sinkhole occurrence in many terranes altered by glaciation. Predictive capabilities would be most significant in the selected terrane (fig. 1).

The most predictable sinkhole development is that resulting from water-level declines due to dewatering by subsurface mines, recessed quarries, and wells. This occurs where the water level, previously located above the top of bedrock during all or most of the year (fig. 1), is maintained below it by pumping (fig. 3). All mechanisms that trigger sinkhole development in unconsolidated deposits are activated by the decline.

Conversely, the unconsolidated deposits are not impacted where the zone in which the water level fluctuates is located below the top of bedrock prior to dewatering. Determining the position of the water table in relation to the top of bedrock aids in predicting whether sinkholes will or will not occur.

Where and when some sinkholes will occur in a dewatered area is also predictable to a limited degree. Many occur where concentrations of surface water are greatest such as streambeds, natural drains, or poorly drained areas. Large numbers occur where natural drainage has been altered and where natural recharge has been increased as a result of activities such as timber removal (Newton, 1976a). Most of the sinkhole activity occurs during or immediately after periods of rainfall, especially deluges, when hydrologic stresses to overburden are greatest.

Prediction (or recognition of an element of risk) in induced sinkhole development due to construction can be made in some instances but not in most. Alteration of the land surface to construct a pond in a natural sinkhole, for instance, would be considered a risk. Similarly, sinkholes associated with major impoundments would be predictable to some degree because all mechanisms associated with their development are triggered by this type of construction. Determining a degree of risk in most instances, however, is rarely accomplished with a significant degree of certainty.

References Cited

Aley, T. J., Williams, J. H., and Masselo, J. W., 1972, Groundwater contamination sinkhole collapse induced by leaky impoundments in soluble rock terrane: Missouri Geological Survey and Resources, Engineering Geology Series, no. 5, 32 p.

Donaldson, G. W., 1963, Sinkholes and subsidence caused by subsurface erosion: Regional Conference for Africa on Soil Mechanics and Foundation Engineering, 3rd, Salisbury, Southern Rhodesia 1963 proc., p. 123-125.

Foose, R. M., 1953, Ground-water behavior in the Hershey Valley, Pennsylvania: Geological Society America Bulletin 64, p. 623-645.

Jennings, J. E., Brink, A. B. A., Louw, A., and Gowan, G. D., 1965, Sinkholes and subsidence in the Transvaal dolomites of South Africa: International Conference Soil Mechanics, 6th, 1965 Proceedings, p. 51-54.

Knight, F. J., 1970, Geologic problems of urban growth in limestone terranes of Pennsylvania: Preprint, Association of Engineering Geologists, Annual Meeting, Washington, D.C., 16 p.

Moore, H. L., 1980, Karst problems along Tennessee highways, an overview: Highway Geology Reprint, 31st Annual Symposium, Austin, Texas, 68 p.

Newton, J. G., 1976a, Early detection and correction of sinkhole problems in Alabama, with a preliminary evaluation of remote sensing applications: Alabama Highway Department, Bureau Research and Development, Research Report no. HPR-76, 83 p.

_____ 1976b, Induced sinkholes--A continuing problem along Alabama Highways: International Association of Hydrological Sciences Proceedings, Anaheim Symposium, 1976, no. 21, p. 453-463.

_____ 1984a, Natural and induced sinkhole development-eastern United States: International Association of Hydrological Sciences Proceedings, Third International Symposium on Land Subsidence, Venice, Italy (In press).

_____ 1984b, Sinkholes resulting from ground-water withdrawals in carbonate terranes-an overview: Geological Society of America Review in Engineering Geology, v. VI (In press).

Newton, J. G., and Hyde, L. W., 1971, Sinkhole problem in and near Roberts Industrial Subdivision, Birmingham, Alabama--a reconnaissance: Alabama Geological Survey Circular 68, 42 p.

Newton, J. G., Copeland, C. W., and Scarbrough, L. W., 1973, Sinkhole problem along proposed route of Interstate 459 near Greenwood, Alabama: Alabama Geological Survey Circular 83, 53 p.

Pennsylvania Department of Transportation, 1971, Geological engineering and geophysical study: LR-1045, Section-E01, Interstate 78, Lehigh County, Pennsylvania, 24 p.

Robinson, W. H., Ivey, J. B., and Billingsley, G. A., 1953, Water supply of the Birmingham area, Alabama: U.S. Geological Survey Circular 254, 53 p.

Sinclair, W. C., 1982, Sinkhole development resulting from ground-water withdrawal in the Tampa area, Florida: U.S. Geological Survey Water Resources Investigations 81-50, 19 p.

Spigner, B. C., 1978, Land surface collapse and ground-water problems in the Jamestown area, Berkley County, South Carolina: South Carolina Water Resources Commission Open-File Report no. 78-1, 99 p.

Warren, W. M., 1974, Retention basin failures in carbonate terranes: Water Resources Bulletin, v. 10, no. 1, p. 22-31.

Williams, J. H., and Vineyard, J. D., 1976, Geological indicators of catastrophic collapse in karst terrane in Missouri: National Academy of Science Transportation Research Record 612, p. 31-37.

Florida karst – Its relationship to geologic structure and stratigraphy

WALT SCHMIDT & THOMAS M. SCOTT *Florida Geological Survey, Tallahassee, USA*

Abstract

Florida is well known for its abundant karst features including sinkholes, springs and caverns. These landforms are most common in two areas; the north-central portion of the panhandle and a broad area in the central and northwestern peninsula. Within these regions sinkhole abundance and morphology are directly related to the geologic structure and resulting stratigraphy.

The dominant positive structures are the Chattahoochee "anticline" in the panhandle and the Sanford High and the Ocala Uplift in the peninsula. These structures have carbonate rocks of the Floridan aquifer system at or near the surface under unconfined or semiconfined conditions near the crests. Increasing thicknesses of younger sediments overlie the carbonates away from the crests. Karst features are most abundant where the carbonates are nearest the surface and are overlain by permeable clastic sediments.

Sinkhole morphology is directly related to the lithologies of the sediments overlying the carbonates. In areas where the carbonates are exposed or very near the surface, the sinks are small and steepsided (such as the numerous "solution pipes" common near Ocala). As the thickness of unconsolidated surface sediments increases, the area affected by sinkhole formation increases and the slope of the sides is significantly shallower (Winter Park Sink of 1981). When more consolidated sediments overlie the soluble carbonates steepsided, deep sinks such as Brooks Sink and Devils Millhopper near Gainesville are the result.

Karst features in Florida also occur in areas where they are not related to carbonate rocks of the Floridan aquifer system. These include features associated with the limestones of the Biscayne Aquifer (Miami Oolite and Key Largo Limestone) and the Tamiami Formation, coquina of the Anastasia Formation and the Plio-Pleistocene shell beds and very shelly sediments. There are also sinkhole features found that are related to volume loss and subsidence due to mineralogic change in non-carbonate clastic sediments.

Introduction

The development and occurrence of karst landforms in Florida is of interest to scientists, homeowners, planners and insurance companies alike. In order to better delineate high risk areas, an understanding of the relationship of structure and stratigraphy to karst development is required. In this paper we define the effects of structure and stratigraphy on the occurrence and morphology of karst features in Florida.

The structural and stratigraphic framework of the State of Florida provides ample opportunities for sinkhole development. These aspects, in conjunction with the hydrologic regime, have created the extensive and variable karst topographies of the State.

The karst features recognized in Florida range in size, from small solution pipe sinkholes to large, closed, karst basins or prairies. Their occurrence is related to the existence and extent of fractures (and faults) in the limestones, the lithology of the limestone and its lithologic variability, and the thickness and lithology of the overburden sediments.

Structure and Geologic Setting

The State of Florida lies in a transition zone between two sedimentary provinces. The

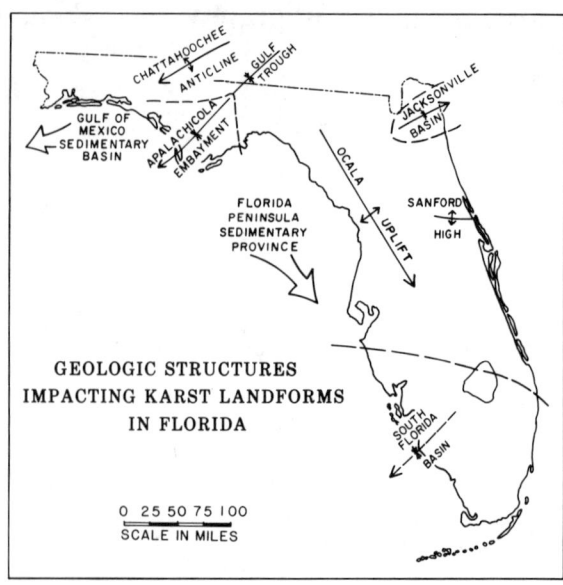

FIGURE 1 - Geologic Structure in Florida

North Gulf Coast sedimentary province consist mainly of clastic sediments derived from the eroding coastal plain, and includes the Apalachicola Embayment and the Jacksonville Basin across north Florida. The Florida peninsula sedimentary province consists of nonclastic sediments predominantly limestones, dolomites and evaporites, and it includes the South Florida Basin (Pressler 1947, p. 1851). These three sedimentary basins surround a positive structural feature known as the Ocala Uplift (figure 1). These features, along with the Chattahoochee Anticline in northwest Florida and the Sanford High in east central Florida, are the dominant geologic structures that influence karst landforms in Florida.

The majority of sinkhole occurrences in the state are associated with the approximate depth below land surface where carbonates of the Floridan aquifer system are encountered. The aforementioned geologic structures are "highs" or "lows" in this Tertiary limestone sequence.

In addition there are a few areas in Florida where karst features occur that are not associated with limestones of the Floridan aquifer system. These include shallow depressions often called dolines, associated with Plio-Pleistocene limestones and shelly sediments in south and east central Florida. There are also shallow depressions that have been called "sinkhole-like" features in noncarbonate clastic sediments in the western Florida Panhandle (figure 2).

Major Positive Structural Features

Chattahoochee Anticline: The Chattahoochee Anticline was first described by Veatch and Stephenson (1911, p. 62-64). This feature was mapped in the tri-state area of Alabama, Florida, and Georgia by mapping exposures of Cretaceous and Eocene rocks along the Chattahoochee River.

It is not clear what this structural high represents. Its presence may be due to surrounding areas subsiding due to sediment loading, or it may indeed be an "uplift". Basement or pre-Mesozoic maps show a high in this area, as do the maps drawn on Mesozoic and Tertiary formations (Applin, 1951; Applin and Applin, 1944; Kwader and Schmidt, 1978; Schmidt and Coe, 1978).

In Florida, Upper Eocene, Oligocene and Lower Miocene limestones are exposed near the crest of the "anticline" in Jackson, Holmes and Washington counties. This area is lower in elevation and has less surface relief than surrounding areas. As a result this physographic region has been called the Marianna Lowlands. This region has been lowered in elevation by removal of material by stream erosion and by solution activity acting on the near surface carbonates. The Marianna Lowlands contain numerous springs, sinkholes and caves predominantly developed in the Ocala Group limestones.

Ocala Uplift: The term "Ocala Uplift" was first used by Hopkins (1920; referred to in Vernon, 1951, p. 54). Since that time the feature was further described by Vernon (1951) and Winston (1976). As is the case with the Chattahoochee Anticline, a mechanism to produce the Ocala Uplift has not been agreed upon. It has been called an anticlinal fold, a gentle flexure, an uplift, a horst and a dome. More recently Winston (1976) proposed the term "Ocala blister dome", with the tilting of the Florida peninsula as the primary cause. Winston points out,... "the structure is present in shallow beds but underlain by beds which do not reflect the structure".

The axis or crest of the Ocala Uplift trends northwest - southeast (figure 1). Regardless of the origin of this feature, the limestones and dolomites of the Middle and Late Eocene, Oligocene, and Early Miocene are near the surface and exposed around its axis (figure 2). These carbonates collectively represent the Floridan aquifer system. In this area, where the aquifer is near the surface (Knapp, 1979), most of the state's numerous karst landforms have developed. This area also is famous for its cave diving, spring-fed rivers and sinkhole swimming spots.

With a generally thin and poorly consolidated overburden covering the limestone, this area has numerous small sinks and sinkhole lakes. The water table is generally near the surface and seasonal changes in rainfall affect the hydrostatic pressure exerted on the overlying soil mantle. In one weekend in the spring of 1982, after about 15 inches of rain, nearly 200 sinkholes opened up in and around the City of Ocala. Most of these were small and many exhibited "solution pipe" morphology.

Sanford High: A semicircular area in east-central Florida (figure 1) called a half-dome was described by Vernon (1951). In this region limestones of the Floridan aquifer system are slightly higher than nearby areas to the north or south (Scott and Hajishafie, 1980). Sinkholes and small depressions are common in Volusia and Seminole counties as a result of this feature.

Major Sedimentary Basins

Gulf of Mexico Sedimentary Basin: This feature encompasses the entire Gulf of Mexico area, of which Florida lies on the eastern rim. This is a depositional center accepting sediment from the entire central part of the United States. The Florida carbonate platform acts as an eastern boundary to this clastic basin.

In the western part of Panhandle Florida, the Tertiary limestones of the Floridan aquifer system dip to the southwest towards this basin. Overlying this southwesterly dipping surface is a clastic wedge of westward thickening sediments. As a result, karst resulting from limestone being at shallow depth is non-existent in this area. There are, however, other "sinkhole-like" features developed in the Western Highlands physiographic province. The hills in the Western Highlands are composed of quartz sands and gravels, and massive silty clays of the Citronelle Formation. Isphording (1984) has established that some of the local depressions in this province are formed by transformation of kaolinite to gibbsite, which can cause a volume reduction by an amount of nearly 35 percent. This mineralogical alteration, along with other material loss due to the porous and permeable nature the sediments, accounts for these "sinkhole" features.

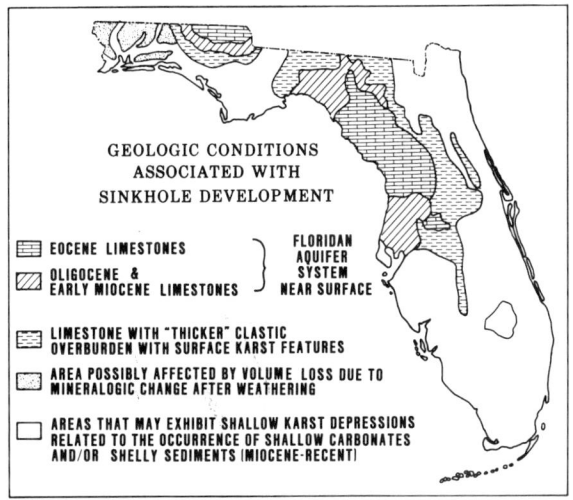

FIGURE 2 - Geologic Conditions

Apalachicola Embayment: The Apalachicola Embayment is a relatively shallow basin between the Ocala Uplift and the Chattahoochee Anticline. The near surface deposits are gently downwarped around an axis that plunges towards the southwest. Filling in this embayment over the limestones are Miocene to Recent clastics which may obtain a thickness of 100 meters or more. As a result of this thickened wedge of clastics, there is little to no karst type landforms developed in this area (figure 2).

Jacksonville Basin: The Jacksonville Basin in northeastern Florida is part of the more extensive Southeast Georgia Embayment. It is separated by a slight high in the Tertiary sediments, located beneath Nassau County called the Nassau Nose (Scott, 1983 p. 24-26). The Jacksonville Basin contains up to 150 meters of Miocene carbonates and clastics over Eocene Limestone, as a result there are no major karst landforms present. Some shallow depressions occur. These are the result of near-surface shell beds and associated carbonate volume loss.

South Florida Basin: The South Florida Basin is a part of the Florida-Bahama Platform. This basin has subsided slowly from Jurassic to Middle Eocene times. After Middle Eocene numerous disconformities imply intermittent tectonic activity (Oglesby, 1965; Winston, 1971). The upper Tertiary (to a depth of about 300 meters) contains calcarenitic limestones overlain by less consolidated shelly limestones, sandy limestone and shelly sands. Some areas are covered at the surface by marine sands, continental sands, or reef rock. Most of the area has very little surface relief. The region south of Lake Okeechobee exhibits "sheet flow" surface drainage as a result.

Karst Landform Development

The development of sinkholes or dolines (Sweeting, 1973) is primarily due to the dissolution of limestone by slightly acidic water. The removal of the dissolved carbonate results in a volume loss in the limestone. The volume loss may be expressed by the development of cavities or caverns, the enlargement of fractures or faults, or the lowering of the limestone surface. Each of these is involved in the development of karst landforms in Florida. Karst depressions in this state also result from dissolving the calcareous shell material incorporated in noncalcareous sediments. If the shell is abundant its removal will be accompanied by a volume reduction and the formation of a shallow depression at the surface.

As previously mentioned karst features (depressions) not related to dissolution of limestone occur in Florida. These depressions result from a mineralogic change in the clay-sized fraction causing a volume reduction (Isphording 1984). Depressions of this type occur in the western Florida panhandle (figure 2).

Stratigraphic Relationships

The oldest rocks involved in surface expression of karst in Florida are the carbonates of the upper Middle Eocene Avon Park Limestone (figure 3). These sediments are exposed in limited areas on the crest of the Ocala Uplift (figures 1 and 2), and are part of the Floridan aquifer system. In these areas the water table is typically very high and there is a thin sand and organic rich overburden. The expression of karst is limited predominantly to small, shallow, swampy depressions. Karst is not generally well developed in this area for two reasons. First, the high water table limits the base line of the dissolution by not allowing acidic waters to percolate along fractures to dissolve the carbonate rock. The second factor is the upper part of the Avon Park consists of mainly dolomite which is not as susceptable to dissolution. Karst depressions may then develop by limited enlargement of fractures and in-filling from overlying sediments (figure 4 A) or by subsidence due to dissolution at the limestone surface and subsequent land surface depression (figure 4 B).

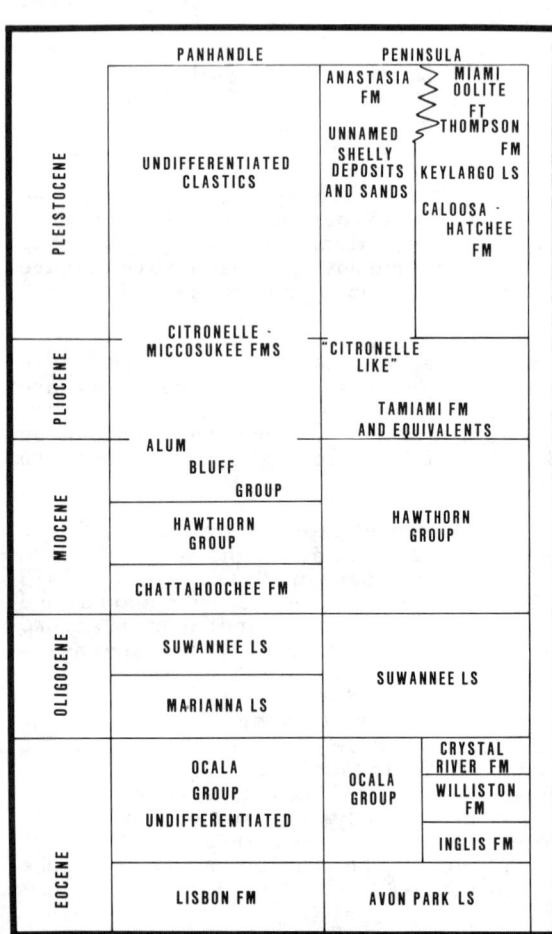

FIGURE 3 - Stratigraphic Chart

The rocks cropping out downdip of Avon Park Limestone on the Ocala Uplift are limestones of the Upper Eocene Ocala Group. The Ocala Group also crops out on the Chattahoochee Anticline in the panhandle. These sediments make up the majority of the Eocene area of figure 2. The Ocala Group (figure 3) is made up of the (in ascending order) Inglis, Williston and Crystal River formations. The Ocala Group is predominantly limestone and exhibits an often well developed karst surface. The limestone plains of the the Western Valley and Gulf Coastal Lowlands (White, 1970) are developed on the Ocala Group sediments.

The limestones of the Ocala Group develop a wide variety of karst features. Shallow depressions such as those discussed for the Avon Park are common (figures 4 A and B) although they may be better developed and approach the configuration shown in figure 4 E. Also encountered are numerous solution pipes (figure 4 C) which are very enlarged fractures or fracture intersections. Roof collapse sinkholes are found, often becoming springs (figure 4 D). Caverns often develop, although few are ever exposed at the surface. A good example of cavern development can be seen in the Florida Caverns State Park in Jackson County (Schmidt, 1982).

It is interesting to note that although roof collapse sinks do occur, they are not thought to be common or the dominant mode of formation for the majority of sinkholes in Florida (Jammal and Associates, 1982). The most common mode of formation is thought to be overburden collapse. In this type of collapse, fractures in the limestone (Ocala Group and Avon Park Formation) are enlarged by percolating ground water forming a network of variable size cavities and caverns (figure 4 E). At some point in time overburden material begins to enter the cavity system leaving a cavity or zone of very loose material in the overburden (figure 4 E). When enough material is removed the overburden collapses creating a sinkhole. This is believed to be the mechanism for the formation of the Winter Park sinkhole of May 1981 (Jammal and Associates, 1982).

Bordering the limestone plains are areas of thicker clastic overburden (figure 2). In the transitional zone between the limestone plains and the Northern Highlands, for example, karst features are very common. The Hawthorn Group sediments overlying the limestone thicken toward the highland areas. As a result the sinkhole landforms have greater relief and often are steeper sloped toward the highland area. The steeper slopes are due to the sinkholes being younger toward the highland therefore less eroded by surface water which reduces slopes. Individual sinks coalesce to form larger karst basins providing the rolling topography of the transitional zone. Geomorphically, the transition zones are erosional remnants of the highlands which are being reduced and removed by karst activity and erosion. Karst features in this area form several ways as suggested by figure 4 A-E. However, as the overburden thickens the true nature of the origin is hidden.

Sinkholes are not common in the Northern or Western Highlands (White, 1970). When sinks do occur they are relatively near the transitional zone. The sinks in the Northern Highlands penetrate through the Hawthorn Group and, in some cases, such as in the Tallahassee Hills also penetrate through the Miccosukee Formation. In areas where the Hawthorn Group lies at or near the surface the sinkholes that form are very steep to vertical sided. They often appear to have been formed by a plug of overburden (Hawthorn Group in this case) collapsing into a cavity as in figure 4 D or E. Devil's Millhopper in Alachua County and Brooks Sink in Bradford County are excellent examples.

Central and northern Florida have large areas where loose to poorly consolidated sands overlie the Hawthorn Group and the Ocala Group. The highlands in this area often show extensive karst development. The highlands include the Mt. Dora, Deland, Crescent City, Orlando, Lake Wales, Lakeland and other ridges (White, 1970). Uplands associated with the ridges include the Marion, Lake and Polk uplands (White, 1970). The topography of these areas typically is very rolling with numerous sinkhole lakes and basins. Karst features appear to form in these areas as shown in figure 4 E. The ultimate size of the feature is due to: 1) the amount of material the cavity system can accept and, 2) the thickness of the poorly consolidated overburden (the thicker the overburden, the broader the sink will be due to the relatively low angle of stability for the sediment). The Winter Park sinkhole is an excellent example of this type of sink.

Large areas of south Florida and east Florida are underlain at very shallow depths by thin limestone units and by units containing abundant calcareous shell material in a clastic matrix. The surface in these areas may be pock marked with very shallow depressions. Solution pitting is a common result of vegetation growth on the limestone surface (Sweeting, 1973). The limestones, belonging to the Hawthorn Group, Tamiami Formation, Caloosahatchee Formation, Ft. Thompson Formation, Anastasia Formation, Miami Oolite, and Key Largo Limestone, exhibit sinks formed similar to those shown in figure 4 A and B. The shell beds, belonging to the Tamiami Formation, Caloosahatchee Formation, and unnamed shelly units of Plio-Pleistocene age, form karst features by dissolution of the incorporated calcareous shell, causing a volume reduction and a resultant depression (figure 4 F).

Summary and Conclusions

The frequency of occurrence and the size of karst features is directly related to the stratigraphy of the area. Sinkholes are most abundant in the areas of outcropping carbonates of Eocene, Oligocene and Early Miocene shown on figure 2. The size of the karst features varies with over burden thickness, affecting a larger area at the surface where overburden is thicker. In the areas shown on figure 2 as having a thicker clastic overburden, individual sinks affect a larger area when the overburden is unconsolidated to very poorly consolidated. However, when sediments of the Hawthorn Group overlie the limestones, and only small amounts of loose sediment exist over the Hawthorn, an individual sink

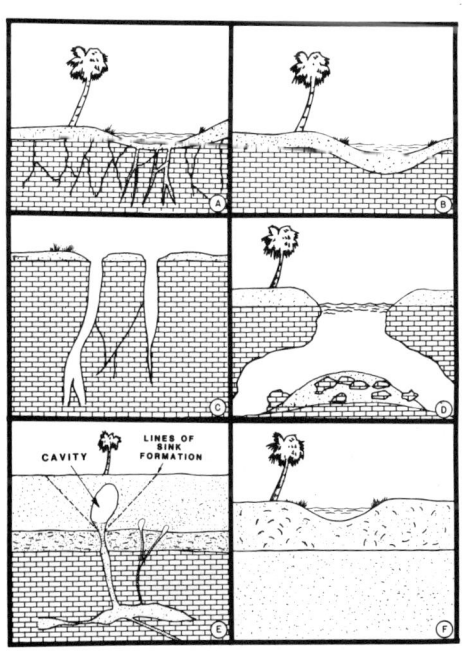

FIGURE 4 - Methods of Sinkhole Formation

15

initially affects a limited area because the Hawthorn sediments will maintain very steep slopes. These slopes will eventually erode to shallower slopes and affect a larger surface area. Fortunately this takes place very slowly and is not an immediate threat to surrounding property.

In the highland areas underlain by a thick sequence of loose sediments lying on limestone the number of sinks recognized at the surface is not as great as in the limestone plains. However, the area affected by an individual sink is many times greater due to the low angle of stability and thickness of the sediments. These sinks also form relatively quickly and therefore pose an increased threat to property.

The potential for property damage by karst is much less in areas underlain by Miocene-Pleistocene carbonates and shell beds. The frequency of occurrence of this type of sink (figure 4 B or F) is lower as can be seen on topographic maps of these areas. Also the resulting area of depression is smaller, limiting potential damage.

References

Applin, P. L., 1951, Preliminary report on buried pre-Mesozoic rocks in Florida and adjacent states: U.S. Geol. Survey Circular 91, 28 p.

Applin, P. L., and Applin, E. R., 1944, Regional subsurface stratigraphy and structure of Florida and southern Georgia: Am. Assoc. Petrol. Geol. Bull., vol. 28, n. 12, p. 1673-1753.

Isphording, W. C., 1984, Sand craters in Gulf Coastal Plain clastic sediments: An extension of the Carolina Bays phenomenon? Geol. Soc. Am., S. E. Section Abstracts with Programs, p. 148.

Jammal & Associates Inc., 1982, The Winter Park Sinkhole: A report of investigation prepared for the City of Winter Park, Florida.

Knapp, M. S., 1979, Top of the Floridan aquifer of north central Florida: Fla. Bur. Geol. Map Series 92.

Kwader, T., and Schmidt, W., 1978, Top of the Floridan aquifer of northwest Florida: Fla. Bur. Geol. Map Series 86.

Oglesby, W. R., 1965, Folio of South Florida Basin, a preliminary study: Fla. Geol. Survey Map Series 19.

Pressler, E. D., 1947, Geology and occurrence of oil in Florida: Am. Assoc. Petrol. Geol. Bull., Vol. 31, p. 1851-1862.

Schmidt, W., 1982, Florida Caverns State Park, a nature-made geologic wonderland: Fla. Bur. Geol. Leaflet 10, 16 p.

_____, and Coe, C., 1978, Regional structure and stratigraphy of the limestone outcrop belt in the Florida panhandle: Fla. Bur. Geol. Report Invest. 86, 25 p.

Scott, T.M., 1983, The Hawthorn Formation of northeastern Florida: Part I - The geology of the Hawthorn Formation of northeastern Florida: Fla. Bur. Geol. Rept. Invest. 94, p. 1-40.

_____, and Hajishafie M., 1980, Top of the Floridan aquifer in the St. Johns River Water Management District: Fla. Bur. Geol. Map Series 95.

Sweeting, M. M., 1973, Karst Landforms: Columbia University Press, New York, 362 p.

Veatch, O., and Stephenson, L. W., 1911, Preliminary report on the geology of the coastal plain of Georgia: Georgia Geol. Survey Bull. 26, 466 p.

Vernon, R. O., 1951, Geology of Citrus and Levy counties, Florida: Fla. State Geol. Survey, Bull. 33, 256 p.

White, W. A., 1970, Geomorphology of the Florida peninsula: Fla. Bur. Geol. Bull. 51, 164 p.

Winston, G. O., 1971, Regional structure, stratigraphy and oil possibilities of the South Florida Basin: Gulf Coast Assoc. Geol. Soc. Trans. 21st Am. Meeting, p. 15-29.

Winston, G. O., 1976, Florida's Ocala Uplift is not an uplift: Am. Assoc. Petrol. Geol. Bull. vol. 60, n. 6, p. 992-994.

Karst progression

DENNIS J.PRICE *Florida Department of Environmental Regulation, Live Oak, USA*

ABSTRACT

There are two areas with different rates of karst activity in the study region. A slight to moderately paleo-karst area in Madison County, Florida and a more active karst area in Suwannee County, Florida. Fractures in the underlying limestone created by the Ocala uplift control and initiatedsinkhole development in areas where the Hawthorn clay units have thinned to 21.3 meters or less. Because of the higher water table fluctuations, exposure of the surrounding limestone to the acid tannic waters enables the limestones surrounding the sink to be put into solution. This encouraged enlargement of the sinks and formed the basins present today. Locally, the land will subside to some base level controlled by the overlying sediment which fill the basin and present day sea level.

Introduction

Humid, subtropical, climatic conditions exist in the study area. The average yearly temperature is 19° C. Rainfall averages 137 centimeters per year. With highest rainfall rates occurring from June through August. During summer months tropical storms and hurricanes can provide heavy precipitation. The winter months usually have longer more sustained rainfall events. The average evapotranspiration rate is 107 centimeters per year.

Two areas have been investigated in this report. Madison County in north Florida, which is bounded on the north by Georgia, on the east by the Withlacoochee River, on the southeast by the Suwannee River, and on the south and southwest by San Pedro Bay, which is a vast lowland swamp. The city of Madison is within the study area. The karst features in Madison County are mostly large basins which appear to be relatively stable.

The Suwannee County study area is approximately 32 kilometers southeast of the city of Madison. The Suwannee River controls the water table in this area. The city of Live Oak is within the study area.

Both the city of Madison and the city of Live Oak are county seats and major centers of population for their respective counties. Both cities and the surrounding areas have flooding problems due, in large part, to the presence of sinkholes and sinkhole basins.

Geology

The following geologic units affect, control, and limit karst features in Madison and Suwannee counties.

The oldest geologic unit is the Ocala group which is composed of three members, Puri (1953). These members in ascending order and depth are the Williston, Inglis, and Crystal River formations. The Crystal River formation is a very pale orange to a very light gray, moderately indurated, biogenic, and very micritic limestone containing many larger foraminifera. The Williston/Inglis is not of particular interest except to note that it is probably cavernous. The Ocala limestone was established as Eocene in age by Cooke (1915).

The Suwannee limestone name was established by Cooke and Mansfield (1936). Its most distinctive fossil is the echinoid Ryncholampus (Cassidulus) Gouldii. The Suwannee overlies the Ocala group and is normally a very pale orange, moderately indurated, very porous calcarenite with numerous foraminifera, moluski, and echinoids present. Another lithology was described by Colton (1978) as "dense, hard, resonant limestone composed of foraminiferal tests completely embedded in dense crystalline calcite." The Suwannee limestone is Oligocene in age and unconformably overlies the Ocala group.

The Suwannee limestone is nonconformably overlain by the St. Marks' limestone, when present, and the Hawthorn formation where the former is absent. The Hawthorn formation overlies the St. Marks' limestone when both are present. The St. Marks' limestone is a very pale orange, sandy, silty, occasionally fossiliferous and micritic limestone. The Hawthorn formation is phosphatic throughout and is composed of sands, clays, and sandy silty calcareous units. The Hawthorn thins to the west and in the study area is dominantly a clay unit which acts as the confining unit restricting karst development.

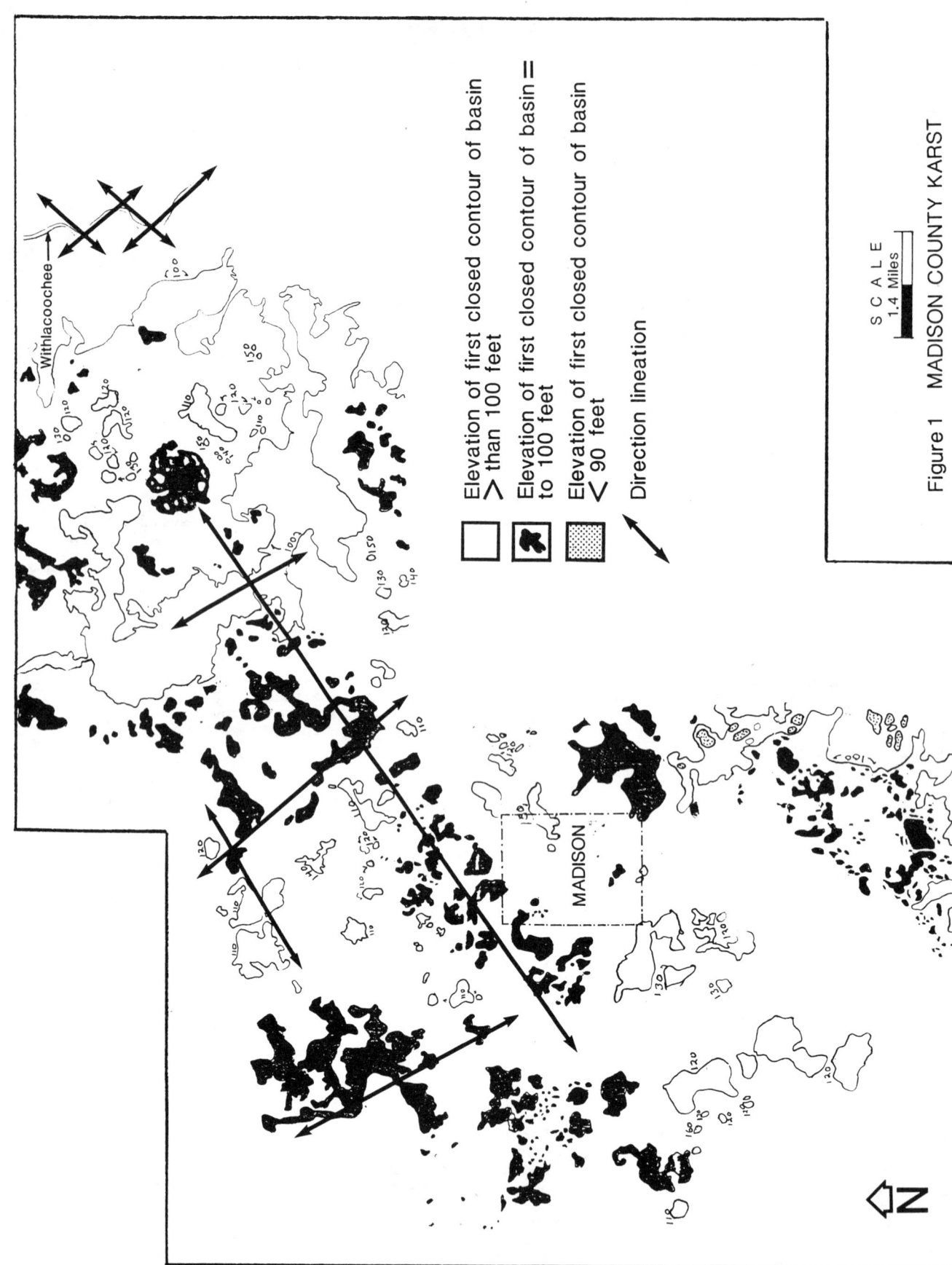

Figure 1 MADISON COUNTY KARST

The Miccosukee formation overlies the Hawthorn and is discontinuous and well represented in the Madison study area. It is moderately to poorly sorted; coarse-to-fine-grained; red and varicolored quartz, sand, and clay. This formation has been placed in the Miocene, Cooke & Mossom (1929) along with the St. Marks and Hawthorn formations.

The Plio-pleistocene age is represented by undifferentiated marine terrace deposits. Sea level fluctuations during this period resulted in these sands being deposited on three terraces. The Wicomico terrace exists from 21.3 to 30.5 meters above Mean Sea Level (MSL); the Okefenokee terrace exists from 30.5 to 51.8 meters MSL; and the Coharie Terrace from 51.8 to 65.5 meters MSL.

The erosional remnants of the Cody Scarp as described by Ceryak (1981) is the boundary between karst and non-karst areas. Throughout the study area the top of the scarp follows the 30.5 and 33.5 meters MSL surface contour line. According to Ceryak the scarp is the transition zone between artesian and non-artesian groundwater conditions in the Floridan Aquifer. The Floridan Aquifer is the main source of water for all uses in the study area. The two study areas in Suwannee and Madison counties have karst activity that occurs near this 30.5 to 33.5 meters contour line.

The Suwannee River is the dominate riverine system present. Many typical karst landforms are present; sinking streams, dry valleys, collapse dolines, solution dolines, dry lakes, and blind valleys. Surface elevations in the study areas range from 10.7 meters to 61 meters MSL. The lower elevations occur in the sinkholes and the river bed.

The Ocala uplift is the most important structural feature present that controls or exerts influence over the location of sinkholes. It was described by Vernon (1951) as "an anticline that developed in tertiary sediments as a gentle flexure, approximately 370 kilometers long and 112.6 kilometers wide." He showed the structure to be active from late Eocene to early Miocene time. The Ocala uplift created the fracture systems that control sinkhole development. The study areas are on the northern end of the uplift.

<u>Madison Karst</u>

Topographically the area is composed of high hills surrounding closed basins which vary from tens of hectares to nearly 500. Elevations range from about 24.4 meters to 61 meters MSL. The geologic units of interest which are present are the Ocala, Suwannee, St. Marks, Hawthorn, and Miccosukee formations and the marine terrace deposits. In and around the basins the St. Marks, Hawthorn, Miccosukee formations, and the marine terrace deposits have disappeared or thinned due to erosion and solution activity. At the highest elevations all of the geologic units are usually present. The first limestone encountered are some carbonates in the Hawthorn and the St. Marks. Both units occur intermittently and both are usually friable. The Suwannee limestone at these higher elevations is well indurated although it and lower limestone units are probably cavernous.

Up the sides of the hills of a basin, borings indicate that the Hawthorn clays thin and the limestones encountered first are soft, friable, and extremely cavernous. Bore holes in the center of the basin did not encounter limestone at a depth of 41 meters. The basins are full of sand and at depth coarser fragmented material is found. The sands which filled the basin are phosphatic except for the first few feet. These phosphatic sands are not represented in the typical geologic column. This indicates a complete removal of a phosphatic sandy unit in the Hawthorn. The sands by appearance are similar to the phosphate units which are mined 48 kilometers to the east and indicates the basin may be Plio-pleistocene.

The basins are recharge areas for the Floridan Aquifer. In recharge areas the water table fluctuations are greatest near the point of recharge. Bretz (1942) indicated that most solution activity in limestone takes place at the water table. Ceryak (1983) indicates water level fluctuations of 9 meters in a nearby recharge area along the Alapaha River in Hamilton County, 32 kilometers to the east. Ceryak also indicates a "high degree of secondary porosity has developed in the upper Suwannee limestone since it is generally the first major carbonate unit encountered by the downward percolating acidic ground water, especially where the aquifer is in a leaky artesian or non-artesian condition."

Personal observations in the phosphate mining region in north Florida indicate rampant solution of the phosphate rock in the first zone of phosphate exposed to the greatest fluctuation of the water table. The extreme is complete removal of the phosphate rock leaving residual phosphatic clay, vugs, and high concentrations of uranium. This is the leach zone of the north Florida phosphates.

This solution of the carbonate rocks provides a mechanism for enlarging the basins after initial sinkhole development. The evidence in boreholes indicates a less indurated and more friable limestone as the sinkhole or basin is approached. This is in the area of highest water level fluctuation. The potential exists for complete removal of a formation if its thickness is in the zone of water table fluctuation. The calcium dissolves leaving residual sands or other minerals. This appears to be what is happening to the St. Marks formation and the upper portion of the Suwannee limestone in this locality.

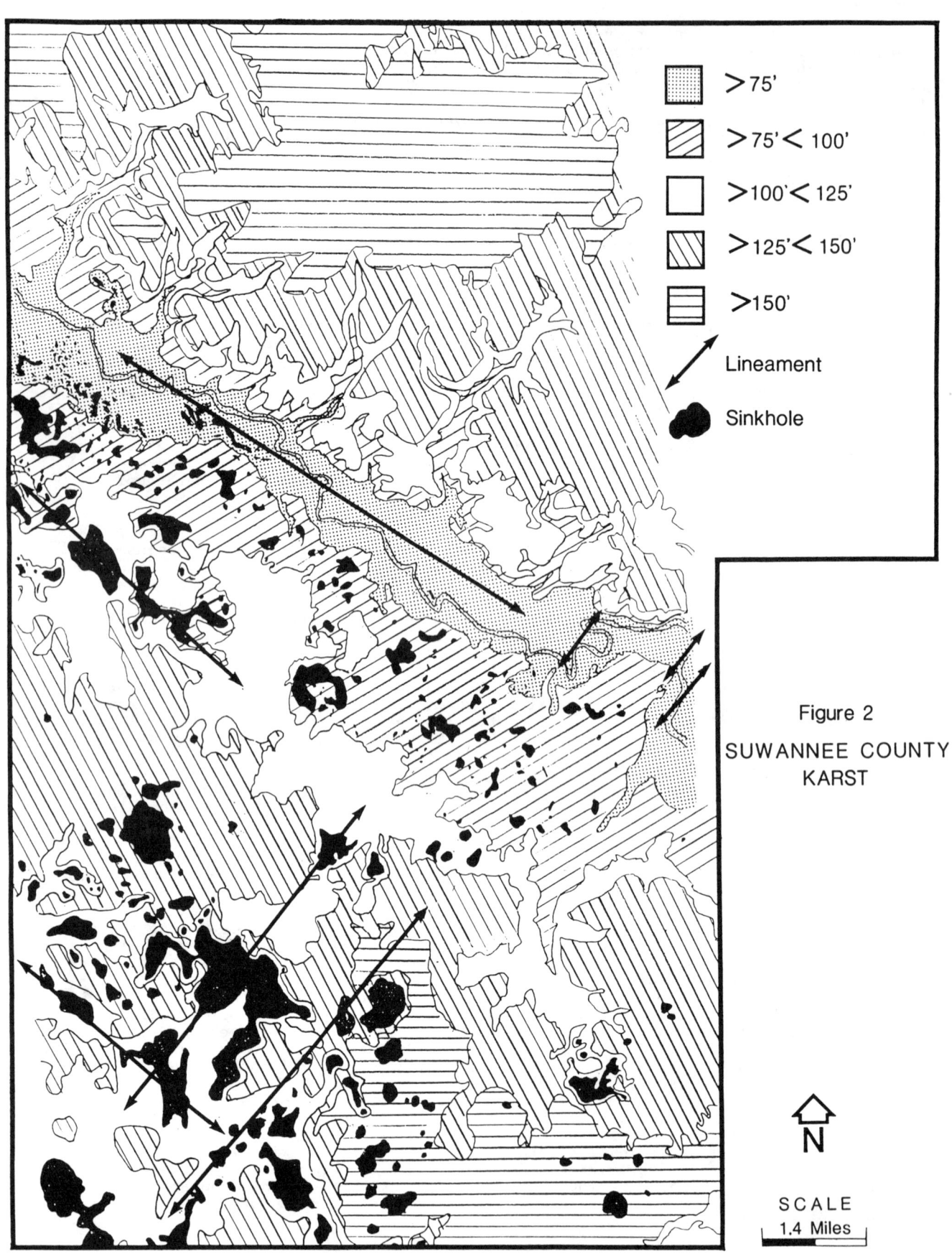

Figure 2
SUWANNEE COUNTY KARST

A result of the Ocala uplift was a regional fracturing of the rock (Vernon 1951). These fractures generally trend in northwest-southeast and northeast-southwest directions.

To determine if the original sinkholes were formed along fracture traces a map of depressions was created (Figure 1). The map shows depressions, a 30.5 meters MSL erosional feature to the northeast, and 30.5 meters MSL contour line to the southeast.

The basins are coded according to the elevation of the first closed contour line surrounding the basin. A large, wide NE-SW lineament is evident between two high ridges which parallel it on the north and south. The depressions on these ridges are higher topographically and when masked the ridges appear as two unbroken parallel features. Between these ridges are numerous basins beginning at 30.5 meters MSL. The basins between the ridges form a lineation and can be followed in the NE-SW direction.

Obscure Lineaments are observed in the direction of elongation of the basin and in conjunction with the proximity of other basins elongated in the same direction. Almost all trend in NW-SE and NE-SW directions.

It is possible to determine what areas are unstable. Major activity will be along the fracture traces or their surface expressions. Minor activity will be along the sides of the basins as they enlarge themselves. It is unlikely that any major collapse will occur in the area. Due to the absence of steep sided sinks, entrenched ravines, exposed limestone, and the amount of fill in the basins it can be reasoned that the area is fairly stable with only minor karst activity occurring. Occasionally a lake will drain indicating a connection to a lower cavernous limestone.

Suwannee County

Sinkholes were mapped to see if alignment along lineaments was obvious (Figure 2). So many sinkholes and sink depressions were present that alignments were hard to define. Several contour lines were traced (22.9, 30.5, 38.1, and 45.7 meters MSL) and erosional patterns became apparent.

Erosional patterns and sinkhole development follows the directions of fracture patterns indicated by Vernon (1951). All lineaments trend in NW-SE and NE-SW directions. The Suwannee River trends in a NW-SE direction in this area and is likely structurally controlled. This NW-SE lineament is also the direction of the dominate fracture zone in the Ocala uplift (Vernon 1951).

Ceryak (1983) stressed on the importance of the thickness of the Hawthorn as a factor in creating artesian conditions. He states that "less than 15.2 to 21.3 meters of basal Hawthorn does not constitute confining beds." Isopach maps (Knapp 1983) indicate a thinning of the Hawthorn to the west and south and a thickening to the east and north. The Hawthorn is absent as a continuous, horizontally mappable unit and only represented by isolated areas of uneroded clays and sinkhole fill in areas with the most numerous karst features.

North of the river where the Hawthorn begins to thicken and phosphatic carbonate units begin to appear there are no sinkholes or depressions (Figure 2). When closed contour lines are encountered; they indicate small, isolated, and shallow cypress heads. It is in this area that the eastern downfold of the Ocala uplift is encountered. Topographically the area is nearly flat.

South of the river the areas above the 45.7 meters MSL line are the most stable, but where surface erosion has thinned the clay unit, sinkholes begin to form. Generally the clay unit is thinner and the competency of the beds are less. Also this area is a transition zone from artesian Floridan Aquifer conditions on the east to non-artesian on the west-northwest. Drillers logs indicate a highly friable less indurated limestone, the first 9 to 15 meters of rock encountered in this area. There is also a large water table fluctuation around the sinks. This allows the acidic tannic waters to come in contact with a greater thickness of limestone. This creates highly weathered areas radiating out from the karst depressions. The combination of the thinning of the clay units, fractures that encouraged sinkhole development and fluctuating water tables causes the solution of the first limestones encountered, widening of sinkholes, and lowering of land surface to a local base level.

Conclusions

In the study areas the initial formation of sinkholes occurs along fracture zones in the underlying carbonates. The fractures occur in northwest-southeast and northeast-southwest directions. The initial sinkholes extend into deeper cavernous zones in the Ocala limestone.

Because carbonate rocks go into solution more readily at the ground water table, the greatest solution activity begins around the original collapse where water level fluctuations are the greatest. The highly acidic, tannic waters dissolve the carbonates down to the lowest water table elevations. This is evidenced by the highly porous carbonates encountered around the sinkholes and provides the mechanism for widening and eventual creation of large depressions. This activity occurs largely in the transition zone between artesian and non-artesian Floridan Aquifer conditions. This transition zone coincides with remnants of the Cody Scarp at about the 30.5 meters MSL contour line.

In areas where sinkhole activity is sparse or non-existent, which coincide with thicker overlying clay units, well drillers generally are able to set casing in the upper section of the Suwannee limestone. In transition zones where the limestone is porous and friable due to solution activity, casing has to be set well into the Suwannee limestone because here the first carbonates encountered are subject to caving and have sand present.

Base level is controlled by the thickness of the calcareous units that will eventually be removed by solution and by the amount of overlying sands and clays that will fill the depressions. Because much of the clay will be carried away by erosion and loss of some sands into the cavernous zones, the resultant base level elevation will be the thickness of units that do not go into solution or are carried away by stream erosion and cavern filling.

REFERENCES

Bretz, J. H., 1942, Vadose and Phreatic Features of Limestone Caverns. Journal of Geology, Volume 50.

Ceryak, Ron, 1981, Significance of The Cody Scarp on the Hydrology of North Central Florida, Southeastern Geologic Society, Guidebood Number 23.

Ceryak, Ron; Knapp, Michael S.; and Burnson, Terry Q., 1983, The Geology and Water Resources of the Upper Suwannee River Basin, Florida. Florida Bureau of Geology Report of Investigation No. 87.

Colton, R. C., 1978, The Subsurface Geology of Hamilton County, Florida with Emphasis on the Oligocene Age Suwannee Limestone. Unpublished Master's Thesis Florida State University, Tallahassee, Florida.

Cooke, C. Wythe, and Mossom, Stuart, 1929, Geology of Florida. Florida Geological Survey 20th Annual Report.

Cooke, C. Wythe, 1915, The Age of the Ocala Limestone. U.S. Geological Survey Professional Paper 95.

Puri, Harbans, 1953, Zonation of the Ocala Group in Peninsular Florida. (Abstract) Journal of Sedimentary Petrology, Volume 23.

Vernon, R. Q. 1951, Geology of Citrus and Levy Counties, Florida. Florida Geologic Survey, Bulletin No. 33.

Impact of ground-water chemistry on sinkhole development along a retreating scarp

SAM B.UPCHURCH *University of South Florida, Tampa, USA*
FRED W.LAWRENCE *McCoy and McCoy, Inc., Lexington, Kentucky, USA*

ABSTRACT

The Cody Escarpment of north Florida is a classic, retreating, karst escarpment. It is the middle of three ground-water chemical domains, which contain: (1) confined, "stagnant", calcite-saturated water under the upland, (2) agressive, organic-rich water from recharge along the scarp, and (3) unconfined, slightly-agressive water characterized by lateral flow below the scarp.

The upland is underlain by the low permeability, Tertiary Hawthorn Formation, which includes upper siliciclastic and lower dolomitic strata. Where the Hawthorn is intact, there is minimal active karst. Along fracture traces and localized zones of increased permeability, the dolostone portion of the Hawthorn is subject to development of interstratal karst. This leads to collapse and subsidence sinks. At the escarpment, Hawthorn sinks coalesce into large uvalas and poljes which relfect the thick overburden where vertical karst develops in the underlying Floridan aquifer.

The Floridan aquifer consists of limestones that have been subjected to multiple karst cycles. Paleosinks are common along the unconformity with the Hawthorn. Recharge is voluminous and highly localized in the large karst basins along the escarpment. The escarpment is, therefore, a zone of vertical karst development with initiation of new conduits and reactivation of old ones. Where the Hawthorn has been removed by scarp retreat, the Floridan is unconfined and recharge is diffuse through small, abundant sinkholes. The pre-Hawthorn paleosinks are exposed in this area and reactivation as alluvial dolines is common.

Water chemistry in the Floridan aquifer reflects the three karst domains. Saturation-index calculations on samples taken from the aquifer where it is confined by the Hawthorn indicate that the water is in equilibrium with respect to calcite and, in some areas, dolomite. This water is not actively recharged and has a high residence time in the aquifer. At the escarpment, saturation-index calculations suggest supersaturation with respect to calcite, but dissolved phosphate and iron concentrations show that, while calcium concentrations are abnormally high, activities are low owing to organic complexing. Thus, recharge is dynamic and the surface-derived water is aggressive to the limestones, which leads to rapid development of karst, especially vertical conduits. In unconfined areas, diffuse recharge leads to the presence of slightly undersaturated waters, which continue slow karstification.

Introduction
There have been many studies that have dealt with the chemistry of karst formation within small-scale cavern systems (e.g., Bogli, 1980). These studies have emphasized the importance of (1) chemical equilibration with the host rock, (2) outgassing of carbon dioxide, (3) water flow path and residence time, and (4) mixing with other waters. There have also been a number of studies of a regional scale that show the importance of flow path and chemical maturation over long distances and suggest that karstification is most intense near the region of recharge (e.g., Back and Hanshaw, 1970; Plummer, 1977).

Studies at an intermediate scale that show the synoptic chemistry of a karst system are lacking, however. Two data sets exist that are unique for the synoptic investigation of intermediate-scale controls on ground-water chemistry in a recharge area and on the development of a karst escarpment. These studies include detailed chemical analyses of over 350 ground-water samples from two, four-township areas that straddle a retreating scarp in north Florida.

Lawrence and Upchurch (1976, 1982) used the two data sets to investigate the utility of factor analysis for identification of the chemical provenance of ground water. They found that the method could be used (1) to identify the rock types with which the water had been in contact during recharge and flow to the sample site, (2) to evaluate the "intensity" of chemical equilibration with host rocks and of waste loading from surface waters, and (3) to determine flow paths. Their studies clearly indicated that maximum recharge was localized along the escarpment and that the karst aquifer water was greatly influenced by position relative to the scarp. This paper describes the ground-water chemistry of the karst aquifer in these two areas and relates the chemistry to the geomorphic features of the scarp system.

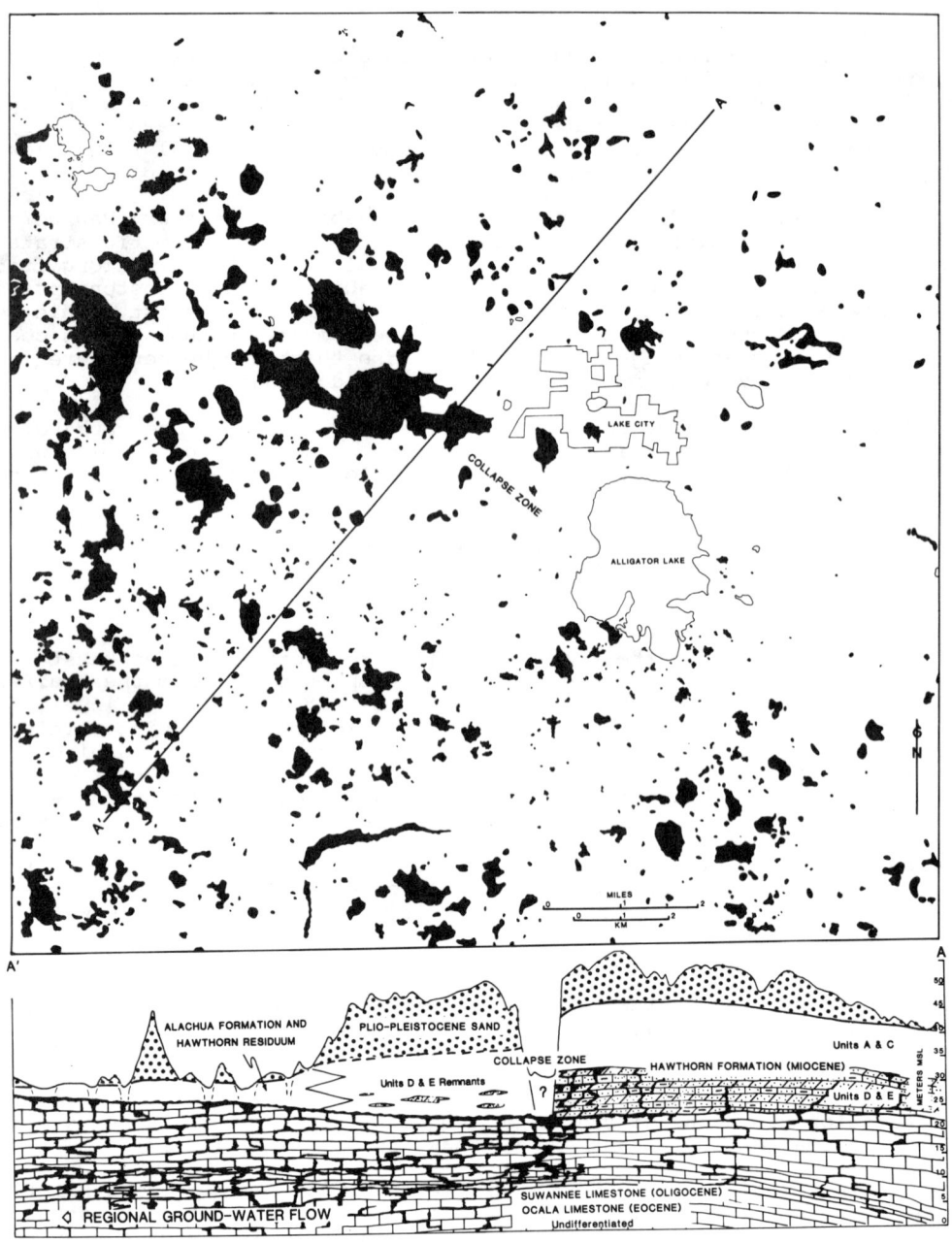

Figure 1: Map and geologic cross section of the Lake City study area. Dark areas on the map are closed depressions.

The two data sets are from four-township areas along the Cody Escarpment in north Florida. One data set is centered about the city of Live Oak, Suwannee County, and the other about Lake City, Columbia County. Both are small communities and have light industry and agricultural economies. Both are located on the escarpment and, owing to significant internal drainage, utilize drainage wells for storm runoff (Hull and Yurewicz, 1979). Chemical influences of these drainage wells are localized (Lawrence and Upchurch, 1976, 1982) and they do not interfere with recognition of regional ground-water processes. For the sake of brevity, only the Lake City data set is illustrated and discussed in this paper. The Live Oak data set shows similar results.

Figure 1 shows the distribution of closed depressions and a geologic cross section in the Lake City area. The northeast half of the study area is characterized by a relatively flat upland, the Northern Highlands (White, 1970), that is underlain by the Miocene Hawthorn Formation. The Hawthorn in the area consists of four lithologic units (Miller, 1978). The upper two (A,C ibid., fig. 1) are dominantly siliciclastic and the lower two (D and E) are dominantly dolostones. The Hawthorn serves as a confining layer for the underlying Floridan aquifer (Ceryak, 1977; Ceryak et al., 1983). Few sinks penetrate to the Floridan and limited leakance is by diffuse flow. The dolostones contain interstratal karst (Wadsworth et al., 1983) near the escarpment (Copeland, 1981; Ceryak, 1981). Interflow through this interstratal karst enhances scarp retreat and doline expansion and coalescence at the scarp.

Lake City is located on the Cody Escarpment in an area of numerous, large sinkhole lakes. The escarpment has a crest at 33.5 m and extends downward to approximately 30.5 m MSL. The sinks and sinkhole lakes along the scarp are aligned along a major fracture trace (Lawrence and Upchurch, 1976) or "collapse zone" (Meyer, 1962). This alignment has been noted in other areas of Florida (e.g., Williams et al., 1977) and seems to be a consistent pattern.

The lakes are organic rich and many are eutrophic. The largest lake, Alligator Lake, drains periodically and nearby wells pump debris from the lake during periods of accelerated drainage. The poljes and uvalas along the escarpment are floored in the Floridan aquifer, which is composed of the Oligocene Suwannee Limestone and the Eocene Ocala Group. Outliers along the escarpment contain Hawthorn Formation siliciclastics and little or no dolostone owing to dissolution in the interstratal karst. The hills within the escarpment and immediately north of the city are capped by a sand ridge known as the Lake City Ridge (Pirkle, 1972; Ceryak et al., 1983).

The southwestern half of the area is a lowland, the Coastal Lowlands of White (1970), overlain by residuum of scarp retreat. This material is a clayey sand, which is partly attributed to the Alachua Formation. The residuum is thin and is penetrated by many small sinks (Fig. 1). Underlying Suwannee and Ocala limestones are riddled by horizontal caverns. Karst is well developed on this plain and many of the small sinks are pre-Hawthorn in age (Price, 1981; Mac Fadden, 1982). These are partly filled with residuum, but many are well drained and represent no confinement to the Floridan aquifer. Sinkhole development in this area is largely through reactivation of the paleosinks and enlargement of karst inherited from scarp-related processes.

<u>Ground-Water Chemistry</u>
In order to evaluate the impact of limestone and dolostone dissolution on karst development, 152 wells in the Lake City area and 109 in the Live Oak were sampled. All were domestic wells completed into the upper Floridan and many were drilled by one contractor, so construction was more-or-less uniform.

At the sample site the well was pumped to remove well bore effects and temperature, pH, and alkalinity were immediately obtained. A complete major element analysis was completed according to standard methods. Chemical-equilibria calculations for major, rock-forming minerals at environmental temperatures were determined using program WATEQF (Plummer et al., 1976). The saturation index (ion activity product divided by the equilibrium constant) was calculated to identify areas where ground waters are aggressive to the host rock.

Figure 2 illustrates the saturation indices for limestone (calcite) and dolostone (dolomite) in the Lake City area. Note that the Floridan waters are at equilibrium with respect to dolomite only in small areas immediately beneath the Hawthorn uplands. These are regions where the interstratal karst developed in the Hawthorn dolostone confining beds is connected to the Floridan.

Saturation indices with respect to calcite and hydroxylapatite show a distinctive pattern that corresponds with three geochemical domains in the Floridan aquifer. Comparison of Figure 1 with the Lake City data (Fig. 2) shows that the northwest-southeast belt of water saturated with respect to both minerals coincides with the Cody Escarpment. Under the

Northern Uplands the water is mostly saturated with respect to calcite, but not apatite. Under the Coastal Lowlands portion of the study area the water is undersaturated with respect to both minerals.

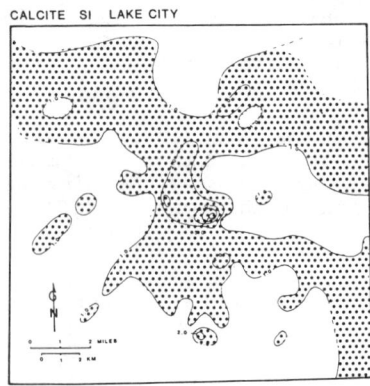

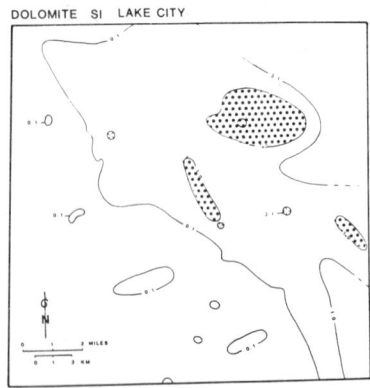

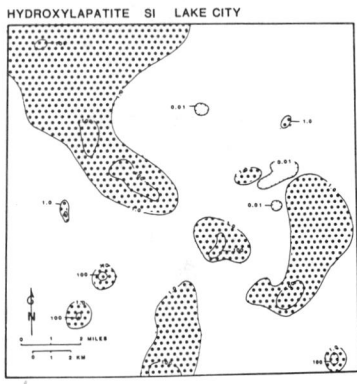

Figure 2: Distribution of saturation indices for calcite, dolomite, and hydroxylapatite in waters from the Floridan aquifer in the Lake City area. Shaded areas are oversaturated with respect to the minerals.

The Floridan aquifer potentiometric surface under the Northern Highlands is nearly horizontal (Ceryak et al., 1983) and Lawrence and Upchurch (1976, 1982) showed geochemically that water in the area is relatively stagnant. In this chemical regime, the water reaches equilibrium with the limestone and regions of supersaturation are lacking. Owing to the bicarbonate present in the water, phosphorus precipitates quantitatively (Odum, 1953) and apparent undersaturation results.

The scarp is characterized by oversaturation with respect to calcite and hydroxylapatite (Fig.2). There is a region of supersaturation with respect to calcite and apatite in the center of the domain. This region coincides with a number of leaky sinkhole lakes, including Alligator Lake. Lawrence and Upchurch (1976) attributed high phosphate concentrations in ground water in this belt to leakance from these lakes and color and iron concentration data support this interpretation. This belt is also characterized by high bacterial counts and lake debris in wells during periods of documented lake drainage. Therefore, it can be concluded that the northwest-southeast belt is a region or recharge along the scarp.

Owing to the insolubility of phosphate minerals in alkaline waters (Odum, 1953), supersaturation with respect to calcite is inconsistent with high phosphate concentrations. Because of the presence of color and iron concentrations in the water, high calcium and phosphate concentrations can be attributed to the presence of chemical complexes with humic substances (Manskaya and Drozdova, 1968; Upchruch et al., 1980, 1983), which is consistent with the color and iron data. Organic complexes are known to be important in the dissolution of calcite and undoubtedly contribute significantly to the formation of the dolines in the scarp. Further work on this problem is underway in a new study at the University of South Florida at the present time. Northeast-southwest zones of saturation reflect other fracture traces and dramatize the importance of fractures in rapid recharge (Lattman and Parizek, 1964).

In the Coastal Lowlands the water is generally undersaturated with respect to calcite and apatite. This reflects diffuse recharge through sandy soils and small sinks with minimal organic content, contact with limestone, or, residence time. Floridan aquifer water is, therefore, slowly dissolving limestone through consumption of small amounts of soil carbonic and organic acids. Karst patterns here are dominated by subhorizontal conduits, and mixing and equilibration with waters recharged from the scarp takes place.

Summary and Conclusions
　　The Cody Escarpment of north Florida is a classic retreating escarpment underlain by semi-permeable, middle to late Tertiary strata. The semi-permeable strata include the middle Miocene Hawthorn Formation and younger, unnamed clastic sediments. The Hawthorn consists of two upper, siliciclastic-dominated and two lower, dolostone-dominated units. A conspicuous, east-west sand ridge, the Lake City Ridge, caps the semi-permeable strata along portions of the edge of the escarpment.

　　Limestone of the Oligocene Suwannee Limestone and the Eocene Ocala Group underlie the area. They constitute the Floridan aquifer, which is confined under the Hawthorn to the northeast, partly confined at the escarpment, and unconfined in front of the escarpment.

　　Ground-water chemical data from over 300 wells in the vicinity of the escarpment show three distinctive geochemical domains characterized by the degree of saturation with respect to calcite, dolomite, and hydroxylapatite. These three domains correspond with the geomorphology of the area and suggest that sinkhole development and morphology are controlled by (1) surface-water drainage and organic content, (2) near-surface rock lithology and structure, and (3) thickness of confining beds.

　　Under the semi-permeable Hawthorn of the Northern Uplands, the Floridan experiences minimal recharge and lateral flow. Recharging water is saturated or slightly undersaturated with respect to calcite and dolomite. Dolomite saturation is a result of contact with the dolostones of the lower Hawthorn. The water is undersaturated with respect to hydroxylapatite because of rapid, quantitative precipitation in the presence of high bicarbonate.

　　Water composition under the Northern Uplands suggests that minimal karst development occurs in the Floridan in this zone. Development of interstratal karst in the Hawthorn dolostones results in collapse sinks and isolated vertical conduits. As the escarpment retreats and the siliciclastic portion of the Hawthorn thins, these sinks become more abundant and are enlarged, especially along fracture zones. These sinks form the nucleus of major karst development along the escarpment.

　　The escarpment domain is characterized by major uvalas and aligned poljes that receive surface runoff and interflow from the interstratal karst in the lower Hawthorn. Vertical conduits develop in the Floridan aquifer at the escarpment and recharge is voluminous and localized in the large sinks. Saturation-index calculations, assuming that all chemical reactions are inorganic in nature, suggest that ground water under the escarpment is highly oversaturated with respect to calcite and hydroxylapatite. Comparison with color and iron data and an understanding of the solubility relationships of apatite in alkaline waters, however, indicate that inorganic equilibria do not apply. Organic complexing of calcium and phosphorus by waters rich in humic substances derived from decay of plants in tributary lakes and streams reduces the activities of calcium and phosphate and greatly increases the aggressiveness of the water to limestone dissolution. Thus, the escarpment constitutes the major zone where aggressive water is localized and karst development is most active. Development of vertical conduits under the rapid recharge of aggressive water results. Where this process coincides with major fracture traces that predispose karst development, poljes are large and aligned. Sinkhole development is maximal in this zone and new sinks, intermittently-draining lakes, and reactivation of paleosinks are expected.

　　In the unconfined, Coastal Lowlands part of the area, a karst plain has developed. Water is slightly undersaturated with respect to dolomite, calcite, and hydroxylapatite. This undersaturation is due to dilution through dispersion of recharge into large numbers of low volume recharge sites in small sinks. Residence time of water in the sinks is minimal compared to that of the escarpment domain, so organic complexing is regionally insignificant. Mixing with saturated waters recharged from the scarp may also lead to undersaturation (Bogli, 1980). Sinkhole development, therefore, consists of infrequent initiation of small collapse sinks, reactivation of paleosinks, and enlargement of alluvial dolines.

Acknowledgements
　　The data collection and chemical analyses for this study were supported by the Suwannee River Water Management District. Computer time was furnished by the University of South Florida's Center for Applied Geology. We thank Dave Strout and Amy Swancar for their assistance in data analysis and drafting.

References Cited

Back, W., and Hanshaw, B.B., 1970, Comparison of chemical hydrogeology of the carbonate
　　peninsulas of Florida and the Yucatan: Jour. Hydrol., 10:330-368.

Bogli, A., 1980, Karst Hydrology and Physical Speleology: New York, Springer-Verlag, 284 p.

Ceryak, R., 1977, Alapaha River Basin: White Springs, FL, Suwannee River Water Mgt. Dist., Info. Circ. 5, 20 p.

Ceryak, R., 1981, Significance of the Cody Scarp on the hydrogeology of north central Florida: Tallahassee, FL, Southeastern Geol. Soc., Guidebook No. 23, pp. 24-29.

Ceryak, R., Knapp, M.S., and Burnson, T., 1983, The Geology and Water Resources of the Upper Suwannee River Basin, Florida: Fla. Bur. of Geology, Rept. of Invest. No. 87, 165 p.

Copeland, R.E., 1981, Mature karst features in north central Florida: Tallahassee, FL, Southeastern Geol. Soc., Guidebook No. 23, pp. 9-14.

Hull, R.W., and Yurewicz, M.C., 1970, Quality of storm runoff to drainage wells in Live Oak, Florida, April 4, 1979: U.S. Geol. Surv., Open-File Rept. 79-1073, 40 p.

Lattman, L.H., and Parizek, R.R., 1964, Relationship between fracture traces and the occurrence of ground water in carbonate rocks: Jour. Hydrol., 2:73-91.

Lawrence, F.W., and Upchurch, S.B., 1976, Identification of geochemical patterns in ground water by numerical analysis: In E.A. Zaleem (ed.), Advances in Ground Wate Hydrology, Minneapolis, Amer. Water Resour. Assoc., pp. 199-214.

Lawrence, F.W., and Upchurch, S.B., 1982, Identification of recharge areas using geochemical factor analysis: Ground Water, 20:680-687.

MacFadden, B.J., 1982, An overview of late Cenozoic terrestrial vertebrates from northcentral Florida: Tallahassee, FL, Southeastern Geol. Soc., Guidebook No. 24, pp. 40-64.

Manskaya, S.M., and Drozdova, T.V., 1968, Geochemistry of Organic Substances: New York, Pergamon Press, 345 p.

Miller, J.A., 1978, Geologic and geophysical data from Osceola National Forest, Florida: U.S. Geol. Surv., Open-File Rept. 78-799, 101 p.

Meyer, F.W., 1962, Reconnaissance of the Geology and Ground-Water Resources of Columbia County, Florida: Fla. Geol. Surv., Rept. of Invest. No. 30, 74 p.

Odum, H.T., 1953, Dissolved phosphorus in Florida waters: Fla. Geol. Surv., Rept. of Invest. No. 9, pp. 1-40.

Pirkle, W.A., 1972, Trail Ridge, a relict shoreline feature of Florida and Georgia: Unpubl. Ph.D. dissert., Chapel Hill, Univ. North Carolina.

Plummer, L.N., 1977, Defining reactions and mass transfer in part of the Floridan aquifer: Water Resources Research, 13:801-812.

Plummer, L.N., Jones, B.F., and Truesdell, A.H., 1976, WATEQF - A Computer Program for Calculating Chemical Equilibrium of Natural Waters: U.S. Geol. Surv., Water Res. Invest. 76-13, 61 p.

Price, D., 1981, Paleokarst features in Hamilton County, Florida: Tallahassee, FL, Southeastern Geol. Soc., Guidebook No. 23, pp. 30-32.

Upchurch, S.B., and Strom, R.N., 1980, Role of humic substances in transportation and deposition of calcium. In P.J. Gleason (ed.), Water, Oil, and the Geology of Collier, Lee, and Hendry Counties: Miami, FL, Miami Geol. Soc., Fieldtrip Guidebook, pp. 5-9.

Upchurch, S.B., Strom, R.N., and Williams, M.J., 1983. Preservation of dolomite in coastal peats of the Ten Thousand Islands area, Florida: In R. Raymond, Jr. and M.J. Andredjo (eds.), Mineral Matter in Peat: Its Occurrence, Form, and Distribution, Los Alamos, NM, Los Alamos Nat. Lab., pp. 215-224.

Wadsworth, J.R., Jr., Brook, G.A., and Carver, R.E., 1983, Surface expression of heavily mantled interstratal karst bordering Okefenokee Swamp, Georgia: Tech. Pap., 49th Ann. Meeting, Amer. Soc. Photogrammetry, pp. 463-470.

White, W.A., 1970, The Geomorphology of the Florida Peninsula: Fla. Bur. of Geology, Geol. Bull. No. 51, 164 p.

Williams, K.E., Nicol, D., and Randazzo, A.F., 1977, The Geology of the Western Part of Alachua County, Florida: Fla. Bur. Geology, Rept. of Invest. No. 85, 98 p.

Sinkhole collapse induced by groundwater pumpage for freeze protection irrigation near Dover, Florida, January 1977

SUSAN J. METCALFE *Seaburn and Robertson, Inc., Tampa, Florida, USA*
LARRY E. HALL *Stauffer Chemical Co., Richmond, California, USA*

ABSTRACT

West Central Florida experienced a period of unusual low temperatures in January 1977, with subfreezing temperatures occurring on five successive nights. The area east of Tampa is heavily agricultural, with one of the primary products being winter strawberries. Growers irrigate their crops with warm groundwater to prevent freeze damage. Simultaneous withdrawals of large quantities of water from the artesian Floridan aquifer resulted in declines in the potentiometric surface of as much as 18m (60 ft) in the vicinity of the strawberry fields. The top of the limestone aquifer is generally at about 15-20m (50-65 ft) below land surface, and is separated from surficial sands by a clay confining unit which is discontinuous. A total of at least 22 new sinkholes formed during this period by collapse of overburden induced by loss of hydrostatic support.

Introduction

During the period of January 17 to 22, 1977, west central Florida experienced freezing weather with lows to -3.9°C (25°F). To prevent freeze damage, the strawberry growers in the Dover area of eastern Hillsborough County irrigated their crops with warm groundwater withdrawn from the Floridan aquifer. The stress placed on the aquifer resulted in a large decline in water levels and in numerous small collapses or sinkholes.

Location and Land Use

The 3100 hectare (12 square mile) study area, shown in Figure 1, is centered around the rural community of Dover, Florida, east of Tampa.

The present land use is predominantly agricultural. Products and crops include citrus, strawberries, vegetables, chickens, and tropical fish. There is some cattle grazing on reclaimed phosphate mining areas.

Geology and Hydrology

The study area is a part of the Gulf Coastal Plain, and is underlain by Tertiary and Quaternary sedimentary deposits which dip to the south and southwest. These sedimentary rocks are largely limestones and are covered with sand of Pleistocene age. Economic deposits of phosphate occur near this area.

The stratigraphy was described in detail by Carr and Alverson (1959), and Applin and Applin (1944). In general terms, the geologic section consists of 3-6m (10-20 ft) of surficial sands; 7-15m (25-50 ft) of clay, phosphate, clayey sand, and clayey limestones underlain by more than a thousand meters of various limestones and dolomites. The upper 300m (1000 ft) of the carbonate section contains the Floridan aquifer, which is a very productive artesian aquifer that supplies the water needs of the area.

Carr and Alverson (1959) reported very deep sand-filled sinkholes in the study area. One of their test holes was drilled to a depth of 46m (150 ft). This hole penetrated through sand to about 30m (100 ft) below the expected top of limestone. Cathcart (1963) noted the presence of buried karst and stated that the age of sinkhole formation ranges from late Miocene to the present.

Another discussion of the solution activity within the study area is in a consultant's report prepared for the City of Tampa (Watson and Company, 1973). This report states that there are several sinkholes within the proposed wellfield and that two new sinkholes formed during aquifer test pumping. The report also states that a test well had to be abandoned when it penetrated a sand column to a depth of 94m (310 ft), about 79m (260 ft) below the expected top of limestone.

Menke, et. al. (1961) noted the presence of sand-filled sinkholes and, after reviewing Tampa's wellfield data, concluded that there is a large amount of sand cavity fill in the study area. The authors of this paper have observed possible cavity fill during the construction of a well at the Dover-Turkey Creek fire station. The well penetrated an interval of chert with loose sand and black sand size phosphate at a depth of 57 to 62m (188 to 205 ft). This sandy unit is overlain by 31m (103 ft) of limestone and the underlying limestone is continuous to the total depth of 91m (300 ft).

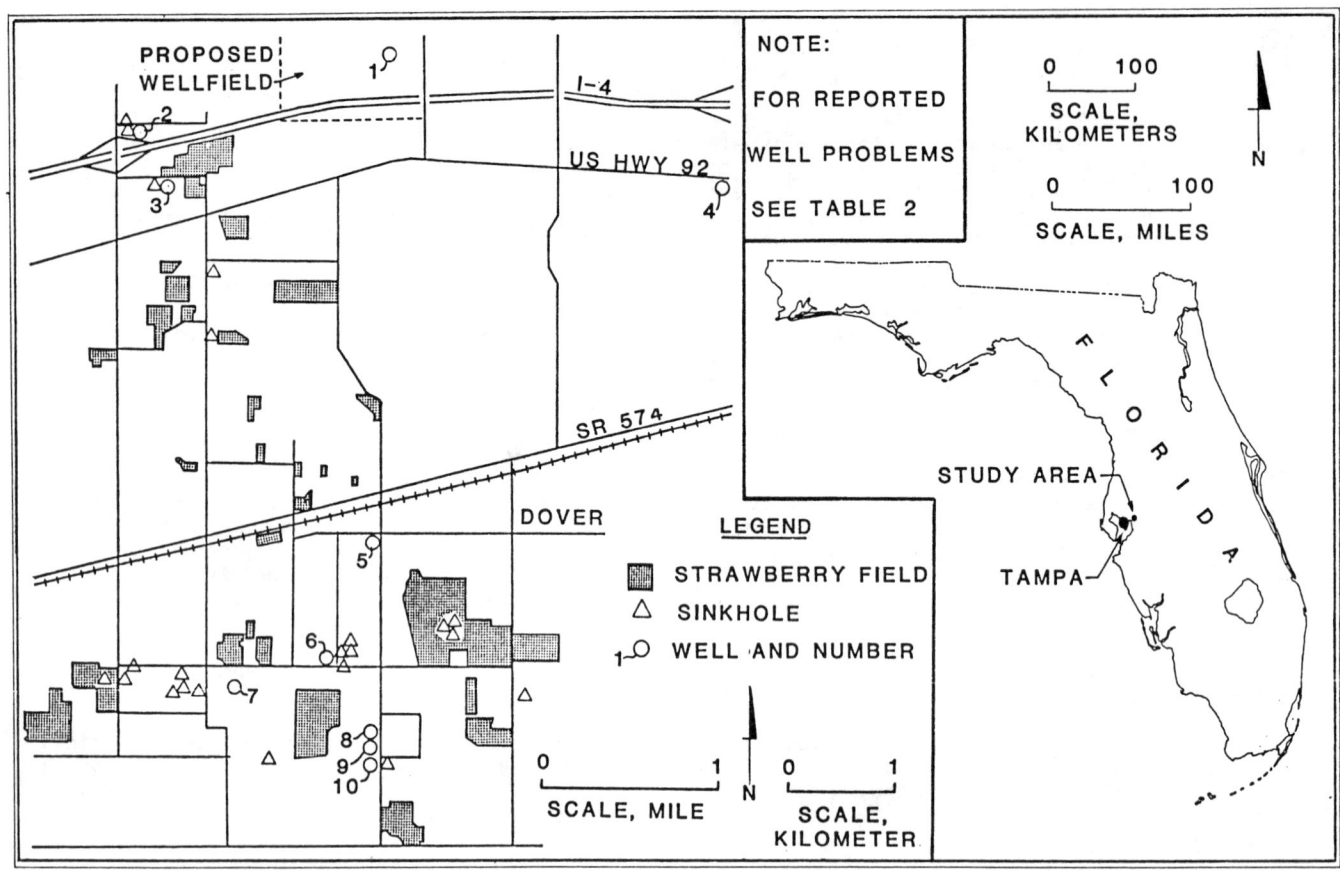

Figure 1: Location of Strawberry Fields, Sinkholes and Wells Reporting Water Level Declines, January 1977, Dover, Florida.

Land surface elevations range from a high of 40m NGVD (130 ft) in the community of Dover to a low of about 20m (65 ft) in the extreme northwest and southwest parts of the study area, averaging about 27.5m (90 ft).

The surficial sands normally contain a water table which occurs from 0-1.5m (0-5 ft) below land surface, depending on location and recent rainfall. The potentiometric surface of the Floridan aquifer fluctuates from wet season (late summer) to dry season (late spring). The normal annual range of fluctuation is about 3m (10 ft). Flow in the Floridan is southwest. Potentiometric surface elevations across the area in the May 1977, dry season ranged from about 7.6 to 16.8m (25 to 55 ft).

Effects of Freeze Protection Irrigation

Strawberry farming is a very important element in the economy of the study area. The plants are set out in the fall and berries are harvested during the following spring. Though Florida winters are mild, there are a few nights when frost or freezing conditions may exist. In order to protect the crop, the farmers spray irrigate their fields with warm, 23°C (73°F), groundwater withdrawn from the Floridan aquifer. Ice forms during certain wind and temperature conditions and coats the plants, however, by constantly applying warm groundwater, the temperature of the continually forming ice stays at 0°C (32°F), and the plants do not freeze.

Three new record low temperatures were established for the Tampa Weather Station during the period of January 17-22, 1977. Table 1 shows the daily minimum temperatures for area weather stations; Lakeland (24km (15 mi) east), Tampa (29km (18 mi) west) and Riverview (16km (10 mi) south.

	January 1977							
Station/Date	16	17	18	19	20	21	22	23
Lakeland WSO[1]	7.8/46	6.1/43	3.9/39	-3.9/25	-2.8/27	2.2/36	1.7/35	4.4/48
Tampa WSO[1]	6.7/44	0.6/33	-1.7/29	-2.8/27	-3.3/26	-0.6/31	-0.6/31	5.0/41
Riverview WTP[2]		1.1/34	-2.2/28	-2.2/28	-2.8/27	0.0/32	0.0/32	3.3/38

1) National Oceanic and Atmospheric Administration - Climatological Data
2) Hillsborough County Utilities, Riverview Water Treatment Plant

Table 1: Daily Minimum Temperatures, °C/°F, January 16-23, 1977, at Lakeland, Tampa, and Riverview, Florida.

The strawberry farmers began to pump warm groundwater for freeze protection irrigation with the onset of the record cold weather, resulting in a drawdown of 2.9m (9.5 ft), as shown by the hydrograph in Figure 2. Drawdowns of up to 18m (60 ft) were reported elsewhere.

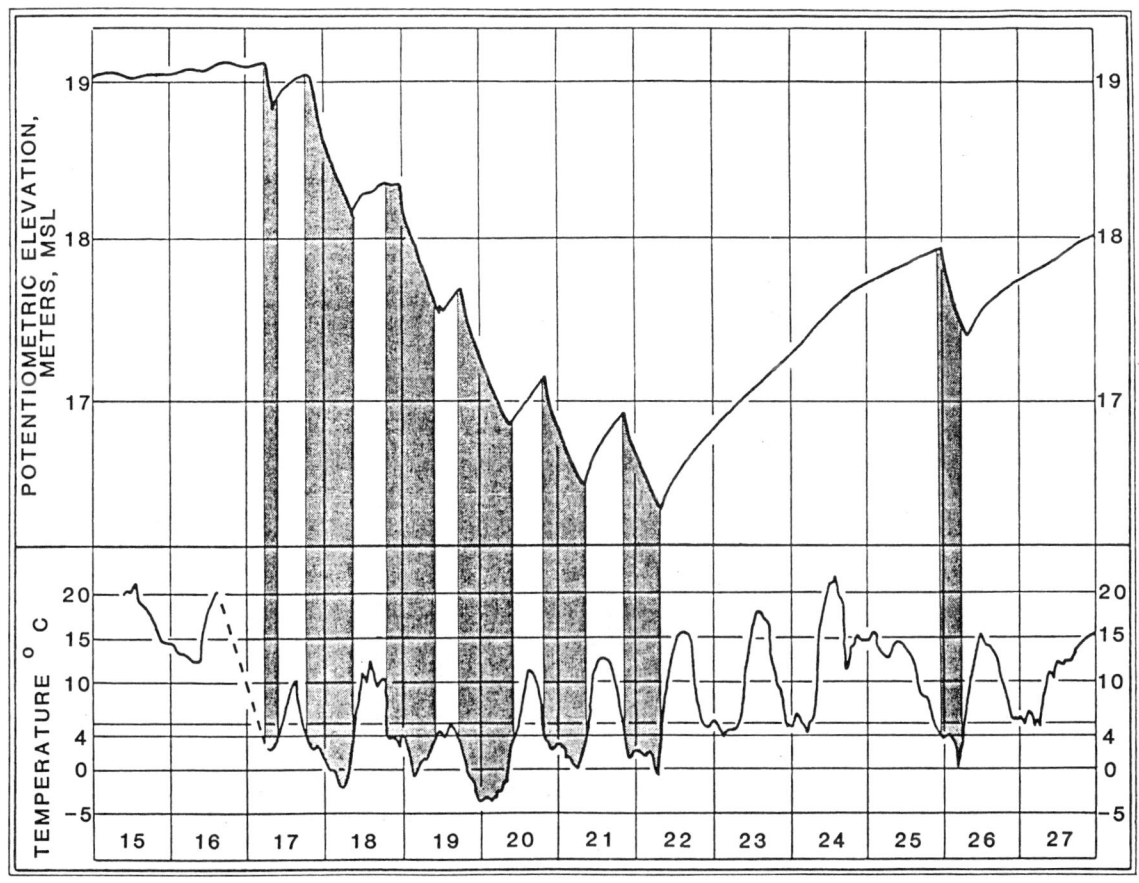

Figure 2: Potentiometric Surface Elevation at Tampa Well #15 and Thermograph at Riverview, Florida, January 1977.

Figure 2 compares the hydrograph of Tampa Well 15 in the proposed Thonotosassa wellfield and the thermograph from the Riverview weather station. It can be easily seen that the drawdown portion of each pumping cycle occurred when the temperature dropped below 4°C (40°F) and continued until the temperature rose above 4°C (40°F). Approximately 95 percent of the pumping time occurred when the temperature was at or below 4°C (40°F).

Residents affected by large drawdowns were forced to modify or repair pumping equipment in order to obtain water for domestic use. Several residents reported that during the cold weather their wells would not pump water; however, as the weather warmed the wells returned to service as water levels recovered. Table 2 lists the reported effects of water level declining in the wells whose locations were identified on Figure 1.

Map # (Figure 1)	Well Depth Data Total m/ft.	Casing m/ft.	Pump m/ft.	Comments and Data Source
1	126/413	20.5/67	None	Drawdown 2.9m/9.5 ft. (see hydrograph Figure 2) (R)eport
2	-	-	183/60	Reset pump at 27.4m/90 ft. during freeze (I)nterview
3	30.5/100	21.3/70	-	Measured 3.7m/12 ft. drop in water level (I)
4	28/92	-	-	No water for about 1 week. Changed to a jet pump and set it deeper (I)
5	36.6/120	-	12.2/40	Reset pump intake to 24.4m/80 ft., did not reach water (I,R)
6	44.2/145	-	-	Reported 18.6m/61 ft. drop in water level (I,R)
7	45.7/150	-	-	61m/20 ft. water level drop. Water heater burned out. No water 2 days (I)
8	33.5/110	-	17.7/58	Without water 4 days, beginning Jan. 19. Water heater burned out (I)
9	52.7/173	-	33.5/110	Without water 1 day (R)
10	-	-	-	Well would not pump (R)

Table 2: Water Level Declines Indicated by Wells, Dover Area, January 1977.

Sinkhole Occurrence

Newspaper and Hillsborough County officials received numerous complaints about the sudden appearance of sinkholes during the period of cold weather. The authors made field inspections and interviewed many of the property owners. In addition to the many sinkholes that were observed, a few sinkholes were reported to, but not observed by the authors. A total of 22 sinkholes were either observed or reported, as described in Table 3. The actual number of sinkholes was probably somewhat higher. For example, there were only two reports of sinkholes on strawberry farms which may have been subject to the greatest drawdowns. Some sinkholes may have occurred in idle fields without being discovered until later.

Sinkholes caused major structural damage to one house, the loss of two citrus trees, and two cases of road damage. Sinkholes occurring in the berm of a fish farm pond, in a cage house of a chicken farm and in a drainage canal would have caused increased property damage if the collapses had been larger or in slightly different locations. Three sinkholes collapsed when vehicles were driven over them; no injuries were reported.

Dimensions Length - Width - Depth m/ft	Remarks and Data Source
Unknown	One hole reported in strawberry field (R)eport
3 x 4.6 x .3+/10 x 15 x 1+	Severe structural damage to home (O)bserved
1.2 x 1.2 x ?/4 x 4 x ?	Edge of road. Repaired by road department (O)
1.8 x 1.8 x .6/6 x 6 x 2	In grove, fell in as truck passed (O)
3.7 x 3.7 x .6/12 x 12 x 2	Old depression, deepened by 2 feet (O)
5.5 x 6.4 x 1.2+/18 x 21 x 4+	Standing water in bottom, lost 15 ft. orange tree (O)
7.6 x 7.6 x .6+/25 x 25 x 2+	In drainage canal, standing water (O)
1.5 x 1.5 x 1.8/5 x 5 x 6	Hole wider with depth (O)
2.4 x 3 x ?/8 x 10 x ?	In grove, had been filled (O)
4.6 x 4.6 x 3.4/15 x 15 x 11	In grove, 3 holes developed Jan. 22-23 (O)
Unknown	3 holes reported in strawberry field (R)
Unknown	Telephone pole unearthed (R)
4.6 x 4.6 x 2.4/15 x 15 x 8	Lost one orange tree (O)
.9 x .9 x .3/3 x 3 x 1	Minor damage, chicken farm cages (O)
2.1 x 2.1 x 1.4+/7 x 7 x 4.5+	Fell in as truck drove over. Bent truck frame. Standing water (O)
Unknown	Cracked pavement, repaired (O)
1.2 x 1.2 x .9/4 x 4 x 3	Berm of fish pond (O)
2.4 x 3.7 x 3/8 x 12 x 10	Standing water 5 feet deep. In grove. Fell in when harvesting equipment passed over (O)

Table 3: Sinkholes Reported in Dover Area, January 1977.

Conclusion

The sudden withdrawal of large quantities of groundwater for freeze protection in January 1977, created severe declines in the potentiometric surface in the vicinity of strawberry fields near Dover, Florida. The large drawdown of the Floridan aquifer also caused inconvenience and minor property damage to nearby residents. The lowered artesian water levels allowed rapid drainage from the water table into the Floridan aquifer. Movement of sandy overburden material through previously existing discontinuities in the clay confining unit into cavities in the limestone below occurred while water levels were lowered. Movement ceased or slowed significantly when water levels returned to normal. The land surface expression of this activity was the formation of numerous sinkholes.

References

Applin and Applin, 1944, Regional Subsurface Stratigraphy and Structure of Florida and Southern Georgia, A.A.P.G. Bull. Vol. 28, No. 12.

Carr and Alverson, 1959, Stratigraphy of Middle Tertiary Rocks in Part of West Central Florida, U.S.G.S. Bull. 1092.

Cathcart, 1963, Economic Geology of the Keysville Quadrangle, Florida, U.S.G.S. Bull. 1128.

Menke, Meredith, and Wetterhall, 1961, Water Resources of Hillsborough County, Florida, Fla. Geol. Survey, Rpt. Inv. 25.

Watson and Company, 1973, Long-Range Study of the City of Tampa Water System.

Influence of a karstified limestone surface on an open-marine, marsh-dominated coastline: West Central Florida

JOAN G.HUTTON, ALBERT C.HINE, MARK W.EVANS & ERIC B.OSKING *University of South Florida, St.Petersburg, USA*
DANIEL F.BELKNAP *University of Maine, Orono, USA*

ABSTRACT

Rising Holocene seas flooding a sediment-starved, low gradient (1:6000), exposed karstified limestone surface has produced an extremely complex, highly digitate, open-marine, marsh-dominated coastline along a portion of the West Central Florida coast. This irregular rock surface, lying beneath an organic-rich, peaty veneer, coupled with numerous, large, fresh water springs discharging from the Floridan Aquifer, is the primary factor controlling the regional, modern coastal geomorphology and sedimentation.

This coastline can be subdivided into: (1) shelf embayments; (2) marsh archipelagos; and (3) berm veneer. Shelf embayments occur at the mouths of the Crystal, Chassahowitzka and Weeki Wachee Rivers and are characterized by shore-parallel, rock-controlled oyster bioherms. Marsh archipelagos have developed between the embayments and probe rod transect data consistently show that: (1) a highly irregular rock surface underlies the marsh veneer; (2) marsh islands and hammocks occur over topographically higher rock areas; (3) tidal creeks, ponds and channels overlie topographically lower areas; and (4) modern marsh sediment, when available, infills and evens out the irregular rock surface. This data strongly suggests that the modern shoreline is controlled by the topography of the underlying rock. Berm veneer is not discussed due to lack of obvious rock control of the shoreline.

As sea level rises, the system is flooded and the following sequence evolves: (1) upland forests on rock highs (soil veneer) become surrounded by high salt marsh or fresh water wetlands; (2) doline-like depressions become isolated, open, circular ponds; (3) tidal creeks begin to form by following fracture lines and by connecting isolated circular ponds together; (4) rock highs become centers of marsh islands; and (5) rock highs become flooded, submerged and are possible nucleation sites for inner-shelf oyster bioherm development.

Introduction

The coastline of West Central Florida from Crystal River to Tarpon Springs (Fig. 1) is part of the "zero" or low energy, non-barrier coast described by Tanner (1960). It is unique because of its marsh-dominated, open-marine character. Factors responsible for its development have been explored (Hine, et al., 1984) and include: (1) the nature of the underlying limestone surface; (2) fresh water discharge through subterranean springs; (3) the extremely low gradient (1:6000); (4) low wave energy and the influence of wind tides on this microtidal coast; (5) the uneven distribution of Quaternary quartz sand deposits; (6) the absence of silici-clastic sediments; and (7) sea-level fluctuations. All of these factors contribute to the formation of this complex, highly irregular and variable coastal system rich in organic and carbonate sediments. However, in the absence of quartz sands, we feel the underlying limestone surface has been of paramount importance to the development of this shoreline.

The coastline can be divided into three coastal types: (1) shelf embayments; (2) marsh archipelagos; and (3) berm veneer. The shelf embayments have formed because of the interaction of spring discharge with Gulf of Mexico waters and are found at the mouths of the Crystal, Chassahowitzka and Weeki Wachee Rivers. The marsh archipelagos are collections of small, rock-cored islands which dominate the northern half of the study area. In contrast, the southern half is a nearly continuous berm veneer in which sandy beach fronts the coastal marsh. For the purposes of this presentation, the berm veneer coastal sector will not be discussed as rock control is neither a dominant nor an obvious factor affecting this part of the modern shoreline.

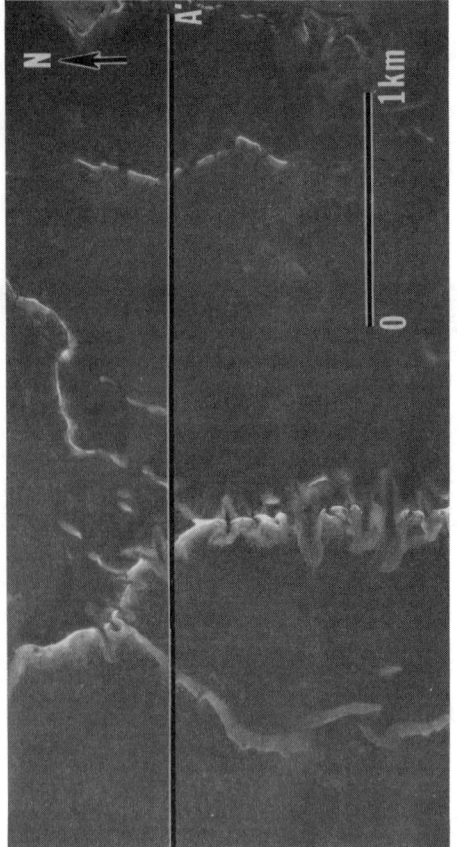

Figure 2A: Aerial photograph (1981) showing inner-shelf location of eastern part of core profile A-A'. Note north-south trending oyster bars.

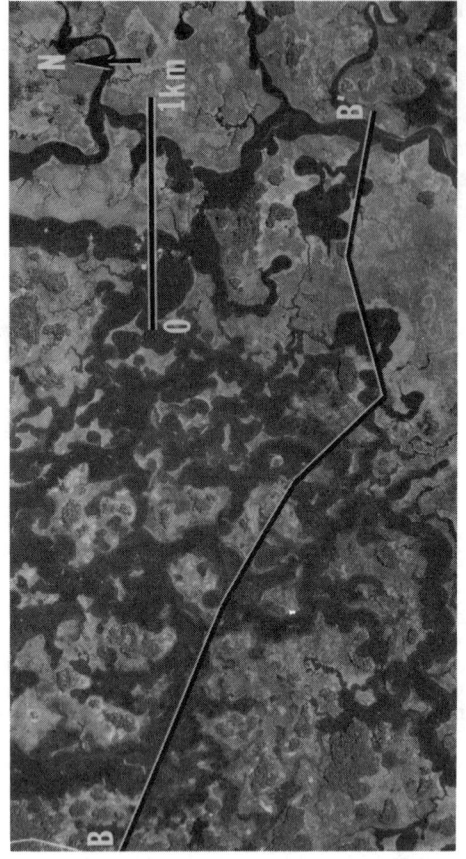

Figure 2B: Aerial photograph (1981) showing location of probe rod profile B-B' in Crystal River marsh archipelago. Note extensive network of lobate tidal creeks separating marsh islands.

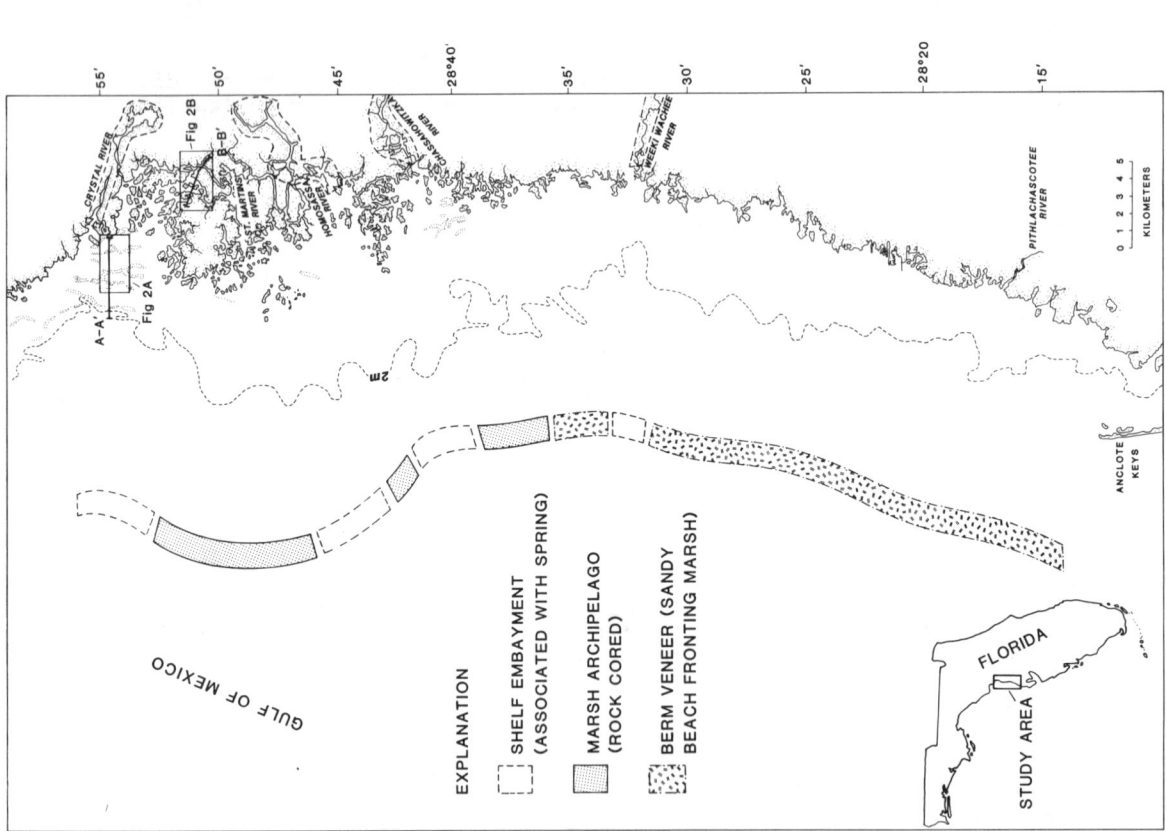

Figure 1: Map showing location of study area with classification of coastline. Profiles A-A' and B-B' also shown in Figs. 4, 5.

General Geologic Setting

Beginning in Lower Cretaceous the Florida Platform was a center of extensive carbonate deposition (McKinney, 1984). It produced an expansive, low gradient carbonate platform which was separated from the terrigenous clastic input of the Appalachians by the Suwannee Channel (Chen, 1965). Carbonate accumulation continued through Late Oligocene at which time the Suwannee Channel began to close allowing an influx of silici-clastics which eventually shunted carbonate production (McKinney, 1984).

Higher Pleistocene sea levels deposited these silici-clastics as large beach ridges paralleling the present shoreline (White, 1970). In the northern half of the study area the ridges lie well inland. Since there is no effective system to transport these sands Gulfward, the coastal area is sediment-starved and the rock topography has become the main factor controlling the modern shoreline. To the south, where these Pleistocene ridges extend further to the west, the sands have been reworked forming a modern barrier-island coastline.

Rocks underlying the northern part of the study area (Citrus and Hernando Counties) are Late Eocene and Oligocene in age. In Citrus County Late Eocene Ocala Group limestones of the Inglis, Williston and Crystal River Formations outcrop extensively (Vernon & Puri, 1964). The Inglis, which forms the base of the Ocala Group, is a fossiliferous, resistant, shallow-water marine limestone which unconformably overlies the Avon Park limestone (Vernon, 1951). It occurs in the study area north of the Crystal River (Vernon, 1951). The Williston, a softer, more porous, detrital, fossiliferous limestone, lies conformably between the Inglis and Crystal River limestones and outcrops north and south of the Crystal River (Vernon, 1951). The Crystal River Formation, a friable, shallow-water marine, fossiliferous limestone, underlies the entire southwestern part of Citrus County and the upper portion of Hernando County (Vernon, 1951).

The Oligocene Suwannee limestone occurs very close to the surface in coastal Hernando County overlying the Crystal River Formation (Yon & Hendry, 1972). It is a granular, shallow-water marine, bioclastic limestone (Vernon & Puri, 1964; Yon & Hendry, 1972) which outcrops at various locations along the coast. During the post-Oligocene or Lower Miocene, large-scale fracturing and faulting of Tertiary sediments occurred forming the Ocala Uplift (Yon & Hendry, 1972; Chen, 1965; Vernon, 1951). With its crest located in Citrus County (Vernon, 1951), the Ocala Uplift caused a regional northwest-southeast trending fracture system to develop. Faulting occurred along its crest and flanks (Yon & Hendry, 1972) and a secondary northeast-southwest trending fracture system was established (Vernon, 1951).

In addition to creating vents for spring flow from the Floridan Aquifer (Rosenau, et al., 1977), these fracture patterns have served as loci of surficial karst topography through solution of the underlying limestone.

Shelf Embayments

The Crystal River, Chassahowitzka and Weeki Wachee shelf embayments are microtidal (1 m; 3.3 ft), low wave energy, fresh water influenced and marsh-dominated. Each is located at the mouth of a river by the same name and all three rivers are spring-fed, discharging from the Floridan Aquifer (Fig. 1). Both the degree of fresh water influence to the nearshore marine environment and oyster population decrease to the south. Linear, roughly shore-parallel, bedrock-controlled oyster bioherms are found just seaward of the fresh water discharge.

In the marsh, hammocks are sites of topographically high bedrock. As sea level transgresses and the marsh erodes, topographic highs become available for bioherm nucleation (Fig. 3A). The most conspicuous populations are found in Crystal Bay, seaward of Crystal River, where oyster bioherms extend 10 km (6.2 mi) alongshore and 3 km (1.9 mi) offshore. Oyster communities are generally restricted to these bars. The oyster bioherms (except the furthest seaward) are emergent at low tide; thus, possibly acting as sedimentological "baffles".

Analysis of cores and surficial sampling yields five main depositional environments/facies observed in the nearshore environment. These are oyster bioherms, inter-bioherm basins, marsh, Pleistocene sands and basal limestone (Fig. 4; see also: Fig. 1, profile A-A'; Fig. 2A). Basal limestone is overlain by remnants of the eroded marsh, unless the marsh is underlain by Pleistocene sands. The inter-bioherm basins are composed of shelly muddy sand, which is probably an admixture of the eroding marsh and shallow marine organic production. Oyster bioherms are observed to accrete vertically, keeping pace with sea level, except for the furthest seaward which are relict. A basal calcilutite (directly above bedrock) has been observed in some cores. It is probably a weathering residuum of the bedrock, or a fresh water marl. Further isotopic and faunal analysis will aid in determining its origin.

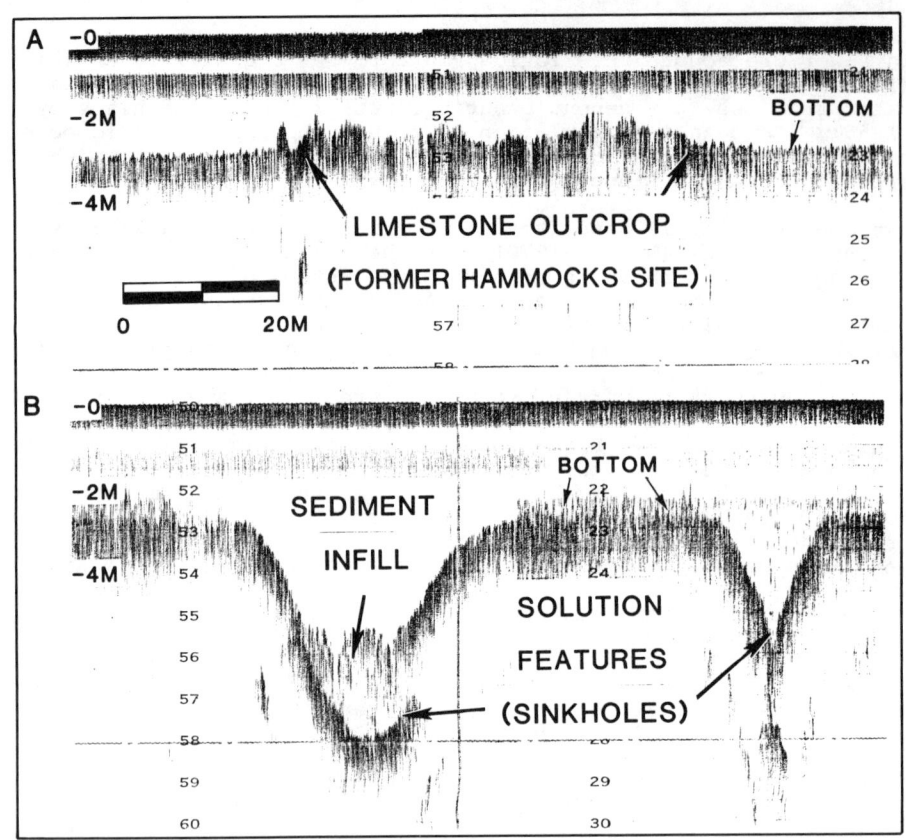

Figure 3: Inner-shelf high resolution seismic profiles using 7 kHz Raytheon RTT-1000. A. Shows rock highs which, when subaerial, were hammocks. They are now potential nucleation sites for oyster bioherms. B. Shows preservation of solution features (sinkholes) on inner-shelf due to lack of modern sediment cover.

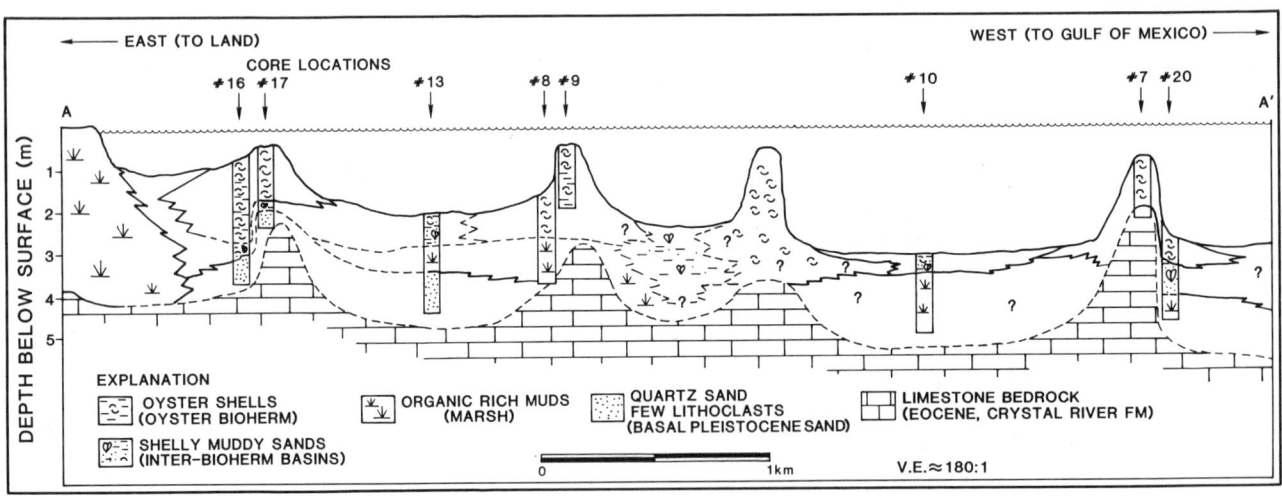

Figure 4: Profile A-A'. A stratigraphic marsh-inner shelf depositional/facies model. Note location of modern oyster bioherms over rock highs.

Observation of bathymetric charts and analysis of the subsurface sediments and bedrock topography via seismic profiling, reveal the absence of deeply embayed river valleys seaward of these embayments. As sea level fluctuates, the spring-fed headwaters of the rivers make a series of discontinuous jumps across the shelf due to the presence of caverns and conduits produced by solution of the underlying limestone (Cooke, 1939). Lowered sea level results in a river discharging further seaward than during a higher sea level stand. Thus, the time necessary to carve out a deep embayment is absent. Seismic profiles of sinkholes offshore suggest that the lack of sediment in this coastal system results in the preservation of these features when "drowned" (Fig. 3B). Since the springs' headwaters are composed of a group of sinkholes (springs), one should be able to track their path across the shelf.

Marsh Archipelagos

In the study area two large marsh archipelagos are prominent and represent headlands between the embayments of the Crystal, Chassahowitzka and Weeki Wachee Rivers. They are composed of numerous flat, low relief islands which front the open Gulf of Mexico and are subjected to mixed, diurnal tidal flooding. Marsh islands are drained by an extensive network of rectilinear (trellis drainage pattern) tidal creeks. Hammocks, which support fresh water vegetation, are abundantly interspersed through the more saline marsh.

More than 40 km (24.9 mi) of probe rod transect data have been taken through this area. Profiles consistently show that: (1) a highly irregular rock surface underlies the marsh veneer; (2) marsh islands and hammocks occur over topographically higher rock areas; (3) tidal creeks, ponds and channels overlie topographically lower areas; and (4) modern marsh sediment, when available, infills and evens out the irregular rock surface (Fig. 5; see also: Fig. 1, profile B-B'; Fig. 2B). This data suggests that the modern shoreline is controlled by the topography of the underlying rock.

The dissolution/degradation of the underlying Late Eocene/Oligocene limestones have produced the irregular karstified surface observed on probe rod profiles. It is possible that solution producing this karst is related to fracturing of these limestones as well as those of the Floridan Aquifer (Rosenau, et al., 1977) which occurred during the Ocala Uplift.

Ocala Group limestones outcrop extensively in the northern archipelago and assume a variety of geometric forms: (1) small circular pits and holes only 10's of cm in width and depth which create a complex honeycomb network difficult to traverse on foot (Fig. 6A); (2) broad, circular depressions up to 1 m in relief and 10's of m in width (Fig. 6B); (3) excavated linear fractures that extend for 10's of m (Fig. 6C); (4) circular, scarped rocky highs up to several m's in relief and 100's of m across (Fig. 6D); and (5) large-scale (km's long) rock-controlled, linear drainage fractures which consist of connected or isolated but aligned circular depressions up to 7 m in relief.

As sea level rises and the present salt marsh retreats, only segments of upland forests which are underlain by rocky topographic highs survive as hammocks which are surrounded by high salt marsh or fresh water wetlands. Doline-like depressions become isolated, circular ponds and tidal creeks form by following fracture lines and by connecting these circular ponds. As transgression continues, the hammocks become centers of marsh islands separated from each other by tidal creeks (Fig. 7A). Finally, the islands are drowned and hammocks eroded leaving exposed, submerged rock highs which may serve as nucleation sites for subsequent inner-shelf oyster bioherm development (Fig. 7B).

Conclusion

The coastline of West Central Florida is geomorphically complex and can be subdivided into: (1) shelf embayments; (2) marsh archipelagos; and (3) berm veneer. Shelf embayments exist at the mouths of the Crystal, Chassahowitzka and Weeki Wachee Rivers and are microtidal (1 m; 3.3 ft), low wave energy, fluvially-influenced systems. Each is characterized by shore-parallel, bedrock-controlled oyster bioherms just seaward of the fluvial discharge.

Marsh archipelagos are collections of low relief islands which are open to the Gulf of Mexico and flooded by mixed, diurnal tides. Probe rod profiles reveal that: (1) a highly irregular rock surface underlies the marsh veneer; (2) marsh islands and hammocks occur over topographically higher rock areas; (3) tidal creeks, ponds and channels overlie topographically lower rock areas; and (4) modern marsh sediment, when available, infills and evens out the irregular rock surface. These data strongly suggest that the modern shoreline has been controlled by the underlying rock topography.

Coastal marsh and shelf embayments represent different stages of geomorphic development of the coastline strongly influenced by rising sea level. As a marine transgression floods the upland forests, they become isolated hammocks situated on topographically higher rock. Doline-like depressions become isolated circular ponds. Tidal creeks begin to form by either following fracture lines or connecting circular ponds. Hammocks become centers of marsh islands separated from one another by tidal creeks. As sea level continues to rise,

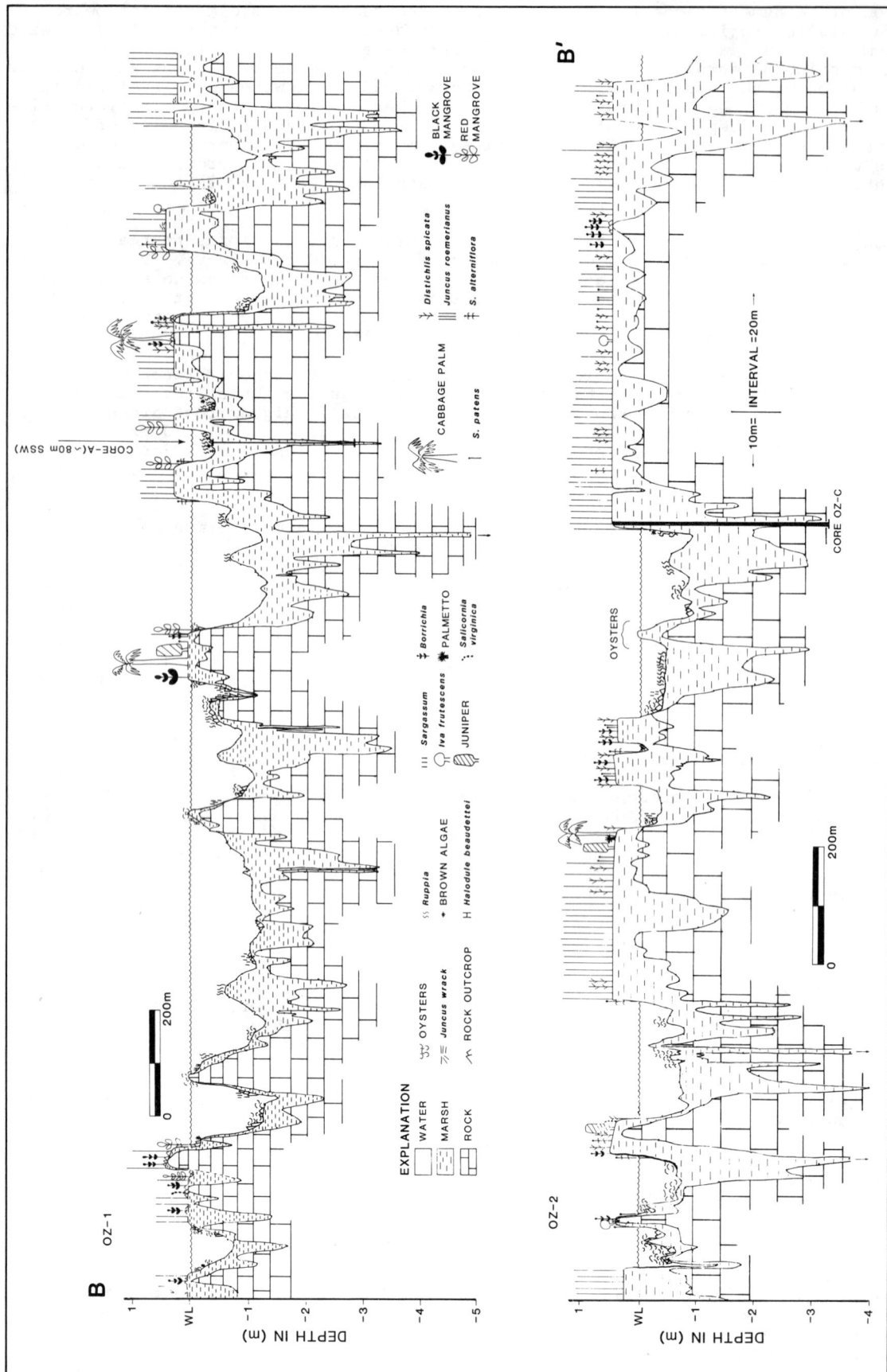

Figure 5: Probe rod profile B-B' showing irregularity of rock surface (V.E. 100:1). Note occurrence of marsh islands and hammocks over rock highs and tidal creeks over rock lows.

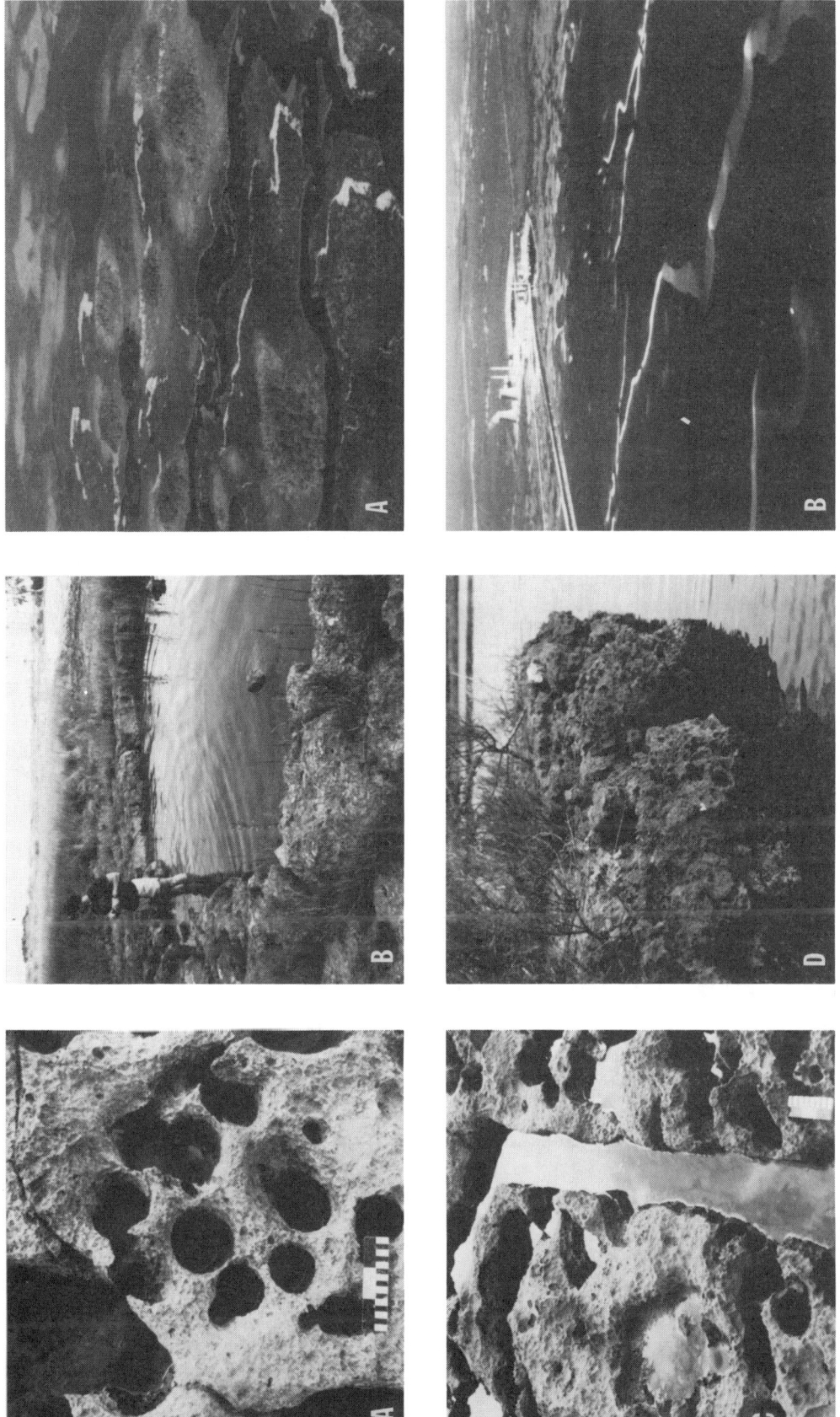

Figure 6: Exposures of Ocala Group limestone in Crystal River marsh archipelago. A. Circular pits and holes are common microtopographies on exposed rock. B. Rock-lined, circular depression connected to tidal flow by small creek (not shown). C. Shows linear, fracture-related solution cavity. Relief along fracture line approximately 10 cm. Note 15 cm scale in foreground. D. Edge of circular, scarped rocky high facing broad tidal creek. Note extensive solution/bioerosion of rock. Scarp relief about 1 m above water level.

Figure 7: Final stages of marsh/inner-shelf evolution. A. As sea level rises, hammocks, situated on rock highs, become centers of marsh islands separated by tidal creeks. B. Further transgression causes marsh erosion and submergence of rock highs. Highs then provide nucleation sites for inner-shelf oyster bioherm development.

islands are drowned and hammocks are eroded to rocky highs submerged on the inner shelf. These rocky highs are then possible nucleation sites for inner-shelf oyster bioherm development.

Acknowledgement

This study has been supported by Florida Sea Grant R/C-8 (NOAA/OSG #NA80AA-D-00038), "Geology of Florida non-barrier shoreline, Tarpon Springs to Crystal River: sedimentary environments, processes, geohazards and resources".

References

Chen, C. S., 1965. The regional lithostratigraphic analysis of Paleocene and Eocene rocks of Florida: Florida Geol. Survey Bull. 45, 105 p.

Cooke, C. W., 1939. Scenery of Florida interpreted by a geologist: Florida Geol. Survey Bull. 17.

Hine, A. C., D. F. Belknap, E. B. Osking, J. G. Hutton and M. W. Evans, 1984. Sedimentation controls along an open-marine, marsh-dominated coastline: NW Florida Gulf Coast, Soc. of Econ. Paleon. & Min. Annual Midyr. Mtg. Abstract Volume, p. 38.

McKinney, M. L., 1984. Suwannee Channel of the Paleogene Coastal Plain: support for the "carbonate suppression" model of basin formation, Geology, 12, (6), 343-345.

Rosenau, J. C., G. L. Faulkner, C. W. Hendry, Jr. and R. W. Hull, 1977. Springs of Florida: Florida Bureau of Geology Bull. 31 (revised), 461 p.

Tanner, W. T., 1960. Florida coastal classification. Trans. Gulf Coast Assn. Geol. Soc., 10, 259-266.

Vernon, R. O., 1951. Geology of Citrus and Levy Counties, Florida: Florida Geol. Survey Bull. 33, 255 p.

Vernon, R. O. and H. W. Puri, 1965. Geological map of Florida: Florida Geol. Survey, Map Series 18, 1 sheet.

White, W. A., 1970. The geomorphology of the Florida Peninsula: Florida Bureau of Geology Bull. 51, 164 p.

Yon, J. W., Jr. and C. W. Hendry, Jr., 1972. Suwannee limestone in Hernando and Pasco Counties: Florida Bureau of Geology Bull. 54.

Seismic-reflection studies of sinkholes and limestone dissolution features on the northeastern Florida shelf

PETER POPENOE, F.A.KOHOUT & F.T.MANHEIM *US Geological Survey, Woods Hole, Massachusetts*

ABSTRACT

High-resolution seismic-reflection profiles show that the shelf off northern Florida is underlain by solution deformed limestone of Oligocene, Eocene, Paleocene and late Cretaceous age. Dissolution and collapse features are widely scattered. They are expressed in three general forms: (1) as sinkholes that presently breach the sea floor, such as Red Snapper Sink and the Crescent Beach submarine spring; (2) as sinkholes that have breached the seafloor in the past but are now filled with shelf sands; and (3) as dissolution collapse structures that originate deep within the section and have caused buckling and folding of overlying Eocene, Oligocene, and to a lesser extent, Neogene strata. The deep dissolution collapse features appear to originate in the Upper Cretaceous and Paleocene rocks. Collapse structures are controlled by both regional joint patterns and by carbonate platform facies. The features are most pronounced over the Late Cretaceous-Paleocene reef and back-reef facies where folding of overlying Eocene and Oligocene rocks has resulted in as much as 80 m of relief. Deformation structures and sinkholes abruptly end at the forereef facies.

Although deformation caused by solution and collapse can be shown to be a continuous process, the major episode of karstification occurred in the late Oligocene and early Miocene when the shelf was exposed to subaerial conditions. Minor deformation has occurred since the early Miocene, and collapse and filling of sinkholes continues today. Collapse features formed chiefly under subaerial conditions, indicating that submarine or fresh/salt-water mixing processes play a minor role in limestone dissolution.

Introduction

The karstic character of the Paleocene, Eocene, and Oligocene limestones of Florida, and Georgia is well known. These rocks contain both the Floridan aquifer system (principal artesian aquifer of Georgia) that supplies water to thousands of municipal, industrial, and irrigation wells, and the deeper, saline, and highly cavernous Boulder Zone (Kohout, 1965). In north central Florida these limestone beds crop out at the surface on the crest of the Peninsular and Ocala Arches where karst features and sinkhole lakes form the aquifer recharge area. In northern Florida the Paleocene, Eocene, and Oligocene limestones are covered in most areas by a veneer of younger impermeable calcareous silts and clays of Miocene, Pliocene, and Pleistocene age that cap the Floridan aquifer system. Because of burial, the limestones in this area appear less karstic. However, in the area between St. Augustine and Daytona Beach, limestone beds are nearer the surface and solution features are clearly seen on seismic records. Offshore large sink holes have breached the sea floor at a submarine spring about 4 km off Crescent Beach (Brooks, 1961) and at Red Snapper Sink (Wilcove, 1975; Kohout and Leve, 1977), a submarine sinkhole on the mid shelf about 40 km from shore.

The U.S. Geological Survey (USGS) has been collecting seismic reflection data along the southeastern U.S. Atlantic margin since 1976. Most data-gathering activity has been concentrated north of the Florida shelf in the Southeast Georgia Embayment or in the deeper offshore Carolina Trough and Blake Plateau Basin; however, a few common-depth-point (CDP) and high-resolution seismic-reflection profiles have been obtained on the Florida shelf. These profiles demonstrate the karstic nature of the Upper Cretaceous and Cenozoic rocks and help to clarify the Tertiary history of northern Florida. A seismic transit in 1979 reported by Popenoe (1983), tied the JOIDES Test hole J-1, 40 km offshore Jacksonville, Fla. to the Red Snapper Sink with high-resolution seismic-reflection data. A small survey was made of the sinkhole itself to establish criteria for karst identification in seismic records (fig. 1) in other areas. Early findings of the JOIDES (Joint Oceanographic Deep

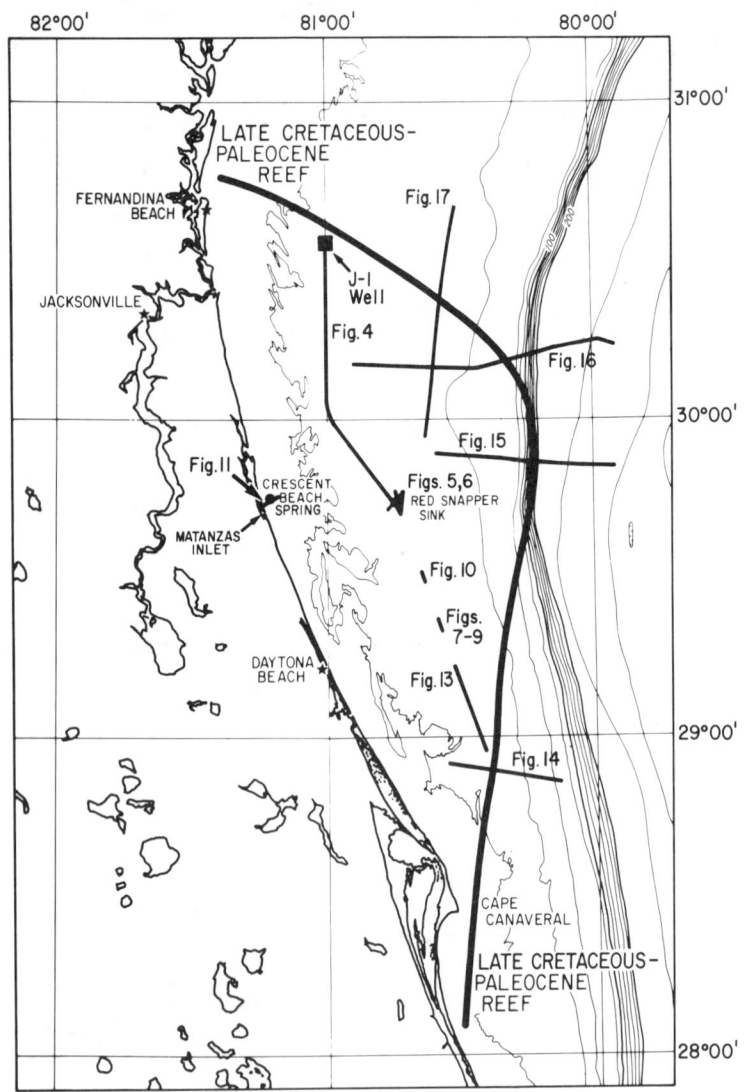

Figure 1: Map showing the location of the J-1 well, Crescent Beach Spring, Red Snapper Sink, and seismic profiles discussed in this paper. Also shown is the surface trace of the buried Late Cretaceous-Paleocene reef.

Earth Sampling) drilling project are given by Emery and Zurudski, 1967; Charm, Nesteroff, and Valdes, 1969; and Poag and Hall, 1979.

Red Snapper Sink

Red Snapper Sink is a funnel-shaped sinkhole on the middle of the Florida-Hatteras shelf 40 km offshore from Crescent Beach, Fla. It was discovered in 1962 by a fisherman who took 5000 lbs. of red snapper (Lutjanus aya) out of it before being driven off by a storm and losing the site. It was rediscovered by fishermen in 1968, where in a period of about 6 weeks 100,000 pounds of red snapper were caught (Wilcove, 1975). The name Red Snapper Sink was applied to the sinkhole and the site was located on chart No. 11486 (U.S. National Ocean Survey, 1976).

In 1970, the U.S. Geological Survey mounted several expeditions to the sinkhole to try to determine its exact location, depth, and hydrologic characteristics. Using a deep-well current meter and a sampling device used for testing water wells on land, they estimated the maximum depth in 1970 at 141 m (464 ft) below sea level. A stainless-steel multiconductor wire operated over a registered sheave reproduced the 141 m depth on several soundings of the hole although depths as shallow as 131 m (430 ft) were also observed.

In 1974, a joint party of USGS and National Oceanic and Atmospheric Administration (NOAA) personnel on NOAA Launch 1257, equipped with a DE-1723D Raytheon fathometer and a Hastings Radist model ZA-67A computerized navigation system, established the location of the sink within about 10 m: lat $29°44°26$ N., long $80°44°52$ W. (Wilcove, 1975). Nine passes were made over the center of the sinkhole following the pattern of the spokes of a wheel. Depth soundings were plotted automatically on Mylar film by an onboard computer and hydroplot. Contours were then hand plotted (fig. 2). Maximum depth recorded in the sinkhole was 133.2 m (437 ft): which appears to be 8 m (27 ft) shallower than the maximum depth measured in 1970.

The sinkhole lies in water depth of about 27 m (88 ft) below mean low water datum. A gentle slope ($\pm15°$) descends from about 27 to 46 m, from which a sharp-rimmed vertical-sided pipe starts and descends to at least 133.2 m (437 ft) depth. The inside diameter of the funnel at the 27 m (90 ft) contour is about 122 m (400 ft).

In 1970, a scuba diver, Robert E. Hill of Merritt Island, Fla., descended into the sinkhole to the 91 m (300 ft) level. While sitting on a ledge at 76 m (250 ft) he drew a

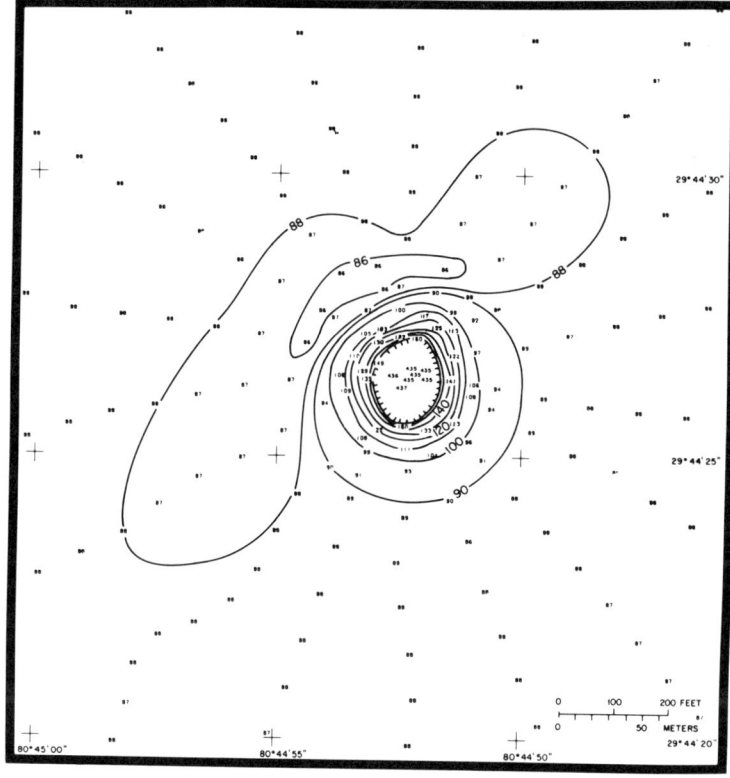

Figure 2: Contoured hydroplot of Red Snapper Sink showing depths in feet below sea level.

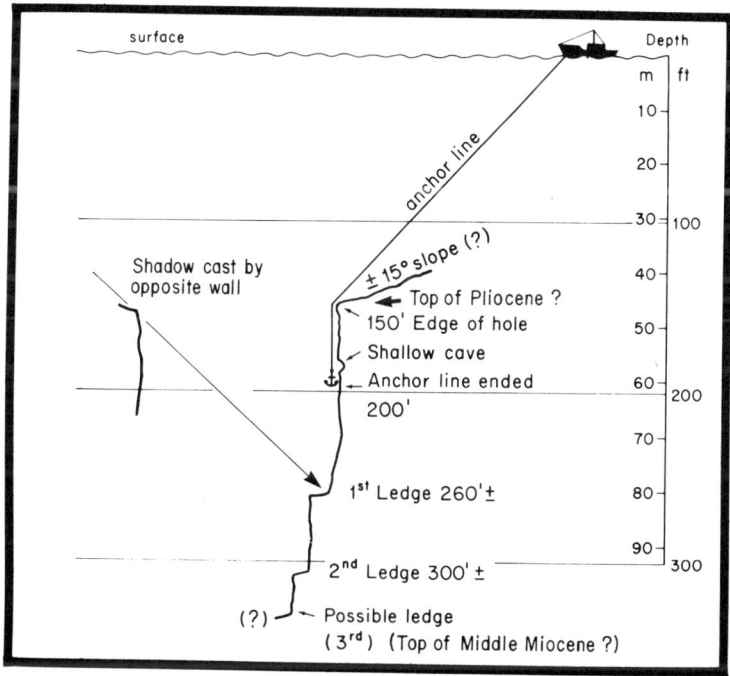

Figure 3: Drawing of Red Snapper Sink made by Robert E. Hill while sitting on the ledge at ±80 m. Age identifications are from this study.

cross section of the sinkhole (fig. 3) and expelled fluorescein dye into the water from a chemical squeeze bottle. The dye cloud dispersed and moved slightly downward but did not move upward (R.E. Hill, written commun., 1970) indicating the possibility of a slight downward velocity of water into the sink and the possibility of salt water intrusion into the Floridan aquifer at this point. Water samples taken by lowering a Foerst water sampler, triggered by the bottom of the sinkhole, showed normal seawater salinity and temperature indicating no freshwater outflow.

A sample of rock was collected by divers Robert Hill and Ray Hyatt at about 61 m (200 ft). A year later, additional samples were collected by Don Serbousek of the Ormond Anchor Chasers Scuba Club at depths of 46 m (150 ft) and 53 m (175 ft). These samples proved to be of Pliocene age with a veneer or encrustation of barnacles, oysters, and other living forms on the side facing the hole. The living encrustation did not appear to be especially thick, suggesting that Red Snapper Sink may be of fairly recent origin. Harbans S. Puri (Florida Bureau of Geology, oral communication, 1971) identified the ostracods Hemicythere cf. H. howei Puri, and Acuticythereis laevissima Edwards in the samples. These two fossils indicate a middle Pliocene age of about 3.4 m.y. (T.M. Cronin, oral commun., 1984) and suggest that the 50 m depth is an outcropping of the Tamiami or Jackson Bluff Formation.

Seismic observations along the profile connecting test hole J-1 with Red Snapper Sink

On August 28, 1979, a high-resolution seismic-reflection profile (fig. 4) was run between test hole J-1 offshore from Jacksonville Fla. and Red Snapper Sink. An airgun (5-in^3) was used as a seismic source, and records were obtained at both a one-fourth and one-half second sweep. Records were obtained simultaneously with a 3.5-kHz hull-mounted echo-sounding system and a EG&G Uniboom seismic reflection system. All of these records showed shallow penetration which did not extend

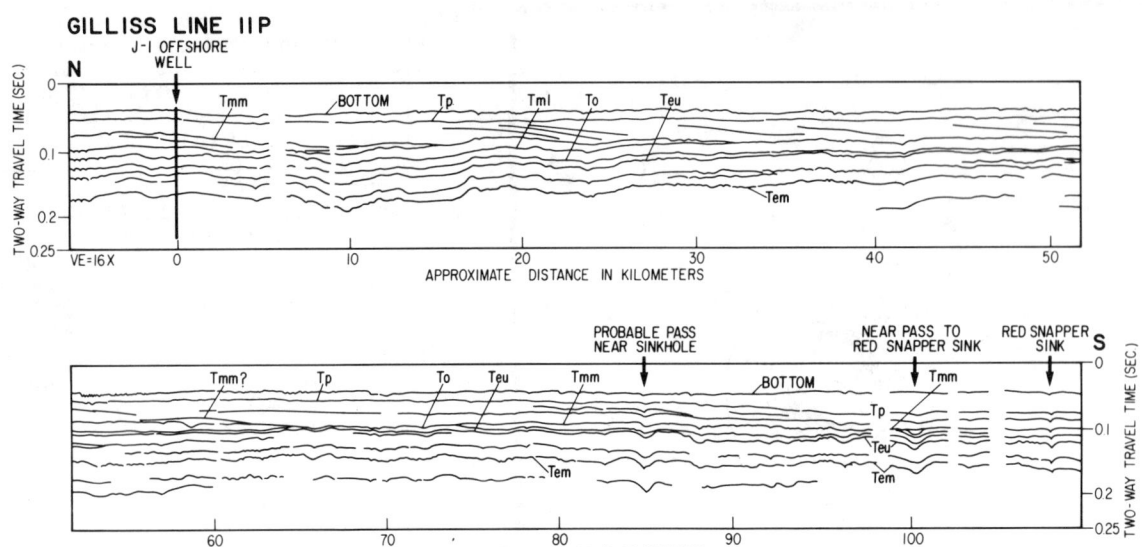

Figure 4: Line drawing of seismic-reflection profile that ties Red Snapper Sink with the JOIDES testhole J-1. The tops of the seismic stratigraphic units are marked as follows: Tem = middle Eocene, Teu = upper Eocene, To = Oligocene, Tml = lower Miocene, Tmm = middle Miocene, Tp = Pliocene, Bottom = Pleistocene and Holocene.

below the Eocene-age rocks. The following velocities, from Emery and Zarudski (1967) were used in correlating seismic reflectors with lithostratigraphic units (Poag and Hall, 1979) picked at the J-1 site: Pliocene and Miocene age rocks = 1.94 km/s.; Oligocene and Eocene age rocks = 3.0 km/s.

At the J-1 test hole, unconformities mark the tops of the Pliocene, middle Miocene (two seismic units), lower Miocene (Langhian), middle Oligocene, upper Eocene, and middle Eocene strata.

The strongest reflector at the J-1 site marks the top of the Eocene rocks. The Eocene section at the test hole consists of massive, dolomitic packstone and grainstone similar to the Ocala Group onshore (Avon Park and Lake City Limestones) (Schlee, 1977). These rocks were deposited in very shallow depths typical of a coastal lagoon (JOIDES, 1965). The hard indurated rocks provide a strong velocity contrast with the overlying softer rocks and on the seismic records produce an internal-reflection pattern of strong subparallel reverberatory type, typical of hard limestones (see figs. 7, 13). Discrete reflectors, which represent depositional or bedding surfaces, can be traced within the limestone units. South of J-1, toward Red Snapper Sink, the seismic data (Fig. 4) show that the Eocene rises to a regional subsurface high so that its top, which is 132 m bsl (below sea level) at J-1, is only 97 m bsl (71 m subbottom) at Red Snapper Sink.

The top of the Eocene section is an erosional unconformity. Most of the relief on this surface however, is due to a pattern of both sharp and broad undulatory flexures and folds that are paralleled by all bedding reflectors within the Eocene strata. Several pronounced downward flexures, which bend all strata, are observed along the profile. The most pronounced flexure is located about 85 km south of J-1, and another occurs as we passed Red Snapper Sink 300 m to the east (fig. 4). On the airgun record, the sinkhole is masked by diffractions from the shelf surface; however, it occurs above an obvious downward flexure on the top of the Eocene that is apparent in all overlying and underlying units. These folds and downward flexures become quite pronounced south of Red Snapper Sink, as discussed later.

At J-1, Oligocene sediments consist of plastic, silty, deep-water, calcareous hemipelagic ooze (Schlee, 1977) that does not produce a strong reflection sequence. They are 9.2 m (30 ft) thick at J-1 and thin southward over the broad Eocene subsurface high. They are very thin at Red Snapper Sink but thicken again to the south where their seismic expression is more pronounced and resembles that of more indurated limestone. These strata also are folded like the underlying Eocene strata.

The early and middle Miocene sections at the J-1 test hole consist mainly of phosphatic grayish-green sandy silts and silty clays that contain phosphate pebbles similar to the Hawthorn Formation of the Coastal Plain (JOIDES, 1965; Schlee, 1977). The top of the early Miocene and the lower unit of the middle Miocene are both strong reflectors. The internal bedding of both units parallel their tops, which are flexed into the folds described previously. However, folding is less pronounced in these units than in the deeper Eocene, particularly south of Red Snapper Sink. Southward from the J-1 site, these units drape the Oligocene top and are folded like the underlying units. They thin and are erosionally truncated over the apex of the Eocene high so that they are thin in the area of Red Snapper Sink. The units occur within a buried syncline-like flexure near Red Snapper Sink where they are buckled downward and their tops truncated by erosion into an angular unconformity.

The lower Miocene and lowest middle Miocene units are overlain at the J-1 site by plastic to compacted sandy, calcareous, quartzose and micaceous silt of middle Miocene age. This Miocene unit progrades gently southward and pinches out on the regional high about 15 km south of J-1. It reappears south of Red Snapper Sink on the southern flank of the regional Eocene high.

The post Miocene sediments consist of sandy phosphatic silt that grades upward to silty, fine- to-medium grained, calcareous, quartzose sand of Pliocene age and to well-sorted, quartzose, shelly sand of Pleistocene age at the surface (Schlee, 1977). Seismic-reflection data show that these units prograde gently southward onto the top of the underlying middle Miocene beds. Although these beds reflect some of the deformation of the underlying units, the deformation is not pronounced except above the sharp downward flexures noted earlier.

Seismic observations at Red Snapper Sink

At Red Snapper Sink, the most notable feature is the sharp downward flexure centered on the sinkhole (fig. 5) and a similar downward warping in the subsurface rocks that was observed 300-m east of the sinkhole (near the 100 km mark, fig. 4). These flexures are not a velocity pulldown caused by the water-column within the sink since they extend some distance from the hole, nor are they due to erosion at the unconformity since reflectors beneath the bent surface conform with it.

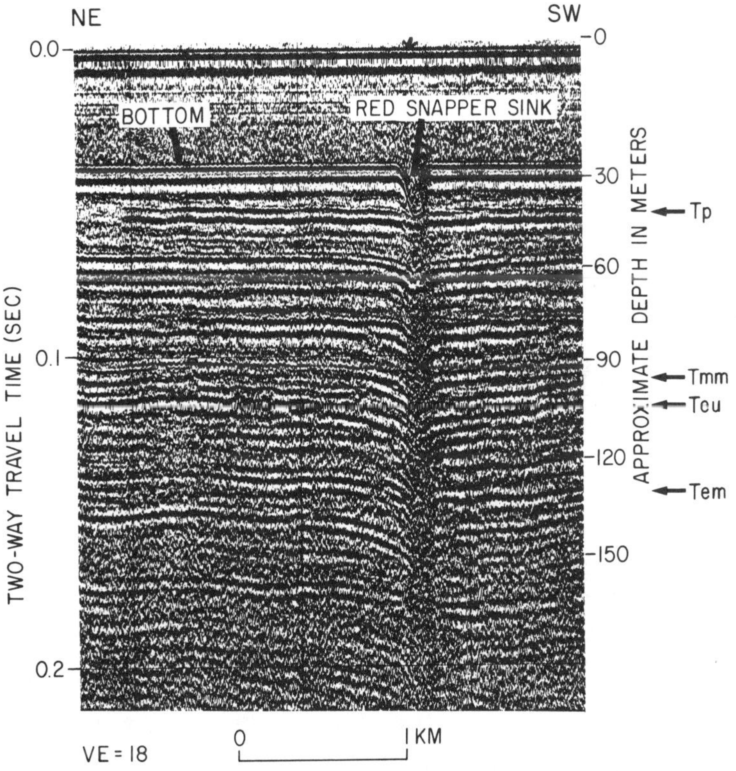

A Uniboom seismic-reflection profile across the sinkhole (fig. 5) clearly shows the surface depression. The bottom of the sinkhole, however, which is masked by diffractions at the edges of the hole, is not seen. A 3.5-kHz record across the sinkhole is shown in figure 6. Again the subbottom is masked by diffractions, but a strong reflector occurs at about .165 to .170 seconds depth directly beneath the surface pipe. This reflection is not a multiple and occurs with a slightly different shape on all crossings of the sinkhole. It appears to be a direct signal return from the bottom of the hole. After correcting for transducer depth, this return indicates that the bottom of the sinkhole, was (in 1979) about 125.9 to 133.2 m bsl, slightly shallower than the depth measured by NOAA in 1974. If the shallower measured depths are correct, they indicate that the sinkhole is being filled with sand. Such filling would be consistent with water motion,

Figure 5: Uniboom seismic profile across Red Snapper Sink. Chronostratigraphic units are carried from J-1 on Uniboom and airgun records.

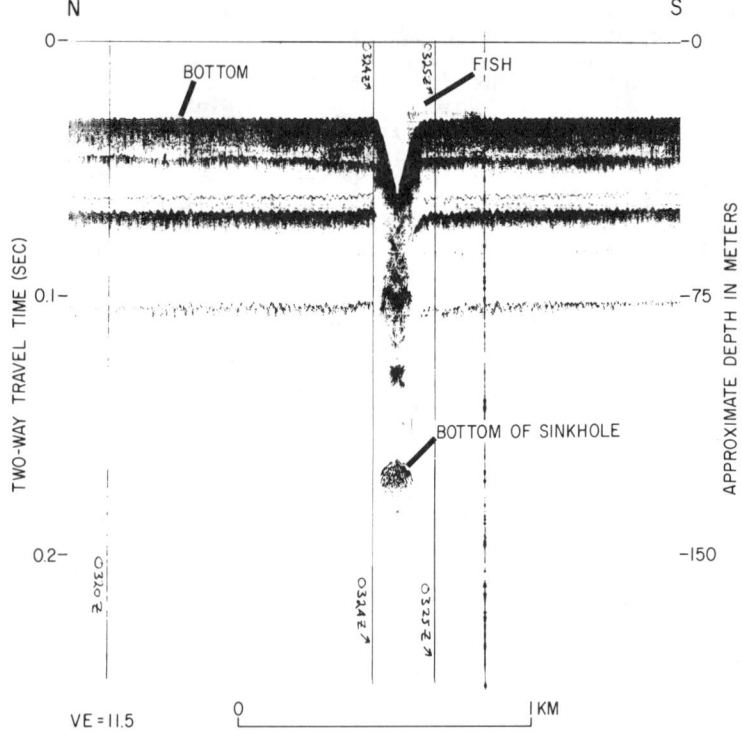

Figure 6: A 3.5-kHz echo-sounder record of Red Snapper Sink. The hyperbolic reflection at .165 to .170 seconds depth is a direct signal return from the bottom of the sinkhole. The strong returns at .1 and .13 seconds mark ledges on sides of hole.

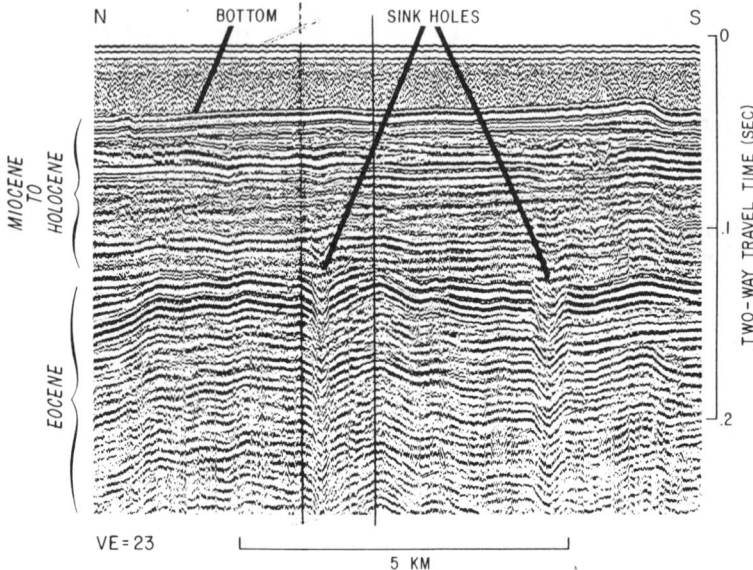

Figure 7: Seismic reflection record showing two large solution features offshore Daytona Beach, Fla. Details of these features are shown in figures 8 and 9. Note the wavey folded character of the Eocene section.

salinity, and temperature measurements that indicate the sinkhole is hydrologically inactive. More depth measurements are needed, however, to verify these preliminary findings.

Seismic character of Eocene and Oligocene limestones near Red Snapper Sink.
Seismic-reflection techniques cannot detect cavernous porosity of subsurface rocks; however, they do define displacements of reflectors or beds caused by solution and collapse. The bending down of reflectors caused by cavernous collapse at the Red Snapper Sink is an example of displacement caused by solution collapse near the surface that appears to follow a deeper collapse trend. The reverberatory reflection character of the Eocene and Oligocene rocks near the sinkhole is typical of limestones; however, the contorted and folded character of these beds in the vicinity of the sinkhole is not typical and is related to solution collapse.

In the vicinity of the sinkhole our profiles crossed a number of collapse features that extend into the Neogene section. The most obvious were found south of Red Snapper Sink on the mid-shelf off Daytona Beach (figs. 7, 8, and 9). In these two large solution features, downbuckling is most pronounced in the Eocene-Oligocene rocks and diminishes toward the surface. Figure 10 shows a Uniboom profile over a sinkhole that can be traced as a collapse structure to within 15 m of the sea floor. The funnel at the top of the sinkhole is similar to that of Red Snapper Sink. This broad funnel shows that the sinkhole was open to the seafloor (or subaerially exposed) at the close of the Pliocene, but it became filled with shelf sands in the Quaternary. Of more importance in this study, however, is the apparent folding of the Eocene and Oligocene strata beneath the originating surface of the sinkholes.

Folding of the limestone formations of northern Florida
The folded character of the limestones that comprise the Floridan aquifer system off

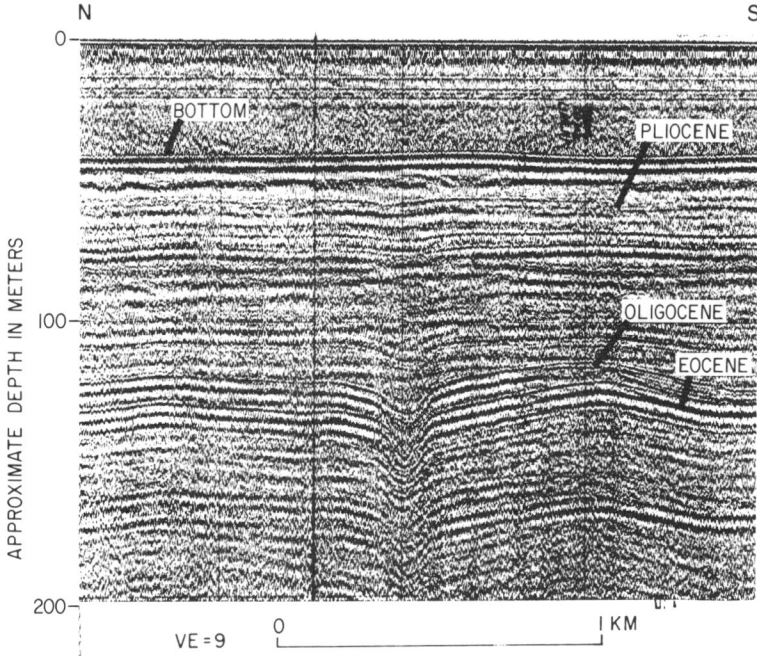

Figure 8: Uniboom seismic profile across the northern collapse feature shown on Figure 7. Note extreme deformation of Eocene and Oligocene rocks and diminishing deformation toward the surface.

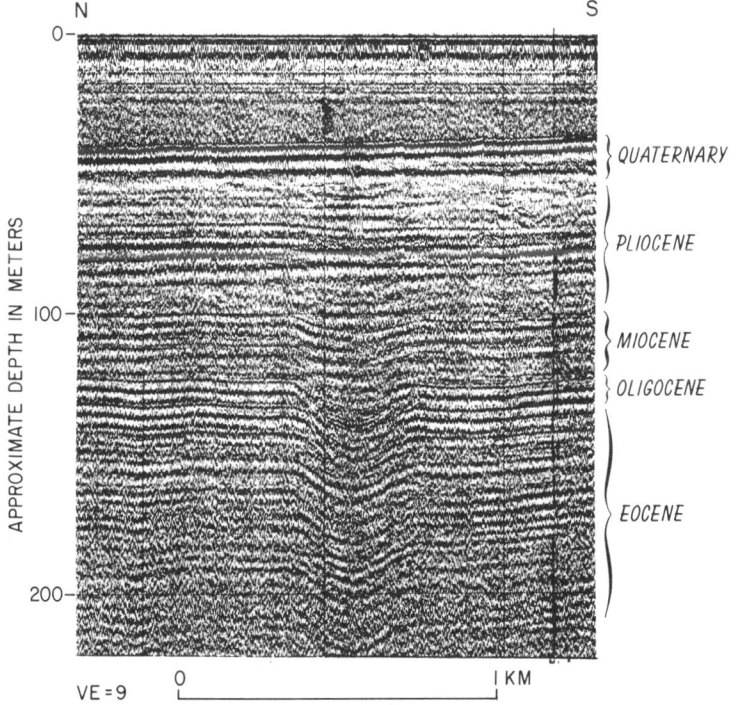

Figure 9: Uniboom seismic profile across the southern collapse feature shown on figure 7. Oligocene rocks are collapsed about 15 m (50 ft.) vertically over a distance of about 240 m (800 ft.) horizontally.

northeastern Florida was first described by Meisburger and Duane (1969), and Meisburger and Field (1975, 1976) from a detailed seismic reflection study along shore from Jacksonville to Miami. These authors noted both local pronounced folding (fig. 11) and broad undulatory flexures of up to 100 m relief that they observed throughout the area but were most pronounced between Vero Beach (South of Cape Canaveral) and Matanzas Inlet. The parallelism of reflection surfaces carried throughout the folded sections suggested to them a structural origin for the features, which they believed were related to the influence of the Ocala "uplift". No faulting was noted in their studies, and detailed mapping showed the features to be either circular, or elongate "folds" of variable strike (fig. 12). Two directions dominate the strike pattern: NNW and ENE. These trends parallel the major fracture trends (Vernon, 1951) observed onshore (fig. 12). The most folded units were below their "green horizon" which they correlated with the top of the Eocene strata (Meisburger and Field, 1976) or with the top of the Floridan artesian aquifer, which consists of highly permeable limestones of both Eocene, Oligocene, and possibly Miocene age. They observed that the upper surface of the folded beds closely matched the top of the aquifer in coastal wells from Cape Canaveral to Flagler Beach, but varied from the aquifer by up to 46 m (150 ft) off Jacksonville and Fernandina Beaches. Many of their flexures affected the entire vertical section, but others were confined to one or more of the lower units "indicating intermittent structural deformation may have occured over a long period of time." They did not recognize solution features in their data, although some of their most "pronounced folding" occurred proximal to the Crescent Beach submarine spring (figure 11).

We were able to examine the folded limestone beds on a more regional basis with both shallow and deep-penetrating seismic reflection data. These data clearly show that the folds are a solution-collapse phenomena of beds deep within the aquifer

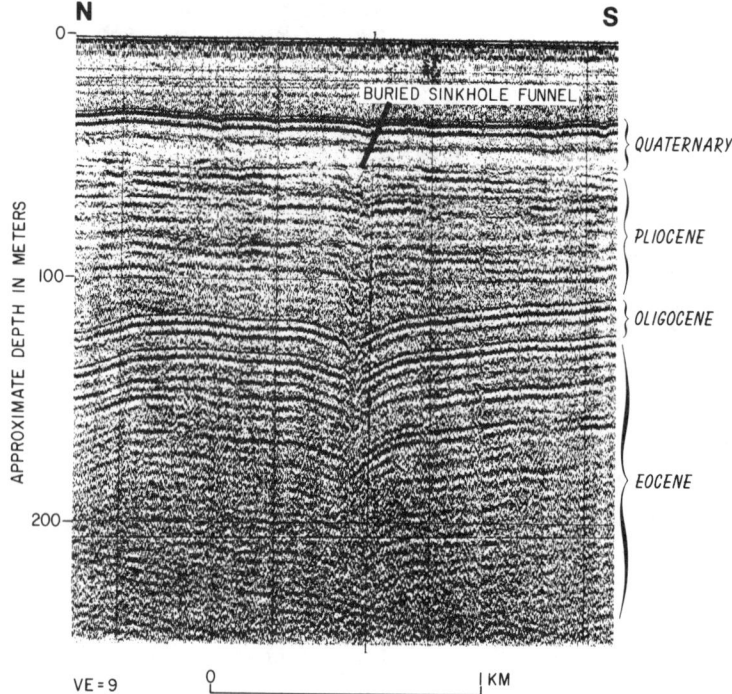

Figure 10: Sinkhole that was breached to the sea floor in post Pliocene time, but is now filled with Quaternary sand. The coincidence of the funnel at the top of the sinkhole with a major unconformity (top of the Pliocene) suggests that collapse was under subaerial conditions.

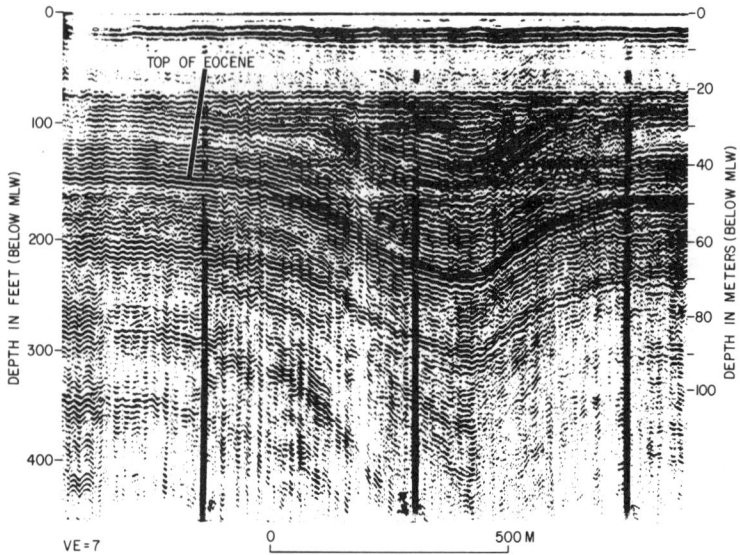

Figure 11: Seismic record (from Meisburger and Field, 1976), showing subsurface folding close to the Crescent Beach Submarine Spring (Brooks, 1961) (fig. 1). The feature is almost identical to the downward flexure observed east of Red Snapper Sink (figure 4), suggesting that both flexures mark dissolution caused collapse along joints.

systems and are not related to tectonic influence. The deep folding controls the placement of the shallower sinkholes, as sinkholes propagate upward from sharp downward flexures located along deep dissolution and collapse trends. The strongest evidence that the folding is not tectonic is that it is facies related and can be considered an identification criterion for the Late Cretaceous-Paleocene carbonate-platform edge and back-reef facies. Four high-resolution records and one Common Depth Point (CDP) seismic-reflection record are shown (figs. 13-17) to illustrate this facies relationship.

The most pronounced folding observed on our seismic-reflection records was found on the mid-shelf off Daytona Beach, Fla. (fig. 13). On this record both short- and long-wavelength folds of up to 80 m relief deform Oligocene and Eocene strata. Several sinkholes originate at sharp downward flexures on the Oligocene unconformity and propagate into the Neogene cover but do not breach the present shelf surface. The Neogene and Quaternary beds only mimic the pronounced folding of the underlying Eocene and Oligocene rocks.

Figure 14 is a deep-penetration CDP seismic-reflection record along a dip section near the record shown on figure 13 which shows the depositional environment of the deformed strata. The folded strata occur near the boundary of a major facies change in both the Eocene and the underlying Paleocene strata and in deeper strata. The discontinuous seismic character of the Eocene and Paleocene rocks on the western side of the profile, which masks underlying reflectors, is typical of high-velocity shallow-water carbonate-bank facies. On the eastern side of the record, strong, continuous reflections separate acoustically transparent seismic sequences that are typical of deep-water pelagic ooze. The folded rocks occurs over the carbonate bank edge of both the Eocene and Paleocene, which overlie a late Cretaceous carbonate bank. These edges mark the transition from the relatively shallow water of

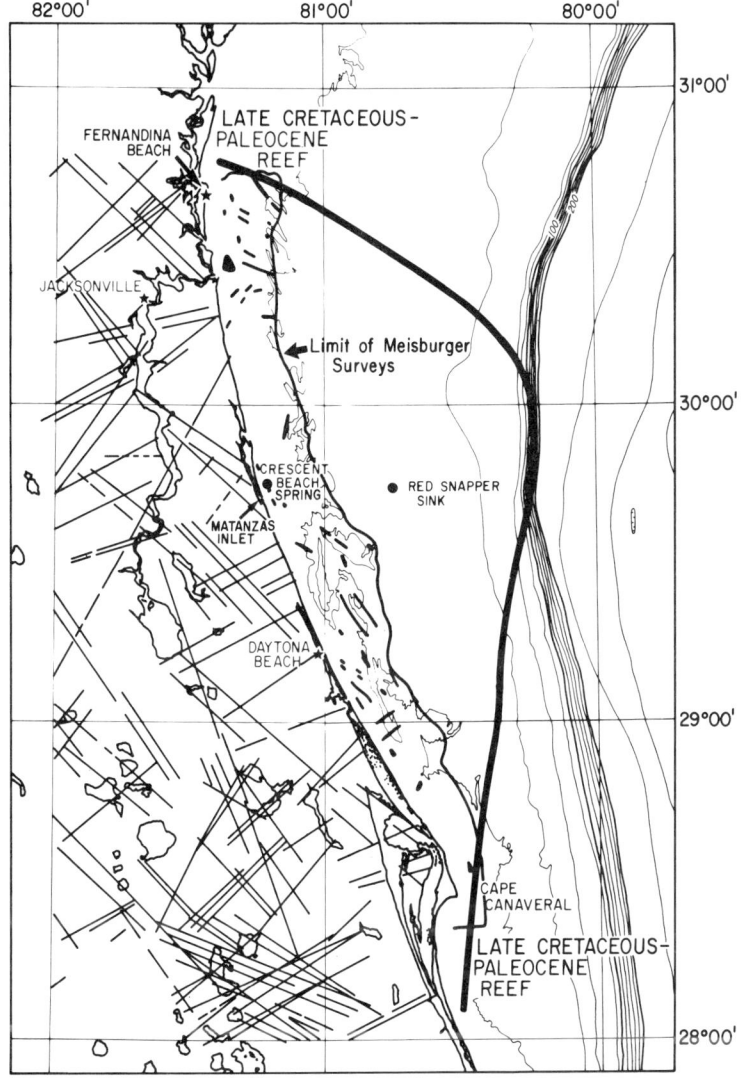

Figure 12: Map showing the largest of the downward flexures or "folds" mapped by Meisburger and Field (1975, 1976) from detailed nearshore seismic-reflection surveys (dark areas offshore) and fracture (jointing) patterns mapped on land by Vernon (1951) from physiographic expressions on aerial photographs (lines onshore). Vernon (1951) described some of the linear features as "trough like depressions and ridges marked by significant vegetation changes and soil colorations. These straight and continuous lines correspond exactly (along the Florida Ship Canal) with the map positions of vertical displacements of 20 to 160 ft distributed along many closely spaced parallel faults." Features such as those described by Vernon are similar to deformation features observed offshore in seismic-reflection profiles and which we attribute to solution collapse.

the Florida carbonate banks to depths of 300 to 500 m offshore in the ancient Florida Straits. Note that the carbonate platform edge is 20 km west off the present shelf edge in this area of central Florida (fig. 1).

A clearer picture of the carbonate-bank edge and its relationship to folding and sinkhole development is shown by the intermediate-penetration high-resolution profiles FAY lines 29 (fig. 15) and 28 (fig. 16). On FAY line 29, the Late Cretaceous and Paleocene reef edge is clearly seen; its top is marked by pronounced folding which extends into the overlying Eocene rocks. Folding abruptly stops at the forereef facies, although some faulting probably due to compaction of the deep-water chalks is evident in the Campanian-Maestrichtian section offshore (Paull and Dillon, 1980). West of the back-reef facies, the folding becomes less pronounced and at least one possible filled sinkhole is observed. On Fay line 28, folding is very pronounced above the Late Cretaceous-Paleocene reef and back-reef facies and diminishes westward. Here, the Late Cretaceous-Paleocene carbonate-platform edge lies west of the Eocene platform edge. No dissolution collapse features are evident in the Eocene rocks where they are not underlain by Paleocene carbonate platform facies.

Figure 17 shows a portion of a intermediate-penetration high-resolution seismic-reflection profile that was run along the mid-shelf from offshore Jacksonville, Fla., into the Southeast Georgia Embayment off Georgia. This line crosses the southern edge of the Late Cretaceous and Paleocene Suwannee Channel (Hull, 1962; Chen, 1965; Popenoe, 1984), which is marked by the westward continuation of the reef seen on figures 15 and 16. The Eocene rocks above the reef are contorted and folded. Southward over the carbonate platform several former sinkholes are evident. Northward within the Suwannee Channel, where Eocene rocks are quite thick, very few dissolution collapse features are evident. Dissolution collapse features may be most pronounced in the reef

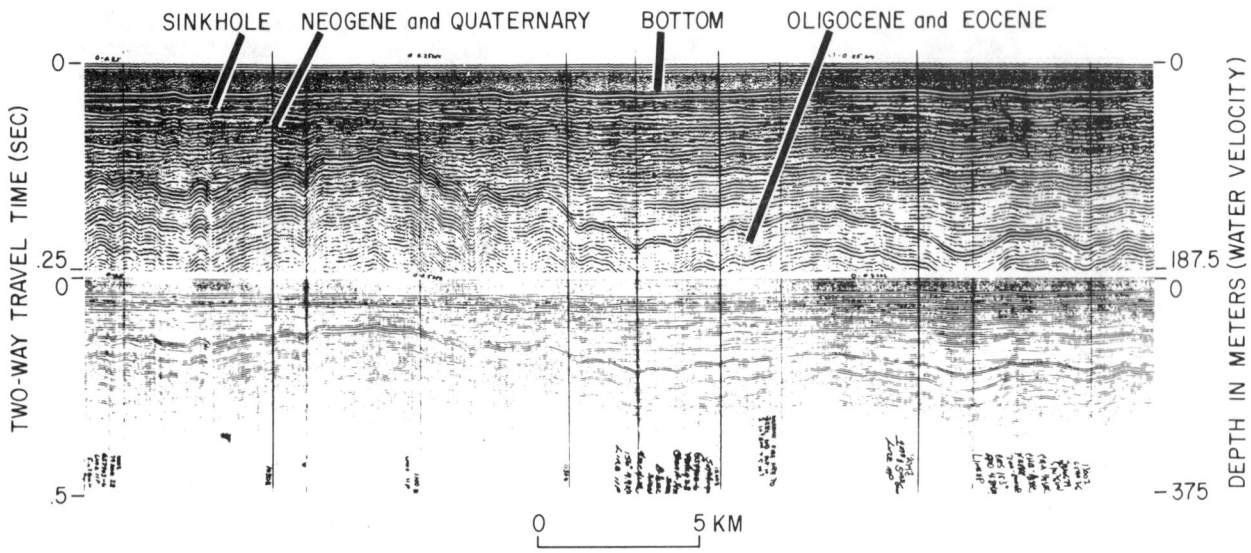

Figure 13: Picture of high-resolution seismic-reflection profile from the mid-shelf off Daytona Beach, Florida. The record is reproduced at two scales, a 1/4-second sweep above a 1/2 second sweep. Vertical exaggeration of the top record is 32, the bottom, 16. The apparent folding of the subsurface Oligocene-Eocene limestones is an effect of limestone solution and karstification of the underlying Paleocene rocks. Similar seismic records were obtained over Eocene rocks on the Miami Terrace directly offshore from known highly karstic areas between West Palm Beach and Fort Lauderdale, Fla. (Freeman-Lynde and others, 1982).

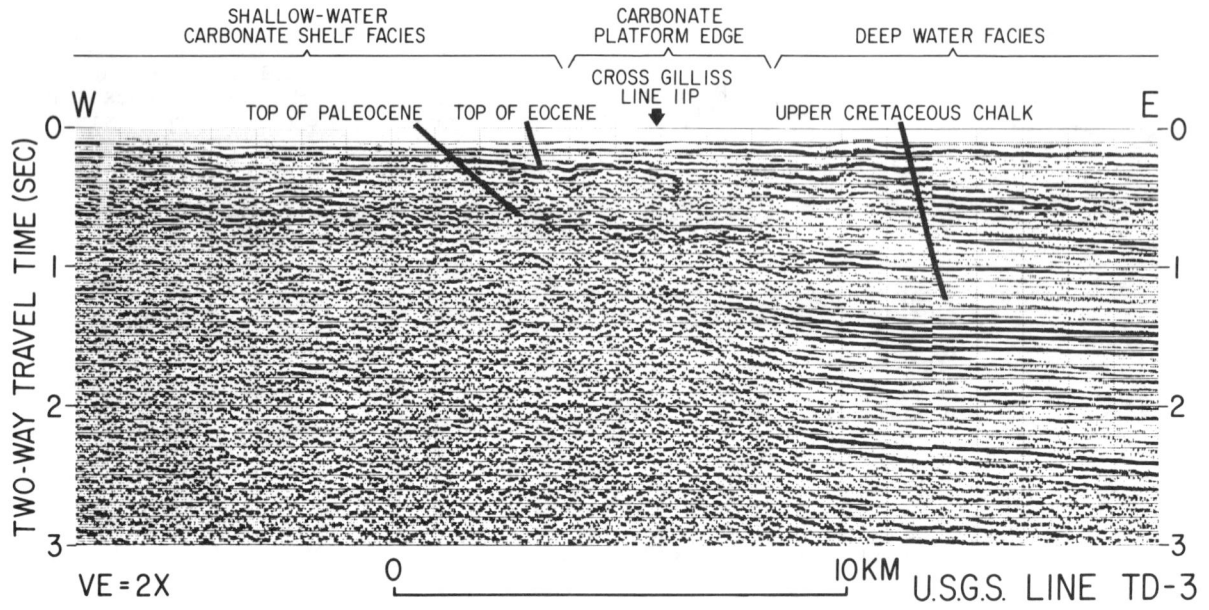

Figure 14: Part of USGS Common-Depth Point (CDP) seismic-reflection profile TD-3 showing major facies change near crossing with GILLISS line 11p (fig. 13). The carbonate platform edge is characterized by folded limestones whose deformation originates in the Upper Cretaceous and Paleocene

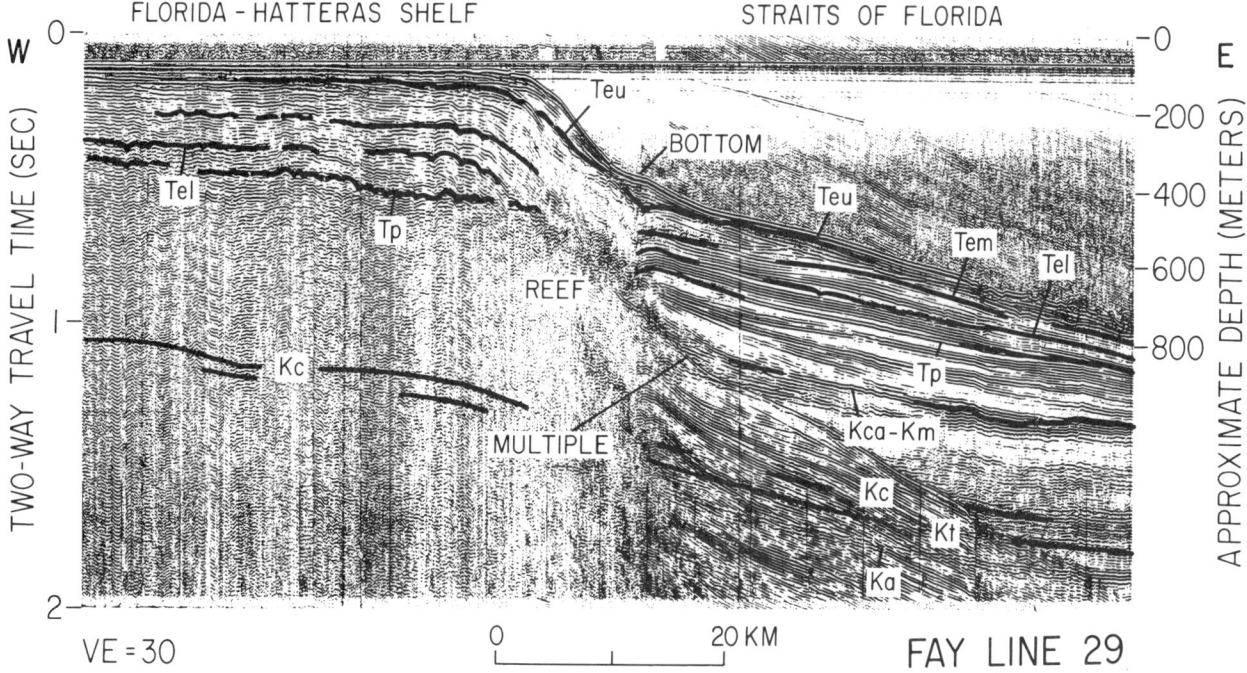

Figure 15: Part of intermediate-penetration high-resolution seismic reflection profile FAY Line 29 showing the Late Cretaceous and Paleocene reef at the edge of the Florida carbonate platform and the overlying folded Eocene strata. Note the abrupt termination of folding at the forereef facies, although the Campanian-Maestrichtian deep-water chalks are cut by many closely-spaced faults believed to be due to compaction of these units. Tops of seismic stratigraphic units are marked as follows: Ka=Albian, Kt=Turonian, Kc-Km=Campanian-Maestrichtian, Tp=Paleocene, Tel,Tem,Teu=Eocene, Tml, Tmm=Miocene.

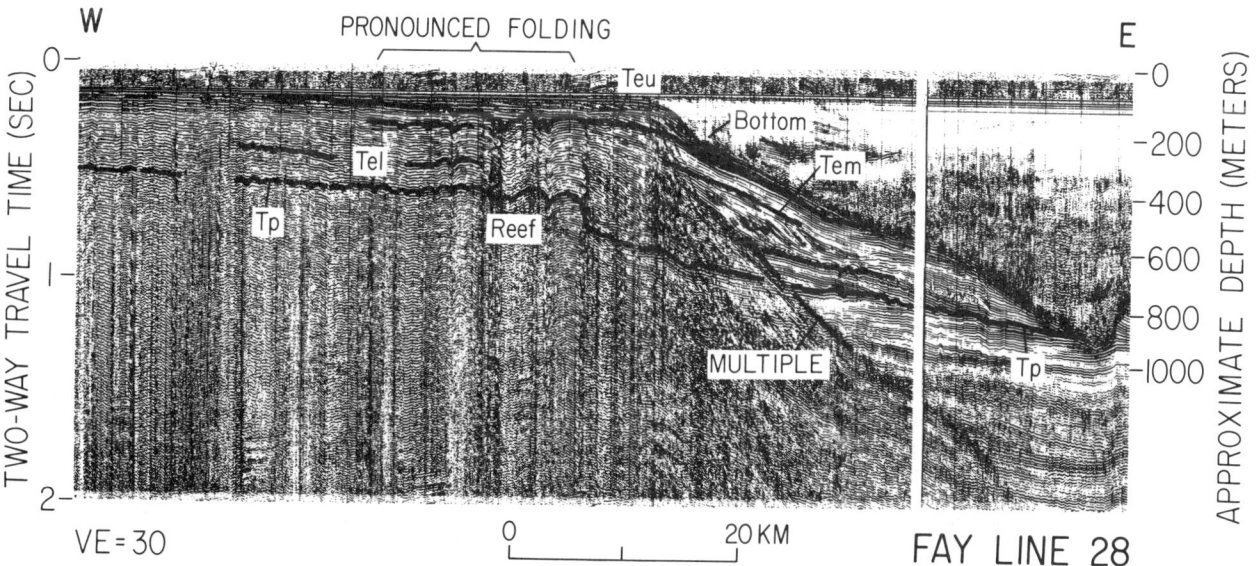

Figure 16: Part of intermediate-penetration high-resolution seismic-reflection profile FAY line 28 across an area where the Paleocene reef edge lies westward of the Eocene carbonate-platform edge. Folding and sinkholes are well developed above the Paleocene reef, but are poorly-developed east of the reef. Seismic-stratigraphic units are marked following the scheme of fig. 15.

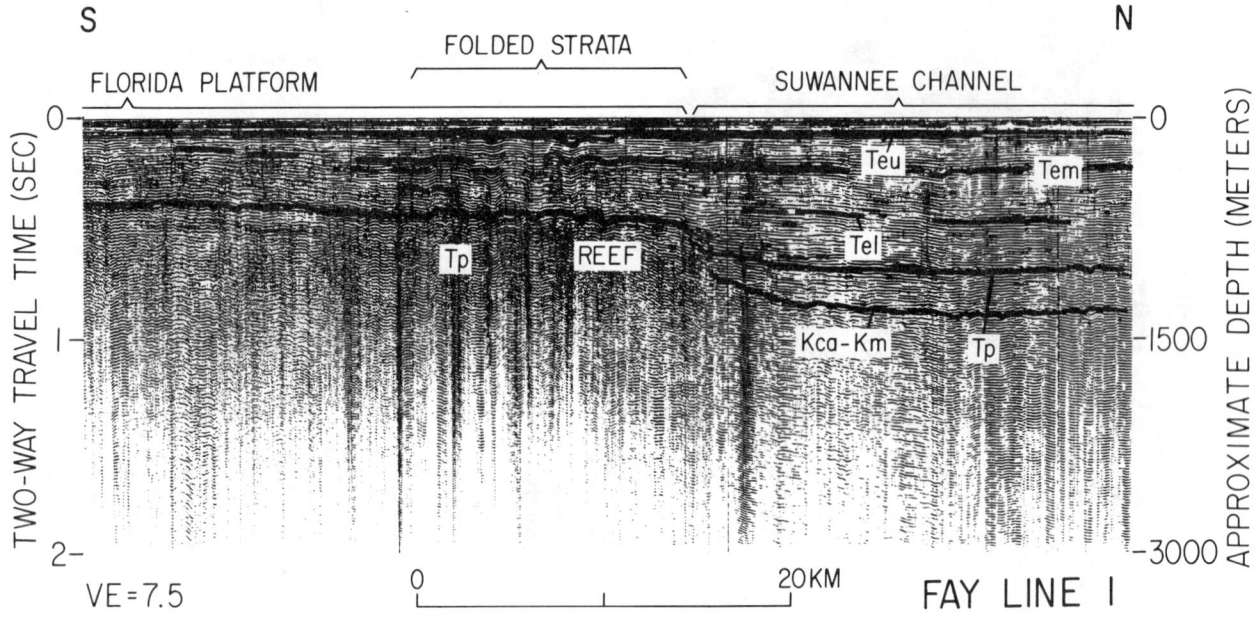

Figure 17: Part of intermediate-penetration high-resolution seismic-reflection profile FAY line 1 showing the relationship of folding and sinkhole development to the Late Cretaceous and Paleocene reef-backreef facies on the southern bank of the Paleocene Suwannee Channel. Although dissolution features are evident in the thick Eocene rocks north of the reef front, they are not pronounced. See figure 15 for explanation of symbols.

and back-reef facies because of the original high porosity of these facies, the high organic and sulfide content of the original sediments, and the easily dissolved aragonite cements that are characteristic of reefs. In addition, these facies would be subaerially exposed more often and longer than the forereef facies. Several other points are relevant to the origin of the collapse features:

1. The pattern of folding is not a tectonic pattern since folds are oriented in many directions. Orientation of folds is compatible with patterns of joints mapped onshore which would have served as avenues for ground-water circulation.

2. Lateral tectonic forces have not affected the Coastal Plain of Florida with which to associate the folding observed on the shelf. The Ocala "uplift" is not a tectonic feature but a depositional thickening of the Lake City Formation (Winston, 1976). Folding cannot therefore be related to the influence of the Ocala "uplift" as advocated by Meisburger and Field (1976).

3. The observed deformation shows a long history of movement. This history is more compatible with continuing solution and collapse caused by ground-water movement than with tectonism.

4. The proximity of some of the most pronounced folding observed by Meisburger and Field (1975, 1976) to the Crescent Beach submarine spring suggests a causal relationship. The deformation between Cape Canaveral and Matanzas Inlet occurs in the area where the Eocene surface is elevated on a regional high and the Miocene is thin to absent (White, 1958). Numerous springs and solution lakes are present onshore and both the Crescent Beach Spring and Red Snapper Sink occur offshore. Along this section of the coast, salt water has encroached into the Floridan aquifer (Vernon, 1951) and because of the lack of confining beds, upward leakages from the aquifer have lowered the piezometric surface (Manheim, 1967; Leve and Goolsby, 1969).

Solution features are pronounced along this section of the coast because during low stands of sea level in the late Miocene, Pliocene, and Pleistocene, the Eocene and Paleocene limestones were not capped by an impermeable layer and would have been subjected to downward percolation of organic and carbonic-acid-charged rainwater.

5. Pronounced folding observed by Meisburger and Field (1975-1976) near Cape Canaveral occurs over the back-reef facies of the Late Cretaceous-Paleocene reef. This reef lies about 7 km due east of Cape Canaveral (Fig. 1).

Timing of the major episode of karstification of strata under the shelf

Meisburger and Field (1975, 1976) noted that many flexures or folds affected the entire vertical section (to the depth visible on their seismic records, or into the Eocene strata), while others were confined to one or more of the lower units, indicating that structural deformation (here attributed to dissolution and collapse) occurred over a long period of time. Deformation continues into the present, as demonstrated by the unfilled sinkholes that breach the present shelf surface. Our data suggest that the open sinkholes may have a short life under shelf conditions if they are not discharging water. However, it can be demonstrated that the most pronounced dissolution and collapse occurred in the low sea level stands (Vail and Mitchum, 1978; Pitman, 1978) of the late Oligocene (Chattian) and early Miocene (Burdigalian) when the shelf was subaerially exposed. The now highly folded Eocene and Oligocene limestones (fig. 13) were no doubt deposited subhorizontally in a high-energy surf zone. Most of the folding that was caused by dissolution and collapse of deeper units occurred after deposition of the youngest unit beneath the unconformity that marks their top (middle Oligocene at the J-1 site: Poag and Hall, 1979) and deposition of the onlapping unit above the unconformity which shows only minor deformation (early Miocene (Langhian) [Poag and Hall, 1979]). This time span limits the major dissolution episode to the two deepest sea-level regressions of the Tertiary (Vail and Mitchum, 1978; Pitman, 1978; Popenoe, 1984): namely the late Oligocene (Chattian) and the early Miocene (Burdigalian) when sea level fell to 80 to 180 m below present (Vail and Hardenbol, 1979). The correlation strongly suggests that the major dissolution and collapse occurred under subaerial conditions and that only minor dissolution and collapse has occurred under submarine conditions. This finding is reinforced by evidence that filled sinkholes terminate at major unconformities (Fig. 10), suggesting that they collapsed under subaerial conditions.

Conclusions

Rocks of Late Cretaceous to Eocene age beneath the Continental Shelf off northern Florida show dissolution features in the form of sinkholes that breach the sea floor, sinkholes that have breached the sea floor in the past and are now filled with shelf sands, and collapse structures that deform overlying units. The major dissolution has occurred deep in the sedimentary section within the Upper Cretaceous and Paleocene rocks. The overlying Eocene and early Oligocene strata were buckled by the collapse of the deeper rocks causing them to appear folded. Solution features are not randomly distributed but are controlled by regional joint patterns, the thickness of overlying Miocene strata, and by facies of the upper Cretaceous and Paleocene carbonate platform. Dissolution of Eocene and Oligocene rocks follow fractures caused by deep collapse, and sinkholes propagate to the surface through the overlying Neogene section along these trends. Sinkholes have a short life on the shelf because they fill rapidly with shelf sands.

The most pronounced deformation from solution follows the reef and back-reef edge of the Late Cretaceous Paleocene carbonate platform. Lesser folding as a result of dissolution and collapse occurs between Cape Canaveral and Fernandina Beach where the Eocene rocks are near the surface on a regional high and where the overlying Miocene strata are thin to absent.

Although dissolution and collapse is a continuous process, the major period of dissolution under the shelf occurred during subaerial exposure in the late Oligocene and early Miocene.

References

Brooks, H. K., 1961, The submarine spring off Crescent Beach, Florida; Quarterly Journal of the Florida Academy of Sciences v. 24, no. 2, p. 122-134.

Charm, W. D., Nesteroff, W. D., and Valdes, S., 1969, Detailed stratigraphic description of the JOIDES cores on the continental margin off Florida; U. S. Geological Survey Professional Paper 581-D, 13p.

Chen, C. S., 1965, The regional lithostratigraphic analysis of Paleocene and Eocene rocks of Florida; Florida Geological Survey Bulletin no. 45, 105p.

Emery, K. O., and Zarudski, E. F. K., 1967, Seismic reflection profiles along drill holes on the continental margin off Florida; U.S. Geological Survey Prof. Paper 581-A, p. A1-A8.

Freeman-Lynde, R. P., Popenoe, P., and Meyer, F. W., 1982, Seismic stratigraphy of the western Straits of Florida (abs): Geological Society of America, Abstracts with Programs, v. 14, nos. 1 and 2, p. 18.

Hull, J. P. D. Jr., 1962, Cretaceous Suwannee Strait, Georgia and Florida; American Association of Petroleum Geologists Bulletin v. 46, no. 1, p 118-122.

JOIDES, 1965, Ocean drilling on the continental margin; Science, v. 150, no. 3697, p. 709-716.

Kohout, F. A., 1965, A hypothesis concerning cyclic flow of salt water related to geothermal heating of the Floridan aquifer; Transactions of the New York Academy of Sciences, Ser.II v. 28, no. 2, p. 249-271.

Kohout, F. A., and Leve, G. W., 1977, Red Snapper Sink and ground water flow offshore northeastern Florida; in Tolson, J.S., and Doyle, F. L., eds., Proceedings of the Twelfth International Congress, Karst hydrogeology, International Association of Hydrogeologists, University of Alabama Press, Huntsville, AL., p. 193, 194.

Leve, G. W., and Goolsby, D. A., 1969, Production and utilization of water in the metropolitan area of Jacksonville, Florida; State of Florida Division of Geology Information Circular no. 58, 37p.

Manheim, F. T., 1967, Evidence for submarine discharge of water on the Atlantic Continental Slope of the southern United States; Transactions of the New York Academy of Sciences, Ser. 2, v. 29, no. 7, p. 839-853.

Meisburger, E. P., and Duane, D. B., 1969, Shallow structural characteristics of Florida Atlantic Shelf as revealed by seismic reflection profiles; Transactions - Gulf Coast Association of Geological Societies v. 19, p. 207-215.

Meisburger, E. P., and Field, M. E., 1975, Geomorphology, shallow structure, and sediments of the Florida Inner Continental Shelf, Cape Canaveral to Georgia; U. S. Army, Corps of Engineers, Coastal Engineering Research Center Technical Memorandum no. 54, 119 p.

Meisburger, E. P., and Field, M. E., 1976, Neogene sediments of Atlantic Inner Continental Shelf off northern Florida; American Association of Petroleum Geologists Bulletin v. 60, no. 11, p. 2019-2037.

Paull, C. K., and Dillon, W. P., 1980, Structure, stratigraphy, and geologic history of the Florida-Hatteras Shelf and inner Blake Plateau; American Association of Petroleum Geologists Bulletin v. 64, p. 339-358.

Pitman, W. C., 1978, Relationship between eustasy and stratigraphic sequences of passive margins; Geological Society of America Bulletin v. 89, no. 9, p. 1389-1403

Poag, C. W., and Hall, R. E., 1979, Foraminiferal biostratigraphy, paleontology, and sediment accumulation rates, in Scholle, P. A., ed., Geological Studies of the COST GE-1 well, United States South Atlantic Outer Continental Shelf area; U.S. Geological Survey Circular 800, p. 49-63.

Popenoe, Peter, 1983, High-resolution seismic reflection profiles collected August 4 to 28, 1979, between Cape Hatteras and Cape Fear, North Carolina, and off Georgia and northern Florida (Cruise GS 7903-6); U.S. Geological Survey Open File Report 83-512, 3p.

Popenoe, Peter, 1984, Cenozoic depositional and structural history of the North Carolina margin from seismic stratigraphic analyses, in Poag, C. W., ed., Stratigraphy and depositional history of the U. S. Atlantic margin; Stroudsburg, PA, Hutchinson Ross Publishing Company (in press).

Schlee, J. S. 1977, Stratigraphy and Tertiary development of the continental margin east of Florida; U. S. Geological Survey Professional Paper 581-F, 25p.

U.S. National Ocean Survey, 1976, St. Augustine Light to Ponce DeLeon Inlet: U.S. Dept. of Commerce, NOAA, NOS navigation chart 11486, 9th Ed.

Vail, P. R., and Mitchum, R. M. Jr., 1978, Global cycles and relative changes of sea level from seismic stratigraphy in Watkins, J. S., Montadert, L., and Dickerson, P. W., eds., Geological and Geophysical Investigations of Continental Margins; American Association of Petroleum Geologists Memoir 29, p. 469-472.

Vail, P. R., and Hardenbol, J., 1979, Sea level changes during the Tertiary: Oceanus, v. 22, p. 71-79.

Vernon, R. O., 1951, Geology of Citrus and Levy counties, Florida; Florida Geological Survey Bulletin 33, 256p.

White, W. A., 1958, Some geomorphic features of central peninsular Florida; Florida Geological Survey Bulletin no. 41, 92p.

Wilcove, Raymond, 1975, The great Red Snapper Sink; NOAA press release reprint v. 5, no. 2, April, 1975.

Winston, G. O., 1976, Florida's Ocala uplift is not an uplift; American Association of Petroleum Geologists Bulletin, v. 60, no. 6, p. 992-994.

Structural and hydrogeologic applications of remote sensing data, eastern Yucatan Peninsula, Mexico

C.SCOTT SOUTHWORTH *US Geological Survey, Reston, Virginia*

ABSTRACT

Landsat and Seasat satellite images and aerial photographs of eastern Yucatan Peninsula, Mexico, were analyzed to delineate geologic controls of ground water. Field investigations are difficult because of low topographic relief, extensive vegetation cover, internal drainage, and limited transportation routes, therefore, synoptic images provide a potential means of studying the structure of the Tertiary and Quaternary carbonate rock terrain. Significant interpretation results include (1) the delineation of linear topographic swales, interpreted as fractures, extending more than 50 km along strike from the previously known limit of the Holbox fracture system, (2) the alignment of sink holes (cenotes) and inlets (caletas) on strike with existing faults and fracture systems, and (3) the identification of tonal anomalies in Ingles Lagoon suggesting fresh-water discharge from a submarine spring.

Introduction

This investigation analyzes existing satellite-image data to determine geologic information that would aid in understanding the hydrogeology of the east coast of the Yucatan Peninsula, Mexico. Research on the hydrogeology of the east coast of Yucatan, a cooperative effort by the U.S. Geological Survey and the University of New Orleans, has been the subject of several geologic field trips (Ward and Weidie, 1982). Satellite image data were analyzed to fulfill the 1982 GSA field-trip objectives of examining geologic control on ground-water discharge and the geomorphic evolution of the Yucatan coast.

Prior to satellite images of the Earth, aerial photographs were the only visual reconnaissance tool available to support field mapping. Aerial photographs are invaluable to geologic mapping; however, international coverage is sparse and often inaccessible. Landsat and Seasat satellite systems have acquired an enormous amount of worldwide data that provide a synoptic base for geologic investigations. Computer-assisted customized digital processing of satellite data provides optimal results; however, costs can be relatively high. An effort has been made in this study to show the results of a low-cost, multi-sensor image interpretation of standard images to support the geologic investigation; however, this approach does not represent the state-of-the-art technique in remote-sensing application for hydrogeology.

General Setting

Flat-lying Eocene, Miocene-Pliocene, and Quaternary carbonate units crop out on the east coast of the Yucatan Peninsula (Ramos, 1975). Known geologic structures on the Yucatan Peninsula are the Rio Hondo fold and fault system, the Ticul fault, and the Holbox fracture system (fig. 1A). The Paleozoic Rio Hondo fold and fault system trends N.$28°$-$36°$E. for approximately 355 km in northern Guatamala and British Honduras. The individual faults are believed to be normal faults bounding horst and graben blocks within the system (Wiedie, 1982). Parallel, northeast-trending horsts and grabens in the Caribbean Sea off the coast of Yucatan, determined by reflection, gravity, and magnetic profiles (Case, 1975), suggest that the zone extends beyond its land-based expression. Gravimetric and magnetic geophysical surveys depict positive magnetic anomalies trending on-strike with the Rio Hondo system as well as with Bouguer gravity contour closures on the trend (Ramos, 1975; fig. 1B). Orthogonal to the Rio Hondo fault system, the Ticul normal fault trends N.$60°$W. for approximately 200 km across central Yucatan. The Holbox fracture system is a relatively unknown structural zone that trends N.$10°$E. for approximately 50 km in northeastern Yucatan (Wiedie, 1982). It is termed "fracture" because of the lack of field data. Photogeologic interpretation and field mapping show near-vertical dipping fractures striking N.$30°$-$40°$E., N.$50°$-$60°$W., and N.$60°$-$70°$E., along the east coast (Wiedie, 1982).

Physiographically, eastern Yucatan, which is termed the Central Plains province, has relatively low topographic relief ranging from 0 to 30 m. No surface drainage exists within the northern and central parts of the peninsula. High porosity and permeability of the carbonate rocks provide internal drainage, which is verified by karst features. A thin veneer of terra rosa soils is caused by the weathering of carbonate rock; the soil supports an extensive vegetation cover of tropical forest, agriculture, and mangrove swamps. Annual precipitation ranges from 50 to 150 cm of rainfall with the heaviest precipitation period occurring between May and September.

Remote Sensing Platforms

Landsat multispectral scanner (MSS), return beam vidicon (RBV), and Seasat synthetic aperture radar (SAR) images and low-altitude black-and-white aerial photographs of eastern Yucatan were analyzed by photointerpretive techniques to delineate geologic controls of groundwater. A rudimentary understanding of the sensors and reflectance properties of surficial materials are required for simple photointerpretation.

Landsats 1, 2, and 3 each carried two sensors, the MSS and RBV. The MSS measured reflected solar radiation in four bands operating from the visible through the near-infrared region of the electromagnetic spectrum: band 4 (.5 to .6 μm), band 5 (.6 to .7 μm), band 6 (.7 to .8 μm), and band 7 (.8 to 1.1 μm). The four wavelength regions have distinct applications: band 4, water penetration; bands 5, 6 and 7, vegetation patterns; and band 7, land/water boundaries. Each MSS scene covers an area 185 km by 185 km with a spatial resolution of 79 m. The Landsat 3 RBV operated in a single broad spectral band of .505-to .750 μm with a spatial resolution of 30 m. Four RBV images correspond to the coverage of an MSS scene.

Although the Seasat experiment was designed for oceanographic monitoring, Seasat SAR data are useful for geologic investigations. In contrast to the Landsat MSS and RBV sensors, which are passive (they rely on the sun for illumination), the SAR is an active system that produces microwave radiation at a wavelength of 23.5 cm (L-band). The SAR illuminates the surface and measures the reflected radiation. Clouds are transparent at this wavelength; thus, radar is considered an all-time, all-weather system. The reflection of the radar data is dependent on the physical properties of the surface, including topography, surface roughness, moisture content, and composition of the surface and near-surface materials.

Satellite images provide regional coverage but lack the spatial resolution for site-specific investigations. Thus, aerial photographs are a more useful tool for field mapping because of their greater resolution (typically 1-5 m). They provide an accurate base upon which to plot field observations and, when viewed in stereo, allow three-dimensional perspectives of topography that often reveal structural and stratigraphic information. Aerial photographs proved to be a good reconnaissance tool for determining the structure of the east coast of Yucatan because of the small areal extent of the fracture sets.

Analysis Procedure and Results

Interpretation, by visual analysis, of Landsat and Seasat images and aerial photographs consists of the delineation of tonal and textural anomalies and the mapping of surface features on the basis of pattern recognition. The analysis of surface features includes the mapping of linear features, which relate to surface drainage, topography, vegetation types, lithology and structure. Because of the low topographic relief, internal drainage, and dominant carbonate bedrock of Yucatan, only surficial structure, karst topography, submarine springs, and vegetation are observed on the imagery. Vegetation can be an indicator of groundwater because of selective habitat, such as phreatophytes, and water-stressed conditions. Vegetation anomalies of Yucatan are extensive, but they will not be discussed as supportive field data is lacking.

Structure

The three major fracture zones on the Yucatan Peninsula are the Ticul fault, the Rio Hondo fault system, and the Holbox fracture system (fig. 1A).

The Ticul and Rio Hondo fault systems are well understood; however, the nature and origin of the Holbox fracture system, which trends N.5°E. to N.10°E., is unknown (Weidie, 1982). To the north, the Holbox fracture system is observed as water-filled swales or linear depressions that are traversed when one heads northeast on Highway 180 toward Cancun. Interpretation of the SAR image reveals a possible extension of the Holbox fracture system approximately 50 km south of the previously known terminus (fig. 2). Irregular and swampy topography, and the characteristics of the SAR aided in the detection of these features.

The Pliocene-Quaternary contact generally coincides with the Holbox fracture system in the north suggesting that a graben in the Pliocene controlled Quaternary deposition (fig. 1A). Well data show that two other stratigraphic boundaries conform to the trend of the Holbox fracture system suggesting an earlier structural control; the contact of the Triassic-Jurassic redbeds with schists, and the trend of the Early Cretaceous land boundary (Ramos, 1975) (fig. 3B and C). In addition, an anomalous concentration of karst topography and cenotes corresponds to the southern terminus of the proposed extension of the fracture system (fig. 2). At the latitude of the southern terminus (N.20°30'), Weidie (1982) identified a change in fracture trend along the coast implying the existance of a conjugate system.

On a regional basis, the Holbox fracture system extends subparallel to a Paleozoic basement structural high which is postulated to be a down to the Gulf fault zone (Weidie and others, 1978) (fig. 3A). The Catoche Tongue, an anomalous submarine valley 50 km wide in the northeast part of the Campeche escarpment, is aligned with this trend and suggests a structural depression (fig. 3A). Although gravity and seismic data suggest a non-structural origin for the Catoche Tongue (Feden and others, 1972), gravity data suggest that a similar feature may be present beneath the shelf of the Florida platform (Uchupi, 1976). The surficial expression of the Holbox fracture zone is interpreted to be extension fractures associated with a buried horst and graben system possibly related to the Rio Hondo and offshore structures. These correlations are speculative; however, field investigations of the 100 km-long Holbox system are warranted.

Karst

The relationship of fracture systems to the origin and development of caletas (inlets) along the east coast of Yucatan has been demonstrated by Back and others (1979). The fractures act as conduits for the mixing of fresh groundwater and marine water with the resultant solution being undersaturated in calcite, thus causing maximal dissolution of limestone (Back and others, 1979). Weidie (1982) concluded that maximum fracture density is closely related to high incidence of caletas, which develop into cuspate bays. Figure 4 presents a sequence of aerial photograph segments depicting the geomorphic development of caletas through dissolution of limestone along fractures. In Figure 5, portions of Landsat MSS images show the development of cenotes (sink holes) along fracture systems.

The alignment of cenotes, trending N.21°E. approximately 15 km west of Akumal (fig. 2), is interpreted to be an extension feature possibly related to the Rio Hondo fault system which is located to the south off the image. The anomalous concentration of karst terrain and cenotes, as seen on the SAR image (fig. 2), corresponds to the southern terminus of the proposed southward extension of the Holbox and the change in azimuth trend of fractures mapped by Weidie (1982), and suggests the existence of a conjugate structural trend.

Submarine Springs

A consistent dark tonal anomaly imaged by Landsat MSS and RBV on several dates is interpreted to be a submarine discharge of freshwater into Ingles Lagoon (fig. 6). Known freshwater springs discharging into brackish water have been observed in Nichupte Lagoon (Brady, 1978; Back, USGS, personal commun., 1982). The dark anomalies of Ingles Lagoon are observed on aerial photographs in Ward (1978) and were confirmed by aerial observation by the author. The contrast in spectral reflectance between the shallow sediment-laden bottom, which is highly reflective, and the deeper freshwater outlet which has low reflectance, is the cause of this anomaly. These features warrant further investigation as a source of freshwater because of the proximity to the airport and resort at Cancun.

Conclusion

Satellite-acquired images are a valuable tool for terrain analysis, especially in poorly charted regions. In this study, visual interpretation of images of the eastern Yucatan Peninsula identified anomalous features that had previously been undetected. These include:

1. Delineation of topographic swales extending 50 km south beyond the previously known limit of the Holbox fracture system. This relatively unknown structure is more than 100 km long, appears to be manifested in geologic and geophysical data, and is interpreted to be extension features related to horst and graben structures.

2. Delineation of karst terrain and the alignment of cenotes and caletas that trend on strike with existing fault and fracture systems.

3. Identification of tonal anomalies in Ingles Lagoon that suggest fresh-water discharge from submarine outlet. Proximity of this locality to the airport and resort at Cancun warrants further investigation of this potential source of groundwater.

References

Back, W., Hanshaw, B.B., Pyle, T.E., Plummer, L.N., and Weidie, A.E., 1979, Geochemical significance of groundwater discharge and carbonate solution to the formation of Caleta Xel Ha, Quintana Roo, Mexico: Water Resources Research, v. 15, no. 6, p. 1521-1535.

Brady, M.J., 1978, Sedimentology and depositional history of coastal lagoons, Northeastern Quintana Roo, Mexico: in Geology and Hydrogeology of Northeastern Yucatan, Ward, W.C. and Weidie, A.E., editors, New Orleans Geological Society, p. 85-112.

Case, J. E., 1975, Geophysical studies in the Caribbean Sea: in The Ocean Basins and Margins, Volume 3, The Gulf of Mexico and the Caribbean, Nairn, A. E. M., and Stehli, F. G., editors., Plenum Press, New York, p. 107-175.

Feden, R.H., Ensminger, H.R., and Massingill, J.V., 1972, Geophysical investigations of the Catoche Tongue, Gulf of Mexico: Geological Society of America Bulletin, v. 83, p. 1157-1162.

Ramos, E. L., 1975, Geological summary of the Yucatan Peninsula: in The Ocean Basins and Margins, Volume 3, The Gulf of Mexico and the Caribbean, Nairn, A. E. M., and Stehli, F. G., editors., Plenum Press, New York, p. 257-282.

Uchupi, Elazar, 1976, Physiography of the Gulf of Mexico and Caribbean Sea: in The Ocean Basins and Margins, Volume 3, The Gulf of Mexico and the Caribbean, Nairn, A.E.M., and Stehli, F.G., editors., Plenum Press, New York, p. 1-53.

Ward, W.C., 1978, Carbonate sand and gravel on the shallow shelf, northeastern Yucatan Peninsula: in Geology and Hydrogeology of Northeastern Yucatan, Ward, W.C., and Weidie, A.E., editors, New Orleans Geological Society, p. 3-29.

Ward, W.C., and Weidie, A.E., 1982, Geology and hydrogeology of carbonate rocks of the northeastern Yucatan Peninsula: in Road Log and Supplement to 1978 Guidebook, 1982 GSA Annual Meeting Field Trip no. 10, 35 p. with additional reprints.

Weidie, A.E., 1982, Lineaments of the Yucatan Peninsula and fractures of the central Quintana Roo Coast: in GSA Field Trip No. 10 - Yucatan, Road Log and Supplement to 1978 Guidebook, Geology and Hydrogeology of Northeastern Yucatan: New Orleans Geological Society, p. 21-25.

Weidie, A.E., Ward, W.C., and Marshall, R.H., 1978, Geology of Yucatan Platform: in Geology and Hydrogeology of Northeastern Yucatan, Ward, W.C. and Weidie, A.E., editors, New Orleans Geological Society, p. 3-29.

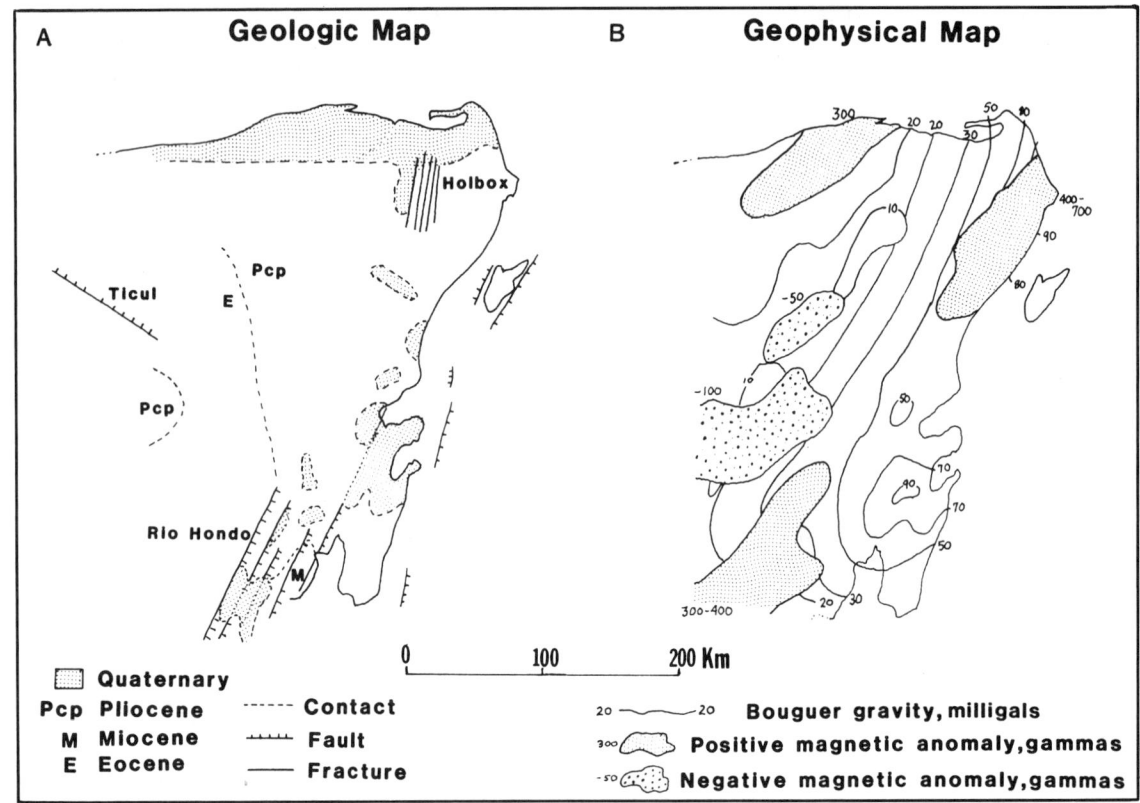

Figure 1. - A) General geologic map of the eastern Yucatan Peninsula, Mexico (modified from Ramos, 1975). Hachures indicate downthrown side of faults. B) Geophysical map (from Ramos, 1975) shows that positive magnetic anomalies and Bouguer gravity closures generally conform to the trend of the Paleozoic Rio Hondo fault zone.

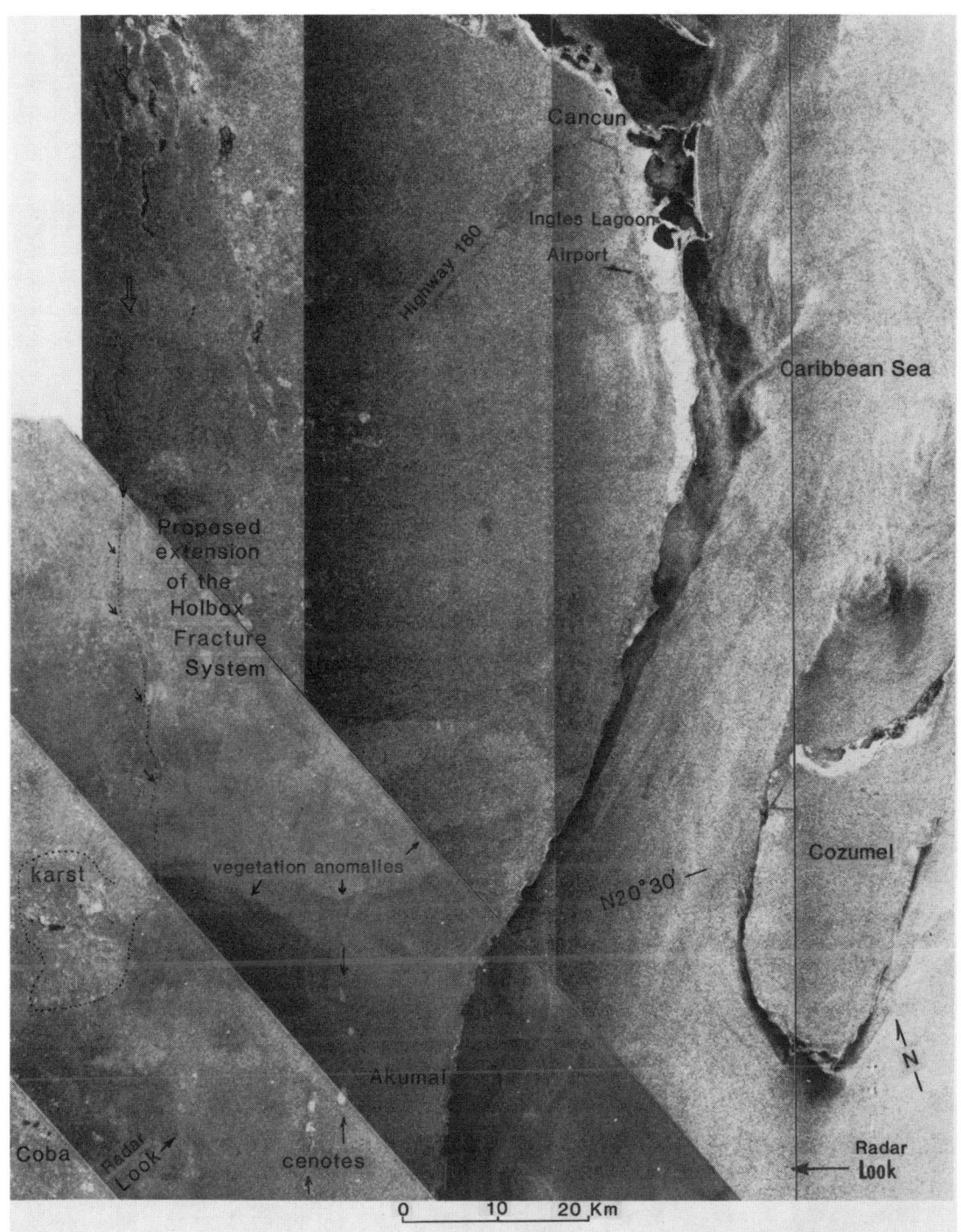

Figure 2. - Seasat SAR image mosaic composed of 7 image swaths acquired 19 August 1978 (I.D. 166, revolution 759) (parallel to length of page) and 21 August 1978 (I.D. 182c, revolution 795) (lower left portion). The hummocky nature of the swales of the Holbox fracture system was imaged by the SAR and traced approximately 50 km beyond the previous known terminus. An anomalous concentration of karst terrain and cenotes corresponds to the extended terminus, as do the large cenotes near Coba. The alignment of cenotes, which trend on strike to the Rio Hondo faults, are seen west of Akumal. The large tonal anomalies in the central to southern portion of the mosaic indicate vegetation differences. Radar look directions are labeled.

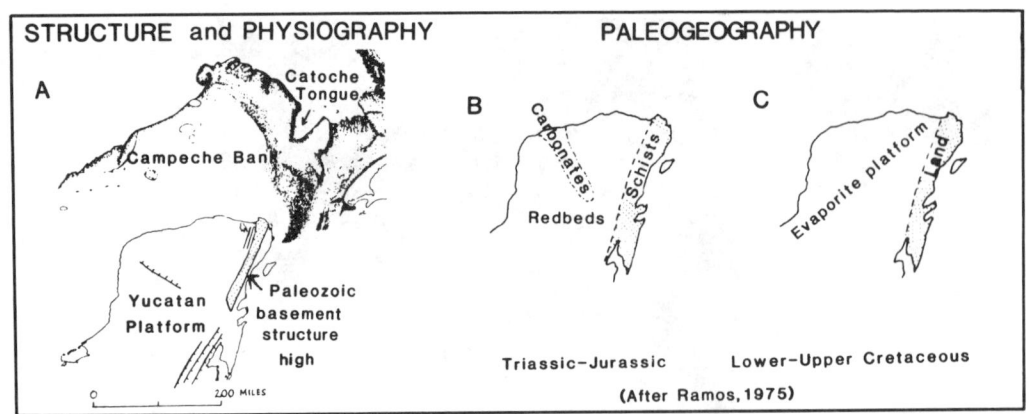

Figure 3. - A) Structure and physiography of the Yucatan Platform and Campeche Bank. Paleozoic basement structural high trends on strike with the Holbox and Rio Hondo systems. The Catoche Tongue, a submarine valley at the north Campeche Escarpment also follows the same trend. B and C) Paleogeography of the Yucatan Peninsula, based on well data, shows that both the contact of Triassic and Jurassic red beds with schist and the Lower Cretaceous land boundary conform to the structural trend (from Ramos, 1975).

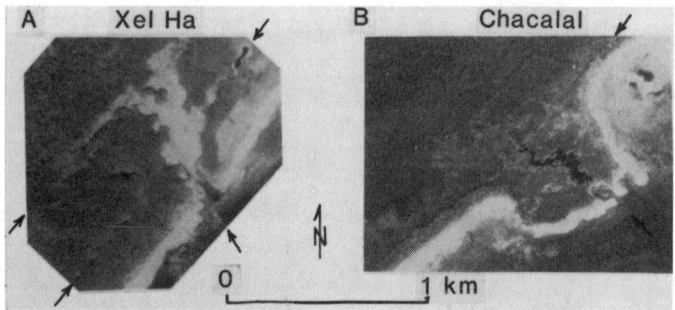

Figure 4. - A-B) Portions of low-altitude black and white aerial photographs, acquired in 1974 by Petroleos Mexicanos, illustrating the geomorphic expression of cenotes and caletas through dissolution of limestone along fractures. Arrows show strike of fractures.

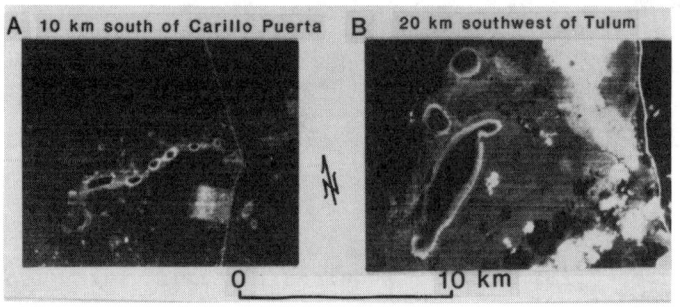

Figure 5. - A-B) Portions of Landsat MSS band 5 image acquired 12 February 1976 (scene ID 2386-15332) showing evolution of cenotes through dissolution of limestone along fractures.

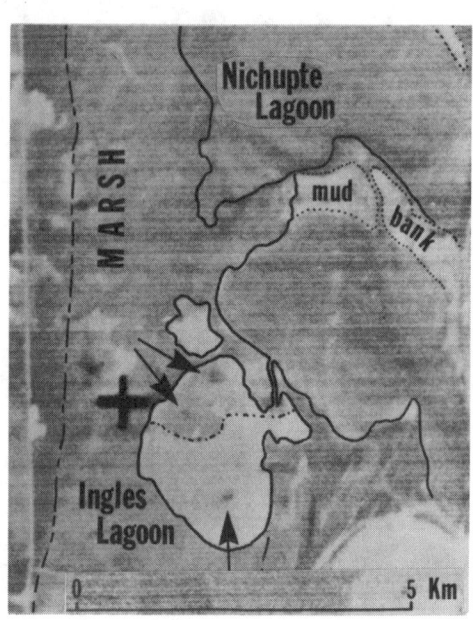

Figure 6. - Portion of Landsat 3 RBV image acquired 11 May 1978 (scene ID 30067-15352D) showing dark tonal anomalies in Ingles Lagoon (arrows). Tonal anomalies on multi-date Landsat images, aerial photographs in Ward (1978) and aerial observations by the author, suggest freshwater discharge from a submarine spring in the muddy lagoon bottom. The large black cross (reseau mark) is for sensor geometric calibration.

A comparison of sinkhole depth frequency distributions in temperate and tropic karst regions

J.W.TROESTER *University of Puerto Rico, Mayaguez, USA*
ELIZABETH L.WHITE & WILLIAM B.WHITE *The Pennsylvania State University, University Park, USA*

ABSTRACT

Sinkholes from several temperate and tropical karst regions were characterized by their depth and density. The temperate karst regions included several Appalachian areas, the central Kentucky karst, Perry County, Missouri, and Northern Florida. The tropical karst regions were the northern karst belt of Puerto Rico and the Cervicos karst region of the Dominican Republic.

The frequency of occurrence of sinkholes decreases exponentially with depth following the equation:

$$(\text{number of sinkholes}) = N_o e^{-Kd}$$

where N_o and K are constants. The N_o coefficient varies depending on the number of sinkholes counted in each region. The K coefficient ranges from 0.22 to 1.19 m^{-1} (0.068 to 0.362 ft^{-1}) in temperate karst regions where it is unaffected by rock type, limestone or dolomite, or structure, folded or flat lying. However the K coefficient in tropical karst regions is significantly different and ranges from -0.09 to -0.11 m^{-1} (-0.027 to 0.034 ft^{-1}). The inverse of K gives a characteristic depth for each population of sinkholes. A plot of the log of the number of sinkholes vs. depth shows a nearly straight line for each karst region. However, each line from temperate karst regions is slightly concave upward while each line from tropical regions is slightly concave downward. These curves probably reflect some more complicated distribution function.

The characteristic depth of dolines in temperate regions range from 0.85 m in Florida to 4.5 m in the Appalachians. The characteristic depth of sinkholes in tropical regions is much greater. In Puerto Rico it is 11.4 m and in the Dominican Republic it is 8.9 m.

In each region, the depression density varies widely, from areas with carbonate rock where no sinkholes are shown on topographic maps to a high of 20.7 sinks per km^2 in one small area. In the Appalachians, this variable is log normally distributed with an average of 1.25 sinks per km^2, whereas Kentucky, Florida, Puerto Rico, and the Dominican Republic all have higher depression densities: 5.41, 7.94, 5.39 and 5.71 sinks per km^2 respectively.

Introduction

The major surface feature of karst topography is the sinkhole, which in most temperate karst regions is a small bowl or funnel-shaped depression called a doline. In tropical karst regions, the scale of the landforms is larger, and sinkholes are often measured in tens of meters. They usually have steeper sides than dolines and are often star-shaped.

White and White (1979) used quantitative morphology to study landforms in drainage basins that contain carbonate rocks in the Appalachians. They found that the frequency of occurrence of sinkholes decreases exponentially with depth, and that this distribution of sinkhole depths is the same regardless of rock type, limestone or dolomite, or structure, folded or flat lying. It was thought that this depth-frequency distribution would exist in other temperate karst regions but it was unknown if the same distribution would exist in the tropics because the most obvious difference between temperate and tropical karst is the amount of relief in the landforms.

Sinkhole depth-frequency distributions were studied by examining topographic maps from three other temperate karst regions, the central Kentucky karst, Northern Florida and Perry County, Missouri, and two tropical karst regions, the northern karst belt of Puerto Rico and the Cervicos region in the Dominican Republic.

Morphometric Techniques Applied to Sinkholes

Of the many measures invented to characterize karst (e.g. Williams, 1966), measures to characterize sinkholes and sinkhole terrains divide roughly into two categories: those that describe the size, shape, and population of sinkholes and those that regard sinkholes as entry points for internal runoff and are concerned with the hydrology of sinkholes.

One of the oldest of the population statistics is the depression density, the number of sinkholes per unit area, introduced by Cramer (1941) and widely used by other authors. Another measure is the doline area ratio, defined as the total area inclosed by closed depression contours divided by the area of carbonate rock outcrop. Shape can be measured by LaValle's (1968) elongation ratio (length/width) or by the product of symmetry proposed by Williams (1971). Average slopes of the sides of sinkholes have also been measured. Some of these are given in Table 1.

When sinkholes become sufficiently large they develop their own internal drainage and these drainage channels can be arranged according to a Horton order or strahler (1952) order. A sinkhole of zero order is one with no internal drainage channels (Figure 1). Instead of defining the boundaries of sinkholes by the highest closed depression contour, one can construct drainage divides around all sinkholes and measure the catchment area that drains internally through the sinkhole. This definition also works for sinkholes that are the swallow points of sinking streams and indeed, much of the catchment area may be on non-carbonate rocks. This definition was used by White (1975) to measure the karstic character of drainage basins and the influence of karst on flood behavior. The area of internal runoff can be used as a parameter directly, or it can also be normalized to the area of carbonate rock outcrop. Williams (1971) defined the inverse of this number as the index of pitting. In highly karstified areas, all drainage is internal, the density of pitting becomes unity, and the area is described as 'polygonal karst' (Williams, 1972-a,b).

In past research on morphometric analysis of sinkholes, most of the attention has gone to the shape and areal distribution. Little has been done with the distributions of sinkhole depths.

Table 1. Measures of Doline Development.

Measure	Definition	Units
Depression Density	$\dfrac{N_D}{A_K}$	L^{-2} (km^{-2})
Doline Area Ratio	$\dfrac{1}{A_K} \sum_i A_{D,i}$	Dimensionless
Index of Pitting	$A_K / \sum_i A_{I,i}$	Dimensionless
Area of Internal Runoff	$\sum_i A_{I,i}$	L^2 (km^2)
Mean Doline Area	$\dfrac{1}{N_D} \sum_i A_{D,i}$	L^2 (m^2)

$A_{D,i}$ = Area of individual dolines.
$A_{I,i}$ = Catchment areas of individual dolines.
A_K = Area of karst
N_D = Total number of dolines in karst area

Methodology

The distribution of sinkhole frequency versus depth for each of the regions was determined by examining each depression as it is shown on the best quality topographic maps available. The depth of each sinkhole was estimated by counting the number of contour lines the sinkhole crosses and multiplying that by the contour interval. The results were tallied for each quadrangle or basin and then summed for the entire region.

The number of sinkholes of a particular depth was plotted against depth for each region on semilog graph paper and the data were fit to an exponential curve

$$n = N_o e^{-Kd}$$

using APL on an IBM-PC. The best fit coefficients for each region are given in Table 2. Note that the contour interval of the topographic maps for the temperate regions is in feet, while the maps of the two tropical areas use a contour interval in meters.

The average depth and the depression density were calculated for each region. The results are in Table 3.

Table 2. Least Squares Coefficients for Exponential Function

	N_o	$K(m^{-1})$	$K(ft^{-1})$	r^2
TEMPERATE KARST REGIONS				
Appalachians				
Dolines on Miss. Ls.	4,276	0.22	0.066	0.98
Dolines on Ord. Ls.	4,000	0.22	0.066	0.98
Dolines on Ord. Dolo.	4,190	0.24	0.073	0.98
All Dolines	12,608	0.22	0.068	0.99
Kentucky	892	0.25	0.076	0.99
Missouri	9,789	0.31	0.094	0.99
Florida	12,299	1.18	0.362	0.99
TROPICAL KARST REGIONS				
Puerto Rico	6,876	0.088	0.027	0.99
Dominican Republic	69,153	0.11	0.034	0.99

Table 3. Average Depth and Depression Density by Karst Region.

	Average Depth (m)	Characteristic Depth, d_e (m)	Carbonate Outcrop Area in km^2	Total Number of Sinkholes	Depression Density Sinks/km^2
TEMPERATE KARST REGIONS					
Appalachians					
Dolines on Miss. Ls.	7.3	4.62	1510	2182	1.45
Dolines on Ord. Ls.	8.3	4.62	1150	1506	1.31
Dolines on Ord. Dolo.	8.0	4.18	1474	1472	1.00
All Dolines	7.8	4.48	4134	5160	1.25
Kentucky	5.4	4.02	153	830	5.41
Missouri	6.8	3.23	*	2217	*
Florida	*	0.85	427	3395	7.94
TROPICAL KARST REGIONS					
Puerto Rico	19	11.37	799	4308	5.39
Dominican Republic	23	8.93	1262	7205	5.71

Temperate Karst Regions

White (1975) and White and White (1979) studied 62 small basins in the Appalachian Highlands from Pennsylvania to Alabama. The average annual temperature ranges from 11 to 17°C and the average precipitation from 900 to 1300 mm. Three different groups of carbonate rocks are exposed in these basins: Ordovician dolomites, Ordovician limestones and Mississippian limestones. Most of the Ordovician rocks are folded into a series of anticlines and synclines in the Valley and Ridge Physiographic Province while the Mississippian limestones are nearly flat-lying in the Appalachian Plateau. The principal surface landforms in the karst of both physiographic provinces are dolines and sinking streams. Two of the 62 drainage basins were eliminated from this study because they did not have complete coverage of U.S. Geological Survey topographic maps at a scale of 1:24,000 with a contour interval of 20 feet.

Depression density varies from 0 in 10 basins which contain carbonate rocks but no dolines that appear on topographic maps, to a high of 20.7 dolines per km^2. The frequency of drainage basins according to depression density was found to be log normally distributed over the three rock types (Fig. 2). The average value for each rock type is given in Table 3. However, they were not found to be significantly different so they were lumped together to produce an overall average of 1.25 dolines per km^2. The distribution of dolines versus depth for each rock type is plotted in Figure 3. The data were fitted to a simple exponential curve over the interval of 20 to 120 feet, and the coefficients are given in Table 2: The K coefficients of the three lines were not found to be statistically different, showing that the distribution of doline depths is independent of rock type and physiographic setting. The breakdown of the simple exponential curve for very deep dolines (>120 ft) can be understood in terms of the small sample size.

The depth-frequency distribution of the Smith Grove, Kentucky 7.5' quadrangle was tabulated by Wells (1973) and is plotted in Figure 4 where it is compared with the total Appalachian data. The Smith Grove map has a scale of 1:24,000 and a contour interval of 10 feet. The quadrangle mostly lies on the Mississippian, St. Louis and Ste. Genevieve Limestones which make up the central Kentucky sinkhole plain. In part of the quadrangle, the limestones are overlain by the Big Clifty Sandstone and then the Haney

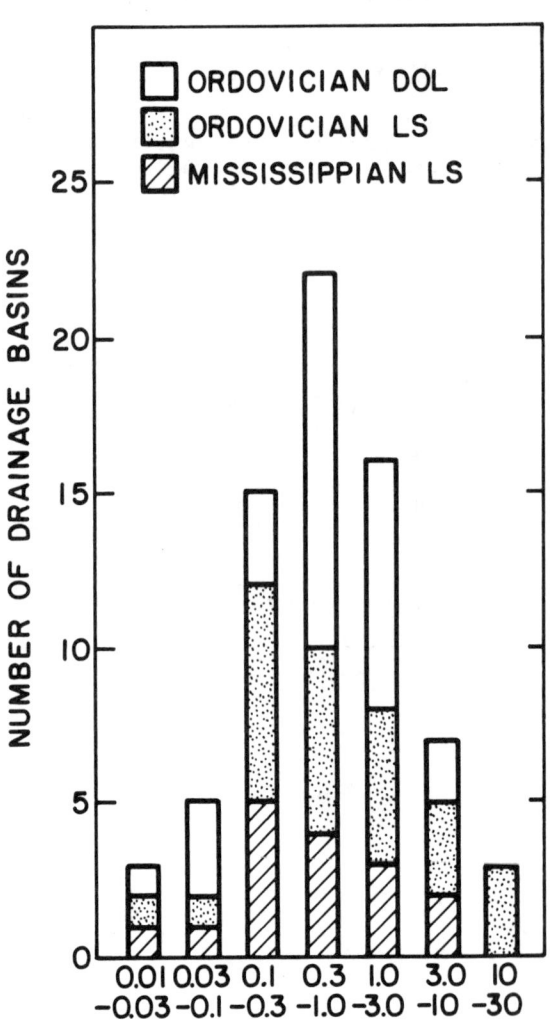

Figure 2: Distribution of depression density in Appalachian drainage basins.

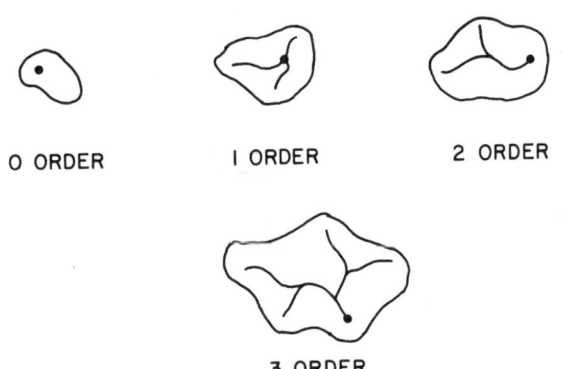

Figure 1: Examples of Strahler ordering applied to sinkholes. Adapted from Williams (1971).

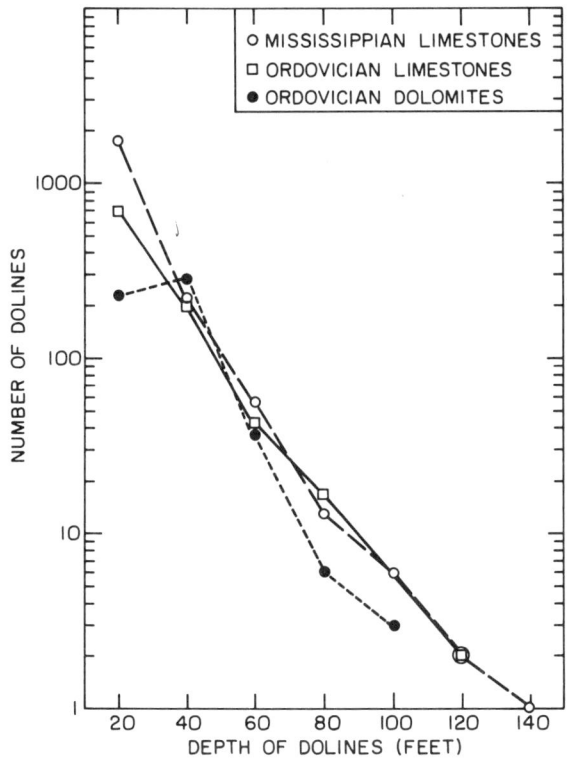

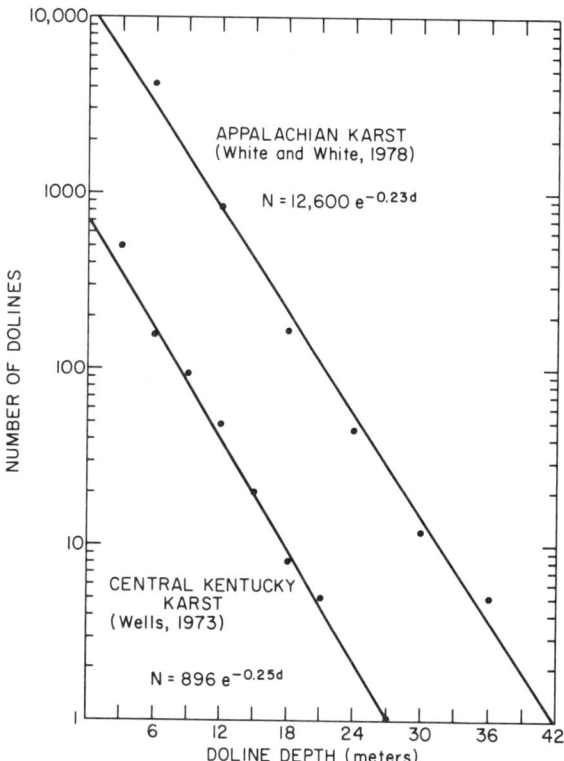

Figure 3: Frequency-depth distribution for Appalachian basins on three rock types.

Figure 4: Comparison of frequency-depth distributions for central Kentucky sinkholes with the total set of Appalachian sinkholes.

Limestone. The sandstone is breached by many sinkholes, and the Haney has many small sinks. The central Kentucky karst has an average annual temperature of 13°C and an annual precipitation of 1300 mm.

The Perry County, Missouri, karst is a sinkhole plain similar to the central Kentucky karst. It contains 577 known caves with four over 20 km long. The average annual temperature is 13°C with an annual precipitation of 900 mm. The karst is developed in the Ordovician Joachim, Rock Levee and Plattin Formations. They are a series of dolomites and limestones. Recent U.S. Geological Survey 7.5 minute topographic maps at a scale of 1:24,000 and a contour interval of 20 feet are available for the entire county. The distribution of sinkholes is similar to that obtained for the Appalachians and Kentucky regions; however, it is slightly steeper. There were no geologic maps available that separated the St. Peters Sandstone from the overlying carbonate formations so it was not possible to measure the area of carbonate rock outcrop or calculate the depression density.

The karst area selected from northern Florida consisted of four quadrangles, Hatchbend, Bell, Wanee, and Fourmile Lake, along the Suwannee River west of Gainesville. The land is of low relief. The highest elevations are upland swamps at about 90 feet above sea level. These slope down to a sinkhole plain developed on the Eocene Crystal River Formation at about 50 feet elevation. The sinkhole plain drops off into the valley of the Suwannee River which here flows at 15 feet above sea level. The contour interval is 5 feet. Sinkholes occur in profusion but are shallow, only a few reaching 25 feet in depth. most are very broad and many are compound sinks. Figure 5 shows the depth distribution. The sinks on Bell quadrangle were the exception to the pattern found in all other regions. There were more 2-contour sinks on Bell Quadrangle than 1-contour sinks. This changes the shape of the distribution function as seen in Figure 5. The reason for this is thought to be the overlapping that occurs among the sinkholes in this area. There is simply no room available for shallow sinkholes to occur. Accordingly, Bell Quadrangle was omitted from the exponential depth distribution analysis but was included in the estimate of depression density.

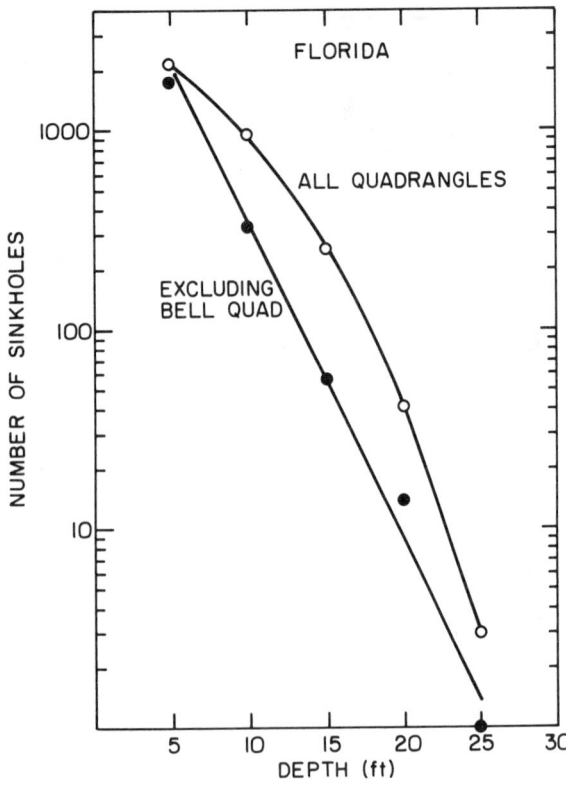

Figure 5: Frequency-depth distribution for shallow sinkholes in northern Florida.

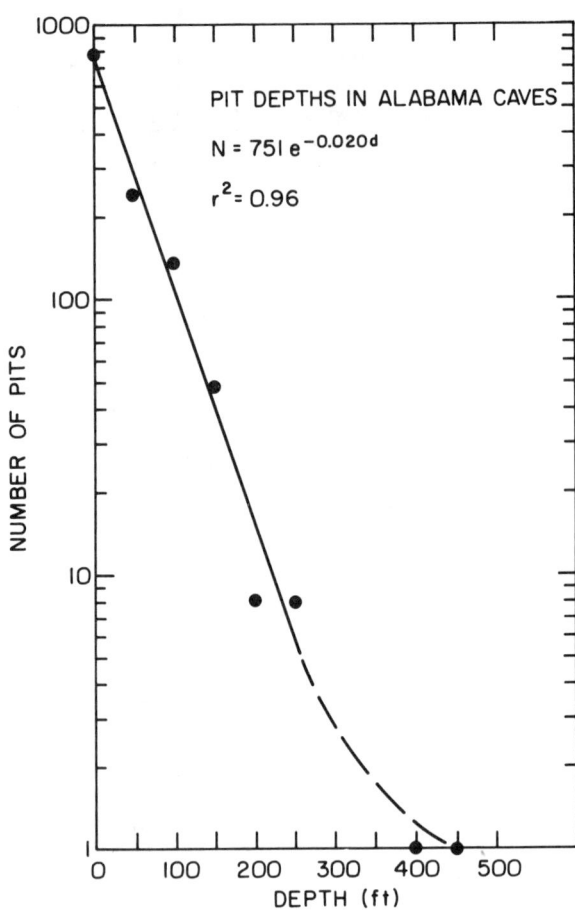

Figure 6: Frequency depth distribution for vertical shafts in Alabama caves.

Vertical shafts are related to sinkholes and often occur as open pits in the Appalachian Mountains. The depth distribution of shafts accessed internally through caves was tabulated by the Alabama Cave Survey (Varnedoe, 1973) and shows the same exponential distribution function as surface sinkholes (Fig. 6). Here one can determine the number of pits of zero depth by counting caves that contain no pits. Remarkably, the zero point also fits the distribution function.

Tropical Karst Regions

The northern karst belt of Puerto Rico is a typical humid tropical cone or cockpit karst developed in middle Tertiary limestones. The geology and hydrology of the area are discussed in Monroe (1976), Monroe (1980), and Guisti (1979). The average temperature is $25°C$, and the average annual rainfall is about 1750 mm. U.S. Geological Survey 7.5 minute topographic maps are available for the entire area at a scale of 1:20,000. The contour interval is either 5 or 10 m. The sinkholes were counted on all nine of the topographic maps with a 10 m contour interval. The results are plotted in Figure 7 and the coefficients for the best fit exponential line are in Table 2. The distribution of the sinkhole depth frequency is different from any found in the temperate regions. The depression density which was calculated for each 2.5 minute quadrangle of the 7.5 minute maps, ranged from 0 in areas with carbonate rocks but no sinkholes shown on the map to 10.6 sinks per km^2 with an average of 5.39 sinks per km^2.

The Cervicos karst region of the Dominican Republic is apparently similar to the northern karst belt of Puerto Rico. The rocks are middle Tertiary limestones. The average temperature is $26°C$ with an annual rainfall of about 2250 mm. The Army Map Service has produced topographic maps at a scale of 1:50,000 for the area. They have a contour interval of 20 m. Sinkholes were counted on the Anton Sanchez, Bayaguana, Cervicos, Cotui, El Valle, Hato Meyor del Rey and Sabana Granda de Boya quadrangles. The distribution is shown in Figure 7; the coefficients are listed in Table 2. Both the K coefficient and the curve of the line are more similar to Puerto Rico's than any of the temperate examples. The area of carbonate outcrop was measured from a geologic map

(Blesch, 1967) at a scale of 1:250,000. The depression density ranged from 0.07 to 5.80 sinks per km² in different areas with an average of 5.71 sinks per km².

Discussion and Conclusions

The exponential fitting function has the form

$$n = N_o e^{-Kd}$$

If all data are replotted as n/N_o, all curves are forced to pass through the y-axis at y = 1 (Fig. 7) and the varying population sizes are normalized out.

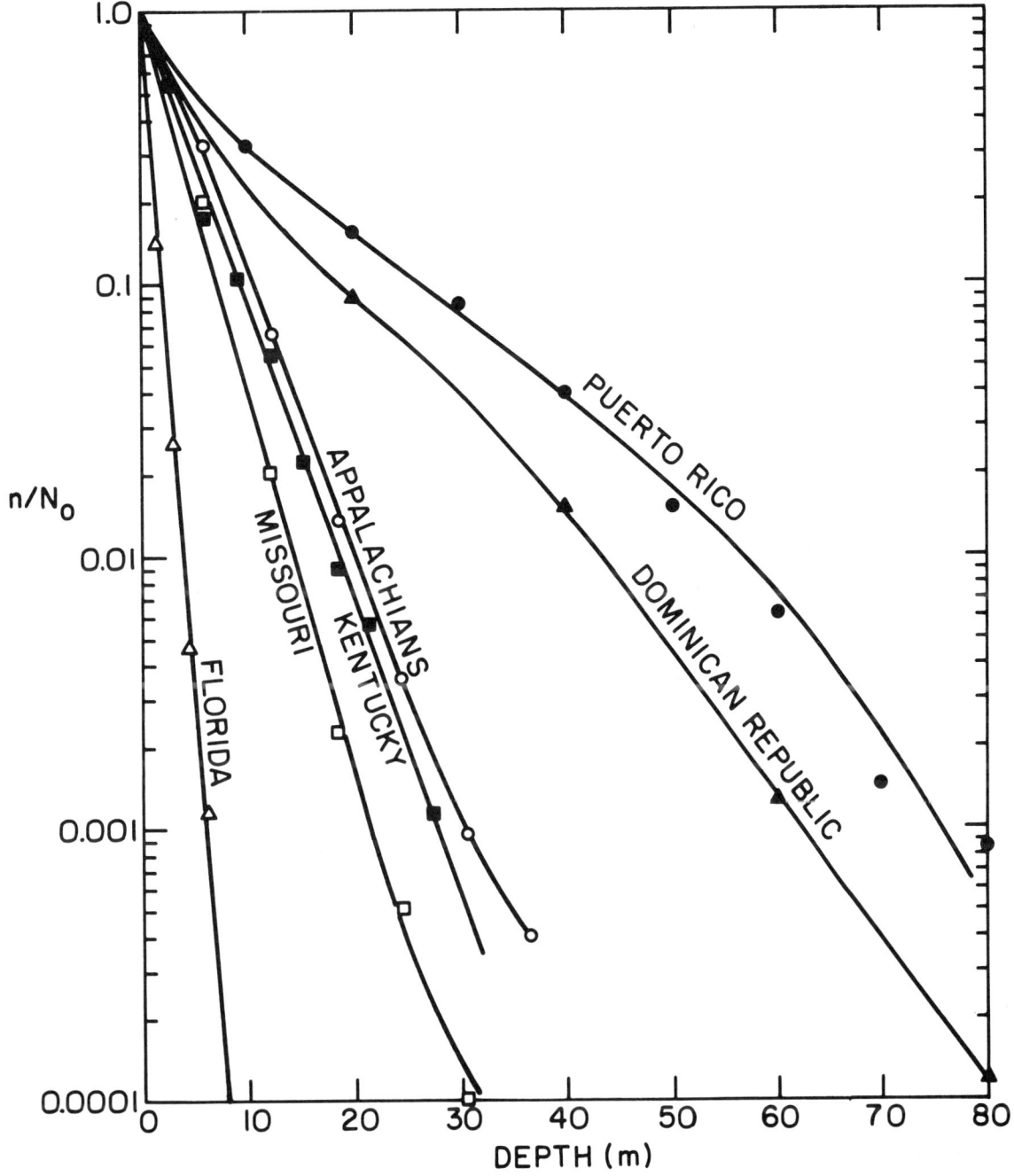

Figure 7: Data from all sinkhole populations in normalized form.

Figure 7 shows that the sinkhole depth-frequency distributions fall into two groups. One group contains the four temperate karst regions, the Appalachians, Missouri, Kentucky, and Florida and the other contains both tropical regions, the Dominican Republic and Puerto Rico. These two groups are also found in Table 2. The K coefficient in the temperate karst regions ranges from 0.22 to 1.18 m^{-1}, while in tropical regions it ranges from 0.09 to 0.11 m^{-1}. The correlation coefficient in Table 2 shows that the fit of the exponential curve to the data is very good, however Figure 7 shows that when the data are plotted on semi-log graph paper, the distributions are slightly curved with the lines from temperate karst regions being concave upward while both lines from tropical regions are slightly concave downward. This probably reflects a more complicated distribution function as has already been demonstrated by Curl (1966) for the distribution of cave lengths and the number of cave entrances.

The coefficient N_o would be an estimate of the number of sinkholes of zero depth if the exponential distribution function was valid over the entire range. Such cannot be the case and the exponential function must break down as the depth approaches zero. One could argue that there are an infinite number of sinkholes of zero depth and that therefore, the distribution function must curve upward, increasing to infinity as depth approaches zero. This cannot be the case either because sinkholes occupy a finite area and as the number of sinkholes increases, they will begin to interfere with each other and the number cannot increase without limit. This may be what is happening on the Bell Quadrangle in Florida. In tropical karsts, much of the landscape is taken up with large sinkholes and positive relief forms such as the cone and tower karst. There is simply no room for small depressions. Likewise, the distribution is terminated at the other end of the scale because of the finite thickness of most carbonate rock units and because of sinkholes eventually reaching the water table or sea level.

The difference in relief in temperate and tropical karst land forms is shown in Table 3. The average depth of sinkholes in temperate regions ranges from 5.4 m (18 ft.) to 8.3 m (27 ft) while in tropical regions it is approximately three times larger. In puerto Rico it is 19 m (61 ft) and in the dominican Republic is it 23 m (77 ft). The differentiation between temperate and tropical karst is not repeated in the depression density data shown in Table 3. Kentucky, Puerto Rico and the Dominican Republic all have similar depression densities 5.41, 5.39 and 5.71 sinks per km^2 respectively while the average for the Appalachians is much lower, 1.25 sinks per km^2.

Exponential curves have a scale factor which requires that x = 1 when y = 1/e. Thus Kd = 1 at the 1/e fraction of sinkholes and 1/K is a characteristic depth of sinkholes at the 1/e point. The characteristic depths are listed with the mean depths in Table 3. These are the characteristic numbers for each of the karst areas examined.

The data set at this point is still small with only six regions. Other temperate and tropical karst regions should be examined to determine if they follow the depth-frequency distributions shown in this paper. A major problem is the lack of good quality topographic maps of tropical karst areas.

Acknowledgements

We would like to thank Mr. Nestor M. Rivera for his assistance in counting the sinkholes in Missouri and the Dominican Republic, and Rafael Rodriquez of the U.S. Geological Survey for his help in obtaining some of the topographic maps.

References

Blesch, R.R., 1967, Mapa Geologico Preliminar Republic Dominicana: Organizacion de los Estados Americanos, Washington, DC.

Cramer, H., 1941, Die Systematik der Karstdolinen: Neues Jahrb: Mineral. Geol. Palaont., V. 85B, p. 293-382.

Curl, R.L., 1966, Caves as a Measure of Karst: Jour. of Geol., V. 74, p. 798-830.

Guisti, E.U., 1978, Hydrogeology of the Karst of Puerto Rico: U.S. Geological Survey, Professional Paper 1012, 68 pp.

LaValle, P., 1968, Karst Depression Morphology in South Central Kentucky: Geografiska Annaler, V. 50, pp. 94-108.

Monroe, W.H., 1976, The Karst Landforms of Puerto Rico: U.S. Geological Survey, Prof. Paper 899, 69 pp.

Monroe, W.H., 1980, Geology of the Middle Tertiary Formations of Puerto Rico, U.S. Geological Survey: Professional Paper 953, 93 pp.

Strahler, A.N., 1952, Dynamic Basis of Geomorphology: Geol. Soc. America Bull., V. 63, pp. 923-938.

Varnedoe, W., 1973, Alabama Caves and Caverns: Privately published, Huntsville, AL.

Wells, S.G., 1973, Geomorphology of the Sinkhole Plain in the Pennyroyal Plateau of the Central Kentucky Karst: unpublished M.S. Thesis, University of Cincinnati.

White, E.L., 1975, Role of Carbonate Rocks in Modifying Extreme Flow Behavior: Ph.D. Dissertation, The Pennsylvania State University, University Park, PA.

White, E.L. and W.B. White, 1979, Quantitative Morphology of Landforms in Carbonate Rock Basins in the Appalachian Highlands: Geol. Soc. of Amer. Bull., Part I, V. 90, pp. 385-396.

Williams, P.W., 1966, Morphometric Analysis of Temperate Karst Landforms: Irish Speleol., V. 1, pp. 23-31.

Williams, P.W., 1971, Illustrating Morphometric Analysis of karst with Examples from New Guinea: Zeits. Geomorph., V. 15, pp. 40-61.

Williams, P.W., 1972-a, The Analysis of Spatial Characteristics of Karst Terrains: in *Spatial Analysis in Geomorphology*, R.J. Chorley, ed., Methuen, London, pp. 135-163.

Williams, P.W., 1972-b, Morphometric Analysis of Polygonal Karst in New Guinea: Geol. Soc. Amer. Bull., V. 83, pp. 761-796.

Sinkhole distribution in the central and northern Valley and Ridge province, Virginia

DAVID A. HUBBARD, Jr. *Virginia Division of Mineral Resources, Charlottesville, USA*

ABSTRACT

Sinkholes caused by the dissolution of Paleozoic limestones and dolomites are present in the Valley and Ridge physiographic province of Virginia (USA). Situated in the northwestern section of the State, the Valley and Ridge province is bounded on the southeast by the clastic, volcanic and plutonic rocks of the Blue Ridge province. Approximately 21,500 square kilometers, two-thirds of the Valley and Ridge province of Virginia, have been examined on aerial photography and field checked.

The carbonate strata have been folded and faulted to form a series of northeast trending belts. Many sinkholes are concentrated along fold and fault structures, a distribution caused by two phenomena: 1) Concentrations of joint and cleavage fractures increase the permeability of carbonate rock units along faults and folds. 2) Inclined carbonate strata are commonly bordered by aquitards or aquicludes which channel surface water and groundwater to and along the carbonate rocks of a structure.

Two types of pseudosinkholes have posed identification problems. Dams and sag ponds created from landslides and/or rockfalls of Silurian sandstones have produced a number of sinkhole-like features. Abandoned lead-zinc and iron ore workings appear as sinkhole-like features in the Shady Dolomite in Wythe and Pulaski counties, Virginia.

Introduction

Sinkholes are saucer- to cylindrical-shaped surface depressions caused by subsidence or collapse as a result of dissolution of bedrock. Exposed carbonate rocks are limited to three general areas in Virginia (USA): 1) Paleozoic limestones and dolomites of the Valley and Ridge physiographic province, 2) indurated Tertiary shell beds (Yorktown Formation) in the Coastal Plain physiographic province, and 3) Precambrian, Paleozoic and Triassic limestones and marbles of the Piedmont physiographic province. In the latter two areas, sinkhole occurrences are rare and are commonly related to water-well development and land modification. This paper will be devoted to the Paleozoic carbonate strata of the central and northern Valley and Ridge physiographic province.

The Valley and Ridge physiographic province, which extends along the northwestern part of Virginia, is bounded by the clastic, volcanic, and plutonic rocks of the Blue Ridge province to the southeast and by the Appalachian Plateaus province to the northwest in adjacent West Virginia and in southwestern Virginia (Figure 1).

In January of 1980, a study of the regional distribution of sinkholes was begun in the Valley and Ridge of Virginia. As of July, 1984, approximately 21,500 square kilometers, about two-thirds of the Valley and Ridge carbonate terrane had been examined (Hubbard, 1983, and Hubbard, in progress) (Figure 2).

Techniques

Sinkholes were located by use of low altitude (4,000 m) panchromatic, stereoscopic, aerial photography. Photographs used were of winter or early spring scenes exposed about midday on generally cloudless days. Sinkholes

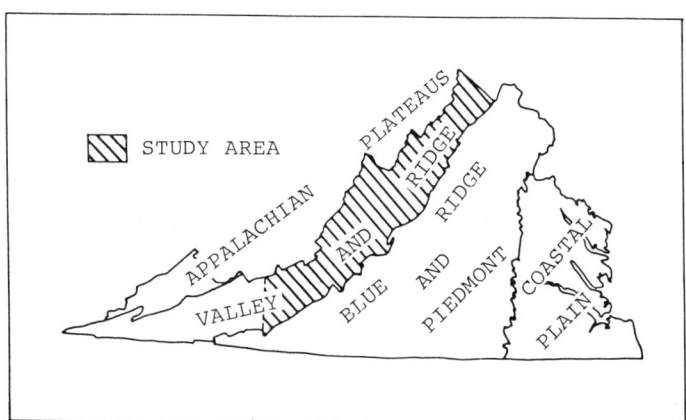

Figure 1: Study location and physiographic provinces of Virginia.

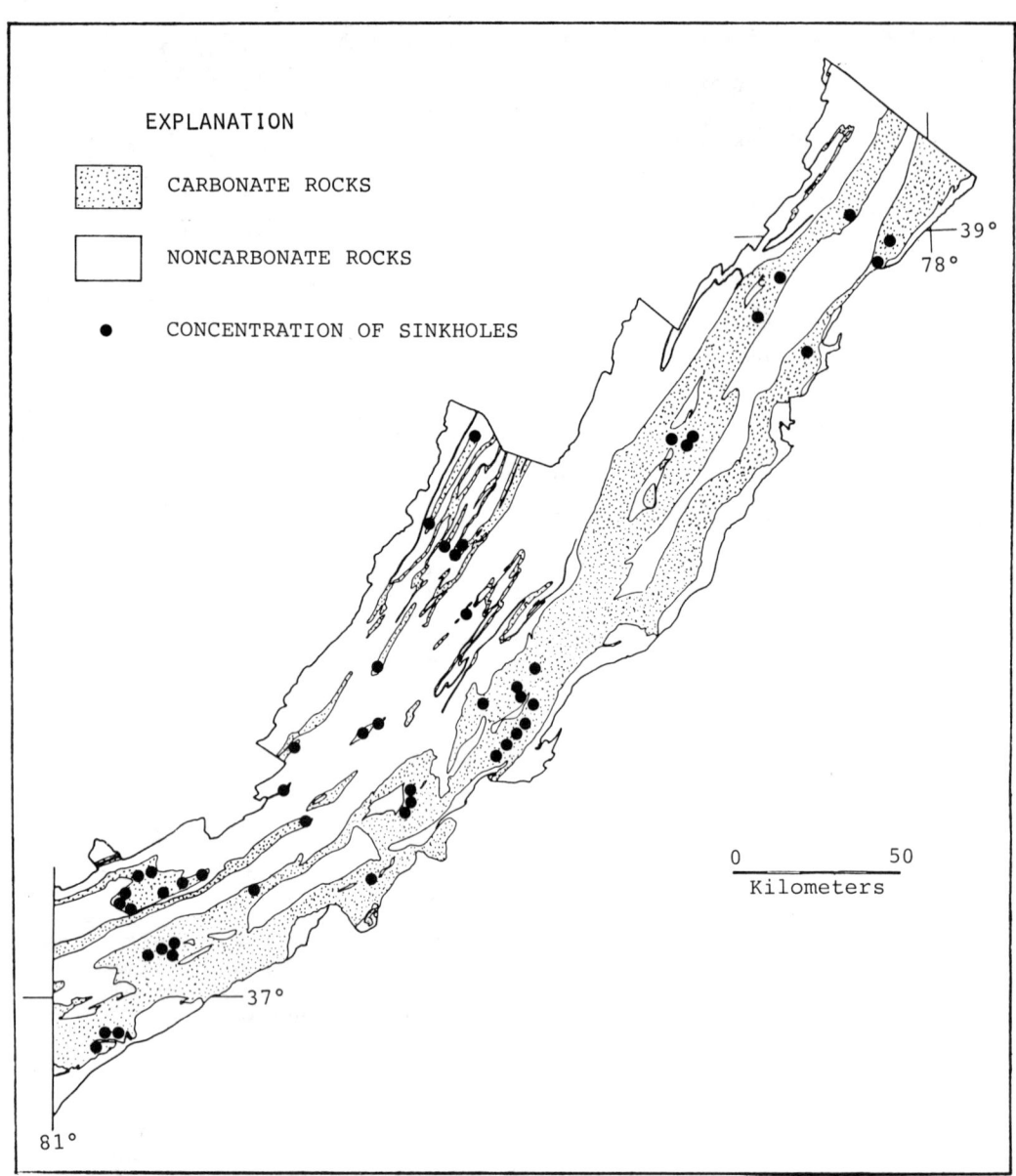

Figure 2: Distribution of sinkholes and carbonate rocks in the study area of the Valley and Ridge province.

were observed on the photography with a Bausch and Lomb Zoom 95 Stereoscope and then plotted on 7.5-minute topographic maps; 17,568 sinkholes were recorded. The sinkholes range from 10 to 3475 meters in diameter. Numerous sinkholes within this range were noted in the field, but were not observed on the photography because of the difficulty in recognizing very low relief features in both fields and woodlands. Sinkholes smaller than 10 meters in diameter could not be recognized consistently.

Geologic data were obtained from published and manuscript maps on file at the Virginia Division of Mineral Resources. Geologic units were combined into map divisions of clastic, limestone or limestone and dolomite rock types and delineated on 7.5-minute topographic maps. Sinkhole and geologic data were then photographically reduced for regional compilation at 1:250,000 scale.

Lithology

Locally carbonate rock lithology is important; however, its influence on sinkhole development is only grossly demonstrated on a regional basis. In the northern segment (Harrisonburg vicinity) of the Valley and Ridge physiographic province, concentrations of sinkholes are found along the contact of the Middle Ordovician limestones and the Cambro-Ordovician carbonates. Limestone beds in the top of the Cambro-Ordovician carbonates (Beekmantown Group) are lithologically very similar to the very soluble Middle Ordovician limestones (New Market Limestone). Other examples of sinkhole distribution patterns related to particular units occur, but on a regional scale the most obvious influence on sinkhole distribution is that of structural position of the carbonate strata.

Structure

The carbonate rock distribution in Virginia's Valley and Ridge physiographic province is largely determined by a number of major thrust faults and folds. These folded and faulted carbonate strata are aligned in a series of belts oriented NE-SW.

Sinkhole distribution is structurally controlled in two ways: by differential development of fracture permeability (cleavage and joints) and by "channeling" of water along inclined beds. Strongly fractured carbonate rocks have greater permeability, thereby enhancing conditions for sinkhole development. Many linear patterns of sinkholes are attributable to fracture patterns. Hubbard and Holsinger (1981) found a relationship between intersections of obliquely viewed lineaments, observed on high altitude (12,200 m) aerial photography, and blind valleys in Rye Cove, Virginia. Good correlation exists between passage orientations in Witheros Cave, Bath County, and the orientations of obliquely viewed lineaments observed on high altitude aerial photography of the area. Cave passage and lineament orientations align with the transverse and diagonal joint sets of the major folds. The longitudinal joint orientation coincides with the major cave passage orientation. Lineaments were not recorded along the strike direction.

"Channeling" is a structurally controlled geomorphic phenomenon. In the temperate mid-Atlantic states (USA), inclined clastic rocks generally form ridges and adjacent carbonate strata tend to form valleys because of their greater solubility. Clastic rocks and less soluble or less permeable carbonate strata may act as aquicludes or aquitards and channel drainage through the more soluble and permeable carbonate strata enhancing sinkhole development. These structural controls (differential fracture permeability and "channeling" in tilted beds) contribute to the observed sinkhole concentrations on some fold structures. Where clastic rocks are exposed in breached folds, fold noses commonly have the greatest concentrations of sinkholes. The Bane dome in Giles County has a fairly uniform sinkhole development around its margin. Uniformity of bed inclination and fracture permeability of the limestone margin and poorly developed channelling in the low relief interlayered carbonate core may account for this.

Hack (1965) notes that sinkhole development favors synclinal structures in the Shenandoah Valley. However, no apparent preferential development was observed, by this author, in the State in general. Locally, either anticlinal (as seen in the Front Royal area, Hubbard, 1981) or synclinal structures may favor sinkhole development.

Other geomorphic influences on the sinkhole distribution include: river entrenchment and the presence of terrace gravels. The meandering course of the Shenandoah River in Shenandoah County is controlled by transverse fracture and bedding trends (Hack and Young, 1959). Locally, entrenchment of this river has formed river bluffs 20 to 40 meters high and greatly steepened the hydraulic gradients in the carbonate rocks near the river. As a result, sinkhole development has been enhanced along these bluffs. Concentrations of sinkholes on or near river bluffs are greater than those in the same rock units away from the entrenched river. Similar sinkhole distribution can be observed along the bluffs of New River.

Terrace gravels diffuse infiltration and result in a deep soil development (Hack, 1960) which appears to inhibit surficial expression of karst activity (Hubbard, 1981).

Pseudokarst Problems

Two types of pseudokarst "sinkholes" have posed identification problems. Dams and sag ponds created from landslides and/or rockfalls of Silurian sandstones have produced a number of sinkhole-like features. One of these, Mountain Lake, is an impressive natural lake in Giles County, Virginia. This 0.85 x 0.3 kilometer lake was apparently formed by landslide-rockfall material that dammed a high stream valley (Sharp, 1933, and Parker, Wolfe and Howard, 1975). A number of sinkhole-like depressions have recently been examined in Giles and Montgomery counties. A. P. Schultz (personal communication) attributes these to landslide-rockfall processes.

Other sinkhole-like features are found in the Shady Dolomite of Wythe and Pulaski counties, Virginia. These pseudosinkholes are the result of mining operations for residual lead-zinc ores in the later 1800's (Case, 1894, and Watson, 1905), and iron ores and pigments (Watson, 1905) mined until 1913. During "open-cut" mining residual ores were removed and buried limestone pinnacles exposed. After years of erosion these excavations resemble sinkholes with some exposed carbonate pinnacles.

References

Case, W. H., 1894, The Bertha zinc mines at Bertha, Virginia: American Inst. Mining Eng. Trans., v. 22, p. 511-536.

Hack, J. T., 1960, Relation of solution features to chemical character of water in the Shenandoah Valley, Virginia: U. S. Geol. Survey Prof. Paper 400-B, p. 387-390.

_____, 1965, Geomorphology of the Shenandoah Valley, Virginia and West Virginia, and origin of the residual ore deposits: U. S. Geol. Survey Prof. Paper 484, 84 p.

Hack, J. T. and Young, R. S., 1959, Intrenched meanders of the North Fork of the Shenandoah River, Virginia: U. S. Geol. Survey Prof. Paper 354-A, 10 p.

Hubbard, D. A., Jr., 1981, Karst development in the Front Royal 7.5-minute quadrangle of Virginia, in Beck, B. F., ed., Proceedings of the Eighth International Congress of Speleology: National Speleological Society, Huntsville, Alabama, p. 511-514.

_____, 1983, Selected karst features of the northern Valley and Ridge province, Virginia: Virginia Division of Mineral Resources Publication 44, single sheet.

_____, in progress, Selected karst features of the central Valley and Ridge province, Virginia: Virginia Division of Mineral Resources Publication.

Hubbard, D. A., Jr., and Holsinger, J. R., 1981, Karst development in Rye Cove, Virginia, in Beck, B. F., ed., Proceedings of the Eighth International Congress of Speleology: National Speleological Society, Huntsville, Alabama, p. 515-517.

Parker, B. C., Wolfe, H. E. and Howard, R. V., 1975, On the origin and history of Mountain Lake, Virginia: Southeastern Geology, v. 16, p. 213-226.

Sharp, H. S., 1933, The origin of Mountain Lake, Virginia: Journal of Geology, v. 41, p. 636-641.

Watson, T. L., 1905, Lead and zinc deposits in Virginia: Geol. Survey of Virginia Bull. 1, 156 p.

Sinkhole distribution in Winona County, Minnesota

JANET DALGLEISH & E.CALVIN ALEXANDER, Jr. *University of Minnesota, Minneapolis, USA*

ABSTRACT

Winona County, located in southeastern Minnesota, is part of a karst region in the upper Mississippi Valley. The karst is developing in flat-lying dolomitic Ordovician rocks. As part of a Minnesota Geological Survey county atlas program, we have systematically field located sinkholes, and prepared a 1 to 100,000 scale map showing sinkhole locations and sinkhole probability. We located 535 sinkholes in Winona County (~ 1600 km^2). Most of these relatively small, geomorphically young sinkholes are not included in the USGS 7.5 minute topographic maps and cannot be readily detected on air photos. The sinkhole density, while low compared to many karst regions was much greater than local, regional, and state land use planners anticipated.

New bedrock, surficial and hydrogeology maps of Winona County were used for correlation with the geographic distribution of the sinkholes. The primary control on the distribution of sinkholes appears to be the bedrock stratigraphy. The secondary controls, not necessarily in order of importance include slope of the land surface, and composition of surficial materials. The depth to the water table does not appear to have an important affect on sinkhole development. Age data indicate that the rate of sinkhole formation has significatnly increased in recent years.

Introduction

Karst is an important geomorphic process influencing the landscape and hydrology of southeastern Minnesota. In Winona County, as in most of the region, the density and distribution of sinkholes varies enormously over small areas. The purpose of this work was to document the density and distribution of sinkholes in Winona County, and to relate that distribution to variations in bedrock geology, surficial materials, and/or hydrogeology. A second goal of this study was to produce a sinkhole probability map of Winona County that would aid state, county and local residents in dealing with the environmental problems associated with the karst region.

The topography of Winona County is characterized by a gently rolling upland plateau incised by deep valleys. The rolling upland, making up about 75 percent of the county, is underlain by dolomite and limestone. Numerous karst features, such as enlarged joints, sinkholes, springs, dry valleys, and caves have formed in the upland as illustrated in Figure 1.

Few perennial streams flow across the upland plateau although there are many small dry valleys. Most of the surface water rapidly percolates into the fractured carbonate bedrock. This water eventually discharges from springs flowing from the Cambrian sandstones exposed in deeply incised valleys or directly into the Mississippi River. Water also emerges from the carbonate bedrock and sandstone escarpments on the upland. Water quality of the Prairie du Chien-Jordan aquifer, underlying the karst plateau shows high levels of surface contaminants as is typical of karst.

Southeastern Minnesota's karst is a fluviokarst in the classification system of Sweeting (1973). The fluviokarst is mantled with residuum and/or glacial tills to an average depth of 6 to 15 m. At the present time, most of the fluviokarst is actively being exhumed.

Stratigraphy: Winona County's karst is forming in gently dipping Lower and Middle Ordovician dolomites and limestones. Carbonate formations cap three bedrock plateaus separated by sandstone and/or shale escarpments. The stratigraphically lowest, and largest plateau is capped by the Prairie du Chien Group, consisting of the Oneota Dolomite, and Shakopee Formation (Hobbs, 1984), and contains the majority of the sinkholes in Winona County. Two smaller plateaus, capped by the Platteville and Galena Formations, are found in southwestern Winona County.

Geomorphology: Since the Cretaceous, the Paleozoic bedrock of Winona County has been exposed to tropical, glacial and temperate climates. Although the age of the karst is

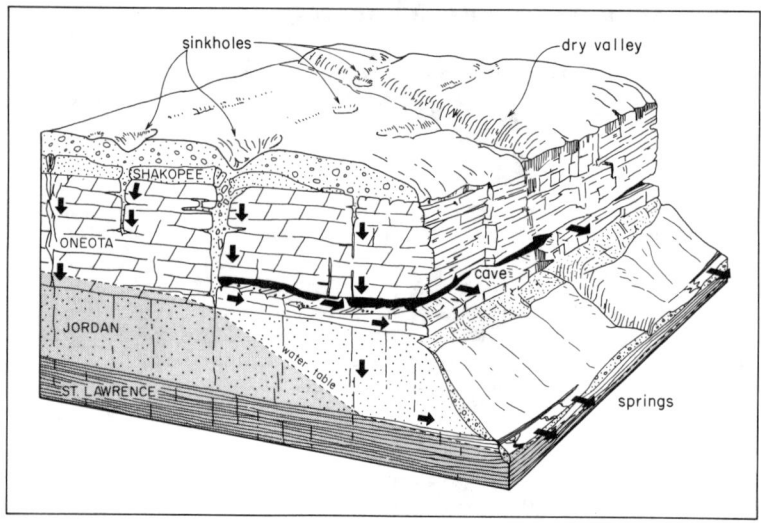

Figure 1. Typical karst and topography relationships in Winona County.

unclear, it appears that karstification began while southeastern Minnesota was exposed to intensive weathering during the warm humid climate of the Late Cretaceous through Tertiary Periods (Sloan, 1964).

The bedrock is mantled with an iron rich residuum which probably formed during the Tertiary as the St. Peter escarpment was eroded to the southwest, and the Prairie du Chien was exposed to weathering (Hobbs, 1984). The residuum appears to be pre-Pleistocene in age since it is stratigraphically lower than the glacial tills found in the western two-thirds of the county. According to Hobbs (1984), the residuum is generally less than 2 m thick, but it reaches 100 m in deep depressions believed to be paleokarst sinkholes.

Karst processes were generally inhibited during the Pleistocene. Pre-Wisconsinan glaciation of the western portion of the county eroded some of the residuum, and smoothed the irregular bedrock surface of the upland plateau, covering the karst with glacial till. According to Hobbs (1984), there appear to be two types of tills, a non-calcareous till in the eastern half of the county, and a calcareous till which is locally leached in the western half of the county. Most of the upland is mantled with loess. There is, however, an area in the west central portion of the county where the loess was either never deposited or has been removed by some process.

During the Pleistocene, the Mississippi River and its tributaries became deeply entrenched. This lowered the regional base level causing the upland plateaus to return to a more erosional stage.

The karst is actively forming during the present temperate climate as the fluviokarst is exhumed. Both surface and groundwater are dynamic geomorphic agents. Surface streams have aided in lowering the bedrock surface as seen by erosion of the St. Peter escarpment which has retreated to the southwest, and by the deeply entrenched streams. Meteoric water, percolating through organic soil and regolith, has dissolved the rock and increased the rock's secondary permeability. The surface topography is returning to its previous irregular form as sinkholes reopen and expand their drainage basins, and surface streams dissect the landscape.

Hydrogeology: The bedrock aquifers underlying Winona County are among the highest yielding aquifers in the United States (Hogberg, 1972). The water quality is generally good, except in the Prairie du Chien-Jordan aquifer where well water frequently exceeds the 10 ppm nitrate-nitrogen drinking water standard. This dolomite-sandstone aquifer is the most widely utilized and shallowest bedrock aquifer over most of the karst plateau. It has characteristics of both fractured flow, and flow through porous media.

The Prairie du Chien Group is mostly dewatered in the northeastern half of the county where the water level is lowered by water discharging into the deeply entrenched Mississippi River drainage system. The Jordan Sandstone contains water throughout most of the county.

Methods of Approach

The initial step in studying the sinkhole distribution in Winona County was to locate sinkholes. Since many of the sinkholes have been filled, and most of the open ones are less than 10 m in diameter (Figure 2), only 20% of the depressions can be clearly identified on USGS topographic maps and air photos. Other sources of information on sinkhole locations included the Minnesota Speleological Survey, and the American Soil Conservation Service, which is currently preparing a Soil Survey of Winona County.

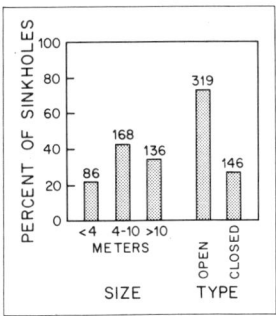

Figure 2. Sinkhole size and type.

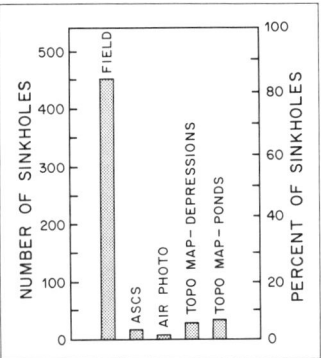

Figure 3. Methods of sinkhole detection.

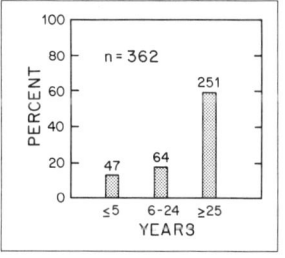

Figure 4. Sinkhole age distributions.

The majority of the sinkholes were identified by systematically canvassing local residents living on the upland plateaus for information on sinkhole locations on their property or in the area. When a sinkhole was located in the field, physical characteristics and historical information was recorded.

A sinkhole map was prepared in conjunction with the Minnesota Geological Survey county atlas program. New bedrock, surficial and hydrogeology maps of Winona County were also prepared at a 1 to 100,000 scale for the county atlas. Information was taken from each of these maps to describe the geology of each sinkhole. A data sorting computer program was used to analyze the physical, historical, and geological parameters collected on each sinkhole.

General Sinkhole Characteristics

Detection of Sinkholes: One conclusion of this paper is that only a small fraction of the sinkholes in Winona County are identifiable on 7.5 minute USGS topographic maps or air photos -- the traditional tools of karst geomorphology. Of the 535 sinkholes identified in Winona County, 85% were initially located through field checking (Figure 3). Prior to this study, Utica Township in western Winona County was widely recognized as having a high density of sinkholes. Even here, only 37% of the sinkholes were located by techniques other than field checking. In contrast, few people working with land use problems in the county realized that Pleasant Hill Township, in eastern Winona County, has sinkhole densities comparable to those found in Utica. All of these sink holes were located through field work.

Age and Rate of Sinkhole Formation: Age information was obtained for 68% of the sinkholes. Sinkhole ages were divided into three age classes: young, < 6 years; intermediate, 6 to 25 years; and old, > 25 years. The number of young sinkholes is the most accurate of the age classes because the sudden appearance of a new sinkhole is noted and remembered by local landowners. The number of sinkholes in the intermediate range was the most difficult age to determine and is underestimated. When sinkholes appeared intermediate in age, the information was omitted unless landowners could verify the age.

The age distribution (Figure 4) implies that the rate of sinkhole formation in Winona County has significantly increased in the last 50 years. Forty-seven of the 535 sinkholes located have formed in the last 5 years. At the present rate, ∿ 9 per year, all of the sinkholes in Winona County could have formed in the last 58 years. Several of the sinkholes are documented as being over 100 years old and the last known geologic event capable of covering all of the paleosinkholes was the loess deposition 10,000 to 12,000 years ago. Unless geomorphic processes fill sinkholes on a 50 to 100 year time scale, the rate of sinkhole formation was much lower in the past.

Aley et al. (1972), Foose and Humphreville (1979), Newton (1976), and Williams and Vineyard (1976) have documented many cases where human activities have locally increased the rate of sinkhole formation through a variety of mechanisms. The only regionally significant human activity on the upland of Winona County is agriculture. Although we are unable to specify the mechanisms, we conclude that on a regional scale human activities associated with agriculture have significantly increased the rate of sinkhole formation.

Sinkhole and Sinkhole Probability Map

Sinkhole locations and five categories of continued sinkhole activity were identified on a Sinkhole and Sinkhole Probability Map (Dalgleish and Alexander, 1984). The regions of varying sinkhole probability were defined primarily by the observed sinkhole density. Bedrock, topographic, surficial, and hydrogeologic conditions were used when geologic controls of sinkhole distribution were clearly defined.

The five regions of sinkhole probability were defined as follows:

No Probability - The surface area exposed in the deep valleys incised below the Prairie du Chien Group is not susceptible to sinkhole formation.

Low - No sinkholes were detected and few if any sinkholes are expected to develop on the steep slopes of the Oneota Dolomite at the top of the deeply incised valleys.

Low to Medium - Areas where sinkholes are widely scattered with only small isolated clusters of 2 to 3 sinkholes.

Medium to High - Sinkholes are generally moderately distributed with several clusters of 4 to 10 sinkholes.

High - Overall the sinkhole density is relatively high with large clusters of more than 8 sinkholes. This region includes some smaller areas where sinkhole densities appear more similar to the medium to high region; these areas were included because there was no indication of geologic or topographic differences from the highly concentrated area to inhibit sinkhole formation. The differences in density are probably random variations in landscape development.

Geological Controls on Sinkhole Formation

Water Table: It appears that sinkhole formation is generally independent of the depth to the water table. Sinkholes are almost evenly distributed above water table depths from 12 to 104 m (Figure 5). Only two small peaks, which do not appear to be significant, are seen at 30.5 to 36.6 m and 79.2 to 85.3 m with 9% and 12% of all the sinkholes studied, respectively. It is possible that the water table depth is an additional factor aiding sinkhole formation where other controlling characteristics are already present.

Stratigraphy: The sinkhole density within each bedrock unit (Figure 6), shows that the Shakopee Formation has an overall sinkhole density of $.68 \pm .07$ sinkholes/km^2. This is about twice the sinkhole density found in the Oneota, and three times the densities found in the other formations. The Shakopee and Oneota are the most susceptible bedrock formations to sinkhole development.

Figure 7 shows the distribution of sinkholes as a function of elevation above or below the New Richmond Sandstone, the lower member of the Shakopee. (The upper member is the Willow River Dolomite.) Most of the sinkholes have developed in the stratigraphic range of the Shakopee and Oneota with a strong peak in the lower strata of the Willow River Dolomite, New Richmond Sandstone, and the upper Oneota Dolomite. Although the bedrock was visible in only six of the sinkholes seen in the field, the New Richmond Sandstone was the exposed bedrock unit in four of the cases. The distribution of sinkholes between bedrock formations within each sinkhole probability area displays similar patterns with the majority of the sinkholes in the Shakopee for each area. Bedrock lithology is evidently the primary control on sinkhole formation, while other factors control the relative density of sinkholes, within the Shakopee and Oneota.

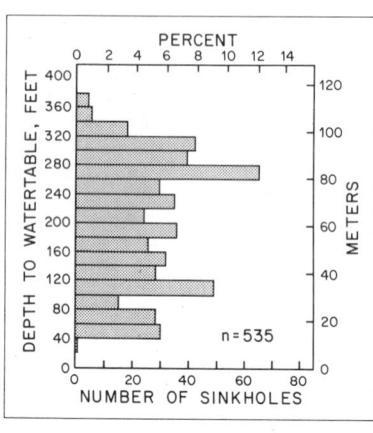

Figure 5. Sinkhole distribution - depth to water table.

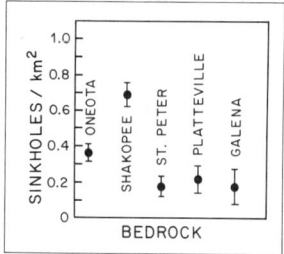

Figure 6. Sinkhole density - bedrock units.

Surficial Geology: Hobbs (1984) has identified five surficial units in Winona County:

Q-RL - Thick loess over residuum.
QRTL - Thick loess over older non-calcareous till and residuum. Residuum and till thins to the east.
QSCL - Thick loess over calcareous till, locally leached. Loess and till and patchy on slopes.
QSCS - Thin sand and till over bedrock with little or no loess cover.
QSCP - Moderately thick, .5 to 3 m, loess over patchy over patchy till on the plateau the St. Peter escarpment.

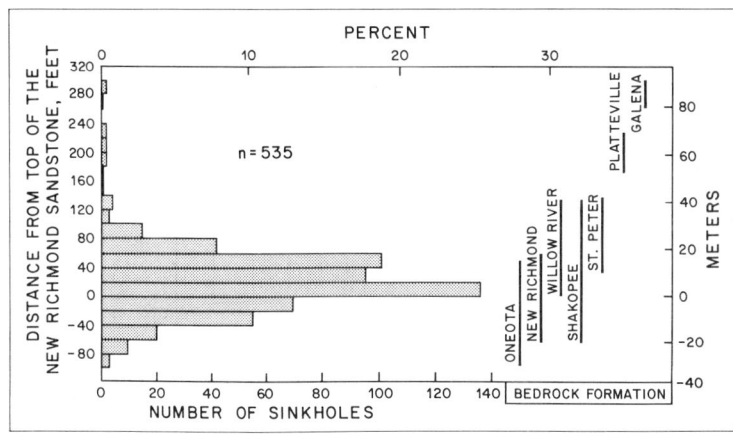

Figure 7. Sinkhole distribution vs. Stratigraphic position.

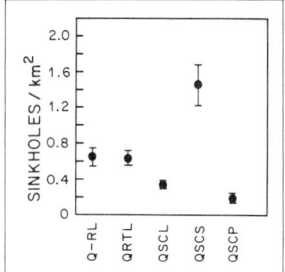

Figure 8. Sinkhole density - surficial units.

Sinkhole density within each surficial unit (Figure 8) shows that the greatest density is found in thin sand and till with little or no loess cover, unit QSCS. The lowest densities are in the thick loess covered calcareous till, QSCL, and moderately thick loess over patchy till on the Platteville and Galena plateaus, unit QSCP.

Stratigraphy and Surficial Geology: As illustrated in Figure 9, not all combinations of surficial and bedrock geology are present in Winona County. Two stratigraphic-surficial combinations yield relatively high sinkhole densities. The first combination is in the Shakopee Formation where it is covered by QSCS, a locally leached till without loess. The density is 2.23 ± .34 sinkholes/km². This area is located on the upland plain within the regional recharge zone which may aid additional sinkhole development. This combination falls entirely within the high sinkhole probability area. The second combination is the Oneota covered with calcareous till and loess, QSCL. The density is 2.38 ± .67 sinkholes/km². This combination is considered less significant since it includes only 5.9 km² and 14 sinkholes. Most of this combination is on gentle slopes near the edge of the Prairie du Chien plateau. The calcium carbonate in this area may be more leached than in central areas of the plateau, due to more surface water flowing toward the plateau's edge.

The Shakopee Formation mantled with residuum and loess has a density of 1.33 ± .21 sinkholes/km². Where the Shakopee is covered with loess, non-calcareous till, and residuum, QRTL, the density is .86 ± .11 sinkholes/km². These combinations have medium sinkhole densities.

	BEDROCK UNIT				
SURFICIAL UNIT	ONEOTA	SHAKOPEE	ST. PETER	PLATTEVILLE	GALENA
Q-RL	0.18 ± 0.05	1.33 ± 0.21			
QRTL	0.36 ± 0.06	0.86 ± 0.11			
QSCL	2.38 ± 0.67	0.31 ± 0.04	0.12 ± 0.09		
QSCS		2.23 ± 0.34	0.23 ± 0.11		
QSCP			0.08 ± 0.08	0.21 ± 0.08	0.17 ± 0.10

Density units - sinkholes / km²

Figure 9. Sinkhole density - Surficial and bedrock units matrix, combinations do not occur where there are blanks.

Discussion

The lower Shakopee, and in particular the New Richmond Sandstone, appears to be the bedrock strata most conducive to sinkhole formation. Considering that the majority of the sinkholes in each sinkhole probability area has formed in these strata, it appears that regardless of other controls on sinkhole formation, stratigraphic position is the primary geologic control.

Water flowing through the New Richmond will not be totally neutralized when it contacts the Oneota Dolomite since the New Richmond Sandstone contains only a small proportion of calcium carbonate. It appears that aggressive water collects in joints of the New Richmond Sandstone which is then channelled to the Oneota Dolomite. This system directs aggressive water, capable of further dissolving the bedrock along joints in the Oneota Dolomite. As the joints in the Oneota are enlarged, the New Richmond Sandstone collapses into the Oneota Dolomite as shown in the block diagram (Figure 2).

The density of sinkholes is not constant over the areas where the New Richmond Sandstone is the uppermost bedrock; therefore, other geologic controls must be affecting sinkhole formation. Surficial geology appears to have an impact on sinkhole development. Another factor to consider with the type of surficial materials is the age of the bedrock surface. In Winona County, the age of the bedrock surface generally decreases to the west, as the carbonate content of the surficial material increases.

The sinkhole density is greater in areas where carbonate concentrations are low or absent in the surficial material. Water percolating through non-calcareous sediments will be more aggressive, when it contacts the Oneota Dolomite than water percolating through calcareous sediments. In contrast, a few sinkholes have developed in calcareous tills, which Hobbs (1984) reports to be locally leached. It remains to be proven if the sinkholes preferentially form in areas that are leached in the calcareous tills. It seems probable that more water has percolated into places where sediments are locally leached than into the surrounding calcareous sediments. If this is true, it is reasonable to suspect that the additional aggressive water will preferentially dissolve the bedrock allowing sinkholes to form.

Since the Prairie du Chien bedrock surface appears to decrease in age to the southwest, the northeast has been exposed to weathering for a longer period of time. This implies that the paleokarst surface is progressively younger to the southwest. The irregular paleokarst surface probably directs water percolating through the sediments into previously enlarged joints and sinkholes.

Paleokarst sinkholes would not only provide additional water for further dissolution of the rock, but the increased volume of water would be capable of transporting sediment deeper into the bedrock. As sediment is transported deeper into the bedrock, cavities will form due to piping, eventually resulting in surface collapse or subsidence. Initially sinkholes may open in sites of paleosinkholes. Once the renewed sinkhole opens surface water may be channelled toward that area causing additional collapses. This process is more evident in eastern Winona County where there are wide shallow depressions with small almost vertical holes in the center.

It appears that the bedrock, paleokarst surface, and surficial material are intertwinned controls on sinkhole formation in Winona County. The relative importance of these factors probably varies throughout the study area.

Conclusions
1. Traditional tools for sinkhole detection, topographic maps and air photos, are not adequate in areas with low to moderate densities of small sinkholes. Detection of sinkholes in such areas must be done from field studies.

2. Age data indicate that the rate of sinkhole formation has significantly increased in recent years.

3. Depth to the water table does not appear to control sinkhole formation in Winona County.

4. Stratigraphic position is a primary control on sinkhole development.

5. Surficial deposits are a secondary control, and significant carbonate contents inhibit sinkhole formation.

Acknowledgements
This study was funded by Winona County, the Minnesota Geological Survey, and a grant from the Legislative Commission on Minnesota Resources.

References

Aley, T.J., Williams, J.H., and Massello, J.W., 1972, Groundwater contamination and sinkhole collapse induced by leaky impoundments in soluble terrain: Missouri Geological Survey and Water Resources Engineering Geology, Series 5, 32 p.

Dalgleish, J.D. and Alexander, E.C., Jr., 1984, Sinkholes and sinkhole probability, Plate 5, Balaban, N.H., and Olsen, B.M., eds., Geological Atlas of Winona County: Minnesota Geological Survey.

Foose, R.M. and Humphreville, J.A., 1979, Engineering geological approaches to foundations in the karst terrain of the Hershey Valley: Bulletin of the Association of Engineering Geologists, v. 16, p. 355-381.

Hobbs, H.C., 1984, Surficial geology, Plate 3, of Balaban, N.H. and Olsen, B.M., eds., Geologic Atlas of Winona County, Minnesota: Minnesota Geological Survey.

Hogberg, R.K., 1972, Ground-water resources in Minnesota, in Sims, P.K. and Morey, G.B., eds., Geology of Minnesota: A Centennial Volume: Minnesota Geological Survey, p. 595-602.

Newton, J.G., 1976, Induced and natural sinkholes in Alabama – A continuing problem along highway corridors: Subsidence Over Mines and Caverns, Moisture and Frost Actions, and Classification, Transportation Research Record 612, National Academy of Sciences, Washington, D.C., p. 9-46.

Sloan, R.E., 1964, The Cretaceous system in Minnesota: Minnesota Geological Survey Report of Investigations, no. 5, 64 p.

Sweeting, M.M., 1973, Karst landforms: New York, Columbia Press, 362 p.

Williams, J.H. and Vineyard, J.D., 1976, Geologic indicators of catastrophic collapse in karst terrain in Missouri: Subsidence Over Mines and Caverns, Moisture and Frost Actions, and Classification, Transportation Research Record 612, National Academy of Sciences, Washington, D.C., p. 31-37.

Tectonics and geology in karst development of Northern Lower Michigan

TYRONE J.BLACK *Michigan Geological Survey, Roscommon, USA*

ABSTRACT

The Michigan Basin appears to be broken up into many different fault blocks of different but simillar low attitudes and generally dipping at less than one degree toward the center. A series of "steps" (growth faulting?) in the structure of the basin in the area has caused at least one trend of fault extensions. Preferential solution of evaporites and some limestone along the trend has produced numerous collapse sinks and valleys. The sink holes range from a few meters to 0.4 km (1320 ft.) in diameter. A second weakly defined trend of karst runs subparallel to the major trend several miles to the north. Additional sink holes plugged with rubble and glacial drift have been discovered by quarry operations. These stagnant sinks appear to be related to structural relief faulting due to solution of the evaporites below.

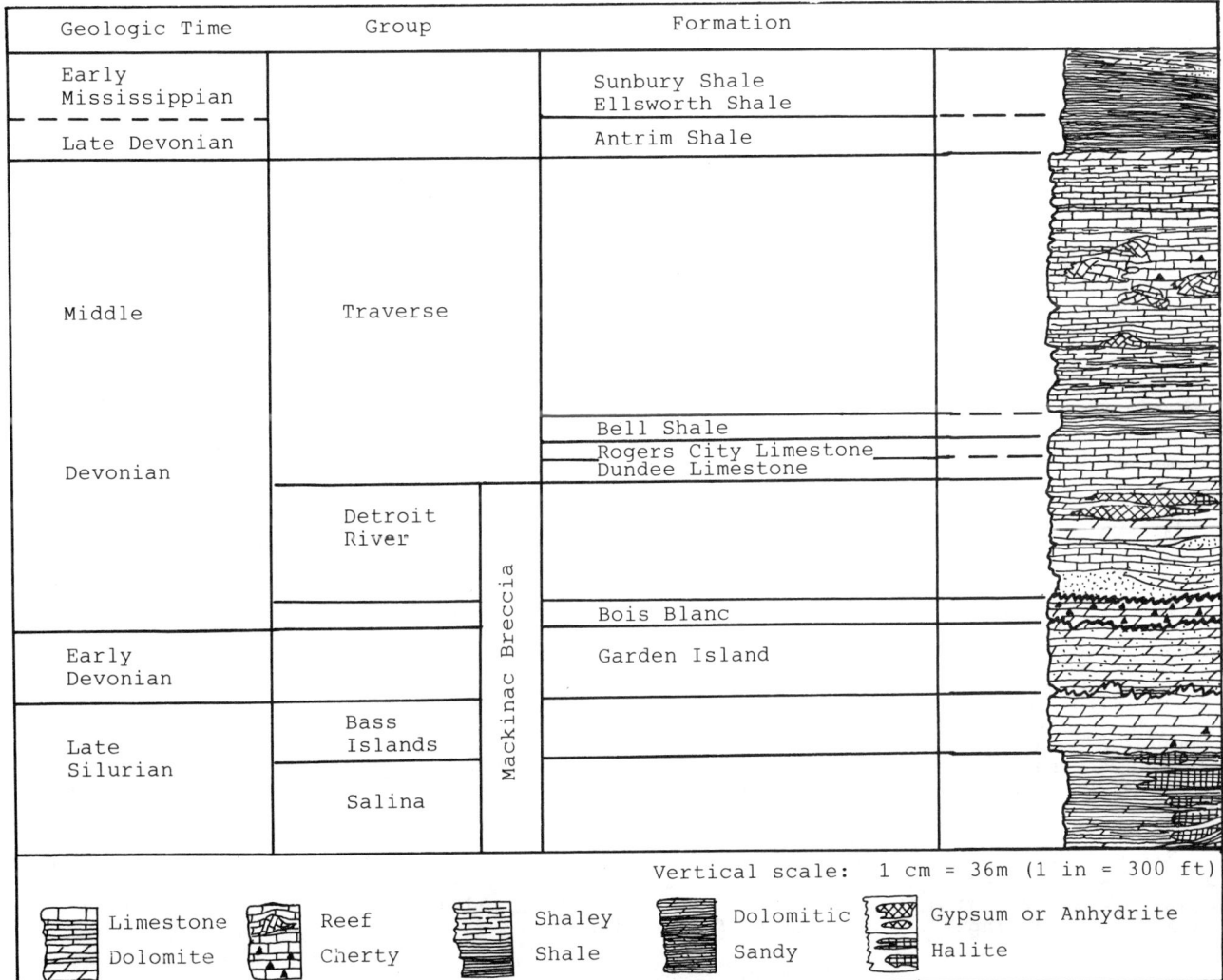

Figure 1: Stratigraphic succession of the Northern Lower Michigan karst area (after Michigan Geological Survey, 1964).

Introduction
    The area studied covers the north-central portion of the Michigan Basin. The Wisconsinan Glacier made it's final retreat approximately 10,000 years ago and left deposits from zero to over 150m (500 ft.) thick. General dip in this area is basinward at less than one degree. But in localized areas the direction of dip due to faulting or folding is variable and may be up to three degrees. The bedrock is Middle to Late Devonian age limestone, shale, claystones, dolomite, and a series of shales and evaporites, (Fig. 1).

    Solution near the outcrop edges of Late Silurian and Middle Devonian evaporites has resulted in a broad area of collapse breccia across the north edge of the penninsula, the Mackinac ("Mackinaw") Breccia (Landes, Ehlers and Stanley, 1945). South of Mackinac Breccia exposures and subcrop, hundreds of collapse sink holes and small swallows lie on a broken arc across the penninsula (Fig. 2). The discovery by quarry operations of drift plugged sink holes suggests that hundreds more exist between the line of active sinks and the exposed breccia.

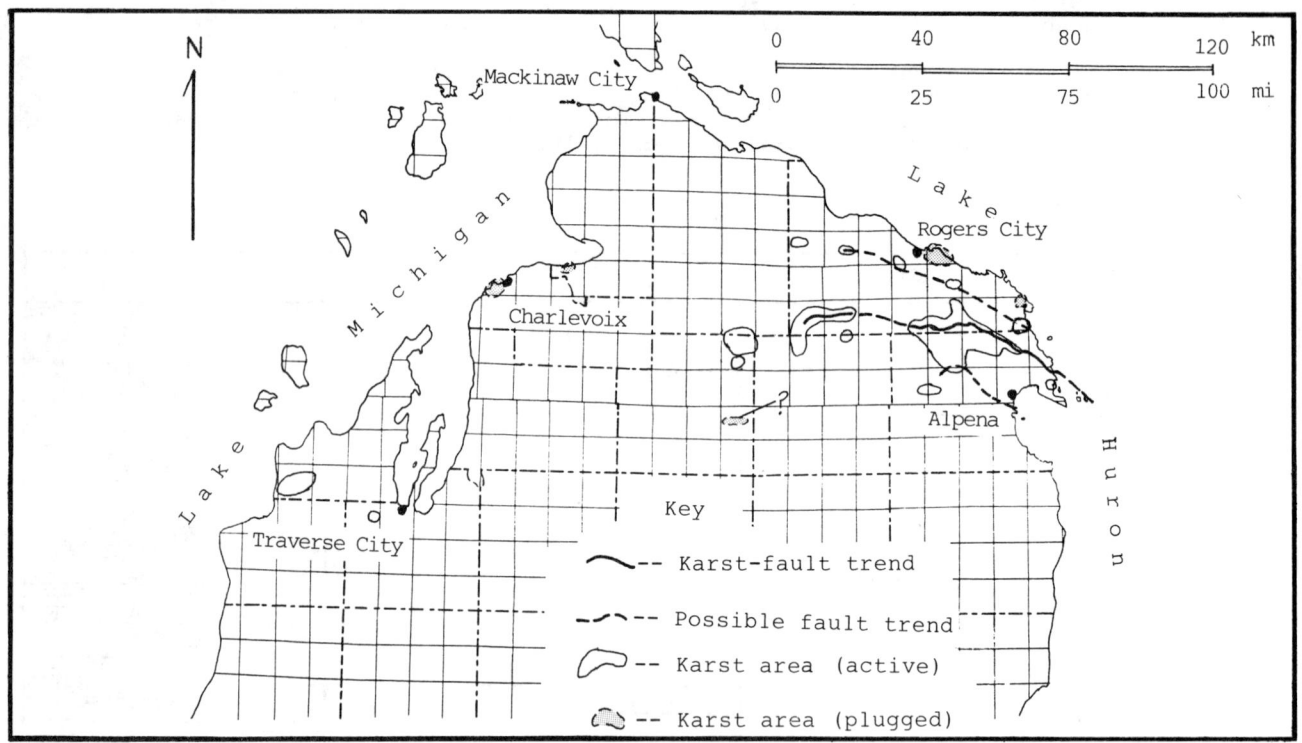

Figure 2.   Karst and fault trends of Northern Lower Michigan (after Black, 1983).

    The author has found that the sinks are controlled not only by solution and brecciation near subcrop edges but also by the exposure of evaporites to solution by ground water circulating in open fault systems. These fault systems probably originate from Precambrian strata and may have been reactivated several times in geologic history. The solution of carbonates above the evaporites is comparatively minor but it does enhance circulation in the system.

Tectonics
    Figure 3 is a tracing of the straight line segments of the reflection pattern from a portion of a seismic line run for oil and gas exploration in eastern Otsego County, Michigan. The seismic line reveals a stair-step pattern on the upper 3 km (2 mi) of strata. This pattern resembles gravity faulting and has been observed in other seismic lines of the northeastern Michigan Basin as well.

    Sanford and McFall (1984) mapped the fracture framework of southwestern Ontario's Paleozoics. An extrapolation of the pattern across Lake Huron to northeast Michigan (Fig. 4) appears to match the occurrence of the parallel "steps" seen on seismic lines and fault trends mapped by the author (Black, 1983).

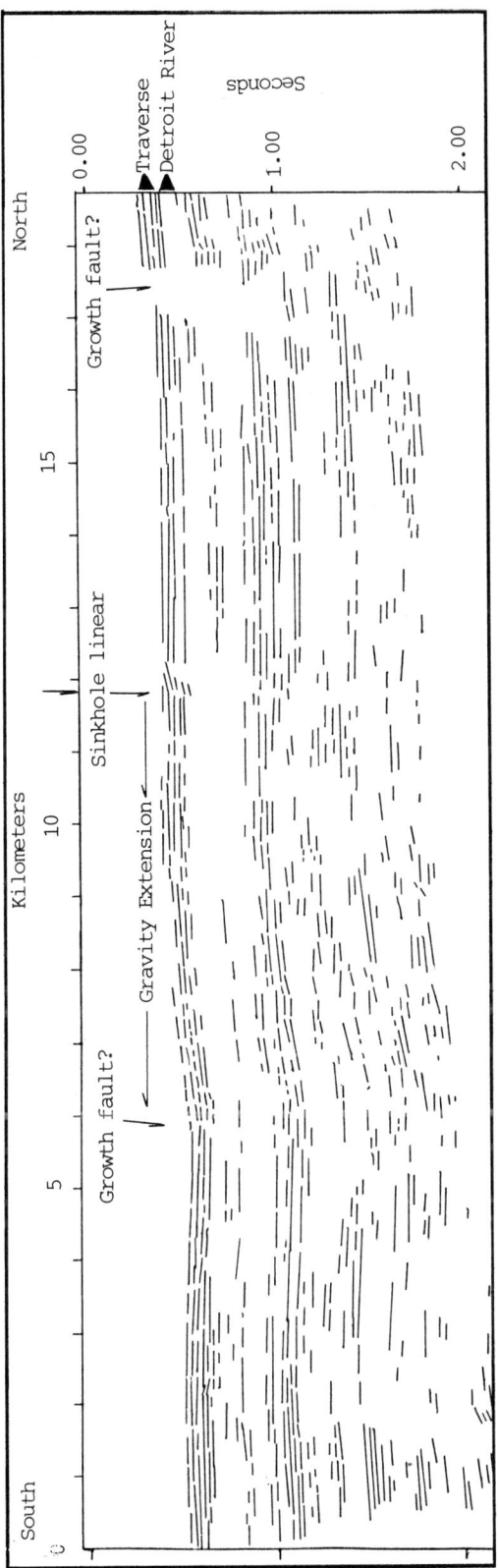

Figure 3: Tracing of straight seismic reflections in eastern Otsego County, Michigan (courtesy of Frontier Exploration, Inc.)

Prouty (1983) has studied lineaments in Michigan as related to shear faulting. Based on his study, he has proposed a shear model for faulting and folding of the Michigan Basin. He applies the concept of one primary direction of force to the regional setting rather than to a simple solid. The result is then numerous orders of magnitudes and directions of shear strains. There have been several episodes of orogenies that have applied disruptive forces to the Michigan Basin. This may explain why the karst-fault trend is made up of segments of numerous faults instead of one or a few faults and why there are erratic patterns of dip in the area. The author also agrees with Landes (1945) that differential settling of younger formations over the solution areas has resulted in structural relief faulting. These shallow faults are expected to be associated with inactive sink holes found by quarry operations.

Sink Hole Geology
The sink holes are most common in Traverse Group exposures, and their location is coincident with most of the fault trends occurring in these exposures. A significant number of sink holes also occur above the Traverse Group in areas of Antrim Shale subcrop and below as in the Dundee Limestone. The Detroit River Group has no known outcrops or sink holes in this area; the only exposures are composed of breccia. Of the sink holes in the Traverse Group that have exposed rock walls, a near vertical shear fault zone striking N40°E through the northeast half of the sink hole is usually found. This zone is highly fractured and is offset downward toward the northeast. The southwest half of these sinkholes reveals a solid but jointed and little weathered limestone.

Around some sink holes and along the more definite sink hole linears are "earth cracks". Earth cracks is used here to refer to openings developed along joints by the creep of blocks bounded by joint planes. This creep includes both motion downward and rotation of the top of the block outward from the formation mass toward a less restricted area such as an open valley or sink hole.

The earth cracks and strong depressional linears influence the volume of water swallowed by the trend much like a gravel pack or fracturing increases the efficiency of a well.

Hydrogeology
There are many observable sites of surface and groundwater discharge into sink holes and swallows. One of these was a disappearing lake and measurements showed it to swallow over 0.283 m³/sec (10 gal/sec) (Stewart, 1981). This figure did not include the surface stream or groundwater drainage into the sink hole which were not estimated. There are five known resurgences. Four are associated with minor systems that flow a distance of less than 0.4 km (0.25 mi) and at less than 0.15 m³/s each. The fifth resurgence is a

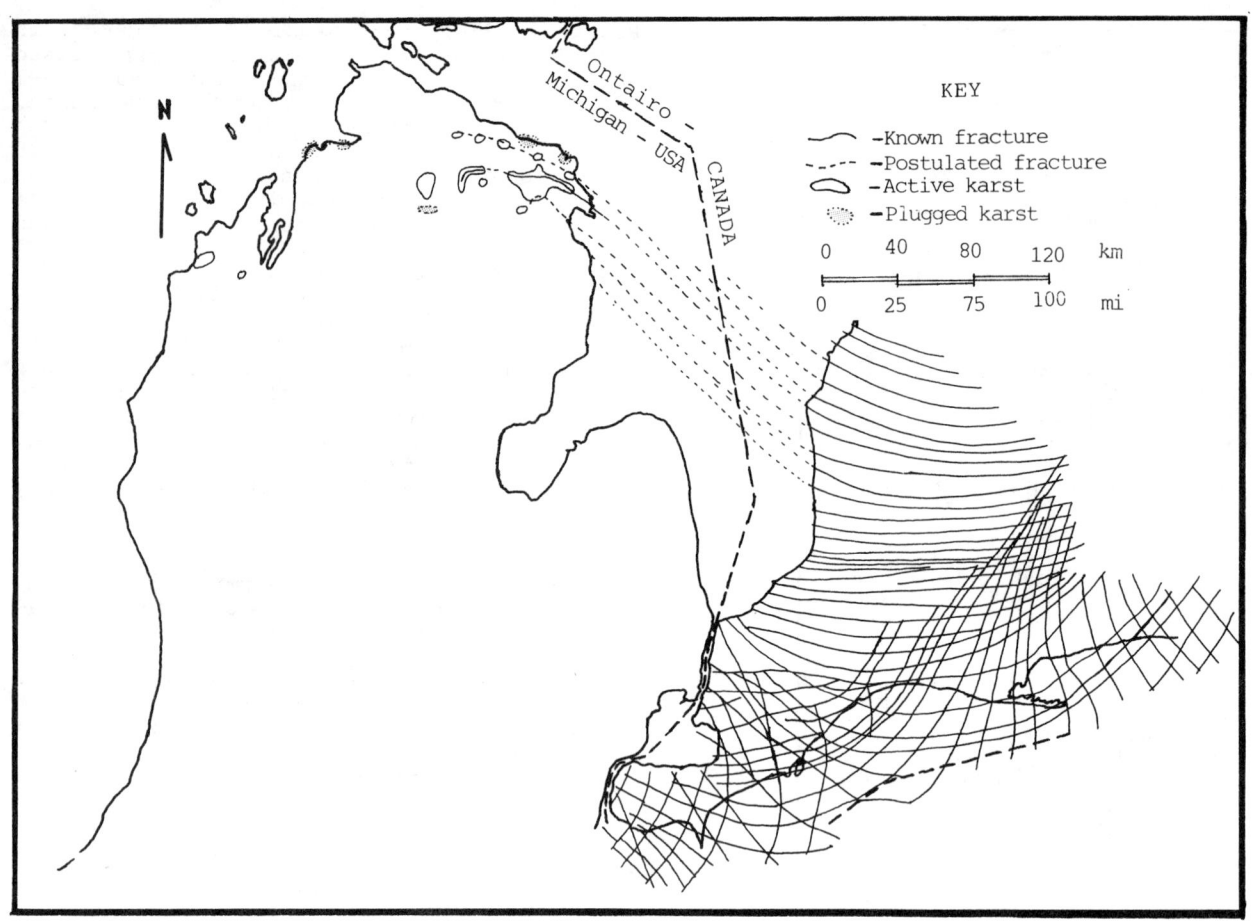

Figure 4. A postulated connection between the fracture pattern of Southwestern Ontario (Sanford and McFall, 1984) and the fault and karst trends of the study area (after Black, 1983).

'blue hole' and associated underwater springs in El Cajon Bay of Lake Huron. It has an outflow of about 3.4 m³/sec (120 gal/sec) and the resurgence water is saturated with respect to gypsum (Moreau, 1983). There are undocumented reports of resurgences further into Lake Huron. Exploratory drilling in the resurgence area and further inland frequently encounters "cavernous" and "lost circulation" zones. The drill holes encounter flows of high sulfate water at depths of 170 to 300m (560 to 1000 ft) coinciding with the Detroit River Group.

References

Black, T. J., 1983; Selected Views of Tectonics, Structure and Karst in Northern Lower Michigan; Tectonics, Structure and Karst in Northern Lower Michigan; Michigan Basin Geological Society, 1983 Field Conference, p. 95.

Landes, K. K., G. M. Ehlers and G. M. Stanley, 1945, Geology of the Mackinac Straits Region and Subsurface Geology of Northern Southern Penninsula: Michigan Geological Survey, Publication 44, Geological Series 37, Figure 2.

Geological Survey, 1964, Stratigraphic Succession in Michigan, Michigan Department of Conservation, Chart 1.

Moreau. 1983, A Review of the Limnological Characteristics of Alpena Michigan Area Flowing Wells and Sinkholes, Tectonics Structure and Karst in Northern Lower Michigan, Michigan Basin Geological Society, 1983 Field Conference, p. 95.

Prouty, C. E., 1983, The Tectonic Development of the Michigan Basin Intrastructures; Tectonics, Structure and Karst in Northern Lower Michigan, Michigan Basin Geological Society, 1983 Field Conference, p. 95.

Sanford, B. V. and G. H. McFall, 1984, Fracture Framework--A Controlling Factor in the Accumulation of Hydrocarbons in Southwestern Ontario, Open file report No. 964, Geological Survey of Canada, Figure 5. (A complete report is due to be published in the Canadian Society of Petroleum Geology in early 1985.)

Stewart, D. C., 1981, Water-loss Problem at Rainy Lake, (Presque Isle County), Michigan, Study done by Department of Geology, Bowling Green University for the Rainy Lake Property Owners Association, p. 8.

Pattern and antiquity of sinkholes along an alluviated karstified valley: Friars Hole, West Virginia

S.R.H.WORTHINGTON & D.C.FORD *McMaster University, Hamilton, Ontario, Canada*

ABSTRACT

Friars Hole is a dry valley incised in Mississippian siliclastic rocks. Along the valley floor a series of limestone inliers are exposed, where streams sink into an extensive cave system. Once abandoned by sinking streams, the alluvial flood-plains which shroud the limestone are progressively dissected by sinkhole development. Solution sinkholes have developed away from the valley sides, while suffosion sinkholes have developed close to the valley margins. U/U and U/th dating and paleomagnetic analyses of speleothems have shown that the cave is more than 1.65 million years old. It has been possible to correlate valley abandonment with underground passage development, and hence the antiquity of the sinkholes has been established.

Friars Hole is a narrow dry valley in Greenbrier and Pocahontas Counties, West Virginia (Figure 1). Along the 11 km length of its floor are a series of inliers of Union Limestone of the Mississippian Greenbrier Group. However, almost all of the catchment (some 97%) is underlain by siliclastic strata of the Pocono and Pottsville Groups. These all drain to the limestone inliers, beneath which some 70 km of cave passages have been explored.

The streams which flow off the impermeable strata sink soon after reaching the limestone. Occasionally, a surface stream flows into an open cave entrance, such as Snedegar North Entrance or Bruffey Creek Cave. In these cases, stream-borne sediments are carried away through cave passages. However, in most instances the streams sink in their alluvial beds and the water drains away into underlying cave passages. Consequently, only fine sediments are carried away, while the coarser fractions remain on the surface to form a mantle of alluvium which may reach 10 m or more in thickness. This alluvium has a mean grain size of 15-20 mm, with over 90% being coarser than 2 mm.

Sinkhole development has taken place along Friars Hole on the inliers of Union Limestone. The sinkholes and modern and former sinkpoints of a section of Friars Hole are shown in Figure 2. In many instances it is possible to follow cave passages to terminations in blockages of sandstone cobbles and boulders lying below surface depressions. However, many former stream sinkpoints have no surface expression. For example, at the Friars Hole Cave inlier (Figure 2), 18 active sinkpoints have been identified from within the cave, yet only four of these are visible on the surface as depressions or stream sinks. There are also 29 inactive sinkpoints, recognized in the cave as abandoned inlet passages. None of these is now identifiable on the surface. Sinkpoints are often underlain by vertical shafts, with typical widths ranging from a few centimetres to several metres. An exceptional example shown in Figure 2 is visible in a cave passage 60 m below the surface, where there is a 25 m wide sandstone boulder blockage. All these shafts became blocked by alluvium before the surface flood plain was abandoned. Subsequently, these blocked shafts formed favourable sites for sinkhole development.

Two populations of sinkholes may be discerned at Friars Hole (Figure 3). Suffosion sinkholes develop relatively rapidly along the margins of the valley where there is some surface runoff down the valley slopes, and in some cases a sinking stream may be present. Typically, the sinkholes are 5-10 m deep, are steep-sided ($20°-35°$), and are irregularly spaced along the valley margin at intervals of 20-300 m.

In contrast, solution sinkholes have developed away from the valley sides, are broader, and rarely have slopes greater than $10°$. They too have an irregular spatial distribution. The shallow sides reflect the predominance of solution on the bedrock by percolation water.

The evolution of Friars Hole has been dominated by Hills Creek, which has a catchment of 50 km^2. There has been a 4 km retreat of the main sinkpoint of the creek since it was first captured underground (Worthington, 1984). The alluviated floodplains abandoned as a result of this retreat have been progressively dissected by the development of sinkholes. Four stages are differentiated in Fugure 4, and the aerial extent of these stages is shown in Figure 1.

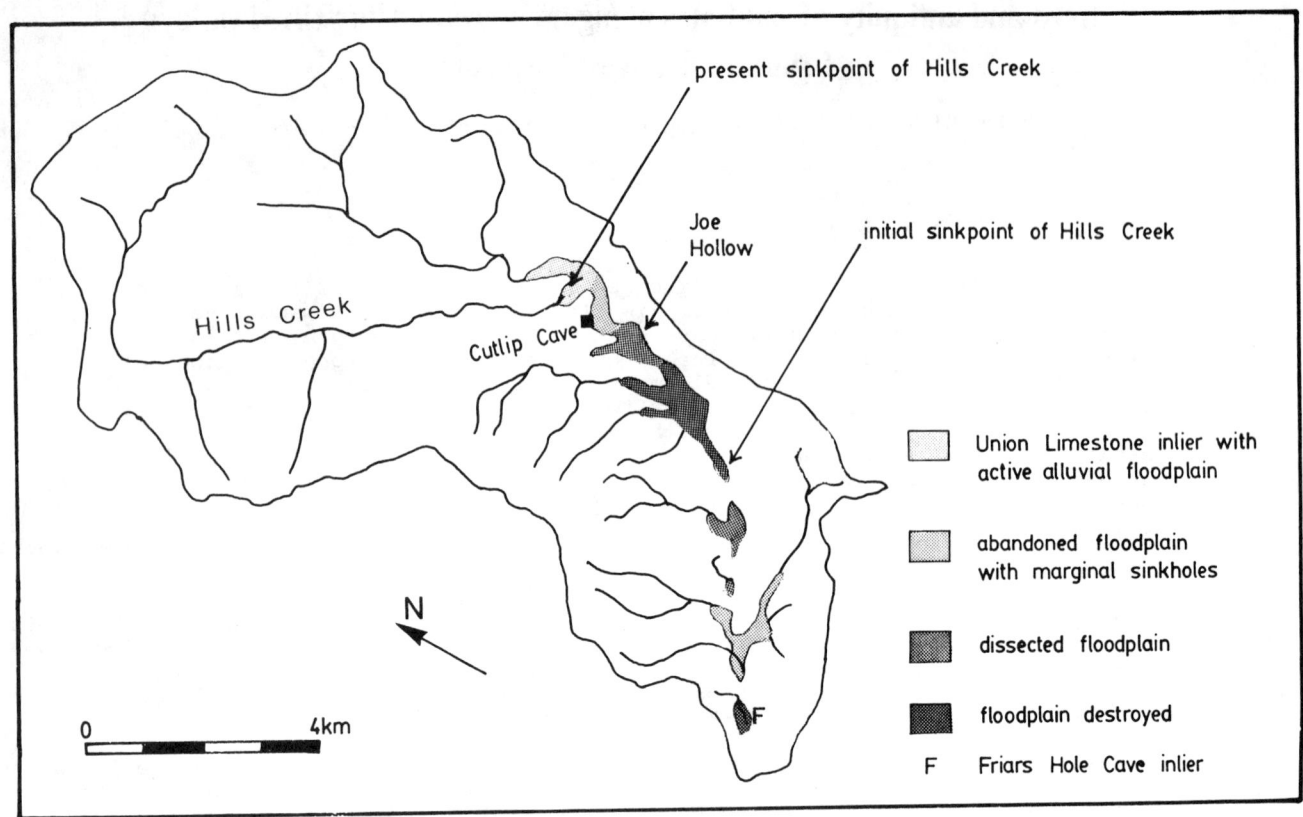

Figure 1   Catchment of Friars Hole showing Union Limestone inliers

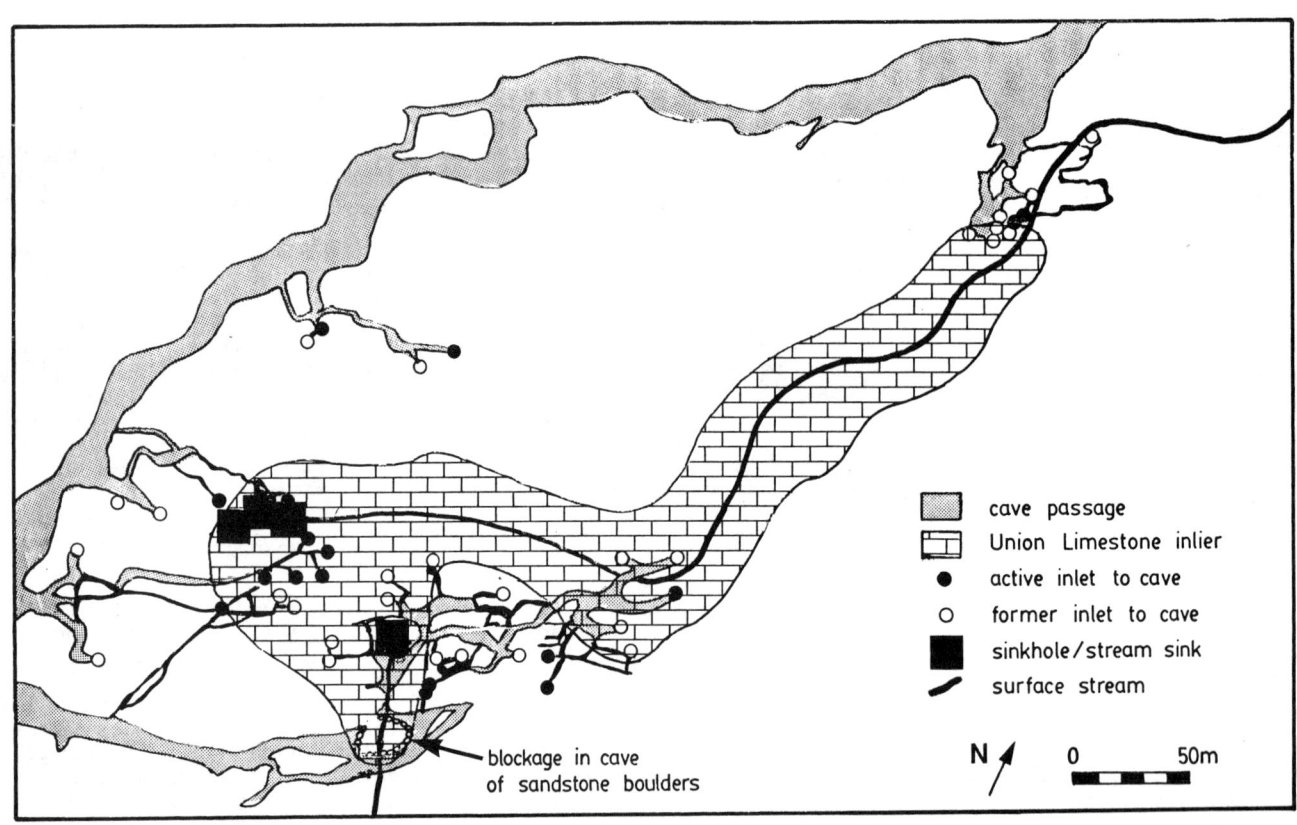

Figure 2   Sinkpoints and cave inlets at the Friars Hole Cave inlier

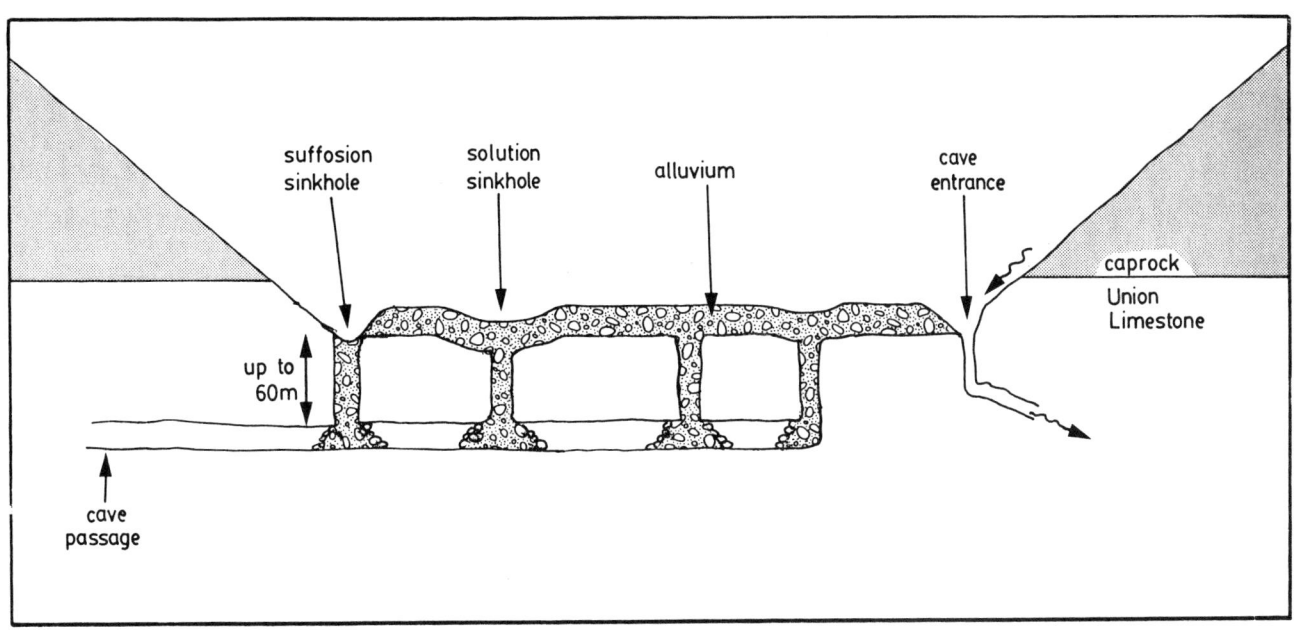

Figure 3 Sinkhole types at Friars Hole

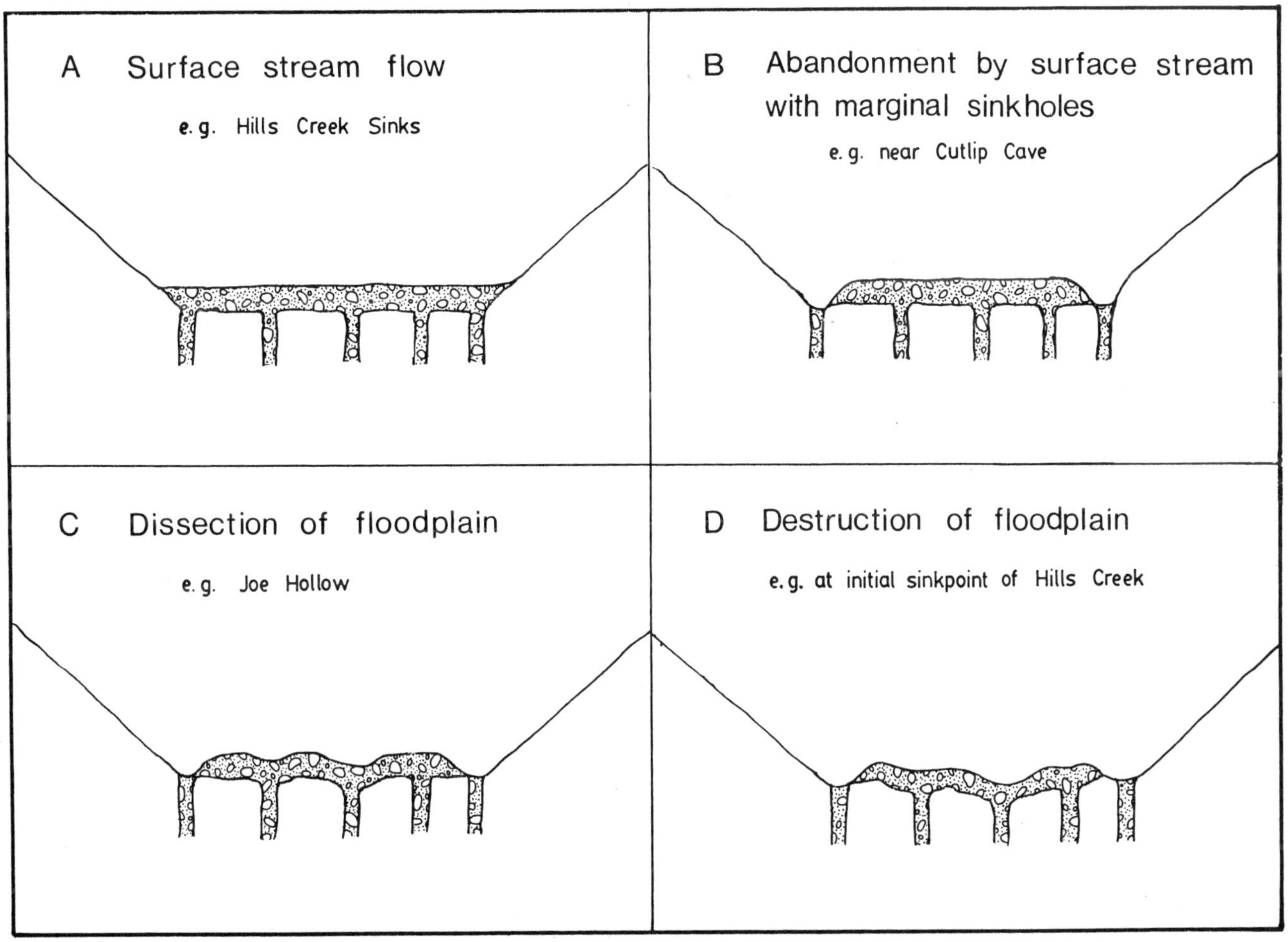

Figure 4 Sinkhole and floodplain evolution at Friars Hole

Nearly fifty uranium/thorium dates have been obtained from speleothems from Friars Hole System, which is the principal cave underlying the valley. These show that almost all the cave is >350,000 years old. The analysis of $^{234}U/^{238}U$ ratios has indicated that three of the sampled speleothems are probably >1.24 million years old. Two of these three speleothems have reversed magnetic polarity, and thus predate the Brunhes Chron (approximately >730,000 years B.P.). The third of these speleothems has normal polarity and is thus probably >1.67 million years old, that date representing the termination of the previous normal-polarity sub-chron.

It has been possible to correlate the retreat of the sink-point of Hills Creek with the development of the cave passages of Friars Hole System. It is thus possible to use the speleothem age results to date the antiquity of the abandoned alluviated flood-plains and gain an insight into the rate of sinkhole development in the area. Results suggest that it has taken more than two million years for the destruction of the alluvial flood-plains at Friars Hole to take place, while the oldest alluviated limestone surface at Friars Hole may be more than four million years old.

The spatial distribution of sinkholes at Friars Hole is thus controlled by two criteria. The availability of surface runoff dictates the type of sinkhole which will develop. Slow-developing shallow solution sinkholes develop where only percolation water is present, while fast-developing, steep-sided suffosion sinkholes result when surface runoff from impermeable rocks is available. Secondly, the maturity of sinkhole development in the valley floor is dependent upon the time that has elapsed since the alluvial flood-plain was last active.

Reference

Worthington, S.R.H., 1984. The paleodrainage of an Appalachian fluviokarst: Friars Hole, West Virginia. Unpublished MS thesis, McMaster University, 218p.

Karst and subsidence in China

ZHANG SHOUYUE *Academia Sinica, Beijing, China*

ABSTRACT

Carbonate rocks are widespread in China, covering more than one fifth of the country and cropping out over more than 1,250,000 km^2. Karst in China occurs in different climatic zones and on various geotectonic units, and has been subjected to various degrees of corrosion producing a wide variety of karstic features of differing ages. Karst corrosion in China can be divided into four regions, eight areas, and fourteen subareas based on climatic, geotectonic and geomorphic factors.

Subsidence has occurred in nearly every province of China. This paper discusses the various mechanisms of karstic subsidence and their distribution, the damage which they cause, and their treatment.

Introduction

Carbonate rocks are widespread in China. They cover over one fifth of the country and crop out over more than 1,250,000 km^2.

Carbonate outcrops range from metamorphosed limestones and dolostones of Archaeozoic age to Cenozoic reef limestone. They are distributed in various climatic zones and developed on various geotectonic units. Karstification of these carbonate formations produced a variety of surface and subsurface features. Some features are produced by several phases of karstification. The development of karst during the Cenozoic Era is the most important in China.

Subsidence, the lowering or collapse of the land surface in either bare or covered karst terrane, has occurred in nearly every province of China.

Geologic Setting in China and Karst Development

Carbonate rocks occur in all the main geotectonic units in China in various degrees (Figure 1). In the Yangtze paraplatform carbonate rocks are several thousand meters thick and range in age from Sinian (PreCambrian---1.7-2.5 billion years B.P.) to Triassic. In the geosynclinal fold belts of South China the thickness of Late Paleozoic carbonate rocks is 2,000-5,000 meters. In the China-Korea paraplatform the thickness is generally 1,000-2,000 meters, ranging from Sinian to Ordovician in age, but in specific areas, such as Yenshan Mountains near Beijing, carbonates may be as much as 6,000-7,000 meters thick.

More than half of the surface karst terrane in China, approximately 680,000 km^2, occurrs in South and Southwest China. For East China we will now compare the carbonate composition of different geotectonic units deposited at the same time.

In Sinian carbonates, dolomites and impurities are relatively more abundant than for other times. In the Cambrian System of the Yangtze paraplatform dolomites are dominant and impurities are higher, but in the China-Korea paraplatform the opposite is true. During the Ordovician Period the composition of carbonate rocks in the Yangtze and China-Korea paraplatforms is similar. Since the Devonian Period, the composition of carbonate rocks in the Yangtze paraplatform and the geosynclinal fold belts of South China has been similar -- calcite is dominant and impurities are low (Figure 2).

All surface and subsurface karst features are caused by the corrosion of carbonate rocks by natural water. Tests of the relative rate of solution and the relative corrodibility of carbonate rocks show that solubility is principally controlled by the composition and that limestone is more soluble than dolomite. Corrodibility includes both dissolution and disintegration and depends on the texture and genesis of carbonate rocks.

Generally, when the content of allocthons is high, the relative corrodibility is low, particularly if the allocthons are clastic. Microcrystalline carbonates, or micrites, are the most corrodible of all carbonate lithologies. If the mineral composition is similar, autocthonous biolithites are more corrodible than micrites. The relative corrodibility of sparry and microcrystalline allocthonous carbonate rocks varies widely due to the amount and composition of the allocthons and the ratio of sparry to microcrystalline texture.

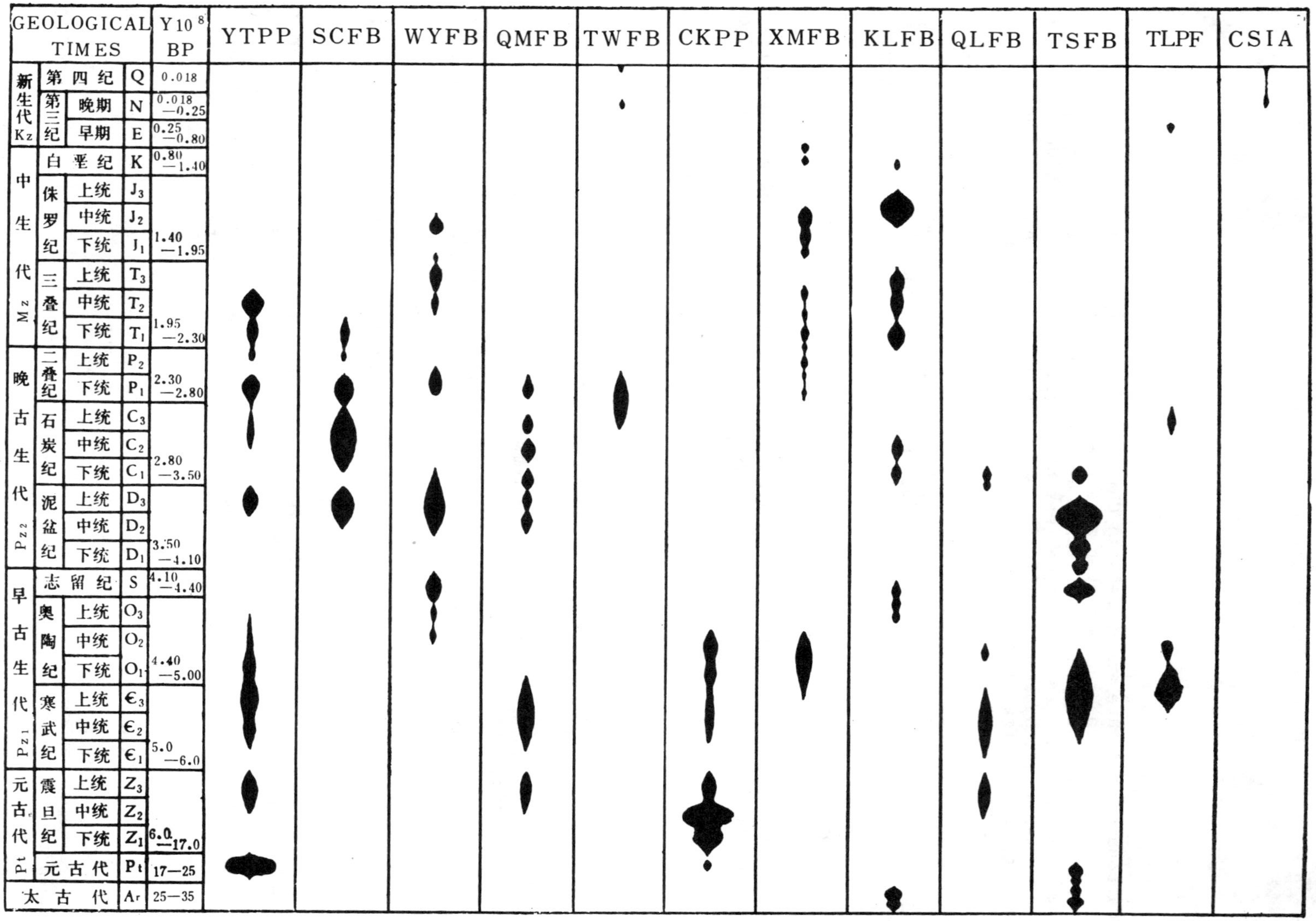

Fig. 1: Ages of carbonate rock formations in various geotectonic units in China. Paraplatform: YTPP=Yangtze; CKPP=China-Korea. Platform: TLPF=Talimu. Geosynclinal fold belts: SCFB= South China; WYFB=West Yunnan; QMFB=Quinling mountain; TWFB=Taiwan; XMFB=Ximalaya; KLFB=Kunlun mountain; QLFB=Qilian mountain; TSFB=Tianshan mountain. Non-geotectonic unit: CSIA=coast, shelf, S. China islands.

Grain sizes of sparry carbonate rocks vary due to the different degrees of diagenetic metasomatism and recrystallization. Their relative corrodibility varies, but it is generally lower than that of micrites with similar composition. In general, coarse grained carbonate rocks are less corrodible than fine grained ones. Metamorphism decreases not only the solubility, but also the relative corrodibility of carbonate rocks.

For the various types of carbonate rocks the relationship between the disintegration/corrosion ratio and the percent dolomite is shown in Figure 3. In many cases dissolution exceeds disintegration. The latter depends on the type of texture and the genesis and is generally less than 40%. For micrites, the rate of disintegration is less than 10% and tends toward zero. For microcrystalline allocthonous carbonates the rate of disintegration is over 20% while for sparry allocthonous carbonates the rate ranges from 5-30% for limestones and from 30-50% for dolomites. The rate of disintegration varies widely for sparry carbonate rocks: 0-30% for limestones and 0-40% for dolomites.

Periodically, natural water passes rapidly through the vadose zone of a karstic massif. During infiltration and downward movement the groundwater becomes of the bicarbonate type, but it is still somewhat corrosive. For this reason, when we discuss karst development in the vadose zone, the rate of solution is more important than the solubility.

In the shallow phreatic zone in a bicarbonate geochemical environment, karst will be developed in either limestone or dolomite. Heterogeneous corrosion produces a network of karst conduits in limestone while homogeneous corrosion of sparry dolomites produces overall karstification. The karstifiability of limestones and dolomites is different.

In general karstic subsidence occurrs on limestones. Karst on dolomite is still very developed, but it results in a fine fissure aquifer.

Karst Zonation in China

In principle, karst zonation will vary for different characteristics. At the first order of karst zonation, China is divided into four karst regions based on climate. Each karst region is a climatic zone with a special aridity.

At the second order of karst zonation, karst areas are divided based on geotectonic units. Each karst area has similar geologic conditions, contains carbonate rocks of a similar age, has a similar geologic history and characteristics of crustal movement. That is, they are similar in how the karstic rocks formed and how the karst and pore spaces developed.

The third order of karst zonation produces karst subareas. The results of various agencies produce different types of karst landscapes which are related to the developmental stage of caves and the hydrogeologic characteristics of the phreatic zone. Geologic structure is considered in a few karst subareas.

Karst in China may be divided into four regions, eight areas, and fourteen subareas (Figure 4). Zonation of karst in China provides a scientific basis for evaluating, modifying, and utilizing karst for national economic growth.

Research on Karstic Subsidence in China

Subsidence has occurred in nearly every province of China (Figure 5). The economic impact of subsidence is unstable building foundations, soil erosion, reservoir leakage, etc. The importance of studying these features and their development is not only a scientific problem but also an engineering, geologic, and exploration problem.

Damage caused by Surface Lowering and Collapse

Although subsidence is usually not spectacular or catastrophic, it causes serious economic damage to construction. For example:
1. Since 1959, subsidence has occurred at 1,811 sites in the Outang mining area of Guangdong Province. Mining and the management of the mine have been affected by the cracking of buildings, collapse, the displacement of roads, and the capture of a river into a sinkhole.
2. In an area of approximately 0.5 km^2 of the rice fields of Yulin County, Guangxi Zhuang Autonomous Region, land subsidence has occurred at over 400 sites.
3. At several coal mines in the central part of Hunan Province, land surface subsidence has occurred at more than 8,500 sites since 1960. The maximum density of subsidence is greater than 500 sites/0.1 km^2.
4. At the Baifang Copper mine, Hunan Province, there is a collapse 80 m in diameter and 30 m deep.

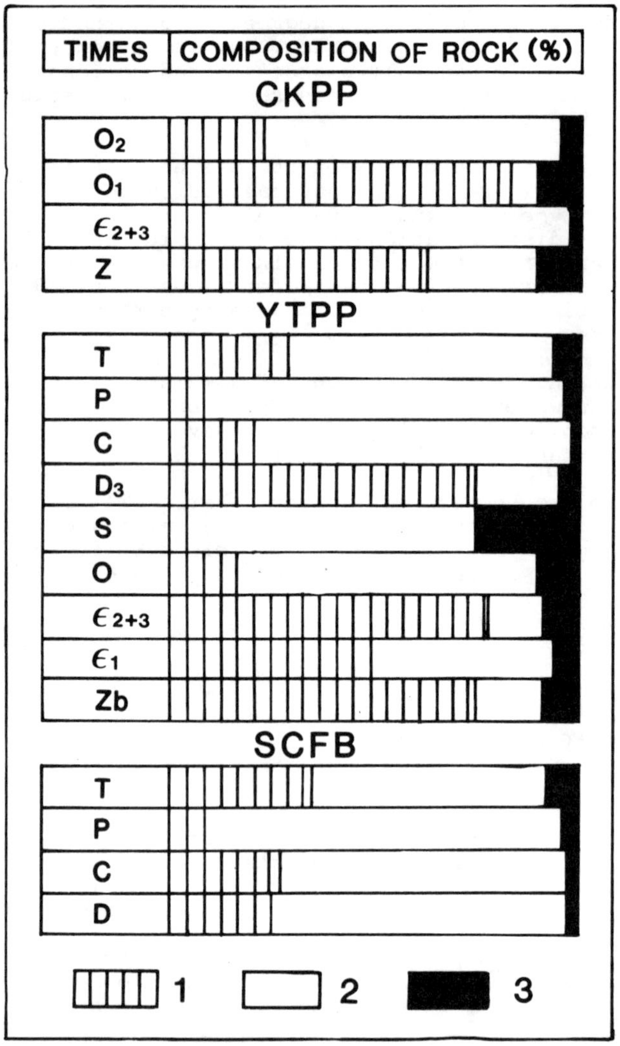

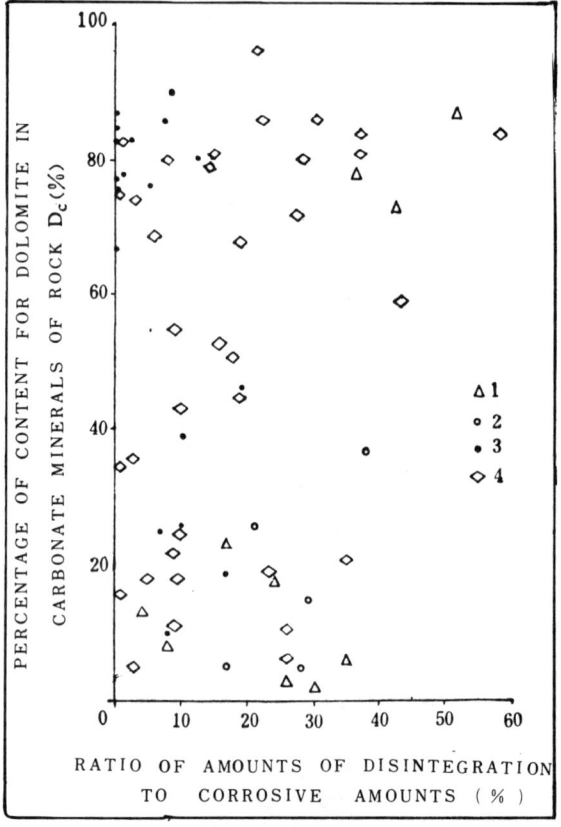

Fig. 2. Generalized composition of carbonate rock formations in Eastern China.
Pattern 1=dolomite;
Pattern 2=calcite;
Pattern 3=impurities
For CKPP, YTPP, and SCFB see Fig. 1.

Fig. 3.
Disintegration/corrosion ratio vs. per cent dolomite for carbonate rocks.
1=sparry allocthonous carbonate rocks.
2=Microcrystalline allocthonous carbonate rocks.
3=Microcrystalline carbonate rocks.
4=Recrystallized (diagenesis, metasomatism) sparry carbonate rocks.

5. A collapse sinkhole developed near the Tianquan Cave at Xingwen County, Sichuan Province, which was 400 m in diameter and 176 m deep.

Distribution of Subsidence

The distribution of subsidence depends on cave development in carbonate rocks or withdrawal of groundwater by natural or manmade activities. Recent research suggests the following conclusions.

1. Subsidence is distributed within the areas underlain by karst. Karst development is related to carbonate rock type and structure. The karst zone is a network of water conduits, at depth, which transmit groundwater from one area to another. The zone of discharge is usually in a valley. Karst always shows a series of depressions on the land surface, such as uvalas and poljes.
2. Subsidence is distributed aver the area with relatively thin cover.
3. As the groundwater cone-of-depression expands, the area of related subsidence also expands.

Mechanisms of Subsidence

At present the study of the mechanisms of subsidence depends on investigation and description without either experimental models or mathematical/physical models. Theoretical understanding is not complete.

1. Subsidence associated with piping: Lowering and collapse are caused when unconsolidated sediments are removed by groundwater flow when the hydrodynamic conditions are changed by natural or man-induced activities.
2. Gravitational collapse: This is the most common type--a cave collapse in bare or covered karst. The roof may be either rock or unconsolidated sediment; it collapses when it is no longer able to support its own weight.
3. Collapse associated with differential pressure: This mechanism of collapse is caused by the fluctuation of the pressure of water in karst conduits. It is called the piston effect. When the pressure head of karst water exceeds atmospheric pressure, the compressed air will blow out of the karst conduit--an air explosion. Inhalation is generated only in a cavity in loose sediment, when the pressure head is less than atmospheric.

Remedial Treatment of Subsidence

An analysis of the origin and regularity of subsidence allows the following remedial treatment methods to be deduced.

1. The method of constructing deep foundations: The building foundation rests on bedrock by the use of shafts, open caissons, pilings, or overburden stripping.
2. By spanning with a beam or concrete slab.
3. By grouting.
4. Methods of ventilation and decompression are used for the treatment of collapse associated with differential pressures. A check valve can be used to prevent collapse due to high pressure caused by a reservoir on karst terrane. A ventilator can also permit air pressure adjustment in either direction.

References

Jiang Yunlong et al., 1982, Ground surface cave in karst regions and its influences on the stability of foundations of industrial or civil structures: Selected Papers from the Second All-China Symposium on Karst Sponsored by the Geological Society of China 1978, Beijing, Science Press. p. 221-230. (in Chinese)

Karst Research Group, Institute of Ccology, Academia Sinica, 1979, Research of China Karst, Beijing, Science Press. (in Chinese)

Su Jiansan et al., 1982, On the problems of air-liquid pressure in tube flow of karst water: Selected Papers from the Second All-China Symposium on Karst Sponsored by the Geological Society of China 1978, Beijing, Science Press. p. 241-250. (in Chinese)

Zhang Shouyue, 1979, The corrosion of carbonate rocks and development of karst: Acta Geologica Sinica, No. 3. p. 247-261. (in Chinese)

Zhang Shouyue, 1980, Carbonate rock formations and karst zonation in China: Scientific Papers on Geology for International Exchange prepared for the 26th IGC, 5, Hydrogeology, Engineering Geology, Quaternary Geology and Geomorphology, Beijing, Publishing House of Geology. p. 91-97. (in Chinese)

Editor's note

This article was completely rewritten, at Dr. Zhang's request, to clarify the English version. It is sincerely hoped that no erroneous misinterpretations occurred in the process.

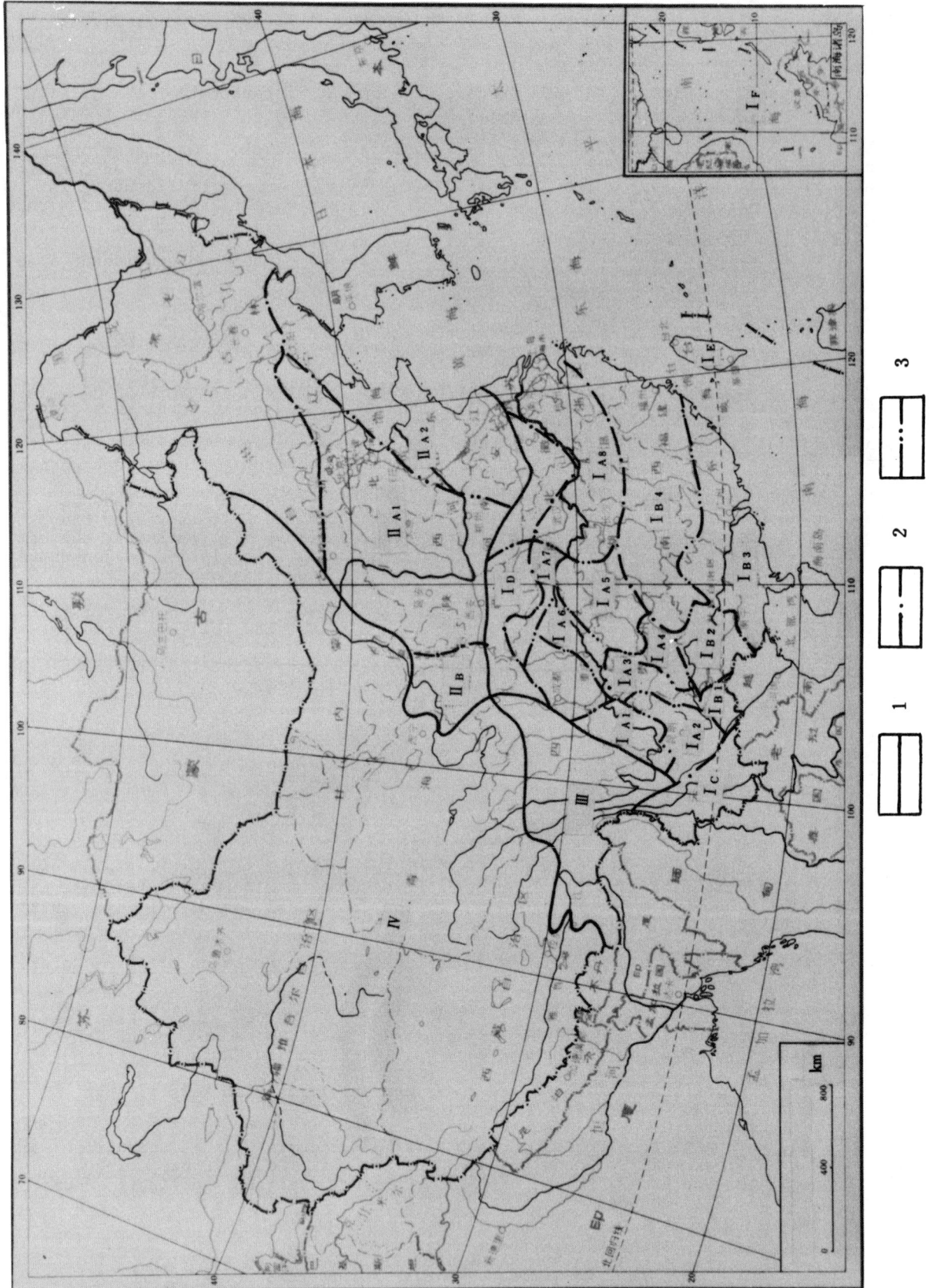

FIGURE 4. Karst zonation in China. Key on next page.

Fig. 4. Karst zonation in China (preceding page).

1. Boundary lines of karst regions.
2. Boundary lines of karst areas.
3. Boundary lines of karst subareas.

I. Corrosion and erosion-corrosion region of tropical and semi-tropical humid climatic type.

  IA. Proterozoic to Mesozoic carbonate rock karst area of Yangtze paraplatform.

    IA1. Canyon-mountain land karst subarea of southwest Sichuan.
    IA2. Karst plain-hilly plateau karst subarea of east Yunnan.
    IA3. Uvala-qiufung* mountain plateau karst subarea of west Guizhou.
    IA4. Karst plain-qiufung and fungling** mountain plateau karst subarea of middle Guizhou.
    IA5. Uvala-qiufung mountain land karst subarea of Guizhou and Hubei
    IA6. Uvala-qiufung mountain land karst subarea of east Sichuan.
    IA7. Uvala-qiufung mountain land karst subarea of Hubei and Sichuan.
    IA8. Karst plain-qiufung and hilly low mountain and hilly land karst subarea of middle course of Yangtze river.

  IB. Late Paleozoic and Mesozoic carbonate rock karst area of geosynclinal fold belts of South China.

    IB1. Karst plain-fungling plateau karst subarea of southeast Yunnan.
    IB2. Uvala-fungling mountain land karst subarea of Guizhou and Guangxi.
    IB3. Karst plain-fungling plain karst subarea of Guangxi and Guangdong.
    IB4. Polje-qiufung mountain land and hilly land karst subarea of Hunan and Jiangxi.

  IC. Paleozoic carbonate rock karst area of geosynclinal fold belts of West Yunnan.

  ID. Late Paleozoic carbonate rock karst area of geosynclinal fold belts of Qinling mountain.

  IE. Late Paleozoic metamorphic carbonate rock karst area of geosynclinal fold belt of Taiwan.

  IF. Cenozoic coral reef limestone karst area of coast, shelf, and islands.

II. Corrosion-erosion region of mid- and warm temperate zone semi-arid and semi-humid climatic type.

  IIA. Archaeozoic to Ordovician carbonate rock karst area of China-Korea paraplatform.

    IIA1. Arroyo-mountain land karst subarea of Shanxi, Hebei and Liaoning.
    IIA2. Arroyo-mountain land and hilly land karst subarea of Shandong and Liaoning.

  IIB. Proterozoic and Paleozoic metamorphic carbonate rock karst area of geosynclinal fold belt of Qilian mountain.

III. Corrosion-denudation region of humid climatic type of Qinghai-Xizang plateau.

IV. Denudation region of arid climatic type of Qinghai-Xizang plateau and temperate zone.

---

*Qiufung: A type of cone karst that is characterized by gently sloping, hemispherical limestone hills, the diameter being several times the height. Qiufung is a transliteration from Chinese.

**Fungling: Tower karst and cone karst. A transliteration from Chinese.

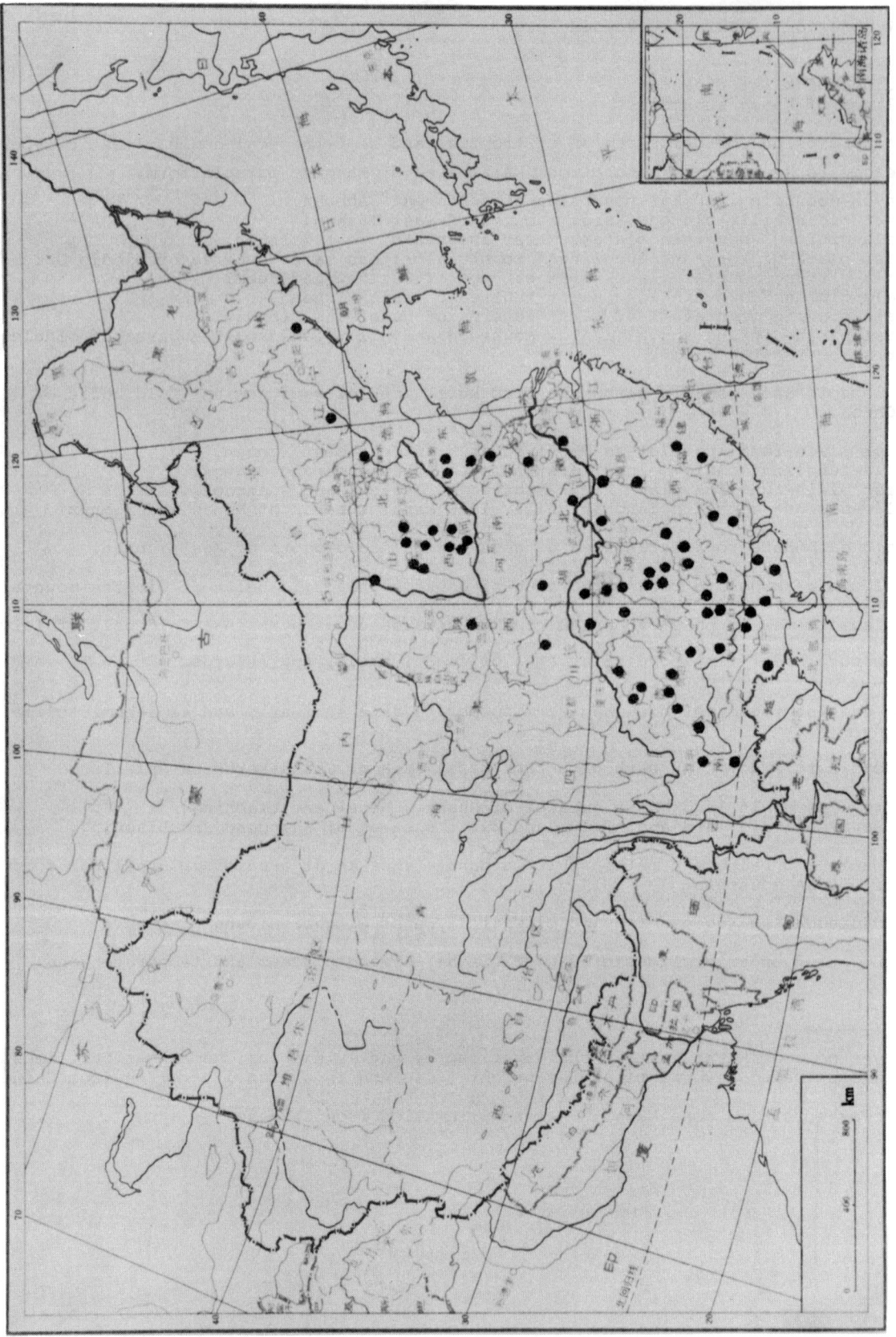

FIGURE 5. Distribution of karst subsidence in China.

A contour map, volume estimate, and description of Teague's Sinkhole

JAMES J. HOLLINGSHEAD *Jammal and Associates, Inc., Tampa, Florida, USA*

ABSTRACT

A contour map of Teague's Sinkhole was produced on February 18, 19, 25, 26, 1984 using a plane table and alidade. The volume of Teague's Sinkhole was estimated at 189,778 M^3 (248,141 Yds.^{3}) by the use of a Lasico Compensating Polar Planimeter, and the contour map, on February 28, 1984. The sinkhole is located in Alachua County, Florida (SW/4, SW/4, Section 14, T.9S., R. 18E.) and is expressed in the Hawthorn Formations and the Crystal River Formation. Teague's Sinkhole lies in the transition zone between the Northern Highlands and the Western Valley of Alachua County. The primary processes of erosion in Teague's Sinkhole are sheetwash and slumping. The orientation of the ravines of Teague's Sinkhole is believed to be fracture controlled. About 95% of the Crystal River Formation that outcrops in Teague's Sinkhole consists of chert.

Introduction

This project was initiated in order to produce a contour map of Teague's Sinkhole which was then used to make an estimate of the volume of Teague's Sinkhole. Geologic and Physiologic descriptions of the area are also provided along with an index map.

The contour map was produced by the use of a plane table and alidade. The field work was performed on February 18 and 19, 25 and 26, 1984. The volume estimate was made using a Lasico Compensating Polar Planimeter, and the contour map, on February 28, 1984.

The region in which Teague's Sinkhole is located has been studied in the past by several authors. The most recent regional studies were done by O.W. Girard (1968), and K. Williams (1974). However, a specific study of Teague's Sinkhole has never been completed. In fact, the sinkhole was nameless until this project. In order to simplify the discussion of the sinkhole, the writer has taken the liberty of naming it in honor of the landowner, Mr. Doyce Teague.

Teague's Sinkhole is located in Gainesville West Quadrangle, Alachua County, Florida (SW/4, SW/4, Section 14, T.9 S., R.18 E.). (See Figure 1)

The sinkhole is covered with numerous oak and pine trees. The slopes are covered with ferns and grasses. The floor consists of sandy clay. The surrounding land is used for cattle ranching.

Physiography

Alachua County lies in the Central Highlands division of Florida. Alachua County is separated into two general topographic divisions: The Northern Highlands and the Western Valley. Teague's Sinkhole lies in the transition zone between the Northern Highlands and the Western Valley. The topography of the western part of Alachua County consists basically of a nearly flat plain (the Western Valley) underlain by the limestone of the Crystal River Formation, mantled by thin sandy soils and residual outliers of the Hawthorn Formation. The area is bordered on the east by a subdued westward facing escarpment (the Transition Zone) and an upland plateau (the Northern Highlands) composed of the less soluble Hawthorn sediments. "Sinks, caves, and typical karst topography occur in the plain, the marginal escarpment, and less commonly on the plateau." (1)

"The primary geologic processes controlling the topographic expression of landforms in this area are, stream erosion and sheetwash along slopes in the marginal zone, groundwater solution in both the Hawthorn and Crystal River Formations, and modification by both higher and lower stands of sea level during the Pleistocene Epoch." (1) In Teague's Sinkhole, active slumping also controls the topographic expression of the southeast ravine.

In the transition zone, small springs result from the channeling of groundwater by impermeable lenses within the Hawthorn Formation. (1) Several springs of this type can be observed in Teague's Sinkhole.

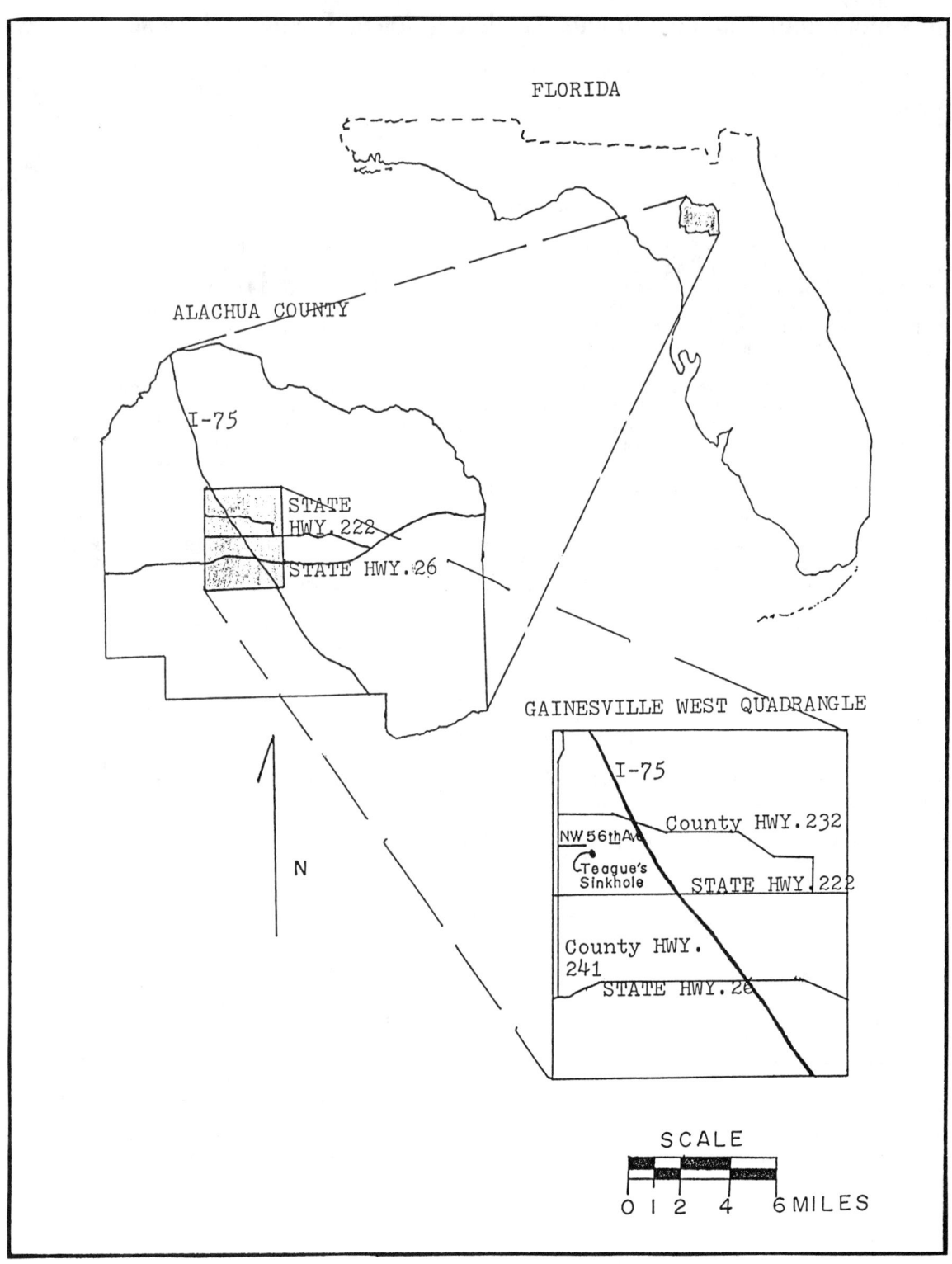

Figure 1: Index Map Locating Position of Teague's Sinkhole.

Geology
    Teague's Sinkhole is expressed in two rock formations: the Hawthorn Formation and the Crystal River Formation.

The Hawthorn Formation
    "The Hawthorn Formation, of early middle Miocene age, was deposited in a transgressing middle Miocene sea." (2) The Hawthorn Formation, in the area of Teague's Sinkhole, consists of thick clays, sandy clays and quartz sands. "Lithologic variability is more the rule than the exception and units may pinch out, interfinger or intergrade both laterally and vertically." (1) The Hawthorn Formation is covered by a veneer of loose sands from Pliestocene terrace deposits. "The identifiable fauna of the Hawthorn Formation in the western part of Alachua County is extremely meager." (1) No fossils from the Hawthorn Formation were found in Teague's Sinkhole. The southeast ravine of Teague's Sinkhole shows a good unweathered exposure of the lower portion of the Hawthorn Formation.

The Crystal River Formation
    "The Crystal River Formation was deposited during the Eocene Epoch on the Ocala Bank, a submarine plateau over which water depth probably did not exceeed 46M (150 feet) at any time." (1) The Crystal River Formation unconformably underlies the Hawthorn Formation. "The most common lithology present in the Crystal River Formation is that of white cream, massively bedded, soft, granular, bioclastic limestone. In places it is almost a coquina of large Foraminifera or of the calcitic shells of Amusium ocalanum or Chlamys spillmani." (1) A loose boulder of this type rock was observed in the northwest ravine of Teague's Sinkhole.

    "Portions of the Crystal River Formation have been locally replaced by silica. Upon erosion, large boulders remain as residual remnants of reworked limestone. They were formed by in situ replacement of carbonate by silica." (1) "The outer surface of these chert boulders is commonly weathered to varying thicknesses and is generally bleached to a white or light grey." (1) In Teague's Sinkhole, the chert is generally a dark grey and is usually covered with mosses. "Chert is commonly found along joints in the Crystal River Formation as seen in several caves in the western part of Alachua County. Horizontal layers and plates of chert are commonly found at the contact between the Crystal River and the Hawthorn Formations. The chert appears to be more common on the surface or close to the contact with the overlying Hawthorn Formation than at depth within the limestone and is also more common in solution cavities than dissiminated randomly throughout the limestone. The above observations suggest that the silica necessary to form the chert boulders was provided by the Howthorne Formation." (1) Observations in Teague's Sinkhole tend to support this suggestion; about 95% of the Crystal River Formation that outcrops in Teague's Sinkhole is made of chert.

    "The Crystal River Formation has a diverse fauna, the limestone being composed principally of the detrital remains of marine organisms. Prominent fossil groups represented are the pelecypods, gastropods, echinoids, bryozoans, crustaceans and foraminifers." (1) Except for the crustacean group, members of all of these fossil groups were observed in the rocks in Teague's Sinkhole.

    Teague's Sinkhole is an example of a ravelled sinkhole in a late stage of erosion. The sinkhole was probably formed by slumpage of material into a solution cavity formed along a joint or fracture. The sinkhole lies in the extensively fractured zone of Alachua County. Also, the outcrops in Teague's Sinkhole are mainly of chert which suggests that it lies upon a joint or fracture. The sinkhole has four main ravines, the axes of which conform to the fracture pattern of the Hawthorn Formation and the Crystal River Formation in Alachua County. The orientation of these ravines is believed to be fracture controlled.

Methods

Contour Map
    The contour map of Teague's Sinkhole was produced using a telescopic alidade and a plane table. Detailed information on alidade and plane table surveying can be found in Robert R. Compton's Manual of Field Geology.

Volume Estimate
    The volume estimate of Teague's Sinkhole was made using a Lasico Compensating Polar Planimeter and the contour map of Teague's Sinkhole. A planimeter is a device that measures irregularly shaped areas, such as areas bounded by contour lines. Details of the operation of a planimeter can be found in the planimeter's instruction manual.

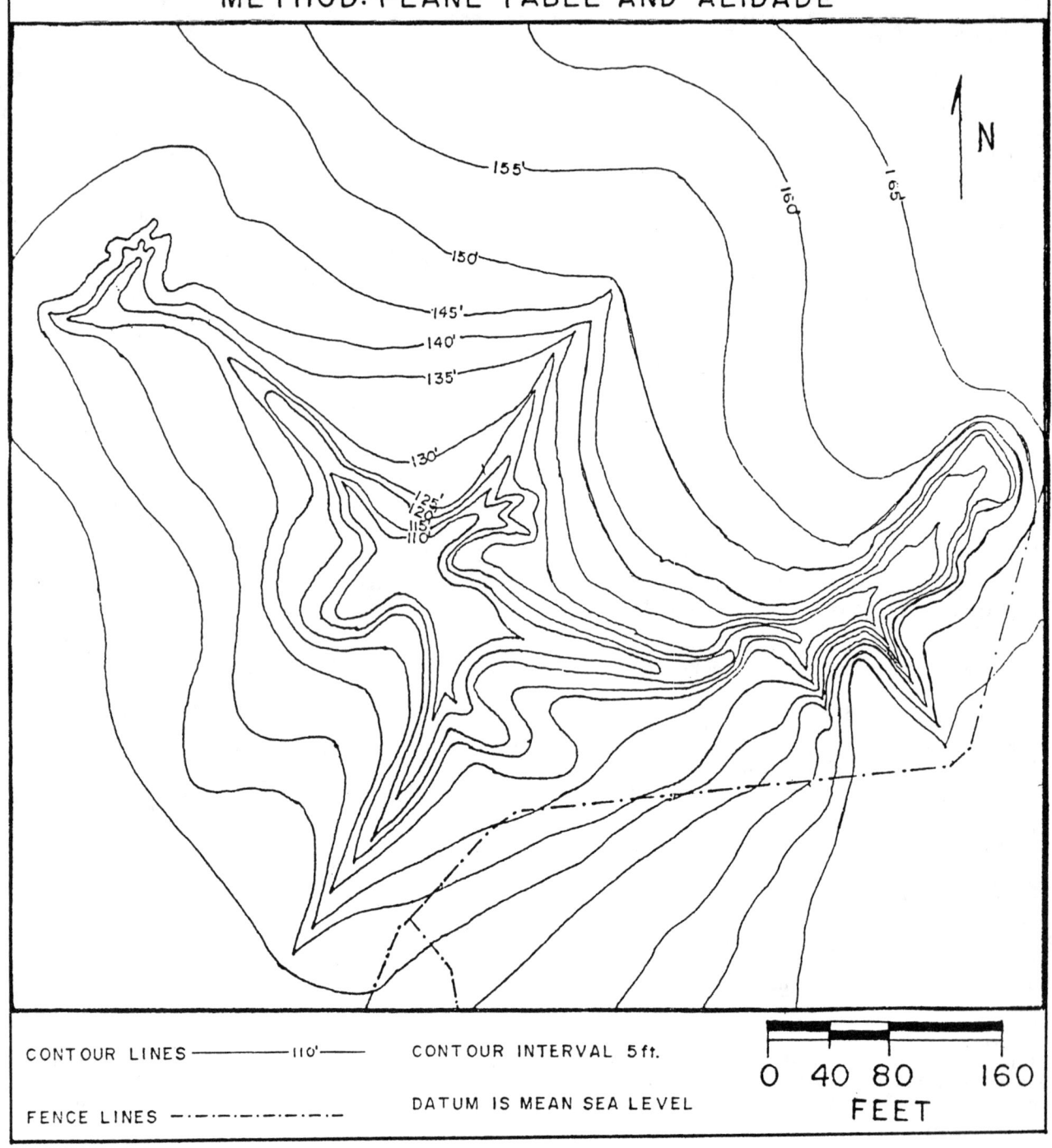

The surface areas bounded by each contour line, from the 33.528 Meter (110 foot) contour line through the 48.768 Meter (160 foot) contour line, were calculated using a Lasico Compensating Polar Planimeter with the arm set on medium. Three planimeter runs were completed for each contour line and the average of these three runs was considered to represent the area bounded by that contour line. This data was used to determine the average surface area between each set of adjacent contour lines. The average surface areas between each set of adjacent contour lines were multiplied by the contour interval of 1.524 M (5 feet) to give an estimate of the volume between each set of contour lines. These volume segments were added together to provide a volume estimate of Teague's Sinkhole from the floor to 15.24M (50 feet) above the floor.

Results

A small contour map of Teague's Sinkhole is included in this report. (See Page 5).

The estimated volume of Teague's sinkhole, from the ground level to 15.24M (50 feet) above ground level, 189,778 M^3 (248,141 yds.^{3}). This is a conservative estimate of the volume of sediment absorbed into the Crystal River Formation through Teague's Sinkhole.

Conclusion

Teague's Sinkhole lies in the Transition Zone between the Northern Highlands and the Western Valley in Alachua County, Florida The sinkhole is expressed in two formations: the Hawthorn Formation and the Crystal River Formation. The orientation of the ravines of Teague's Sinkhole is believed to be fracture controlled. The volume of Teague's Sinkhole from the floor to 15.24M (50 ft.) above the floor is estimated at 189,778M^3 (248,141 Yds.^{3}).

References

Kenneth Edward Williams, The Geology of the Western Part of Alachua County, Florida (Unpublished Master's Thesis, University of Florida, 1974).

Oswald Woodrow Girard Jr., The Geology of the Gainesville West Quadrangle, Alachua County, Florida (Unpublished Master's Thesis, University of Florida, 1968).

Robert R. Compton, Manual of Field Geology (John Wiley and Sons, Inc., New York, 1962), p.p. 88-112

Development, occurrence, and triggering mechanisms of sinkholes in the carbonate rocks of the Lehigh Valley, eastern Pennsylvania

PAUL B.MYERS, Jr. *Lehigh University, Bethlehem, Pennsylvania, USA*
MICHAEL PERLOW, Jr. *Valley Fdn Consultants, Inc., Bethlehem, Pennsylvania, USA*

ABSTRACT

The southern half of the Lehigh Valley in eastern Pennsylvania contains over 1500 meters of Cambro-Ordovician carbonate rocks most of which are susceptible to the development of sinkholes. The sinkholes develop in overlying residual soils and frost-churned glacial drift in response to the undermining and removal of material into a solution-enhanced drainage system in the underlying bedrock. This undermining is accomplished primarily by percolating ground water in the vadose zone. Approximately 2000 sinkholes in the Lehigh Valley have been cataloged as to principal triggering mechanism, rock and soil type. These statistical data greatly facilitate the prediction, prevention, and/or mitigation of this problem.

Introduction

The Lehigh Valley is a broad NE-SW-trending valley in eastern Pennsylvania (Figure 1). Physiographically, it represents the Great Valley Section of the Appalachian Valley and Ridge Province in this area. It is bounded on the north by a relatively high, linear ridge held up by resistant quartzites of Lower Silurian age and on the south by the Reading Prong, a series of hills composed of Lower Cambrian quartzites and Precambrian crystalline rocks, primarily pyroxene granulites to amphibolites. The rocks underlying the Lehigh Valley consist of a thick (1585-2000m) series of Cambro-Ordovician carbonate rocks overlain by approximately 3600 meters of Middle Ordovician graywacke siltstones and claystone slates. The carbonate rocks occupy an outcrop belt approximately 12 kilometers wide by 51 kilometers long that forms the southern half of the valley. This portion of the valley is also the lowest in elevation as might be expected.

The area has been glaciated and the eastern two-thirds of the Lehigh Valley is covered by intensely weathered, frost-churned drift of Illinoian age. The thickness of this glacial cover is highly variable ranging from absent to over 10 meters.

Numerous sinkholes resulting in millions of dollars of property damage have developed in the carbonate rock outcrop belt in the southern half of the valley. Although these sinkholes have been a local problem for over a century, serious study into their characterization and mitigation has been relatively recent (Wheeler and Myers, 1976; Perlow, 1984).

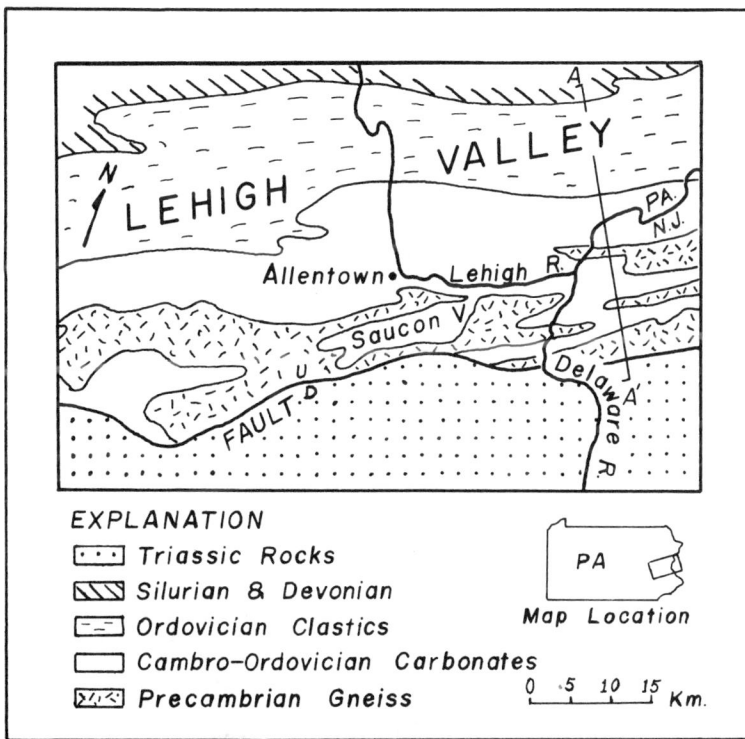

Figure 1: Map showing location and general geologic setting of the Lehigh Valley, Pennsylvania. Line A-A' shows approximate position of section in Figure 2.

Geologic Setting

The rocks of the Lehigh Valley represent a Lower Paleozoic carbonate shelf facies of Middle Cambrian to Lower Ordovician age which gave way to a deeper water foreland flysch facies, the Martinsburg Formation, by Middle Ordovician times. The carbonate rock units and their characteristics are shown in Table 1.

TABLE 1. Characteristics of Lehigh Valley Carbonate Rocks

Formation (Age)	Thickness (m)	Formation Description and Weathering Characteristics
Jacksonburg Fm. (M. Ord.)	170 - 460	Dark-gray shaley limestone grading downward into crystalline, high-calcium limestone. Low to moderate porosity and permeability; thin soil mantle; relatively few solution features.
Ontelaunee Fm. (L. Ord.)	0 - 200	Med.-gray, finely crystalline dolomite; cherty at base; missing at many locations. Solution enhanced porosity and bedrock pinnacles characteristic. Moderate to thick soil mantle.
Epler Fm. (L. Ord.)	270±	Interbedded v.f. grained, med.-gray limestone and gray dolomite. Solution enhanced porosity; few bedrock pinnacles; v. thick soil mantle.
Rickenbach Fm. (L. Ord.)	220	Gray, fine to coarse dolostones, thin bedded at top to thick bedded toward base. Solution enhanced porosity and bedrock pinnacles characteristic; mod. thick soil mantle.
Allentown Dol. (U. Camb.)	575	Alternating thick beds of light- and dark-gray weathering dolomite; stromatolites and oolites common; some orthoquartzite beds. Solution enhanced porosity and bedrock pinnacles characteristic; soil mantle generally thin.
Leithsville Fm. (Uppermost L. and M. Camb.)	350	Interbedded fine- to coarse-grained dolostones and tan phyllite; few thin sandstone beds. Solution enhanced porosity; bedrock pinnacles common; commonly covered with thick colluvium near uplands.

The rocks of this area have been subjected to intense deformation during the Late Ordovician (Taconic Orogeny) and during the later Paleozoic (Alleghenian Orogeny). They have been thrown into a series of large, basement-cored nappes upon which a later series of more open folds that are asymmetric to overturned to the northeast have been superimposed (Figure 2). The area has been cut by a series of faults that range in age from lower Paleozoic to Triassic. The rocks have been exposed at the surface since the Mesozoic and the only materials deposited in the valley since the Late Ordovician consist of Pleistocene glacial drift, alluvial gravels and thick aprons of colluvial material that have been shed off of the uplands that border the valley both on the north and the south. The colluvial deposits are over 60 meters thick along the uplands and thin markedly as one moves away from the ridges out into the valley.

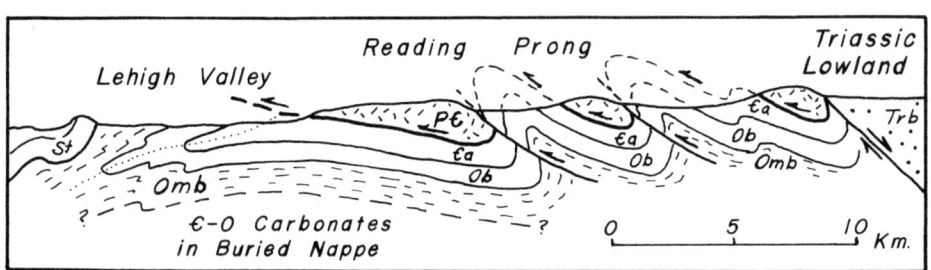

Figure 2: Generalized cross-section of the Reading Prong and Lehigh Valley (modified from Drake, 1970). Trb, Triassic rocks; St, Silurian rocks; Omb, Ordovician clastics; Ob, Ordovician carbonate rocks; €a, Cambrian carbonate rocks; P€, Precambrian crystalline rocks.

Development of Sinkholes

Because of the structural complexities in the region, no one stratigraphic unit persists in surface outcrop over a broad area. Consequently, until recently, little effort has been directed toward a systematic study of the relationship of sinkhole activity and lithology. Attitude of bedding planes in the various units ranges from horizontal to vertical and structural units may be upright to completely overturned. Solution of the carbonate rocks has occurred along bedding planes, fault and joint planes, and, to a minor extent, along cleavage planes. Most, if not all, of the sinkholes have developed within the materials overlying bedrock. Percolating waters move along the bedrock-soil interface to sites of solution enhanced openings in bedrock and the soil material is carried downward into the underlying bedrock by a combined process of erosion and spalling. This creates a hole that propagates upward in the soil to the point where the overlying materials are unable to support themselves over the opening and collapse occurs. No sinkholes in the area are related to cavern collapse. In many areas throughout the valley the sinkholes exhibit an allignment that parallels fault zones, joint sets, or vertical bedding. In other cases the sinkholes appear to exhibit a more random pattern suggesting a region underlain by a particularly solution-prone rock unit.

Evaluation of Sinkhole Occurrence

Recently, Perlow (1984), Perlow, et al. (1984) undertook a systematic study of the sinkholes of the Lehigh Valley in an attempt to gain a better insight into their cause and possible mitigation. Over 2000 sinkholes have been identified within the area by means of reconnaissance surveys, remote sensing techniques (primarily infra-red and standard aerial photography) and geophysical techniques (primarily terrain conductivity and electrical resistivity). An extensive search of township highway records, engineering reports, and records of utility companies has revealed numerous sinkholes in the area that have been active over long periods of time. A sinkhole classification system has been established based on the most probable triggering mechanism and correlations between the various types of sinkholes and geologic formations and soil types has been established (Perlow, 1984). Five principal types of sinkholes have been recognized based on the probable triggering mechanism. These are as follows:

1. Naturally occurring sinkholes - These include sinkholes that are located in undeveloped or farm areas that exhibit no systematic pattern. They are the result of percolation of surface waters concentrated by natural overland flow and surface runoff.

2. Fault/structure-related sinkholes - In many localities, the sinkholes are distributed in linear patterns or systems of intersecting linear patterns. In some cases these sinkholes can be shown to parallel zones of highly faulted or fractured bedrock. In other cases a structural control is suspected because of the linear distribution of the sinkholes. Zones of highly faulted or fractured bedrock are especially susceptible to weathering and solution thereby creating a situation where the associated sinkholes tend to be recurring, relatively large, and deep.

3. Construction-related sinkholes - Excavation and removal of the fine-grained, upper portions of the soil mantle commonly exposes an underlying material that is more granular and consequently more permeable and with less shear strength than the overlying materials. Surface waters, which can be introduced naturally or during the construction process, can rapidly erode these silty, granular materials and, in areas where the underlying bedrock is highly pinnacled and riddled with solution enhanced openings, sinkholes will readily develop. Perlow (1984) documents over 184 sinkholes that have developed along a 7 km. section of PA Route 33 since highway construction began in 1972. These sinkholes are all located in cut areas along natural or highway drainage courses.

4. Sinkholes in urban land areas - Breaks or leakage in water or sewer lines, subsurface stormwater disposal, and on-lot sewage disposal all introduce sources of percolating water that can rapidly trigger the development of a sinkhole. Leaky utility lines or other sources of introduced water can accelerate the subsurface erosion of soil materials and the creation of voids which in turn put increased stress on utility lines. These unsupported lines eventually fail commonly resulting in catastrophic development of sinkholes. Since this type of sinkhole is common in urban land areas, the resulting void is commonly bridged by asphalt, concrete or other similar materials. This permits an unusually large subsurface opening to develop and, when the surface cap fails, it is commonly an instantaneous collapse. Where gas lines have been ruptured in this manner, dangerous explosions can, and do, occur.

5. Sinkholes caused by dewatering or mine pumpage - Significant changes in the elevation of the water table brought about by heavy pumping to dewater mines or quarries can lead to increased sinkhole activity. The drop in the water table permits the increased development of percolation in the vadose zone which creates new sites of subsurface erosion and the development of voids.

An example of this can be found in the Saucon Valley (for location see Figure 1), a carbonate-floored window in a basement-cored nappe. Since 1953, pumping of 20-40 mgd out of a zinc mine at Friedensville has lowered the water table up to 145 meters in the vicinity of the mine. This has led to an increase in sinkhole activity throughout the area of the cone of depression. However, lack of accurate statistical data on sinkhole activity in this area prior to 1953 makes documentation of this observed increase in activity difficult.

Correlations

Correlations between the 2000 documented sinkholes and various geologic and cultural parameters are beginning to provide insights into the conditions that lead to the development of sinkholes in the Lehigh Valley area. These correlations are summarized in Table 2. As can be seen from the data, the carbonate rock units that are most prone to sinkhole development are the Rickenbach and Allentown formations. The Jacksonburg formation is an argillaceous

TABLE 2. Average Sinkhole Density for Various Geologic Formation and Sinkhole Type

Formation	Total Area (mi^2)	Total No. of Sinks	Avg. Sinkhole Density (No/mi^2) (all occurrences)	Average Sinkhole Density (sinks/square mile)			
				Naturally Occurring	Construction Related	Utility Related	Structure Related
Jacksonburg	24.0	54	2.2	1.8	0.3	0.1	--
Ontelaunee	6.4	28	4.2	1.4	1.4	1.4	--
Rickenbach	18.7	174	9.3	4.5	4.0	0.8	0.05*
Epler	74.5	518	6.9	4.0	2.5	0.3	--
Allentown	85.0	731	8.6	2.2	2.2	4.2	--
Leithsville	32.0	69	2.1	1.0	0.2	0.9	--

*Includes only those occurrences where a clear-cut relationship between structure and sinkhole exists (modified from Perlow, 1984).

carbonate unit and, as might be expected, it is less soluble and sinkhole activity is low compared to the more pure carbonate units. The low frequency of sinkholes within the outcrop area of the Leithsville formation can be attributed to two factors. First, the Leithsville formation has numerous phyllitic interbeds and movement of ground-water and consequently solution is relatively inhibited. Second, and more important, the Leithsville formation is the oldest carbonate rock unit and consequently crops out close to the Precambrian unlands on the south side of the valley. Colluvial material shed from the Precambrian rocks commonly covers the Leithsville outcrop area with a thick blanket of material thus armoring the Leithsville from extensive weathering and erosion.

The data in Table 2 indicate that almost 50% of the documented sinks have been associated with natural drainage in undeveloped areas, whereas construction-related sinks account for nearly 30% of the occurrences and utility-related sinks comprise over 20%. The very few occurrences of fault/structure-related sinks is due to the fact that only those sinks in which a relationship to a fault or fracture zone could be documented are included in this category, its percentage would be much larger. Also, sinkholes related to dewatering and mine pumpage have not been included in Table 2 although dewatering appears to be a significant factor in accelerating sinkhole activity to the south in the Saucon Valley and this phenomenon should be anticipated in any area where extensive dewatering is undertaken in a carbonate rock terrain.

The high rate of construction-related sinks in the outcrop area of the Rickenbach formation can be tied to the fact that PA Route 33 cuts across a wide outcrop belt of the Rickenbach and the high number of sinks that developed along Route 33 introduces a strong bias to the statistics dealing with this formation. However, despite the fact that local circumstances can introduce a strong bias to some of these data, long-term statistics on sinkhole activity in an area can be extremely useful for land use planning and anticipation and mitigation of sinkhole problems.

Conclusions

Sinkholes in the Lehigh Valley are related almost exclusively to the creation of voids by the subsurface erosion and removal of residual and transported soil material into solution enhanced openings in the underlying bedrock. The triggering mechanism for over 50% of the 2000 documented sinkholes in this area has been some sort of construction/utility-related activity.

Accumulated, long-term statistical data on sinkholes in an area such as this can be of use in land use planning and the anticipation of special problems during and after construction.

Also, long-term statistical data could form the basis for the development of a sinkhole insurance program that could markedly ease the financial burden commonly encountered by individuals, municipalities, and private corporations in areas of known sinkhole activity.

References

Drake, A. A., Jr., 1970, Structural Geology of the Reading Prong; in Fisher, G. W., et al., eds., Studies of Appalachian Geology - Central and Southern: New York, Interscience Publishers, p. 271-291.

Drake, A. A., Jr., 1978, The Lyon Station - Paulins Kill Nappe - The Frontal Structure of the Musconetcong Nappe System in Eastern Pennsylvania and New Jersey: U.S.G.S. Prof. Paper 1023, 20p.

Perlow, M., Jr., 1984, Evaluation of Sinkhole Occurrences, Engineering Construction and Maintenance Problems in Lehigh Valley Limestone Areas: AEG-ASCE 1984 Joint Symposium, Geologic and Geotechnical Problems in Karstic Limestone of the Northeastern United States, Frederick, Maryland.

Perlow, M., Jr.; Schadl, S. M.; Fang, H. Y.; Chancy, R., 1984, Sinkhole Case Histories and Site Development Guidelines for Limestone Bedrock Areas in Eastern Pennsylvania: Transportation Research Board Symposium; Special Problems in Karstic Limestone Bedrock, Washington, D. C.

Wheeler, J. R. and Myers, P. B., Jr., 1976, Carbonate History and Foundation Problems in Carbonate Rocks of the Lehigh Valley: in Engineering, Construction, and Maintenance Problems in Limestone Regions Symposium; Lehigh Valley Section ASCE, Lehigh University, Aug. 3-4, 1976, p. 51-77.

Submarine 'sinkholes': A review

MARCO TAVIANI *CNR, Bologna, Italy*

ABSTRACT

The exploration of the submarine realm has led to the discovery that structures closely resembling subaerial dissolution karstic features are relatively frequent at great depth. Rounded, subcircular or oval depressions from a few meters up to 30 km in diameter have been reported from different basins (Mediterranean, North and Red Seas, Gulf of Mexico). The objective difficulty in making detailed studies of these depressions does not allow in most cases to inequivocally understand the nature of these structures and the mechanisms controlling their formation. The possibility of submarine karstification of carbonates leading to large scale structures seems unlikely but has not yet been investigated. The formation of sinkholes due to subsurface solution of salt does not necessarily require a subaerial environment. Important saline basins, such as the Gulf of Mexico, Mediterranean, Red and North Seas, display large depressions interpreted as salt collapse-structures due to the solution of underlying evaporites. The role that they play as mini-sedimentary basins appears to be important. In the Red Sea these depressions are likely to be connected to the formation and maintanance of hypersaline brines fundamental for the formation of metal-enriched precipitates of economic interest.

INTRODUCTION

The intense investigation of the deep-sea during the last decades has substantially advanced our knowledge of the morphology of the ocean bottom. Thus, morphological structures analogy to the better known structures of the subaerial environment have been discovered. A case in point is given by the detection of karst-like features in many basins of the world to depths as great as 2000 m. This paper will give a brief account on this attractive topic by discussing some of the theoretical and applied problems connected with it.

The first report of structures on the ocean bottom thought to be the product of large scale dissolution and collapse of soluble rocks is given by LOHMAN (1972) who provides convincing arguments of subsurface erosion (leaching) of the Permian (Zechstein) salt on the basis of seismograms recorded in the North Sea. Some of the circular to subcircular depressions (Chain, Discovery, Valdivia, Kebrit and Oceanographer Deeps) acoustically determined on the bottom of the Red Sea are suspected by SCHOELL et al. (1974: p.309) and BÄCKER et al. (1975: p.102-103 and 119) to be ultimately collapse structures due to removal of underlying salt. To support their view, BÄCKER et al. (1975) recall the morphological similarity between these structures and those shown by the diapirs at Gubbet Mus Nefit (Dahlak islands, southern Red Sea) reported by HOOPER in FRAZIER (1970). Salt collapse is also connected, according to TRABANT & PRESLEY (1977), to the formation of the Orca Basin, an elbow-shaped depression located in the Gulf of Mexico. Finally, ROSS & UCHUPI (1977), LORT (1977: p. 195), BELDERSON et al. (1978) and KASTENS & SPIESS (1984), independently argue that many depressions observed on seismograms, Gloria and Deep-Tow records from the eastern Mediterranean are due to collapse linked to salt leaching of underlying evaporites. It is noteworthy to record that HINZ (as quoted in SIGL et al., 1973) was the first to advance the hypothesis that "collapsing of underground caverns" due to chemical erosion of Miocene evaporites is one of the probable origin of the hummocky and rolling landscape of the Eastern Mediterranean.

Thus, it appears that karst-like structures are widespread phenomena in the oceans and not just local accidents.

It can be noted that the evidence available at present is limited to solution of salt. In the subaerial environment true karst is a phenomenon primarily linked to solution of carbonatic rocks (limestone, dolomite) through ground water circulation. Even if suspected (see for example, BIJU-DUVAL et al., 1982, 1983a, 1983b), the occurrence of deep ocean karstic dissolution of carbonatic rocks is yet to be proven. On the other hand, it is important to point out that dissolution of very soluble rocks such as Na,Mg,K salts need not take place only in subaerial conditions. Salt could be infact easily removed by seawater circulation when outcropping or within the subsurface (suberosion according to the German literature). This kind of leaching is therefore somewhat independent from the surrounding environment in the sense that the most important requirement seems to be water (either fresh or marine) circulation within the salt-bearing evaporites. Dissolution and collapse features in salty rocks are well known also on land (see references in KASTENS & SPIESS, 1984). As observed by LOHMAN (1972), who wisely discriminates between "true" karst and salt leaching the latter phenomenon can eventually evolve in collapse structures, the sinkholes, when salt leaching on the top of a diapir is faster than the flow of salt from downwards. Although the final result of the process may produce features resembling dolines, poljes and other subaerial karst morphologies, this does not strictly imply that they formed under subaerial karst conditions. If compared to the detailed morphological, lithological and structural analyses that can be carried out on the karst landscape on land, any approach to study the underwater environment still appears very primitive. Basically all the information we have about the underwater sinkholes derives from indirect, acoustic observations which supply little more than a rough morphology. LOHMAN (1972) has studied seismograms which give a vertical section of the sinkholes. The hypothesis that the Red Sea deeps are effectively collapse structures is not supported by any definitive direct observation but rather their peculiar morphology, detected by a traditional bathymetric survey. The same approach is shared by ROSS & UCHUPI (1977) and by TRABANT & PRESLEY (1978). Even the use of more sophisticated devices such as the GLORIA (BELDERSON et al., 1978) and Deep-Tow (KASTENS & SPIESS, 1984) systems provide only morphological information.

The morphological studies derived from the acoustic methods alone can not be considered diagnostically sufficient. The geological and structural setting of a given area must also be carefully analyzed. This factor is well understood by the above mentioned investigators which derive their final interpretation from a thorough analysis of all the geological and geophysical evidence at their disposal. In fact, circular or oval depressions on the ocean bottom can be associated with other mechanisms other than salt dissolution, for example, the collapse of the roof of pillow-lavas (SARTORI, pers. com.), subsidence related to methane escape from the sediment (Missisippi River Delta: PRIOR & COLEMAN, 1980), dynamic sedimentary processes (Malta-Siracusa escarpment: BIJU-DUVAL et al., 1983b) etc. Furthermore, karstic features originally produced under subaerial conditions can be encountered on rocks later subsided below the sea.

THE TERMINOLOGICAL PROBLEM

The use in marine geology of terms introduced to describe the subaerial karst is misleading when there is no definitive proof of the genesis of a given underwater "karstic" feature. In order to avoid confusion between a superficial, morphological appearance and the mechanism which originated it, it is prudent to use terms as "doline-like", "polje-like" or phrases as "resembing dolines" and so on. An uncorrect terminology can generate more than a wrong interpretation. For example, KASTENS & SPIESS (1984) use the terms doline and zanjones referring to some collapse and dissolution features present on the Eastern Mediterranean Ridge. However, from their paper it appears that their dolines (typically a karst structure of subaerial carbonate rocks) are collapse structures due to salt leaching and that their zanjones (originally described by MONROE, 1964, as a local karst feature of carbonates in Puerto Rico, and by the same author, 1976: p.4, defined as a local term for corridor) are solution features again in evaporites. Accordingly, it can be argued that the mechanisms known to produce those structures on land are also responsible for the origin of those discovered at sea. As a consequence, one can conclude that dolines and zanjones (an unnecessary term) may form also in evaporitic sequences (namely halite and/or K,Mg salts) exactly as they do in carbonate rocks; or, that the soluble bedrock was not made up by evaporites but by carbonates. Moreover, a zanjon is thought to form under special conditions through the action of "acidic waters derived largely from decay of forest vegetation" (Monroe, 1964). Similar conditions are highly unlikely in the deep ocean. It could be logically argued that the "karstic" morphology of the Eastern Mediterranean bottom is inherited by an ancient subaerial karst landscape. It is evident that the data at our disposal do not allow us to objectively recognize the mechanism(s) which formed these underwater structures and a less rigid use of the terminology would have generated less confusion between the observed morphology (the objective datum) and its interpretation (subjective).

In the present paper I shall use the informal term sinkhole when refering to the underwater morphological expressions recognizable as collapse structures due to subsurface dissolution of soluble rocks whatever (karst or salt leaching) is their likely origin.

SUBMARINE SINKHOLES: THEIR GEOLOGICAL ASPECTS

The submarine sinkholes buried in the North Sea subsurface (LOHMAN, 1972) range from a 1 m up to 5 km in diameter. Those of the Red Sea have a comparable magnitude (1-4 km about) as well as the collapse structures of the Eastern Mediterranean (1-6 km). The maximum diameter of the Orca Basin is about 30 km.. The vertical relief varies from a few tens to many hundreds of meters. It is immediately evident that these structures collect and preserve sediment acting as mini-sedimentary basins (LOHMAN, 1972) whose geological importance can be regional depending upon their size and setting.

The most common mechanism proposed by the authors implies that the collapse takes place at the top of a salt diapir; however, subsurface dissolution of the North Sea Permian evaporites is also thought (LOHMAN, 1982) to take place in the absence of diapirism.

Limiting our consideration to the first mechanism, we observe that sinkholes generated by salt-leaching pass through the following steps: 1) primary deposition of salt-bearing evaporites; 2) subsequent deposition of a sedimentary cover; 3) halokinesis and/or halotectonics (TRUSHEIM, 1960) due to the effects of sedimentary loading and/or tectonic activity leading to diapiric uplift; 4) subsurface removal of salt greater than its inward flow within a diapir with consequent formation of caverns; 5) formation of the sinkhole for the collapse of the cavern's roof.

In the Red Sea the widespread occurrence of a thick Miocene salt-bearing evaporitic sequence is well known from seismics and drilling (BONATTI et al., 1984, with references therein). Although the Plio-Quaternary sedimentary cover rarely exceeds the 100-200 m, extensive salt movement is favoured by active tectonic movements connected to the basin's oceanization. Furthermore the regional heat flow is high implying high subsurface temperature; high temperature is known (GUSSOW, 1968) to enhance the plasticity of salt which can, therefore, move also in absence of considerable loading. "Diapiric" subbottom configurations can be seen in many seismic reflection profiles of the Red Sea (ROSS & SCHLEE, 1973; BONATTI et al.,1984). BÄCKER et al. (1975) observe that the Red Sea deeps which they think to be collapse features, are connected with presumed transform faults. This observation seems to be correct since the Oceanographer Deep is situated in the area of the Brothers Fracture Zone (CRANE & BONATTI, in press), Valdivia, Discovery and Chain Deeps are situated in the Atlantis Fracture Zone (CRANE & BONATTI, in press) and Kebrit Deep can well be related to a transform parallel to the Dead Sea-Aqaba transform as suggested by some bathymetric alignments (see plate 5 of BÄCKER et al., 1975). Whether this connection is purely casually or directly connected to the sinkhole formation is still to be investigated.

Miocene (Messinian) evaporites are present also in the Mediterranean Sea where they form an extensive subsurface formation; the presence of Na,Mg,K salts within the evaporitic formation has been demonstrated by DSDP drills (RYAN, HSÜ et al., 1973). Diapiric structures have been extensively reported from the Eastern Mediterranean (LORT & GRAY, 1974; SMITH, 1976, 1977;), although not without criticism (e.g., GOGUEL, 1978; HIEKE, 1982). It is interesting to note that at present collapse features at the top of presumed diapirs are reported only from the Eastern Mediterranean while they are apparently absent from the Western basin where diapiric structures are far numerous (ERICKSON et al., 1977: p.270). This fact could be related to the different geologic and tectonic framework of the two basins. In the light of the previous discussion, it is interesting to re-examine the conclusion of HSÜ et al. (1973) which advance the hypothesis that the hummocky landscape of the Eastern Mediterranean largely derives from the karstification of the evaporitic deposits exposed to subaerial conditions when the basin completely dried up. They correctly reason that the solution of sulphates and carbonates, originating collapsing caverns, occurred as a result of ground water circulation. They also argue that the collapse of caverns formed during the subaerial phase may have continued also after marine conditions re-established in the Mediterranean. However, if the "karstic" morphology is due to dissolution of salt and not of less soluble products (as gypsum, anydrite and carbonates) there is no need of subaerial exposure. Lacking definitive information, none of the hypothesis can be ruled out with the possibility that both processes may have occurred.

The Gulf of Mexico, where the Orca Basin is situated, is an area of thick accumulation of salt which was deposited in Jurassic time in a geological/tectonical setting closely resembling the Miocene Red Sea (HUMPHRIS, 1977). Sediment loading triggered important halokinetic movements resulting in salt tectonics. TRABANT & PRESLEY (1977) argue that Orca Basin, the largest "sinkhole" until now, reported in the deep sea, is linked to massive collapse due to salt removal from an underlying salt mass or exposed salt diapir.

THE ECONOMIC IMPORTANCE

The disruption of salt-bearing evaporites in a tectonically active setting can facilitate intraevaporitic water circulation; salt within the evaporites can be easily dissolved forming brines which can eventually accumulate in nearby depressions. The sinkholes emplaced on saline formations, as those treated in the present paper, are an ideal place where brines can be formed and preserved from dissipation. The more or less prolonged stagnation of these brines could bring the bottom of the basin to anoxic conditions. The Red Sea Deeps and Orca Basin are two excellent examples of this phenomenon. The chlorinity (‰ Cl) of the Red Sea brines of the deeps reported in this paper ranges between 136.6 to 156 (BÄCKER & SCHOELL, 1972) comparable to that filling the bottom 200 m of Orca Basin that is 149.5 (TRABANT & PRESLEY, 1977). Anoxic conditions at bottom allow the preservation of organic matter; thus, under favorable geodynamic (low heat flow, absence of volcanism...) and hydrographic (especially high surface productivity) conditions, a sinkhole can become a source of hydrocarbons, as suggested by TRABANT & PRESLEY (1977). These geodynamic processes, active in the Red Sea since at least the time of the evaporite deposition (e.g., BONATTI et al., 1984), prevent the formation and preservation of this kind of hydrocarbons. However, in the Red Sea deeps metal-enriched sediments accumulate in relation to the intense hydrothermalism accompanying the separation of the Nubian and Arabian plates from each other (BÄCKER, 1982; THISSE et al., 1983). In this context, the sinkhole may have a two-folded importance: first, as a site where the brine (in which useful base metals are dissolved) can easily form through seawater circulation within the steep walls exposing evaporites, and second as a collector able to keep the brine itself which would otherwise be quickly dissipated.

Considering the striking geological, lithological and tectonic similarities between the Red Sea Miocene to Recent setting and the early phase of separation of the Atlantic margins, it is likely that fossil deeps which acted as sinkholes in the past (eventually enriched in base metals as the Red Sea ones) lay more or less deeply buried at places of the Eastern and western Atlantic margins where evaporites of that age has been reported.

CONCLUSIONS

1) Submarine collapse structures (sinkholes) probably due to leaching of underlying salt have been detected in many oceans where thick salt-bearing evaporites are present (Gulf of Mexico, North, Red and Mediterranean seas).

2) In order to avoid confusion between a simple morphological configuration of the ocean subbottom and the ultimate nature and genesis of these features, it is recommended to avoid as much as possible the use of a terminology strictly connected to subaerial geomorphologies.

3) Sinkholes form in a different sets of tectonic environments. The presence of diapirism seems to be an ideal pre-requisite, although exceptions to this rule have been reported.

4) The submarine sinkholes could act as mini-sedimentary basins and can be an important source for ore formation and/or preservation (young passive margin: Red Sea) as well as for hydrocarbons (mature passive margin, Gulf of Mexico).

ACKNOWLEDGEMENTS

Manuscript reviewed by M. Frignani, J. LaBrecque, R. Sartori, G.B. Vai and N. Zitellini. Contribution no. 459 of Istituto di Geologia Marina del C.N.R.

REFERENCES

Bäcker, H., 1982, Metalliferous sediments of hydrothermal origin from the Red Sea, in: P.Halbach & P.Winter eds., Marine Mineral Deposits, Verlag Gluckauf GmbH, Essen, 102-136.

Bäcker, H. & Schoell, M., 1972, New deeps with brines and metalliferous sediments in the Red Sea, Nature Phys. Sci., 240, 153-158.

Bäcker, H., Lange, K. & Richter, H., 1975, Morphology of the Red Sea Central Graben between Subair Islands and Abul Kizaan, Geol. Jb., D13, 79-123.

Belderson, R.H., Kenyon, N.H. & Stride, A.H., 1978, Local submarine salt-karst formation on the Hellenic Outer Ridge, eastern Mediterranean, Geology, 6, 716-720.

Biju-Duval, B., et al., 1982, Données nouvelles sur les marges du bassin ionien profond (Méditerranée orientale). Résultats des campagnes Escarmed, Revue de l'IFP, 37(6), 713-731.

Biju-Duval, B., et al., 1983a, Exemples de sédimentation condensée sur les escarpements de la Mer Ionienne (Méditerranée orientale). Observations à partir du submersible "Cyana", Revue de l'IFP, 38(4), 427-438.

Biju-Duval, B., et al., 1983b, Dépressions circulaires au pied de l'escarpement de Malte et morphologie des escarpements sous-marins. Problèmes d'interprétation, Revue de l'IFP, 38(5), 605-619.

Bonatti, E., Colantoni, P., Della Vedova, B. & Taviani, M., 1984, Geology of the Red Sea transitional region (22° N-25° N), Ocean. Acta, 7(4), in press.

Crane, K. & Bonatti, E., 1984, Structural analysis of the Red Sea margins using the space shuttle imaging radar (SIR-A): implications for a propagating rift, submitt. to Tectonophysics.

Erickson, A.J., Simmons, G. & Ryan, W.B.F., 1977, Review of heatflow data from the Mediterranean and Aegean seas, In: B.Biju-Duval & L.Montadert eds., Intern. Symp. on the Struct. Hist. of the Medit. Bas., Split (Yugosl.), Ed. Technip, Paris, 263-280.

Frazier, S.B., 1970, Adjacent structures of Ethiopia: that portion of the Red Sea coast including Dahlak, Kebir Island and the Gulf of Zula, Philos. Trans. Roc. Soc. London, A267, 131-141.

Goguel, J., 1978, Diapiric structures in the eastern Mediterranean Cilicia basin: comment and reply, Geology, 6(8), 452-453.

Gussow, W.C., 1968, Salt diapirism: importance of temperature, and energy source of emplacement: in: Diapirism and diapirs, AAPG Mem. 8.

Hieke, W., 1982, "Reflector M" and diapiric structures in the Ionian Sea (eastern Mediterranean), Mar. Geol., 46, 235-244.

Hsü, K.J., Cita, M.B. & Ryan, W.B.F., The origin of the Mediterranean evaporites, in: W.B.F.Ryan, K.J.Hsu et al., Init. Repts. DSDP, 13 (2), 1203-1231.

Humphris, C.C. Jr., 1978, Salt movement on continental slope, northern Gulf of Mexico, AAPG Stud. in Geol.,7, 69-85.

Kastens, K. & Spiess, F.N., 1984, Dissolution and collapse features on the eastern Mediterranean ridge, Mar. Geol., 56, 181-193.

Lohman, H.G., 1972, Salt dissolution in subsurface of British North Sea as interpreted from seimograms, AAPG Bull., 56(3), 472-479.

Lort, J.M., 1977, Geophysics of the Mediterranean Sea basins, in: The Oceans basins and margins, 4(A), Nairu, Kanes and Stehli eds., Plenum Press, New York, 151-213.

Lort, J.M. & Gray, F., 1974, Cyprus: seismic studies at sea, Nature, 248, 745-747.

Monroe, W.H., The zanjon, a solution feature of karst topography in Puerto Rico, U.S. Geol. Surv. Prof. Pap. 501-B, B126-B129.

Monroe, W.H., 1976, The karst landforms of Puerto Rico, Geol. Surv. Prop. Pap. 899, 69 p.

Pautot, G., 1983, Les fosses de la Mer Rouge: approche géomorphologique d'un stade initial d'ouverture océanique réalisée à l'aide du Seabeam, Ocean. Acta, 6(3), 235-244.

Prior, D.P. & Coleman, J.M., 1980, Sonography mosaics of submarine slope instabilities, Missisippi River Delta, Mar. Geol., 36, 227-239.

Ross, D.A. & Schlee, J., 1973, Shallow structure and geologic development of the southern Red Sea, GSA Bull., 84, 3827-3848.

Ross, D.A. & Uchupi, E., 1977, Structure and sedimentary history of southeastern Mediterranean Sea-Nile Cone Area, AAPG Bull., 61(6), 872-902.

Ryan, W.B.F., Hsü, K.J., et al., 1973, Initials Reports of the Deep Sea Drilling Project, volume XIII (Part 2), Washington, U.S. Governm. Print. Off., ix + 930 p.

Schoell, M., Bäcker, H. & Baumann, Q., 1974, The Red Sea geothermal systems.New aspects on their brines and associated sediments, in: Problems of ore deposition, Fourth IAGOD Symp., Varna, 1, 303-314.

Sigl, W., Hinz, K. & Garde, S., 1973, "Hummocky and rolling landscape" in the Ionian sea. A contribution to the "cobblestone" problem, Meteor Forsch.-Ergebnisse, 14, 51-54.

Smith, S.G., 1976, Diapiric structures in the Eastern Mediterranean Herodotus Basin, Earth Pla. Sci. Lett., 32, 62-68.

Smith, S.G., 1977, Diapiric structures in the eastern Mediterranean, Cilicia basin, Geology, 5, 705-707.

Thisse, Y., Guennoc, P., Pouit, G. & Nawab, Z., 1983, The Red Sea: a natural geodynamic and metallogenic laboratory, Episodes, 3, 3-9.

Trabant, P.K. & Presley, B.J., 1978, Orca Basin, anoxic depression on the continental slope, northwest Gulf of Mexico, AAPG Stud. in Geol., 7, 303-311.

Trusheim, F., 1960, Mechanism of salt migration in northern Germany, AAPG Bull., 44(9), 1519-540.

A brief review of the South African sinkhole problem

ANTHONY B.A.BRINK *Watermeyer, Legge, Piésold and Uhlmann, Rivonia, South Africa*

ABSTRACT

The Transvaal dolomites, and particularly those overlying the auriferous reefs of the Far West Rand where the richest gold mines in the world are situated, have a notorious reputation. Three ground water compartments have been dewatered here during the past 25 years to enable shafts to be sunk through the water-bearing dolomites and down to the gold-bearing conglomerates underlying them. The lowering of the water table, which is the base level of subsurface erosion, initiates a new cycle of subsurface erosion which results in an accelerated development of both sinkholes and dolines. Not only have 38 people lost their lives by being interred in sinkholes in the Transvaal, but damage to buildings and other structures, largely as a result of doline development, has amounted to tens of millions of rand. Research during the sixties showed that sinkholes develop as a result of successive arch-collapse in the soil overlying solution-widened joints which lead down to large caverns, and that dolines result from consolidation of highly compressible manganese dioxide "wad" following water table lowering. As geophysical techniques failed to predict the location of subsurface arch-collapse, the telescopic benchmark (TBM) was developed as a safety device to give timely warning of sinkhole development. Sinkholes may develop in dolomites originating from any of the tidal zones of the epeiric Proterozoic basin in which they were deposited, but dolines are largely confined to the manganese-rich dolomites of the subtidal zone.

INTRODUCTION

Figure 1:

The Transvaal Province of South Africa

Accelerated sinkhole development has taken place during the past 25 years in the Transvaal Province of South Africa (Figure 1).

Dolomite occupies six thousand square miles in the Transvaal, but accelerated sinkhole development is confined to the goldmining area of the "Far West Rand" - the richest gold mines in the world are situated here (Figure 2). Gold is mined from the Witwatersrand conglomerates at a depth of about one kilometre or more beneath the surface. In large parts of the Far West Rand the water table has been lowered through the dolomites in order to facilitate sinking of shafts down to the gold-bearing ores.

It is in the dewatered ground water compartments that catastrophic sinkholes have caused loss of life and damage to property in the last two and a half decades. In December 1962, a three-storey crusher plant building was swallowed into a sinkhole on the West Driefontein Mine. None of the twenty-nine occupants of the building was seen again.

Two years later a family of five were buried in a sinkhole which swallowed their house during the night. There was nothing to be seen of their house after the collapse. Nor indeed of Mr. Oosthuizen's rose garden on which, we subsequently learnt, he sometimes used to leave the hosepipe running all night.

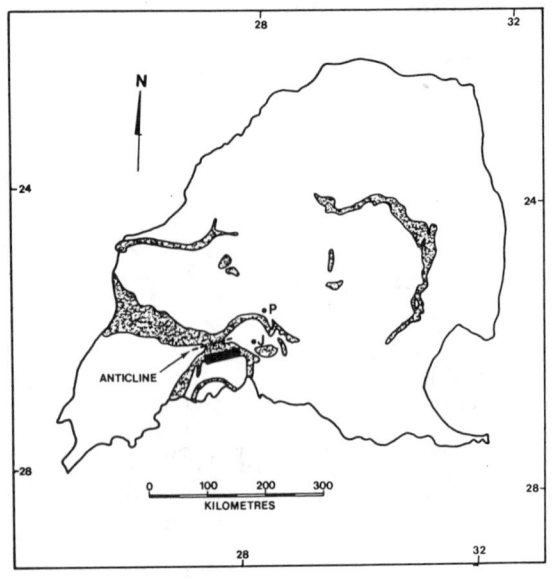

Figure 2: The distribution of dolomite in the Transvaal.
P = Pretoria   J = Johannesburg
Black rectangle = the Far West Rand

One Sunday afternoon in 1970, Mr. Nortjie sat at a pavilion watching four friends play a game of tennis. His friends heard a loud noise like a pistol shot, and when the cloud of dust had subsided over the sinkhole, there was nothing to be seen of part of the pavilion or of their companion.

Although few have been catastrophic, in the sense of causing loss of life, hundreds of sinkholes have developed in the dewatered dolomitic "compartments" during the past twenty-five years.

The other type of surface subsidence encountered in the area is the doline. It develops as a gradual subsidence of the surface over a period of years. Dolines may be several hundred metres in diameter.

The Mechanism of Sinkhole and Doline Development

During the early sixties we embanked upon a major research project to determine the mechanism of sinkhole and doline development, in an attempt to make predictions and provide warning systems. This research involved exploring the geometry of a great many sinkholes and caves.

In some cases it was possible to enter a near-vertical "slot" (or grike) beneath a sinkhole. In many cases we found that the slots led down to large caverns which had developed in the phreatic zone beneath a long-static water table.

The dolomite is divided into water-tight ground water compartments by a series of intrusive igneous dykes. In the Far West Rand three such compartments, numbered 1, 2 and 3 in Figure 3, were dewatered during the fifties, sixties and seventies respectively; it now seems possible that the compartment numbered 4 may be dewatered during the eighties.

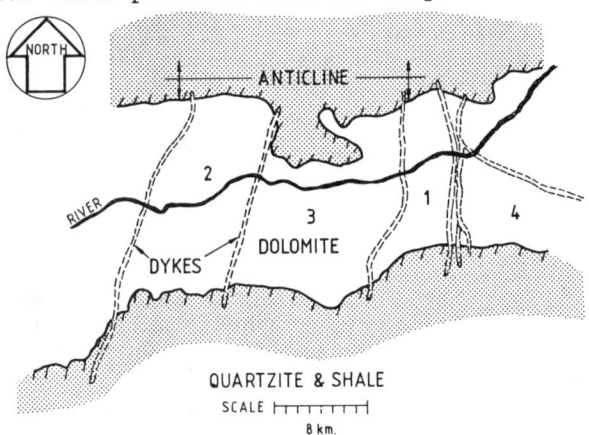

Figure 3: Ground water compartments in the Far West Rand:

1 = Venterspost

2 = Oberholzer

3 = Bank

4 = Gemsbokfontein

The first large sinkhole to appear after the commencement of dewatering in the Venterspost Compartment was one hundred metres in diameter and forty metres deep: surely one of the largest sinkholes in the world. At the bottom of this sinkhole we could see a small "slot" or "chimney" between two pinnacles of bedrock. It was difficult to understand how more than three thousand cubic metres of soil had apparently disappeared through this small opening in a matter of seconds.

But further research subsequently led us to an understanding of the mechanism of sinkhole development as a result of successive arch collapse (Figure 4). After lowering of the water table, water from perhaps a leaking pipe or canal will find its way down a slot and will erode the soil within the slot into the underlying cavern. Successive collapse of the arch so-formed above the abutments of the slot will eventually break through to the surface as a sinkhole.

The mechanism of doline development as a result of lowering of the water table involves a gradual subsidence of the surface as consolidation takes place in the "wad" and other highly compressible materials occupying deep subsurface valleys (Figure 5). These subsurface valleys, or "bogaz" features, generally represent solution-widened fault zones. Wad, a highly compressible black material, is insoluble manganese dioxide which is left as a residue when manganese-rich dolomite dissolves.

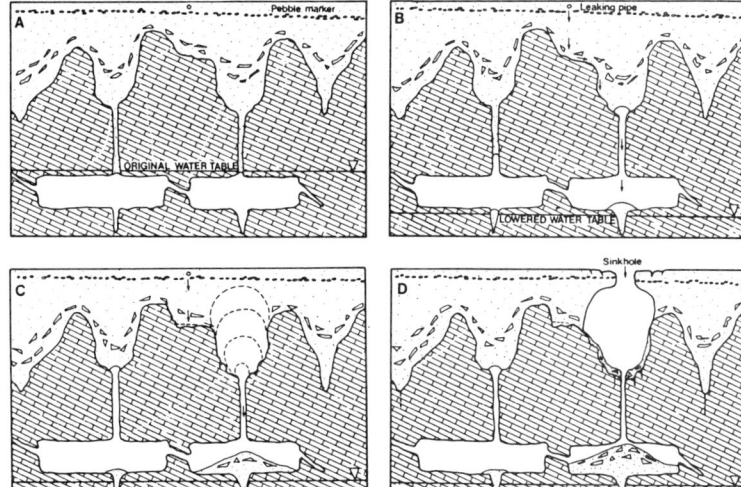

Figure 4: The mechanism of sinkhole development:

"A" shows the equilibrium situation before the lowering of the water table and "B" the position after lowering of the water table. There is active subsurface erosion, and the slot is flushed out by a process of headward erosion. "C" shows the progressive collapse of the roof of the vault which has formed above the "slot", possibly arrested temporarily by the ferruginised pebble marker. "D" shows the collapse of the last arch to produce a sinkhole surrounded by concentric tension cracks (Brink, 1979).

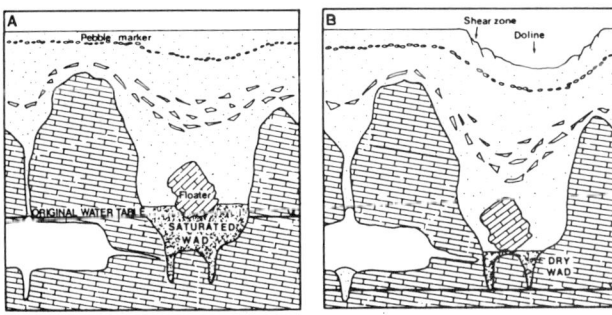

Figure 5: The mechanism of doline development:

"A" shows the equilibrium situation before the lowering of the water table. The palaeo-doline is not apparent at the surface but is indicated by sagging in the chert rubble and the pebble marker. "B" is the position after lowering of the water table. Reactivated doline-development becomes apparent when the consolidation of wad leads to surface subsidence. The periphery of a doline is characterised by a shear zone and tension cracks (Brink, Partridge & Williams, 1982).

A number of geophysical methods have been tried out in an attempt to predict locations in which sinkholes or dolines may develop. With the exception of the gravimeter method they have all proved useless. The gravimeter had proved useful in determining the depth to solid bedrock, and hence the localities of "bogaz" zones in which dolines may occur if the water table is lowered: but it does not indicate where sinkholes are likely to occur.

The TBM: a Warning Device

In an attempt to provide a warning device to indicate whether a sinkhole was developing beneath any existing structure, the telescopic benchmark (TBM) was devised. This consists of a number of rods or pipes cemented into boreholes at different depths. To save money these are generally placed at different depths within one borehole (Figure 6). If observations show that the length of the borehole is increasing with time, this indicates that a sinkhole is developing. If the borehole length is decreasing with time, a doline is developing.

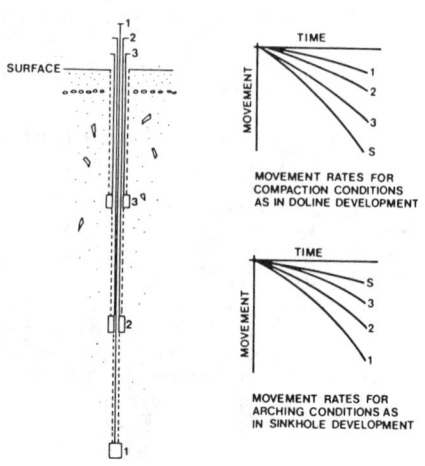

Figure 6: Telescopic benchmarks give early warning of the development of a doline or a sinkhole (Jennings, 1966).

Enviroments of Deposition of Dolomite

The enviroments of deposition of the dolomites have some significance in relation to the engineering behavior of the strata.

The dolomites were deposited in Early Proterozoic times, about 2 200 million years ago, in an epeiric basin: a shallow, tropical, inland sea. Certain of the strata were deposited in the relatively deep waters of the subtidal zone (Figure 7). Here the algal colonies which thrived in the tropical waters developed very large dome-shaped stromatolites, and the carbonates which were deposited contained a high percentage of manganese. As the water was alkaline no chert was deposited. In the intertidal zone, however, wave action prevented the formation of large stromatolites so here we find only small ones; and the admixture of fresh river water decreased the alkalinity so that chert was also deposited as layers within the dolomite. In the supratidal zone deposition took place in very shallow water and the chert-rich dolomite is characterised by the presence of evaporites, such as gypsum and anhydrite.

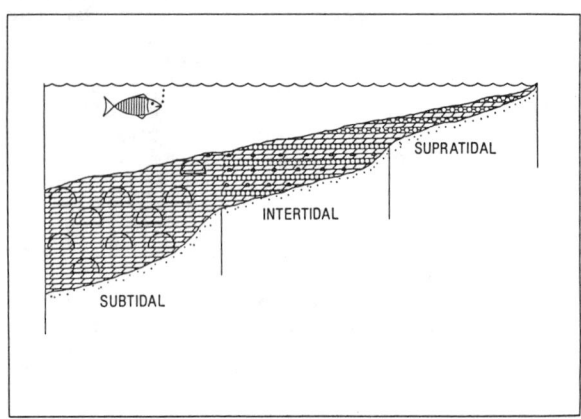

Figure 7: Environments of deposition of the dolomites in an epieric Proterozoic sea.

The stratigraphic column of the dolomite succession (Figure 8) shows that the thin Oaktree Formation at the base had a mixed, or rapidly changing environment; the thick Monte Christo Formation, rich in chert and with small stromatolites, was deposited in an intertidal environment; the Lyttelton Formation was subtidal, and the Eccles Formation supratidal.

Sedimentologists study similar enviroments today, for example in the Bahamas, at Abu Dhabi in the Persian Gulf, and at Shark Bay, Australia (Tankard et al, 1982). Their studies magnificently illustrate, once more, the truth of the Uniformitarian Principle - "the present is the key to the past".

The subtidal Lyttelton Formation with its large stromatolites, lack of chert, and high manganese content provides a good source of pure dolomite for use as concrete aggregate. When it weathers, however, the high manganese content gives rise to the formation of compressible wad which makes the formation highly susceptible to doline formation. Perhaps the past is also the key to the present?

But sinkholes have been found to develop in any of the dolomitic formations, regardless of their environment of deposition. and although engineers have made spectacular advances in devising ways of building more safely on dolomite (Wagener, 1982), the prediction of future sinkhole locations still remains largely unsolved.

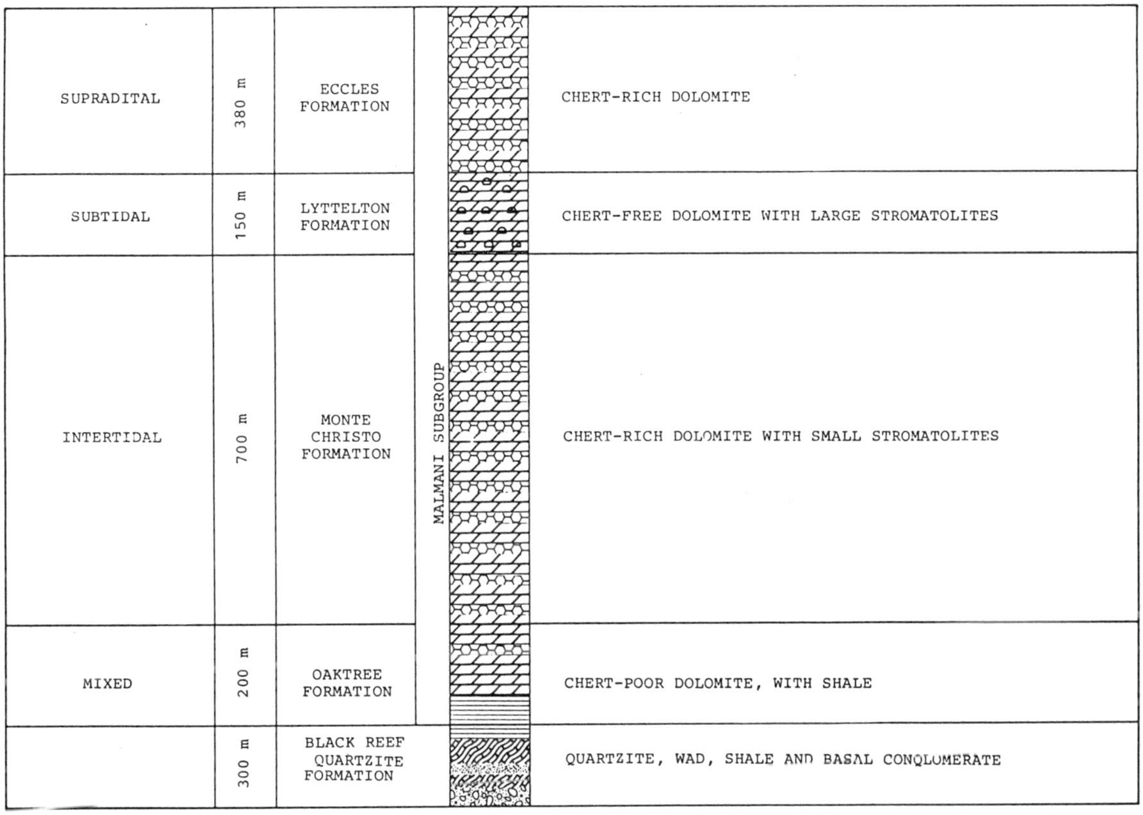

SUPRADITAL	380 m	ECCLES FORMATION		CHERT-RICH DOLOMITE
SUBTIDAL	150 m	LYTTELTON FORMATION		CHERT-FREE DOLOMITE WITH LARGE STROMATOLITES
INTERTIDAL	700 m	MONTE CHRISTO FORMATION	MALMANI SUBGROUP	CHERT-RICH DOLOMITE WITH SMALL STROMATOLITES
MIXED	200 m	OAKTREE FORMATION		CHERT-POOR DOLOMITE, WITH SHALE
	300 m	BLACK REEF QUARTZITE FORMATION		QUARTZITE, WAD, SHALE AND BASAL CONGLOMERATE

Figure 8:   Stratigraphic column of the dolomites in the Far West Rand.

References

Brink, A.B.A. (1979). Engineering Geology of Southern Africa. Volume 1. Building Publications, Silverton.

Brink, A.B.A., Partridge, T.C. and Williams, A.A.B. (1982). Soil Survey for Engineering. Oxford University Press.

Jennings, J.E. (1966). Building on dolomites in the Transvaal. Transactions of the South African Institution of Civil Engineers, Volume 8, Number 2, pp. 41-62.

Tankard, A.J., Jackson, M.P.A., Eriksson, K.A., Hobday, D.K., Hunter, D.R. and Minter, W.E.L. (1982). Crustal Evolution of Southern Africa. Springer-Verlag.

2. Site studies and evaluation of sinkhole-susceptibility

Catastrophic subsidence, Shelby County, Alabama

PHILIP E.LAMOREAUX *P.E.LaMoreaux and Associates, Inc., Tuscaloosa, Alabama, USA*

ABSTRACT

Recent subsidence and collapse of the land surface in at least three areas underlain by carbonate rocks in Jefferson and Shelby Counties, Alabama, have dramatically demonstrated the need for greater understanding of the causative mechanisms involved. Buildings, highways, utility lines, oil and gas pipelines, and constructions of all types are threatened by these collapses. In many cases, owing to the high cost or lack of sufficient technology, repair of the resulting damage is not feasible. It is far more economical and much safer to apply corrective measures during the planning and development of projects in carbonate terranes than to wait until a problem manifests itself at the surface. Subsidence and collapse are not new to Shelby County, but subsidence activity has intensified in recent years and has caused problems to transportation, communication, pipelines, and water supplies.

Techniques that are currently being used to study the source, occurrence, and movement of ground water in karstic areas include a great variety of remote sensing techniques, satellite and high altitude imagery, high and low level air photography, as well as sonar, down-hole television cameras, deep-well current meters, down-hole pH samplers, conductivity (profile) traverses, and a variety of geophysical and mechanical logging devices. These investigations are directed at defining geological structure, evaluating the degree and trend of jointing and faulting; locating and determining the extent and size of solution cavities; discovering concealed discharge areas; and studying the directions and rates of flow of waters with depth, and associated changes in temperature and chemical composition of ground waters in carbonate rocks.

This paper describes in detail some of the remote sensing methods used for these investigations. In each instance, for best results, the more exotic studies are combined with sound standard hydrogeological investigations because of the complex nature of occurrence of ground water in carbonate rocks.

This paper describes in some detail some of the principal land-use problems studied by remote-sensing techniques including: foundation, water-supply development, mining, agricultural activity, location and construction of dams, oil and gas pipelines, highway construction and maintenance, disposal of solid and liquid wastes, and land-use in ground-water discharge areas. Sinkhole formation in areas of extreme dewatering for mining have been studied extensively in South Africa (Jennings, 1966), Pennsylvania, U.S.A. (Foose, 1953), and Alabama, U.S.A. (Powell and LaMoreaux, 1969; Newton and others, 1973).

Introduction

On December 2, 1972, Hershel Byrd, a rural resident of Shelby County, Alabama was startled by a house-shaking rumble and the sound of breaking trees. Two days later, a crater 325 feet long, 300 feet wide, and 120 feet deep was found. This giant collapse, and associated features, has been studied in great detail over the past years. One of the most useful remote-sensing methods is low-altitude color infrared photography. Satellite imagery and high altitude photography have also been used to identify major regional structures and lineaments. More conventional air photographs, including photographs taken as low as 500 feet above land surface, have been used for detailed mapping of geologic features, subsidence, drainage alignments, vegetation stress and correlation with the surface mapping of geologic units, faults, jointing, and the development of subsidence features associated with solution cavities.

Based on ground reconnaissance and aerial photographs, over 2,000 collapses or related features have formed in a 16-square-mile area in Dry Valley near Calera, Shelby County, Alabama. Most of these subsidence features in Dry Valley are underlain by the Lenoir, Newala, and Longview limestones in the valley and the Chepultepec and Copper Ridge dolomites of the Knox Group underlie the low, eroded ridges where the large collapse --- "Golly Hole" or "December Giant" --- is located.

Catastrophic collapses or subsidence, can be related to many natural phenomena such as heavy rainfall, large seasonal fluctuations in the water-table, earthquakes, or other changes in the hydrogeologic regime affecting residuum stability. Man-imposed effects such as artificial drainage, dewatering, seismic shocks (blasting), break in water or sewage pipes, or watering by irrigation may result in a collapse.

Sequential air photographs taken at regular intervals are an invaluable tool in tracing the history of subsidence problems. They provide a basis for interpretation of geologic, hydrologic, and climatologic features, as well as man-induced events.

Use of Remote Sensing Methods (LaMoreaux, 1984)

A broad definition of remote sensing includes all methods of collecting information about an object without being in physical contact with that object. However, a more restrictive definition is used; it includes only those methods that employ electromagnetic energy, including light, heat, and radiowaves, as means of detecting and measuring target characteristics (Sabins, 1978). The major types of remote sensing used in carbonate hydrology are aerial photography, satellite imagery, thermography, and radar. Sonar, down-hole television cameras, and other remote sensing techniques are also used by carbonate hydrologists.

The determination of the optimum remote sensing band or band ratios (i.e., range of detected wave lengths) and the type of remote sensing to be employed depends on the objective of the study and what features are sought to be enhanced. This optimum band or band ratio selection can be done by statistical methods (which generally require use of a computer) or by manual techniques such as the coincident spectral plot methods described by Shourong (1982).

The steps in aquifer mapping using remote sensing as listed by Moore (1980) are: 'image analysis, image interpretation, geologic interpretation, and groundwater interpretation. Image analysis consists of objective detection, classification, delineation, and identification of land cover and physiography. Image interpretation is the visual and subjective mapping of landforms, drainage characteristics, lineaments, and curvilinears. A geologic interpretation begins with surficial lithology and structure, and then proceeds to surficial geomorphic processes, subsurface geologic relationships, and geologic processes. A groundwater interpretation builds on the conceptual geology by inferring aquifer characteristics and water quality.'

Information derived from the use of these techniques is particularly valuable in the study of carbonate terranes and the problems associated with subsidence and collapse. Moreover, because large areas can be examined in a with very short period of time, remote-sensing technology can be viewed as a time-saving tool. Possible application include: (1) inventory of sinkholes; (2) monitoring sinkhole development; (3) mapping sinkhole alignments; (4) investigating the relationships among sinkhole development groundwater movement, fracture traces, and lineaments; (5) preparing and updating base maps; (6) delineating incipient collapse zones; (7) detecting areas of abnormal surface drainage; (8) mapping submarine and surface karst springs; (9) locating potential water well sites; (10) mapping regional geologic structures; (11) locating exposures of bedrock; and (12) aiding in general project planning.

Satellite imagery and high-altitude photography have been used to locate major regional structure and lineaments in a study of a sinkhole prone area in Alabama, USA (LaMoreaux, 1979). More conventional air photographs, and photographs taken at 500 feet above land surface, have been used for detailed mapping of geologic features, subsidence, drainage alignments, and vegetation stress and for correlation with the surface mapping of geologic units, faults, jointing, and the development of solution cavities.

Photographs taken at regular intervals are an invaluable tool in tracing the history of subsidence problems. Often inaccurate information, based on personal interviews, can be eliminated by a complete photographic history. Thus, historic remotely sensed data can provide information on subsidence during past, as well as present active stages.

The pattern of subsidence offers many clues to the causes of its occurrence in one area as opposed to other nearby areas and may give a direct indication of underlying geologic structures. Distribution of sinkholes also can be used effectively in choosing the best location for a well of large capacity in a karstic area.

Surface drainage features are an important aspect of the formation of recent sinkholes because surface water, moving through the overburden into bedrock openings, aids in the process of forming sinkholes. Black and white infrared photography or thermography taken during the leafless season and after rainfalls will reveal where water is entering the subsurface through jointing or sinkhole collapses. This will save considerable field time in locating and plotting the exact location of surface-water loss.

Previous investigations have shown the usefulness of remotely sensed data in detecting changes in the soil moisture content and surface temperature anomalies that are caused by subsurface development of uncollapsed cavities (Coker, Marshall, and Thomson, 1969; Newton, Copeland, and Scarbrough, 1973).

Voids in the residual clays that have little or no surface expression can sometimes be detected by specific signatures on photographs that record vegetative vigor or lack of vigor. In open fields, circular vegetative patterns indicate that water may be concentrated on the surface due to subsidence over a subsurface cavity, or that water is draining into openings in the ground. The vegetation may show increased vigor due to extra water. Vegetative anomalies may also result from evaporation beneath the land surface where cavities in residuum have progressed upward into the root zone. The borders of these circular vegetative patterns sometimes appear as ring-like features. Trees may show a lack of growth vigor as a result of subsurface collapse exposing their root zones to excessive evaporation. Such a collapse could also cause trees to topple and die. Subsurface voids in the residuum beneath or near a tree could cause a decrease in moisture in the soil that would be detected by vegetation stress and thus recorded on infrared film. Some correlation has been observed on color-infrared photographs between pine tree "kills" by insects and the occurrence of collapses in Shelby County, Alabama, USA. The death or weakening of trees caused by collapses or the formation of subsurface voids in root zones results in infestation by beetles or other insects because of their affinity for attacking trees in a stressed condition.

Formation of sinkholes in areas of extreme dewatering for mining has been studied extensively in South Africa (Jennings, 1966), Pennsylvania, U.S.A. (Foose, 1953), and Alabama, U.S.A. (Powell and LaMoreaux, 1969; Newton and others, 1973). Remote-sensing is an excellent beginning tool for hydrologists and geologists to better understand the perplexing problems of ground water exploration, development, stream flow analysis, dam building, highway construction, and water pollution in limestone terranes throughout the world. Some of the most useful remote sensing techniques used in the Shelby County where extensive dewatering for mining has caused catastrophic subsidence are described as follows.

Thermal Infrared Imagery
Thermal infrared imagery and infrared photography have been useful in predicting the development of active subsidence along proposed highway routes in Jefferson County. Photographs obtained during and after a study of subsidence problems in and near Roberts Industrial Subdivision, Birmingham, Alabama (Newton and Hyde, 1971) indicated that infrared photography is most useful in locating areas of potential collapse. Panchromatic, color, and infrared photography has been used effectively to locate fracture traces. These fractures control the development of solution cavities which become the preferential flow zones for movement of ground water.

In Florida, Coker, Marshall and Thomson (1969) used multispectral scanning equipment, several types of photography, and a special-purpose computer to detect stressed vegetation and terrane-surface temperature anomalies that were associated with areas of impending collapse. They concluded that such an approach could potentially detect hydrogeological phenomena as much as 100 feet below the land surface. Matalucci and Abdel-Hady (1969) pointed out the usefulness of infrared surveys in the delineation of highway construction problems in carbonate terranes. In Pennsylvania, Lattman and Parizek (1964) investigated the relationship between solution cavities and photogeologic fracture traces in a carbonate area within the Valley and Ridge physiographic province. In each of these examples a relationship was found between ground-water movement and fracture traces.

Imagery, especially the near-infrared band, is an excellent aid in regional structural interpretation. Powell, Copeland and Drahovzal (1970) found this also to be true of Apollo 9 photography in the same area in eastern Alabama. The Shelby County sinkholes do not show up as discernible features, as the diameter of the largest sink (about 300 feet) approaches the limit of the smallest object that the M.S.S. can detect; however, large-scale lineaments, to which solution activity may be related, do show up well, as do regional geologic structural trends. The Shelby County studies related regional structure features to site specific features such as dip, strike, jointing, stream alignments, trends of sinkholes, stream alignments, and quarry joint studies.

Hydrogeology of Dry Valley Area

Dry Valley is within the Cahaba Valley District of the Valley and Ridge physiographic province which is characterized by northeast-southwest trending valleys and ridges. The Cahaba Valley was formed by differential erosion of folded and faulted rock formations composed primarily of limestone and dolomite.

Rock formations underlying the Dry Valley area outcrop in northeast-southwest trending parallel bands. The rocks dip to the southeast at 20 to 60 degrees and range in age from Cambrian to Mississippian. From northwest to southeast, the rock formations include the Copper Ridge Dolomite of Cambrian age; the Chepultepec Dolomite, Longview Limestone, Newala Limestone, Lenoir Limestone, and Athens Shale of Ordovician age; the Chattanooga Shale of Devonian age; and the Fort Payne Chert and Floyd Shale of Mississippian age.

The Copper Ridge and Chepultepec Dolomites form the western boundary of Dry Valley and support a stream-dissected ridge that is locally more than 100 feet above Dry Creek. The valley of Dry Creek is underlain, for the most part, by the Longview and Newala Limestones. The Newala is mined from recessed quarries and underground mines in the valley as a source of raw material for the manufacture of cement. The Athens Shale and Fort Payne Chert --- a sinuous, narrow ridge --- forms the eastern boundary of the valley.

A mantle of unconsolidated material consisting mainly of residual clay covers the carbonate bedrock in the area, obscuring surface exposures of geologic contacts and faults. This unconsolidated material, or residuum, resulting from the solution of underlying carbonate rocks, commonly contains varying amounts of insoluble chert debris. Some of this unconsolidated material is carried by water into openings in bedrock and fills solutionally-enlarged fractures and other openings underlying the valley floor. The buried contact between the residuum and the underlying bedrock, because of differential solution, is highly irregular. Pinnacles of bedrock extending upward into the residuum, and boulders of "floating" rock within the residuum, are common and may easily be mistaken for the bedrock surface.

Studies in the Dry Valley area are concerned primarily with ground water that occurs under water table conditions, the configuration of which conforms, in general to topography, but is effected by geologic structure and variations in the rates of recharge and discharge by groundwater withdrawals.

The generalized hydrologic conditions that would characterize Dry Valley under natural conditions would be one in which ground water moves from higher land surface areas to lower elevations. Where the water table intersects land surface, ground water discharges into streams, lakes, and ponds. During the "dry season" ground water discharge maintains the flow of streams. Streams that do not intersect the water table are intermittent and flow only during and for a short time after, heavy rainfall periods.

The hydrogeologic conditions that presently occur in the Dry Valley are characterized by a water table that has been lowered by extensive groundwater withdrawals. All surface water runoff flows into Dry Creek, which discharges into Spring Creek. Natural drainage patterns have been extensively modified by road construction and mining operations. Dry Creek is intermittent north of Shelby County Highway 16 because of the small drainage area and downward infiltration of water from the main channel to the water table. A tributary of Dry Creek which originates from Simpson Spring and an impoundment on the Floyd Shale, has appreciable flow during the "dry season". This tributary is at the contact between the Athens Shale and Fort Payne Chert and is separated from Dry Valley by the impermeable shale. Discharge measurements made by the U.S. Geological Survey in October 1973 (low flow period) indicated the runoff from Simpson Spring and the impoundment at 0.07 cfs (cubic feet per second).

South of Highway 16, flow is maintained in Dry Creek by discharge of ground water pumped from Dry Valley. Discharge measurements made by the U.S. Geological Survey at Dry Creek on Shelby County Highway 23 in October 1973, indicated the discharge was about 32 cfs. At the time of this measurement, the main channel of Dry Creek north of Highway 16 was dry. Therefore, all natural runoff originating in the Dry Valley drainage basin does not flow through the basin as surface water. Instead, streams in the area commonly discharge directly into sinkholes. Runoff from the Simpson Spring tributary of approximately 5 cfs on March 4, 1977, was discharging into a recent collapse near County Highway 16 in the SE¼ NW¼, Section 18, T. 22 S., R. 2 W. On April 3, 1977, the runoff was observed to be flowing into a second collapse farther upstream at about 6 cfs. Both collapses have been filled, but are in the process of reopening.

Surface water discharging into sinkholes in the area, enters the solution cavity system in the underlying carbonate rocks, and from this part of the karst system is pumped by dewatering wells to Dry Valley, and is subsequently discharged back into Dry Creek south of Highway 16.

Groundwater withdrawals amounted to more than 14,000 gpm (gallons per minute), or about 31 cfs, approximately the same discharge measured in Dry Creek at Shelby County Highway 23. Water-table maps for the area show that a cone of depression from mine dewatering has formed beneath that part of the Dry Valley area.

The water-table decline in this area is the result of extensive groundwater pumping from wells, recessed quarries, and underground mines in Dry Valley. Water levels at the center of the cone of depression are more than 400 feet below land surface, the center of the cone of depression corresponding closely with the location of the deepest underground mine in the area.

As a result of the extensive water-table decline, all wells north of Alabama Highway 25 that are less than 100 feet deep in Dry Valley, are dry. Previously perennial springs west of Dry Valley along Spring Creek Road are reported to flow only during the "wet season". This is the result of a cone of depression expanding during dry periods, causing downward percolation of water from the Spring Creek basin.

Sinkhole Occurrence in the Dry Valley Area

Catastrophic or collapse sinkholes in Dry Valley are near the center of a cone of depression and are associated with a decline in the water table. The development of sinkholes in Dry Valley has increased since 1967 and several thousand have occurred to date in the area. They range in general diameter from 2 to 100 feet and in depth from a few to 30 feet. The "big sink" which occurred in December 1972 and is approximately 325 feet long, 300 feet wide, and 120 feet deep is not in the valley but in the ridge to the northwest underlain by dolomite.

The western margin of Dry Valley is underlain by the Copper Ridge and Chepultepec Dolomites while the interior is underlain by the Longview, Newala, and Lenoir Limestones. It is in the area underlain by the fractured Longview Limestones that the majority of sinkholes, and the large collapses, have occurred. In areas underlain by other formations, sinkholes are generally smaller and less numerous. The disparity in size of those occurring in the Longview Limestone and those in the rest of Dry Valley appears to be related to differences in thickness of the overlying unconsolidated material as well as to bedrock lithology and solution-cavity enlargement along fractures. In addition, outside the area of the Longview Limestone it is also apparent that other factors have contributed to their development. Twenty-eight of the 33 sinkholes identified just north of Highway 16 are in areas where timber has been removed or in areas near drains along the Highway, causing a change in surface drainage characteristics.

Several collapses occurred on and adjacent to the road during and immediately after a period of heavy rainfall in the spring of 1977, leading to the closing of the Highway. In the valley, however, by discharge of surface water into the sinkholes, they are enlarged laterally through erosion of unconsolidated material forming their sides. The continuation of this process will result in much larger holes, but it is unlikely that sinkholes as large as the "big sink" will occur in the interior of Dry Valley.

In general, there are two types of sinkholes in the area. One is the shallow, small-diameter hole that may be successfully filled with crushed rock; the other is the large, catastrophic sinkhole that may be hazardous to transportation, communication and transmission facilities, property, and lives. The potential for each must be identified and proper planning and precaution taken to guard against a catastrophic event that would cause unnecessary deaths or damage to property.

References

Alabama Geological Survey and the U.S. Geological Survey: Published reports and files, Tuscaloosa, Alabama.

Allum, J.A.E., 1973, Infrared color aerial photography in mineral exploration shows promise as experience in interpretation grows. Northern Miner., V. 58, No. 51, page 35.

Cato, G.A., 1972, Basic principle of earth resource sensors, in: Shahrokri, F., ed., Remote Sensing of Earth Resources. Tennessee University Space Inst., pp. 64-82.

Coker, A.E., Marshall, R. and Thomson, N.S., 1969, Application of computer processed multispectral data to the discrimination of land collapse (sinkhole) prone areas of

Florida, in: 6th Internat. Symposium on Remote Sensing of Environment, Proc. 1: Michigan Univ., Inst. Sci. and Technology, pp. 65-77.

Foose, R.M., 1953, Ground water behaviour in the Hershey Valley, Pennsylvania. Geol. Soc. America Bull., 64, pp. 623-645.

Graves, Stanley, Graves Well Drilling Company, Sylacauga, Alabama.

Jennings, J.E., 1966, Building on dolomites in the Transvaal. The Civic Eng. in South Africa, V. 8, No. 2, pp. 41-63.

LaMoreaux, P.E. and others, Guide to the hydrology of carbonate rocks. Unesco, 1984.

LaMoreaux, P.E. and Associates, Inc.: Unpublished reports to the Southern Natural Gas Company, Tuscaloosa, Alabama.

Lattman, L.H. and Parizek, R.R., 1964, Relationship between fracture traces and the occurrence of ground water in carbonate rocks. Jour. Hydrol., V. 2, pp. 73-91.

Matalucci, R.V. and Abdel-Hady, M., 1969, Surface and subsurface exploration by infrared surveys in: Remote Sensing and its Application to Highway Engineering, Highway Research Board Spec. Rept. 102, Washington, D.C. pp. 1-12.

Newton, J.G., Copeland, C.W. and Scarbrough, W.L., 1973, Sinkhole problems along proposed route of Interstate Highway 459 near Greenwood, Alabama. Alabama Geol. Survey Circ., 83, 63 pages.

Newton, J.G. and Hyde, L.W., 1971, Sinkhole problem in and near Roberts Industrial Subdivision, Birmingham, Alabama: a reconnaissance. Alabama Geol. Survey Circ., 68, 42 pages.

Newton, John, Hydrologist, U.S. Geological Survey, Tuscaloosa, Alabama.

Powell, W.J., Copeland, C.W. and Drahovzal, J.A., 1970, Delineation of linear features and application to reservoir engineering using Apollo 9 multispectral photography. Alabama Geol. Surv. Inf. Serv., 41, 37 pages.

Powell, W.J. and LaMoreaux, P.E., 1960, A problem of subsidence in a limestone terrane at Columbiana, Alabama. Alabama Geol. Surv. Cir., 56, 30 pages.

Sabins, F.F. Jr., 1978, Remote sensing. San Francisco, W.H. Freeman, 426 p.

Shourong, Shu, 1982, The coincident spectral plot method for selecting the remote sensing bands of carbonate rocks. Carsologica Sinica, V. 1, no. 2, p. 158-166.

Southern Natural Gas Company: Infrared aerial photography, furnished by, Birmingham, Alabama.

Sinkhole development in North-Central Puerto Rico

MIKOLAJ WEGRZYN & ALEJANDRO E.SOTO *University of Puerto Rico, Mayaguez, USA*
JUAN A.PÉREZ *US Army Engineering Topographic Laboratory, Fort Belvoir, Virginia, USA*

ABSTRACT

Extensive karst formations cover approximately 125 km (80 miles) of Northern Puerto Rico from the capital of San Juan to the west coast town of Aguadilla. Near the coast the limestone formations are overlain by blanket sands and alluvium. Numerous sinkholes develop continuously in the blanket sands, causing serious damage to housing development and industrial plants located in this region. The study of the sinkhole related terrain failures was conducted in 1982-83 by authors as a project of the Research Center, University of Puerto Rico. Two most devastating cases were of particular interest: one resulting in the total loss of a one-family house and the second in multiple sinkhole collapse on the grounds of a pharmaceutical plant. The study methods comprised aerial photography, reconnaissance examination of geological, hydrogeological and topographical maps, inventory of selected sinkholes, borings and laboratory soil testing, seismic refraction and electric resistivity survey. It was found that the development of the sinkholes in the blanket sands is directly related to the drainage of rain water through this sand layer to underlying cavernous limestones. Groundwater level fluctuations even sudden and large, have no bearing on the sinkhole development in the area of study because of the great depth (60 + meters) of the water table. The potential of sinkhole related terrain failure was evaluated and recommendations were given on how to deal with this problem in future.

Regional Setting

The island of Puerto Rico features complex physiographic and geologic settings. The three main geographic regions consist of the mountainous area in the central part of the island, extensive karst belt at the north and northwest, and a discontinuous fringe of coastal plains (Monroe 1976), Fig. 1.

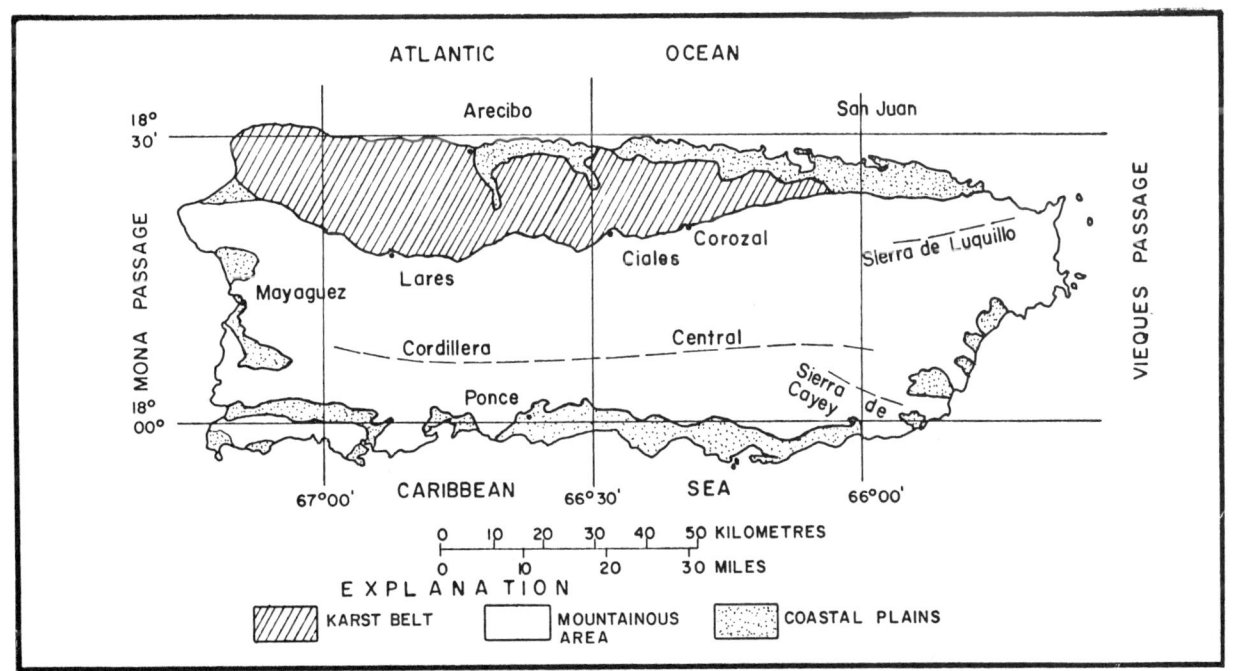

Figure 1. Principal physiographic divisions of Puerto Rico (after Monroe, 1976).

Numerous sinkholes, causing serious damage to housing developments and industrial plants, develop continously in the blanket sands overlying the limestone formations of the karst belt.

The karst features have developed on Oligocen to Miocene limestone formations which rest unconformably on volcanic, volcaniclastic and intrusive rocks of lower Cretaceous to Eocene age. Five limestone formations are recognized. From youngest to oldest they are (thickness range given in parentheses; Monroe 1976):

$$\begin{array}{ll}\text{Camuy Formation} & (\ \ 0{-}170\text{ m}) \\ \text{Aymamón Limestone} & (190{-}200\text{ m}) \\ \text{Aguada Limestone} & (\ 70{-}110\text{ m}) \\ \text{Cibao Formation} & (250{-}280\text{ m}) \\ \text{Lares Limestone} & (\ \ 0{-}280\text{ m}) \end{array}$$

The entire sequence dips north at angles of 6° or less (Fig. 2). Near the coast the limestone formations are overlain by blanket sands and alluvium. The most spectacular and abun-

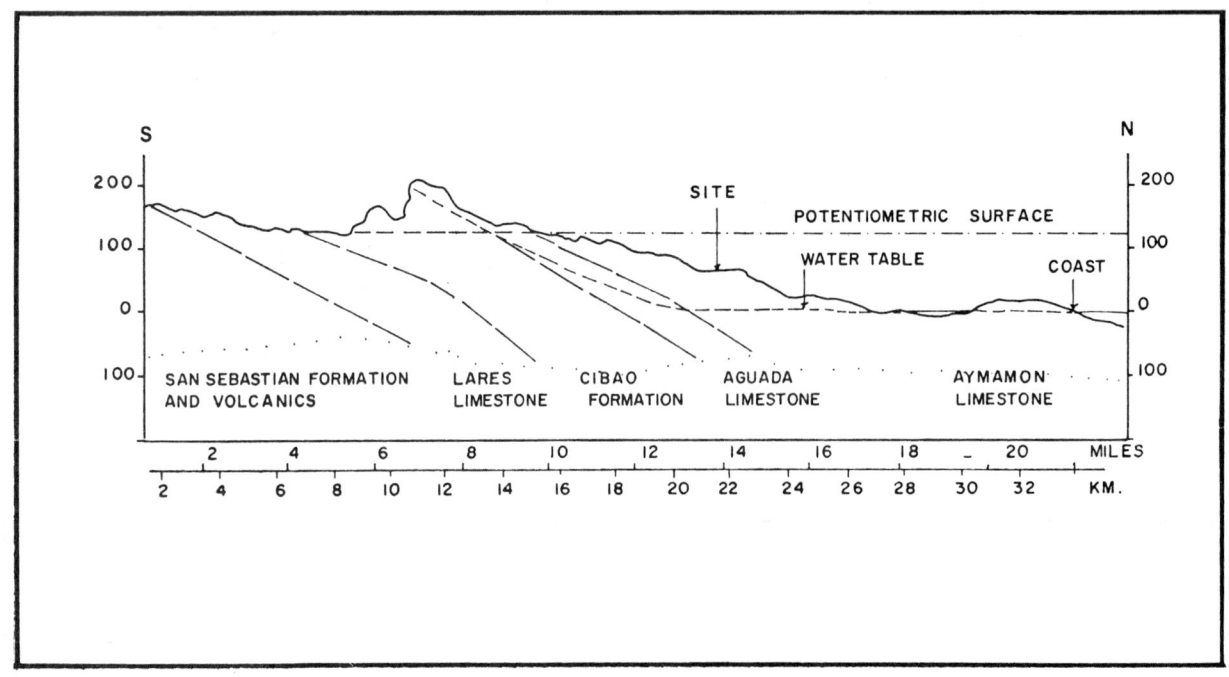

Figure 2. North-South cross section through the middle Teriary limestone formations, showing the position of groundwater table and the site of the study area (after Giusti and Bennet, 1976).

dant karst features occur in the Lares, Aguada and Aymamón formations. These include closed depressions, conical hills, numerous caves, sinkholes and other solution features.

Drainage in the limestone outcrop area is predominantly underground. The runoff is quickly transferred to several aquifers by numerous solution-enlarged passageways. Several large rivers which originate in the mountainous interior of the island cross the limestone belt. Two of these, the Río Camuy and the Río Tanamá have part of their present course underground. The Río Guajataca flows in a deep, narrow canyon, which probably represents a cave system.

The Groundwater table in the Aguada and Aymamón Limestones is only 5m to 10m above mean sea level but it steepens abruptly in the vicinity of the Aymamón-Aguada contact and then flattens off within the Cibao and, presumably, the Lares Limestones (Giusti and Bennet 1976). (Fig. 2). Artesian aquifers occur in some members of the Cibao Formation and in the upper part of the Lares Formation.

Figure 2 shows the approximate location of two sinkhole sites, described later, in relation to local geology and the position of the groundwater table. Four other sinkhole sites in this area exhibited very similar geologic and hydrologic setting. Several wells in the vicinity of the sinkholes tap the water table aquifer at depths in excess of 75m (250 ft).

Puerto Rico has a semitropical climate, characterized by temperatures from about 20°C (68°F) to 32°C (90°F) and rainfall scattered throughout the year (Fig. 3). The most intense precipitation occurs usually between May and October, about twice as much as between January and March (Calvesbert 1970). Hurricanes, tropical disturbances and torrential rainstorms generate heavy rains and floods. A definite link was established between the formation of sinkholes and heavy rainfall. Between December 12 to 15 of 1981, gaging stations in north-central Puerto Rico registered one of the most intense storms on record. Over 74 cm (29 in) of rain fell in Toa Baja during this period, of which 42.3 cm (16.6 in) occurred in 24 hours on December 13. Both amounts exceeded that of the 100-year storm. Many sinkholes which opened after this rainstorm prompted the study of sinkhole related terrain failures in Puerto Rico (Wegrzyn et al., 1983)

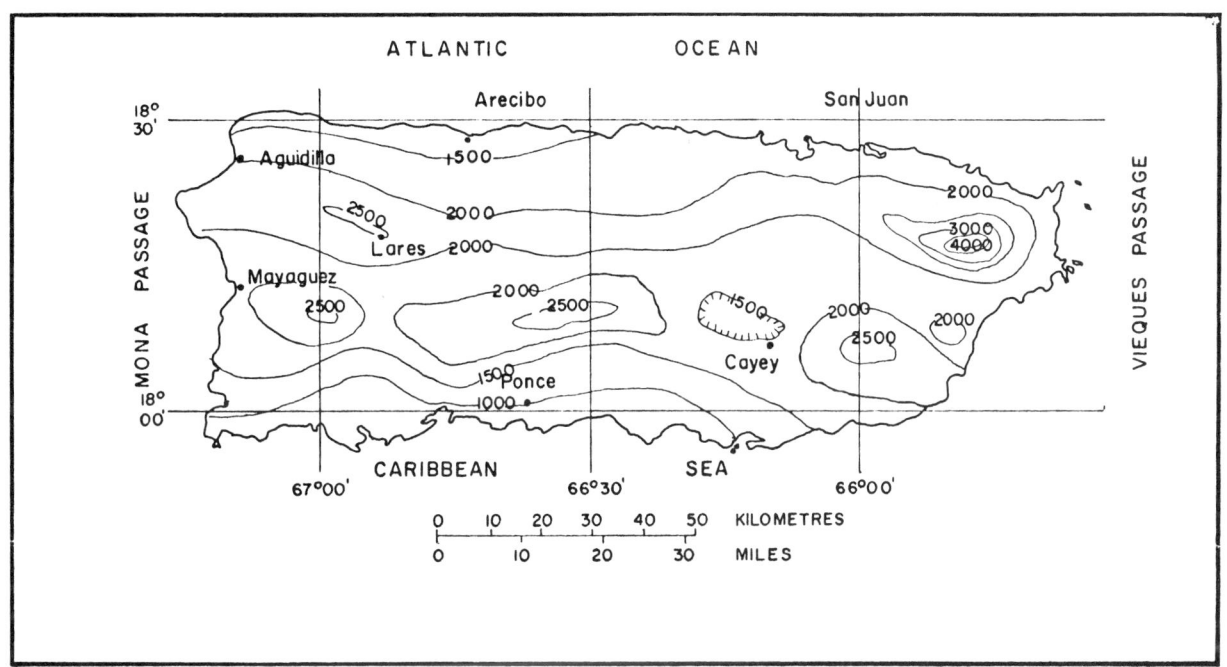

Figure 3. The annual rainfall in Puerto Rico, contoured in millimeters. Data from U.S. National Weather Service.

Sinkhole Development in Puerto Rico.

Most sinkholes in northern Puerto Rico result from the collapse of the surficial blanket sands into cavities in the underlying limestone. The cavities derive from solution processes which began thousands of years ago and are still in progress. Calcium carbonate, the main constituent of limestone is readily soluble in acid waters, the rate of solution increasing with acidity. Under normal circumstances surface and ground water acidity depends mainly on its carbonic acid content. Carbonic acid is formed by the solution of carbon dioxide in water, a common reaction in the atmosphere. The atmosphere and dry soil both contain 0.03 to 0.08 percent of carbon dioxide. Miotke (1973) measured much higher carbon-dioxide concentration, 0.9 to 1.3 percent, in the air on the sides of the residual limestone hills of the Puerto Rico karst belt and even higher concentration, 1.5 to 7.0 percent, in the blanket sands. As rainwater percolates through such soils it becomes strongly acidic. Upon reaching the underlying limestone, at depths of up to 30 meters solution begins. The solution process is particularly fast in closed depressions, along joints and in weathered zones. With time, randomly oriented passages are formed. These openings are subsequently enlarged resulting at times in the formation of cavern system but more often producing a rock mass locally riddled by small interconnected solution passages.

It is in this setting that collapse sinks develop. Surface collapse generally occurs during or shortly after periods of intense or extended rainfall, causing the ponding of water in shallow surface depressions anywhere from several to hundreds of meters in diameter. Eyewitnesses report that the collapse begins without warning as if a plug and had been removed from the bottom of the pond. The water quickly drains through unseen subterranean passages leaving a gaping hole on the ground surface. The remaining depression is deeper and usually wider than before. Depth may vary from one to several meters. The sinkhole walls are usually inclined at angles in excess of 45°. There may be one or several open holes on the bottom generally less than 1.0 meter wide through which the water drained. Flotsam, trash and soil accumulate in openings sometimes completely sealing them. A strongly solution-pitted limestone invariably forms part of the drain hole wall. Usually entirely recrystallized, solution-pitted limestone may also occur in the pit walls, either as massive, irregular blocks or as smaller limestone fragments surrounded by reddish blanket sand soil. The sinkhole may refill to drain suddenly several times, or it may become inactive and drain slowly. New sinks may coalesce with the older one, or be connected to them along narrow ravines. Over time, however, the sink becomes inactive and soil slumping and sloughing at the walls begins to fill and widen the pit. Eventually all that remains is a shallow depression which continues to become ponded during heavy rains.

Study of the Sinkholes.

First systematic study of sinkhole related terrain failures in the blanket sand plains of North-Central Puerto Rico was conducted in 1982-83, through the Water Resources Research Institute, Research Center, School of Engineering, University of Puerto Rico. The study methods comprised aerial photography, reconnaissance examination of geological, hydrogeological and topographical maps, inventory of selected sinkholes, borings and laboratory soil testing, seismic refraction and electric resistivity survey. Two of the examined sinkhole failures were of particular interest: H.R. Robbins Sinkhole and Parcelas Márquez sinkhole.

H.R. Robbins pharmaceutical plant is located at kilometer 63.0 on the north side of highway PR-2, about 20 km (12 miles) east of Arecibo. East of the main entrance gate and parallel to PR-2 there is a large drainage pit (85 m. x 45 m.; 280 ft x 150 ft.). The purpose of the pit is to drain storm water which flows in from the surrounding hills and accumulates on the factory grounds. During the December 13 to 15, 1981 storm, described before, the drainage pit was filled to capacity with the water being 2 m. (6.5 ft.) deep in places. A large part of the factory grounds were covered with 0.6 to 1.2 m. (2 to 4 ft.) of rainwater. According to the report of a guard who was in the guard house the water level in the drainage pit dropped suddenly with a roaring sound. In about 45 seconds 5,500 m^3 (1.2 million gallons) of water emptied through four sinkhole openings formed at the bottom of the drainage pit. Their diameter varied from 1.5 m. to 12 m. (5 ft. to 40 ft.) with depths ranging from 1.5 m. to 9 m. (5 ft. to 30 ft.). The case of H.R. Robbins sinkhole exemplifies the suddenness of the failure related to the extemely high rainfall of over 74 cm (29 in) in the period of four days with peak intensity of 42.3 cm (16.6 in) during 24 hrs. Twenty-seven (27) test soundings were made in the failure area to depths ranging from 5 to 11 meters (16.4 to 36 ft.) at which depth limestone was encountered. Only two open cavities were found, one at a depth of 1-2 meters and another one at depths of 3-4 meters (9.8-13.1 ft.). Extensive repairs were carried out in December of 1981 and through the first months of 1982 to minimize the potential of future terrain failures so that the sinkhole is no longer evident. The construction of the drainage pit over the unexplored calcareous bedrock proved to be erroneous. Continously repeated drainage of the surface runoff through the bottom of the pit contributed to the enlargement of the solution channels and finally to the numerous sinkhole failures. Ground water on this site is located more than 60 m. (197 ft.) below the ground surface and therefore had little, if any, effect on the sinkhole development.

The Parcelas Márquez study area comprised an area of approximately 4.1 km^2 (1.6 mi^2) located south of state Highway PR-2, about 43 km. (26 miles) west of San Juan. The site is located on a moderately undulating plain with large number of broad, gently-sloping closed depressions, having maximum relief of 2 to 3 meters (6.5 ft. to 9.8 ft.). These become ponded during frequent periods of intensive rainfall. There are no stream channels in the area. The ponded water evaporates or seeps down through the blanket of soil to a subsurface drainage system consisting of a net of interconnected solution channels.

The most devastating sinkhole in Parcelas Márquez developed beneath a carport and column of a concrete house, causing partial collapse of the structure and forcing the owner to abandon the site. A Second smaller sinkhole was located on the neighbouring property, partly below the floor of a reinforced concrete-masonry house, also abandoned for this reason. The larger sinkhole was almost circular in shape, with a diameter of about 12 m. (39.4 ft.) and a depth of 2.25 m. (7.4 ft.). The site slopes were irregular with inclinations varying from about 45° to 60°. According to the owner of a nearby lot the sinkhole occured in May 1981, following the period of heavy rains, which completely flooded the area.

The failure was very sudden and water drained through the sinkhole in a period of a few hours. Observations made in 1982-83 indicated that the sinkhole, although still active, was partly clogged, so that the crater required over 48 hours to drain after heavy rains. Numerous smaller sinkholes were discovered all over the study area.

Borings, the seismic refraction survey and electric resistivity exploration placed the limestone bedrock at depths between 12 m. and 14 m. (40 ft. to 46 ft.), covered with decomposed limestone and blanket sands. Water wells in the area tap the water table aquifer at depths in excess of 75 m. (245 ft.)

Shallow examination pits were cut along the rim of the large sinkhole. Field and laboratory soil tests were conducted on surficial soil. These included granulometric, plasticity, permeability (remolded), consolidation and vane shear tests. The surficial soil contained 51% of particles larger than 0.425 mm (sieve No. 40) and only 5.7% of fines. It classified as A-2 soil according to AASHTO System or as clayey sand according to USC System (ASTM D-2487). Shear strength determined in site with a Torvane probe was slightly lower than 100 kPa (14 psi), permeability of samples compacted to the same density as in field (13.7 kN/m^3; 87.4 pcf) tested with the falling head permeameter was about 2×10^{-4} cm/sec (7.9×10^{-5} in). The drainage properties of blanket sand were examined according to the criteria established by Sowers (1979), Fig. 4. The gradation curves for the samples from Parcelas

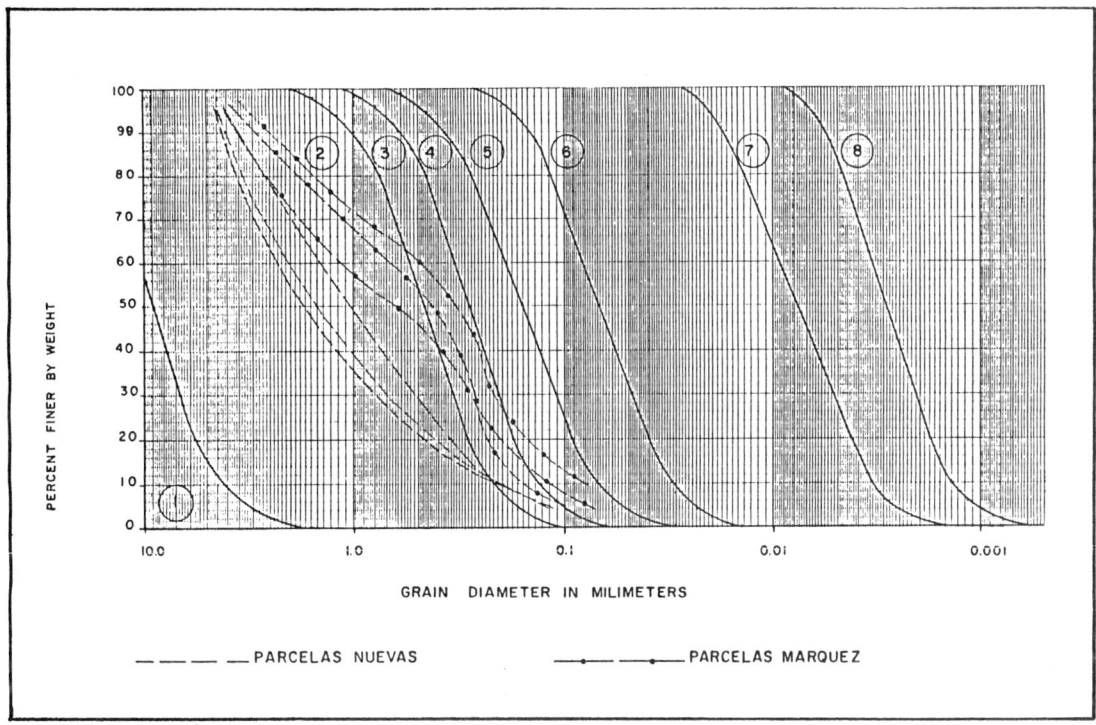

1. Drainage difficult because of large flow. Cutoffs, void filling, blankets helful.

2. Excellent operation of open drains, simple gravity well points. Large flow likely.

3. Good operation of open drains, simple well points.

4. Good to fair open drains. Sanding of well points and vacuum helpful. Erosion in open drains.

5. Gravity drainage slow and erosion may be serious. Sanding of well points and vacuum needed.

6. Gravity drainage impossible except for fissures, sand seams. Vacuum well point usually effective.

7. Sanded well points with vacuum sometimes successful. Electro-osmosis will increase drainage.

8. Drainage by consolidation, accelerated by sand blankets and vertical sand drains.

Figure 4. Drainage capabilities of soils from Parcelas Nuevas and Parcelas Márquez. Drainage criteria after Sowers, 1979.

Márquez indicated good drainage characteristic and rather high erodibility of blanket sands. The results of borings indicated that the fines were continuously washed out of the top soil and deposited at greater depth. Much higher contents of fines (50% to 80%) were found in the borings than in the top soil (about 2% to 8%). Erosion by surface runoff contributed probably also to the decrease of fines in the top soil. Only one small cavity 45 cm (1.5 ft.) high was detected in one of the drillholes to a depth of 2 m. (6.5 ft.)

Conclusions

(1) All sinkhole failures examined in the recent study resulted from the collapse of overburden - blanket sands - into underlying open passages in limestone bedrock.

(2) The collapse of sinkholes related directly to the drainage of rain water through the blanket sand layer to underlying cavernous limestones.

(3) Groundwater level fluctuations, even sudden and large had no bearing on sinkhole development in the area of study, because of the great depth (more than 60 m., 197 ft.) of the water table.

(4) There is a definite link between the heavy rainfall floods and the formation of sinkholes in the karst region of Puerto Rico.

(5) The risk of sinkhole failures can be considerably reduced with proper surface drainage measures. Ideally, surface runoff should be diverted away from depression areas and open sinks, unless these are located in land of no particular value, in which case they may be used to transfer runoff to the underlying limestone safely.

(6) Aerial photointerpretation is extremely useful in identifying depression areas susceptible to collapse failures in the blanket sand plains, particularly when photographs taken on widely separate dates are used.

Acknowledgment

The authors gratefully acknowledge the funds provided for this study by the Office of Water Research and Technology, U. S., Department of the Interior and the support offered by the U. S. Geologic Survey, Caribbean District in performing the electric resistivity and seismic refraction surveys.

References

Calvesbert, R. J., 1970, Climate of Puerto Rico and U. S. Virgin Island: U.S. Environmental Sci. Services Admin., Climatography of the Unites States 60-52

Giusti, E. V., and Bennett, G.D., 1976, Water resources of the north coast limestone area, Puerto Rico: Puerto Rico Water - Resources Bull. 15.

Miotke, F. D., 1973, The subsidence of the surfaces between mogotes in Puerto Rico east of Arecibo (translated from German by W. H. Monroe): Caves and Karst, Vol. 15, No. 1, pp 1-12.

Monroe, W. H., 1976, The karst land forms of Puerto Rico: Geol. Survey Professional Paper 899

Sowers, G. F., 1979, Introductory Soil Mechanics and Foundations: Geotechnical Engineering, 4-th edition, MacMillan Publishing Co., Inc.

Wegrzyn, M., Pérez J. A., and Soto A. E., 1983, Rainstorm related terrain failures in Puerto Rico: Final Technical Report to U. S., Dept. of the Interior from the Research Center, College of Engineering, Univ. of Puerto Rico, Mayaguez.

Collapse sinkholes in the blanket sands of the Puerto Rico karst belt

ALEJANDRO E.SOTO & WANDA MORALES *University of Puerto Rico, Mayaguez, USA*

ABSTRACT

Collapse sinkholes in the northern Puerto Rico karst occur within the outcrop belt of surficial deposits known as blanket sands. These are former beach sands laid down unconformably on middle-Tertiary limestones since middle-Miocene time. As the karst terrain formed, the blanket sands subsided into developing karst depressions and intrakarst plains. Subsequent weathering has modified the original sand composition leaving a quartz-kaolinite residue. Aerial photointerpretation is useful in identifying potential sinkhole sites. Comparison of air photos taken in 1936 and 1982 (1:10,000) shows most present sinkhole problems could have been avoided. Zones evidencing subsidence activity today are essentially the same as those recognizable in the 1936 photography. Statistical analysis of sinkhole and depression long axes from a 80km² area shows significant concentration of orientations in a NE direction. These suggest possible structural control of sinkhole development. A model for Puerto Rico collapse sinkholes is illustrated.

Introduction

Collapse sinkholes in the north Puerto Rico karst occur mainly within a 400km² area where mid-Tertiary limestone is covered by a mantle of blanket sand. These are former beach sands laid down unconformably on the limestone. Yearly losses attributed to collapse sinks and related ground subsidence add up to hundreds of thousands of dollars. Much of this loss could be reduced by proper planning which considers the spatial and temporal occurrence of collapse related ground subsidence. The work reported herein is part of an ongoing project aimed at identifying areas susceptible to collapse failure and providing a better understanding of collapse mechanisms, particularly as they relate to soil and bedrock compositional and structural variations.

Blanket Sands

The geologic-hydrologic setting of the blanket sands is summarized in another paper of this volume (Wegrzyn and others). The blanket sands occur close to the north coast within the outcrop belt of the three youngest mid-Tertiary limestone formations (the Aguada Limestone, Aymamon Limestone and Camuy Formation). They consist predominantly of angular to subrounded, fine to medium grained quartz sand and reddish brown to orange to white kaolinitic clay. Quartz clay ratios generally range between 80:20 and 30:70 with an average of 65:35 (Briggs, 1966). Secondary constituents, including organic matter, feldspar and iron oxides generally make up less than 3% of the total soil volume.

Up to 30m thick, the blanket sands mantle roughly one fifth of the northern Puerto Rico limestone belt. They occur in sinkholes and in broad undulating plains usually rimmed by residual limestone hills (mogotes) and ridges. On close inspection the blanket sand plains are seen to contain many broad, gently sloping closed depressions having a maximum relief of several meters and a maximum diameter of several 100 meters. The low parts of the depression may contain swallow holes, swamps, or ponds. Natural or man-made drainage channels often lead towards depression low points. In places the surface topography resembles a scaled down version of the hummocky terrain which characterizes areas of prolonged landslide activity. This is most noticeable near many of the collapse pits or craters which dot the plains.

According to Briggs (1966) the blanket sands were originally deposited on a surface devoid of karst features and subsequently lowered into depressions resulting from differential dissolution of the underlying limestone. The sands were derived from the erosion of volcanic, sedimentary and plutonic rocks of the pre-Eocene central highland complex which makes up the backbone of the island of Puerto Rico. They were laid down as beach deposits by regressing seas beginning in mid-Miocene as arching of the central part of the island raised the limestone formations above sea level.

Runoff on the newly emerged land quickly seeped underground through the porous sand deposits initiating the dissolution and karstification of the underlying limestone. Once a subterranean drainage system became established there were few surface streams to erode and transport the sands towards the new coastline, much as occurs today. Dissolution proceeded most rapidly where the limestone was fractured or purest leading to the development of relief. The blanket sands began to subside into the developing karst terrain. Trapped within the subsiding basins the less resistant mineral components weathered rapidly, with only the original quartz grains remaining unaltered. Blanket sands from the developing topographic highs was washed into the deepening depressions and today is absent from residual limestone hills and ridges.

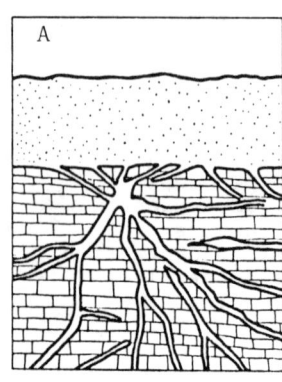

A: Stage one

B: Stage two

C: Stage three

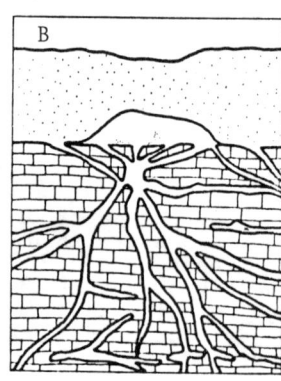 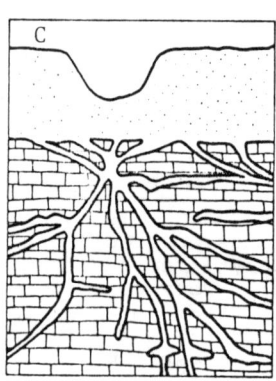

Figure 1: Schematic representation of the development of a collapse sinkhole.

Locally, subsidence rates were increased by washing away or collapse of blanket sand into solution enlarged passages in the underlying limestone, the process responsible for collapse sinks. The development of a collapse sink is illustrated schematically in Figure 1. Figure 1-A shows blanket sand overlying limestone locally riddled by small diameter (usually less than 1m), solution enlarged passageways. Initially the soil bridges the openings. Gradually soil falls or is washed into the limestone channels and a cavity begins to form above the bedrock contact (Fig. 1-B). This stage may be accompanied by sagging of the ground surface. In the final stage (Fig. 1-C) the cavity roof collapses and a sinkhole is formed. This sequence of events is essentially that proposed by Sowers (1975) and by Newton (1976). It is significant, however, that water table fluctuations do not affect the process except possibly where perched groundwater exists locally.

After the initial collapse the crater may flood during heavy rains. Further collapse may ensue, either at the site of the initial failure or in adjacent ground. Slumping of the crater walls enlarges the pit. The slumped material may disappear during subsequent collapse events or it may accumulate in the crater if the underlying openings become temporarily plugged with soil or plant debris. When this happens the crater is smoothed out so that with time a shallow depression or sag marks the site of the collapse. Preliminary findings suggest that sinkholes tend to recur at such sites.

Airphoto Mapping of Collapse Sinkholes and Related Depression Features

Aerial photographs dating from 1936 to 1982 (scale approximately 1:10,000) are being used to identify sinkholes and related subsidence features. To date the area covered on the more recent photographs encompasses some 80km² ESE of Arecibo (see Fig. 1 in paper by Wegrzyn and others, this volume); the area mapped on the older photos is about one eighth of this. The area mapped (1982 photos) contains over 650 collapse sinkholes, undrained depressions, and elongate depressions having drainage outlets. Their position will eventually be transferred to standard 1:20,000 U.S.G.S. topographic maps to be used to delineate sinkhole susceptibility. Most features are assymetric. The width and length of the 569 sinks and undrained depressions was measured. The results are tabulated in Figure 2 which shows the frequency of occurrence of subsidence features grouped into 10 width-length ratio categories. The long axis orientation of 574 assymetric features was determined. These were then subdivided into 10° arc classes and plotted on a rose diagram (Fig. 3).

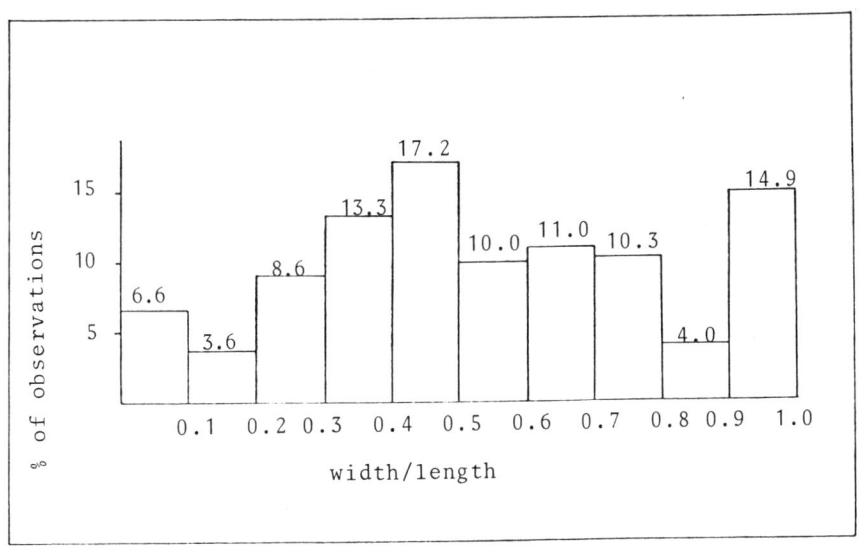

Figure 2: Histogram of sinkhole related depression with-length ratios.

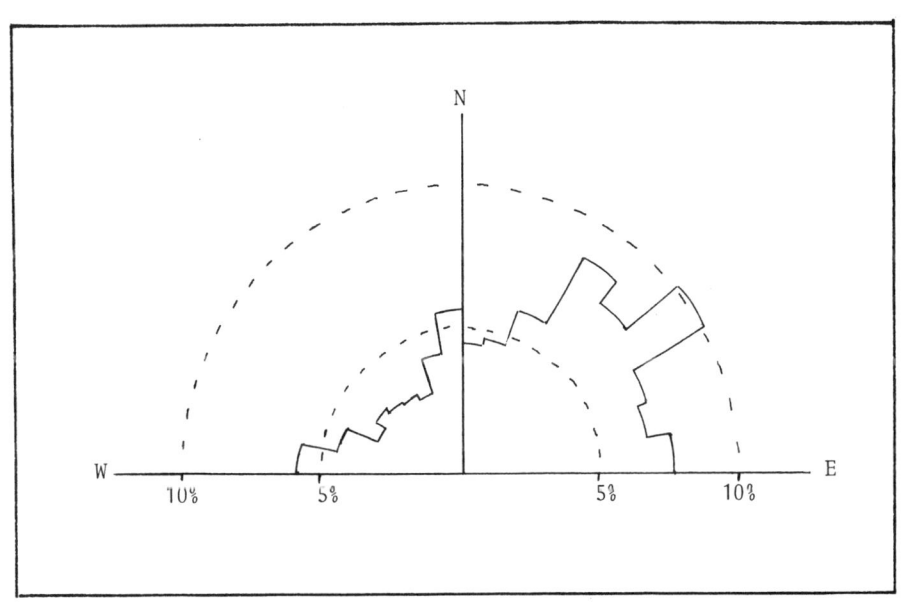

Figure 3: Depression long axis orientations

Discussion

Aerial photointerpretation is useful in identifying potential sinkhole sites. Comparison of airphotos taken in 1936 and 1982 shows zones evidencing subsidence activity today are essentially the same as those identified in the 1936 photographs. Figure 4 shows airphoto interpretation maps of sinkhole related features in a small area east of the town of Manati. Note that the site of several elongate depressions in the 1936 photos are occupied by sinks or undrained depressions in the 1982 photos. Similar persistence of subsidence features in certain areas is evident in all areas where 1936 and 1982 photos have been compared.

Figure 3 shows a pronounced concentration of long axis directions oriented between N30E and EW. This suggests some external factors may be influencing sinkhole development. Such controls may be structural or lithological, such as fracture patterns or lateral facies changes, or they may relate to such things as regional land slope or dominant wind direction.

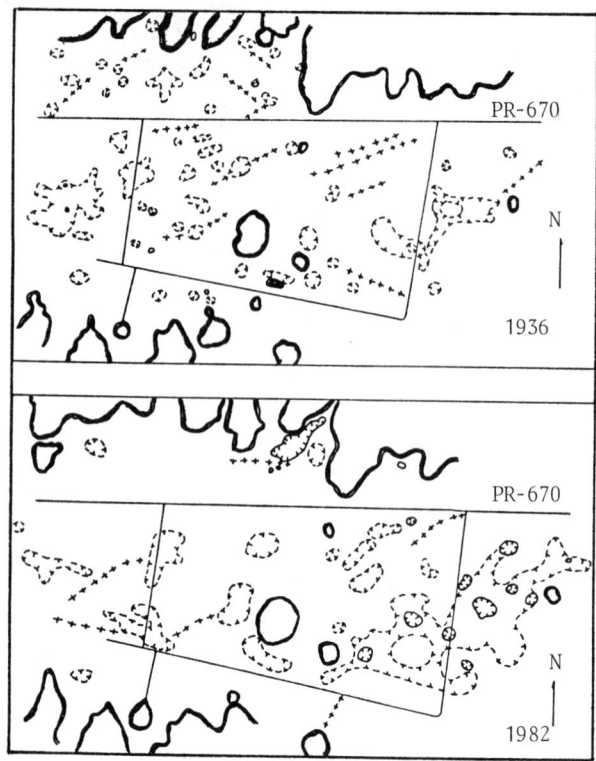

Figure 4: Photointerpretation maps of an area near Manati, 1936 and 1982. Heavy dark lines mark the limit of residual limestone hills and ridges. Elongate depressions shown by line of crosses. Straight lines are roads. Area shown is approximately 1.8 km. long and 1.0 km. wide.

A statistical analysis of orientation significance, the chi-square test (Day, 1978, Gray, 1974, Hay and Abdel, 1974, Rahmann, 1974, Williams, 1974) was performed on the orientation data. It shows that only the concentrations of orientations between N31E-N40E and that between N51E-N60E exceed that to be expected to arise from random variations (chance) in the sinkhole forming processes at the 0.05 confidence level. Whether or not these orientation concentrations reflect some sort of structural or other type of control is still unclear. It should be noted that the N51E-N60E class was one of three which Day (1978) found statistically significant in an analysis of residual limestone hill orientation. He concluded this was suggestive of some type of structural control in hill development.

Acknowledgements

Funding for the initial phases of this investigation was provided by the Water Resources Research Institute, School of Engineering, University of Puerto Rico - Mayaguez.

References

Briggs, R.P., 1966, The blanket sands of northern Puerto Rico: Caribbean Geol. Conf.,3rd, Kingston, Jamaica, 1962, Trans., p. 60-72.

Day, M.J., 1978, Morphology and distribution of residual limestone hills (mogotes) in the karst of northern Puerto Rico: Geol. Soc. Am. Bull., v.89, p.426-432.

Gray, J.M., 1974, Use of chi-square on percentage orientation data: Discussion: Geol. Soc. Am. Bull., v.85, p.833.

Hay, A.M., and Abdel Rahmann, M.A., 1974, Use of chi-square for the identification of peaks in orientation data: Comment: Geol. Soc. Am. Bull, v.85, p. 1963-1965.

Newton, J.G., 1976, Early detection and correction of sinkhole problems in Alabama: NTIS Report PB-257561, Alabama State Highway Dept., Montgomery, Ala., 83p.

Sowers, G.F., 1975, Failures in limestones in humid subtropics: Jour. Geotech. Eng. Div., Proc. ASCE, GT-8, p. 771-787.

Wegrzyn, M., Soto, A.E., and Perez, J.A., 1984, Sinkhole development in north-central Puerto Rico: Proc. First Multidisciplinary Conference on Sinkholes, Orlando, Fla.

Williams, P.W., 1974, Use of chi-square on percentage orientation data: Reply: Geol. Soc. Am. Bull., v.85, p.833-834.

Predicting the location of surface collapse within karst depressions: A Jamaican example

MICHAEL DAY *University of Wisconsin-Milwaukee, Milwaukee, USA*

ABSTRACT

Shallow, extensive karst depressions in northern Jamaica are punctuated by collapse pits that pose a threat especially to livestock. To identify high risk sites the locational attributes of 108 collapses within 37 depressions were analysed. Although the overall collapse distribution approximates that predicted by a random set of processes, certain sites within individual depressions are especially prone to collapse. In particular, 55% of collapses occur between 20m and 50m from depression bases and 57% occur in internal topographic lows. By contrast, only 25% of collapses occur within centripetal drainage channels and only 22% occur along depression long axes. Such analysis provides a simple, useful index of site hazardousness.

Introduction

Detailed site analysis, employing test boreholes, geophysical survey and monitoring of near-surface water movements, is the surest way to identify collapse-prone locations in karst terrain, but such investigation is too time-consuming and expensive to be universally viable. Detailed survey is necessary where major construction projects are proposed or threatened but at risk more usually is the property of individuals who cannot meet or justify the expense.

In these situations it is useful to have a quick, cheap way to determine where surface collapse is most likely to occur. Although it will not be completely dependable, a determination based upon surface observation and measurement has much potential value in determining which sites are particularly hazardous and should be avoided.

It is clear that closed depression development, and by inference surface collapse, is the result of a set of random solutional and mechanical processes whose distribution is influenced by geological and geomorphological factors. Paramount among these spatial controls are variations in secondary permeability and actual surface-subsurface water movement (see for example Palmquist, 1976, 1979 and Day, 1983). Closely associated with these are variables such as bedrock purity, strength and fracturing, regolith character, and proximity to surface drainage channels. Additionally, because of feedback variations depression formation may be independent (e.g. McConnell and Horn, 1972), contagious (e.g. Drake and Ford, 1972, Palmquist, 1976, 1979, Kemmerly, 1982) or competitive (e.g. Williams, 1972).

Studies of karst depressions and collapse in tropical areas are relatively few, although collapse there constitutes an appreciable hazard (Day, 1978a). Also, few studies have considered collapse within well-developed, extensive depressions although much collapse occurs in this situation. *A priori*, one might anticipate that in such a terrain collapse would be subject to two sets of controls-those operating at the regional scale and those which are a function of internal processes within individual depressions. The aim of this study is to examine which of these controls are important and to apply these results to the identification of particularly collapse-prone sites.

The Study Area: Geology, karst landforms and collapse

The karst under discussion covers 13km^2 north of Browns Town in St. Ann Parish, Jamaica (Figure 1) and has been the subject of previous morphometric and hydrologic studies (Day, 1976, 1978b).

Bedrock is Upper Eocene to Lower Miocene White Limestone. Members present, in order of areal dominance, include the Browns Town-Walderston, Somerset, Claremont and Troy. These range petrographically from rubbly biomicrites to compact dolomitic limestones

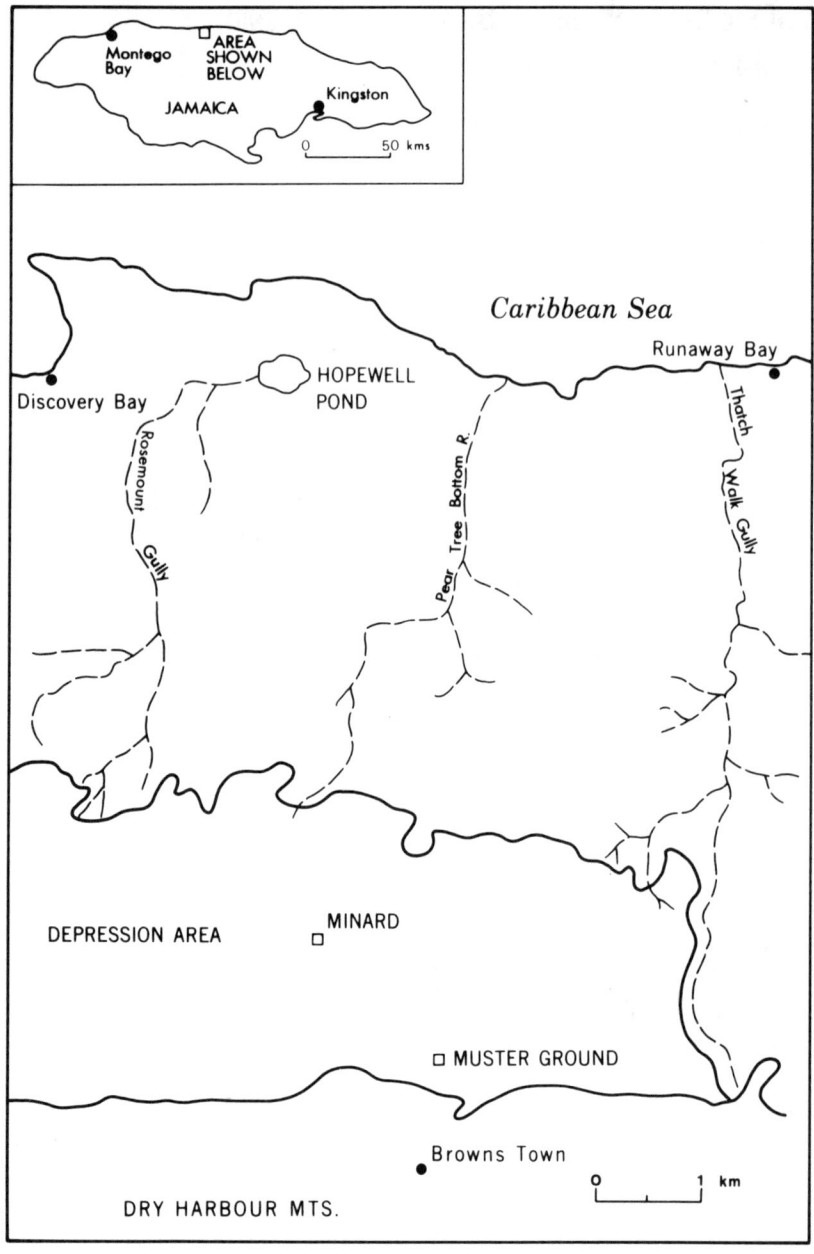

Figure 1. Location of Study Area.

but all are extremely pure, containing less than 1% insoluble residue. Regional dip is towards the north at less than 5°. The limestones are mantled by thin red-brown bauxitic loams (Barker, 1968) whose depth exceeds 1m in some closed depression bases. Soils are acidic; soil moisture retention and carbon dioxide content are highest in depression bases.

The study area is bounded to the south by the eastward extension of the Duanvale fault scarp which rises about 100m to the cone karst of the Dry Harbour Mountains. North of the karst depressions a series of ephemerally active valleys trend north across the Montpelier Limestone.

The topography is gently rolling, consisting of shallow, extensive depressions separated by low divides and residual hills. Surface slopes are generally less than 10°. Depressions occur both singly and in complexes which might technically be termed uvalas. Thirty seven individual depressions have been identified; mean density is $2.85/km^2$ and the depressions occupy 41.6% of the total surface area. Mean depression area is $0.15km^2$, mean diameter is 219m and mean depth 18.4m. The depressions exhibit a distribution which approximates uniformity although depression long axes show a significant alignment approximately eastwest, concordant with local fault directions (Day, 1976).

Centripetal drainage characterises the depressions. Both surface runoff and shallow throughflow accumulate in depression bases where water may pond for several weeks following heavy rain. Surface runoff is conducted along poorly-defined channels and during intense storms, discharges into basal ponds may be of the order of 5 to 10 cumecs. Water which accumulates in depression bases, that which is not lost by evaporation or other means, sinks below ground and has been traced to springs up to 5km distant in the valley systems to the north.

Although there is no evidence that the depressions resulted primarily from collapse, rapid subsidence occurs currently within the landscape. Steep-sided pits, up to 10m deep and wide appear sometimes overnight, although in most cases collapse is preceded by slow surface sinking or cracking. Although the majority of holes formed are less than 1m deep and wide, monitoring since 1974 reveals that an average of about 2 major collapses occur each year. The number of small holes developing may well be 5 to 10 times that. These collapses represent a hazard to structures, people and livestock. Although no person has yet been injured, and no buildings have been damaged seriously, several cattle have been killed or destroyed after injury. The study area is used extensively for grazing of Brahman and other cattle and injury or death in a collapse event of a Brahman bull worth U.S. $10,000 or more represents a grave potential loss. The typical response following detection of a hole is fencing and infilling.

Hypotheses

On the basis of previous studies of this and other areas, the broad hypothesis to be tested is that there exists a spatial distribution of collapses reflecting variations in near-surface solution and cavitation. More specifically, it is expected that collapse will be associated with structural weakness and with foci of surface and sub-surface water movement. Additionally, individual collapses may be associated spatially. Two groups of hypotheses are to be tested:

- A) Collapses throughout the area are related to regional variations.
    1. Collapses reflect a non-random distribution mirroring that of the large depressions.
    2. Collapses are more numerous on certain of the limestone members.
    3. Collapses are more numerous adjacent to faults or other fractures.

- B) Collapses are related to position within depressions.
    1. Collapses are more numerous in depression bases.
    2. Collapses are more numerous within topographic lows in depressions.
    3. Collapses are more numerous adjacent to depression long axes.
    4. Collapses are more numerous adjacent to internal stream channels.

Methodology.

To identify collapse-prone sites within the study area, existing collapses were located during fieldwork between 1974 and 1984. The error in identification of collapses is estimated at up to 20%. Holes less than 0.5m wide and deep may have a number of origins and were not included in the sample. Many larger holes have been infilled and cannot be identified reliably.

Using field measurement in combination with geological maps (1:12,500) and air photographs (1:25,000), collapses were analysed to provide information on density and location. In order to test the hypotheses proposed on the basis of previous studies, collapses were examined with particular respect to:

- A) Position within the landscape
    1. Distance to nearest-neighbor collapse
    2. Identity of underlying limestone member
    3. Distance from fault or other fracture trace

- B. Position within individual depressions
    1. Distance from depression base
    2. Location relative to other low points in depression
    3. Distance from depression long axis
    4. Distance from internal stream channels.

Results.

108 collapse pits were identified and recorded during the study. Mean collapse pit density is $8.3/km^2$ and the 37 depressions within the study area contain an average of nearly 3 pits each. At least 20 of the pits have formed during the study period. Ages of the others are unknown, but since most holes are filled within a few years of formation, and since the contemporary rate of formation is estimated at between 2 and 20 per year, it is probable that all are less than 50 years old.

Analysis of nearest-neighbor distances provides no reason to suggest that collapse distribution is the result of other than independent random processes. $\overline{La}$, the mean distance between collapses, is 160m and $\overline{Le}$, the expected distance between collapses in an infinitely large, randomly-located population with the same density ($8.3/km^2$), is 174m. The ratio $\overline{La}/\overline{Le}$ is 0.920 which is not significantly different (at the 0.05 level) from that predicted as the result of a random set of process (where $\overline{La}/\overline{Le}= 1.0$).

Comparison of collapse densities on the different White Limestone members shows a minor but statistically insignificant variation. Densities on the Browns Town-Walderston are slightly higher, at $8.6/km^2$, but the area underlain by the other members is so limited as to make meaningful comparison difficult.

Distance from fault or other fracture trace has little bearing upon collapse location. Less than 5% of collapses occur on traces inferred from geological maps and air-photo interpretation and less than 10% occur within 20m of a trace line.

Of the 108 collapses, 59 (55%) occur between 20m and 50m from depression bases. By contrast, 10(9%) occur within 20m of depression bases, 28(26%) occur 50m to 100m from bases and 11(10%) occur more than 100m distant. Since the mean depression radius is about 110m, the zone of maximum collapse is a band approximately one fifth to one half of the distance up the slopes to the depression edge.

62(57%) of all collapses are located within topographic lows or sub-basins within depressions. The remaining 46(43%) do not correspond to internal low spots. Only 27(25%) of collapses occur along or within 5m of drainage channels which direct runoff to depression bases. Similarly only 24(22%) of collapses occur along or within 10m of the line representing the depression long axis.

Conclusions.

The results support the second group of hypotheses rather than the first and suggest a heirarchy of controls over surface karst development. In a karst terrain where extensive depressions occupy much of the surface area, position within a depression, rather than within the overall landscape, is the more reliable indicator of collapse potential. Regional trends appear to be of little importance at the individual collapse scale although they may exercise considerable influence over the general nature of the large depressions within which the collapses occur.

In practice, the study provides a useful, general set of guidelines that identify sites where collapse is most likely. Low spots, one fifth to one half of the distance up depression sides are the most hazardous and these should be avoided as sites for construction and cattle penning Additionally, these sites deserve most scrutiny in order to locate collapses promptly and institute remedial measures. Collapse patterns elsewhere may differ, although in similar terrain an analogous pattern might be expected, and for practical purposes in other areas similar studies are recommended.

Geomorphologically, the study indicates that much change in large karst depressions is dependent on internal form and position. The concentric zone of maximum collapse may correspond to a subsurface zone of maximum solution with major implications for long term depression evolution. Further research on the internal distribution of forms and processes within depressions can but benefit both landowners and karst geomorphology.

Acknowledgments.

This study was funded in part by a scholarship from the Natural Environment Research Council (U.K.) and by travel grants from the University of Wisconsin-Milwaukee.

References

Barker, G. H., 1968. Soil and land-use surveys, No. 24, Jamaica, Parish of St. Ann. Regional Science Centre, University of the West Indies, Trinidad.

Day, M. J., 1976. The morphology and hydrology of some Jamaican karst depressions. Earth Surface Processes, 1, 111-129.

Day, M. J., 1978a. Engineering hazards in tropical karst terrain. Applied Geography Conferences, 1, 288-298.

Day, M. J., 1978b. The morphology of tropical humid karst with particular reference to the Caribbean and Central America. Ph. D. thesis, Department of Geography, Oxford University, England.

Day, M. J., 1983. Doline morphology and development in Barbados. Annals of the Association of American Geographers, 73(2). 206-219.

Drake, J. J. and D. C. Ford, 1972. The analysis of growth patterns of two-generation populations: the example of karst sinkholes. Canadian Geographer, 16(4) 381-384.

Kemmerly, P. R., 1982. Spatial analysis of a karst depression population: clues to genesis. Bulletin of the Geological Society of America, 93, 1078-1086.

McConnell, H. and J. M. Horn, 1972. Probabilities of surface karst. In: Spatial Analysis in Geomorphology, ed. R. J. Chorley, 111-133.

Palmquist, R. C., 1976. Distribution and density of dolines in area of mantled karst. In: Hydrologic Problems in Karst Regions, ed. R. R. Dilamarter and S. C. Csallany, 117-129.

Palmquist, R. C., 1979. Geologic controls on doline characteristics in mantled karst. Zeitschrift fur Geomorphologie, Supplement Bande, 32, 90-106.

Williams, P. W., 1972. The analysis of the spatial characteristics of karst terrains. In: Spatial Analysis in Geomorphology, ed. R. J. Chorley, 135-163.

Investigation techniques on dolomites in South Africa

PETER W.DAY & FRITZ VON M.WAGENER *Jones and Wagener, Inc., Rivonia, South Africa*

ABSTRACT

The investigation techniques used on dolomites are, of necessity, different to those employed on other geological strata. This is partly due to the nature of the material and due to the added requirement of investigating the risk of subsidence settlements such as sinkholes and dolines. The investigation techniques used in South Africa include photo interpretation, thermal line scanning, landsat imagery, gravimetrics, seismic and resistivity surveys, percussion drilling, backactor trenching and variable frequency vibration amongst others. This paper discusses the above methods and presents case studies of two investigations conducted by the authors.

INTRODUCTION

As much as 14 percent of the highly developed Witwatersrand area, the heart of South Africa's gold mining industry, is underlain by dolomite. This has led to ever increasing residential and industrial development on subsidence prone areas (van Schalkwyk 1981). The situation is complicated by the extensive dewatering of the dolomites due to deep level gold mining which has given rise to numerous large sinkholes and extensive doline developments (Brink 1979, Wagener 1982).

The dolomites, which were deposited during the Precambrian era, reach a thickness of 2000m in places. They are underlain by amygdaloidal andesitic lava followed by the shales and auriferous quartzites of the Witwatersrand Supergroup. The dolomite has been extensively intruded by thick, continuous syenite dykes which divide the dolomite and underlying strata into watertight compartments.

Conventional investigation techniques employed on other geological strata are often poorly suited for use on dolomites. This is due to the highly irregular, pinnacled nature of the dolomite rock surface, the variable nature of the overburden and the frequent abundance of chert gravel in the residuum. (Refer paper by Wagener and Day in these proceedings for typical soil profile on dolomites in South Africa.) Unlike most other formations, the consistency of the overburden generally deteriorates with depth and frequently competent chert gravel layers are underlain by wad and even cavities before solid rock is reached. Consequently, the investigation of dolomite frequently requires the use of specialised techniques not used on other geological formations.

INVESTIGATION TECHNIQUES IN SOUTH AFRICA

The investigation of dolomitic sites has a two-fold purpose. Firstly the investigation should evaluate the risk of subsidence settlements and give recommendations for reducing this risk to acceptable levels. Secondly the investigation should provide the parameters required for the economical design of foundations. The latter purpose is common to most site investigations whilst the former is unique to sites underlain by dolomite.

Amongst the factors which influence the risk of subsidence are the surface topography and drainage, the nature and thickness of the transported soils and residuum, the nature and topography of the underlying rock, the depth and anticipated fluctuations of the water table, the presence of structural features such as faults, fracture zones and dykes and the nature of the proposed development. The parameters required for foundation design generally include the compressibility, shear strength, permeability and extent of the various strata on the site. Any investigation on dolomite should aim to identify and define the above variables.

The investigation techniques used in South Africa may be divided into three categories namely remote sensing, geophysical methods and direct methods.

Remote Sensing

This includes photo-interpretation, thermal line scanning and satellite imagery.

Photo interpretation is a very popular and highly effective tool. Aerial photography at a scale of 1:20 000 is generally used for this purpose and is readily available. During interpretation, particular attention is paid to surface topographical features such as raised areas and linear depressions, vegetation, surface texture and other features which may be indicative of the nature and thickness of the overburden and the weathering of the underlying rock. An experienced interpreter can generally delineate pinnacle and boulder outcrop zones, areas rich in chert gravel, sandy areas, palaeo and recent sinkholes, drainage patterns, fault zones and dykes.

Thermal line scanning measures the thermal radiation of the ground in the 10 - 14 micron wave-length range using a scanner mounted in an aircraft. In South Africa, the best results are obtained in the dry winter months and the area is flown just before sunrise when atmospheric conditions are stable. The imagery from the scan is made into a mosaic at a convenient scale. Cold areas such as deep sandy zones and other areas of moisture accumulation show up as dark patches on the imagery whereas zones of near surface rock which retain the warmth from the previous day, show up as light patches. The depth to which the nature of ground influences the thermal radiation from the surface is however limited to one or two metres. Thermal imagery is particularly valuable in areas of thin overburden where its ability to delineate fault zones and regions of deeper weathering are often superior to that of photo interpretation.

Landsat imagery, although less readily available, is also used on occasions. This is generally only the case where large areas are being investigated and a macroscopic overview is required. Enhanced imagery has been used to map the boundaries of the five main formations which make up the Chuniespoort Group of dolomitic rocks which differ principally in manganese and chert content.

Geophysical Methods

The geophysical methods used include gravity, seismic refraction, resistivity and magnetometer surveys and variable frequency vibration.

Gravity surveys, which make use of the large density contrast between the overburden and the rock to give an estimate of the depth to rock, are widely used in dolomitic areas. The resolution which can be obtained depends on the size of the feature relative to the depth of overburden and the station spacing. For normal site investigations a grid spacing of 15 to 30m is used depending on the depth to bedrock. Interpretation is generally based on residual gravity anomalies, that is the measured Bouguer values minus the regional gravity gradient. Gravity low trends and steep gravity gradients representing deeply weathered features and steeply sloping bedrock respectively are generally considered high risk areas. It should be borne in mind that gravity surveys are unable to distinguish between small features near the surface and large features at depth.

Seismic surveys are used on occasions but suffer from the drawback of not being able to detect softer material such as wad below a competent layer. Seismic surveys on dolomite often make use of a high energy source such as a small explosive charge or drop weight of 200kg or more. In addition, the response is sometimes enhanced by "stacking" or accumulating the signal generated by repeated drops. Using these methods, investigation to depths of 30m or more are possible.

When interpreting the results of either a gravity or a seismic survey, it must be realised that the methods are only capable of modeling the general topography of the dolomite rockhead. The resolution which can be anticipated diminishes with depth and station spacing. These methods are not capable of accurately depicting the pinnacled nature of the dolomite surface. As a result, all anomalies detected by seismics or gravity should be investigated in more detail using percussion drilling or other suitable techniques prior to drawing conclusions as to their significance.

Resistivity and magnetometer surveys are generally used for detecting variations in composition of the overburden or bedrock. These methods are used extensively in the location and delineation of dykes in the dolomite.

Variable frequency induced vibration is a relatively new technique which has been tried during the past year in South Africa. A variable frequency hydraulic vibrator is placed on the ground surface and the frequency is swept repeatedly through a range of 2 to 10 or 10 to 50 hertz in a period of about 30 seconds. The response of the ground to this vibration is sensed by accelerometers on the ground surface set out in a grid around the vibrator. In theory (Savage, 1978) the presence of a void in the ground will give rise to increased response over a narrow frequency range due to the resonance of the void. The resonance frequency is indicative of the size and depth of the void, larger voids at shallow depth having low resonance frequencies. To date little success has been achieved by the authors with variable frequency vibration techniques. This is partly ascribed to the complex nature of the dolomite soil profiles on which the trials were carried out. The method may indeed have merit for the detection of cavities in a deep profile with homogeneous overburden but has been found wanting in profiles comprising near surface pinnacles and variable overburden. Additional trials should be carried out in more favourable conditions prior to reaching conclusions on the value of the technique.

Direct Methods

This category includes all methods in which the soil and rock are examined directly by coring, drilling, penetration testing or other such methods. In contrast to the geophysical methods described above, these methods generally provide point information with a relatively high degree of confidence.

Of these methods, percussion drilling is the most popular due to the relatively low cost and ability to penetrate most materials. Down-the-hole hammers are preferred to top hammer rigs. Common hole diameters are 165mm (6 $\frac{1}{2}$ inch) reducing to 150mm and if necessary to 140mm (5 $\frac{1}{2}$ inch). Drilling air pressures of 700 to 1000kPa (100 - 150 psi) are used. During drilling, chip samples are collected at the top of the hole and penetration rates are recorded over each metre, or occasionally each half metre, drilled. Records are also kept of estimated water inflow, cavities, difficult penetration due to clays and the like. Chip samples are logged under a binocular microscope and the degree of reaction of the chips with dilute hydrochloric acid and hydrogen peroxide are recorded. This method provides "point" information only and a hole drilled a mere metre away may indicate a vastly different depth to bedrock due to the pinnacled nature of the rock. This observation should be borne in mind before attempting to draw contours of depth to rock dolomite.

Piezometers and telescopic benchmarks are installed in percussion holes where the monitoring of the water table and ground settlements is required (Jennings 1966).

Backactor trenching is frequently used to assess the nature of the near surface strata. Even where bedrock is not exposed in the trenches, much valuable information can be gained by observing the slumping of chert bands or extent of infilling with transported sands. Trenches are generally dug at right angles to the strata and trenches should be of the order of 20m long to allow observation of the variability of the strata along the length of the trench.

In the absence of chert gravel and particularly in deep wad profiles, auger drilling using a 750mm (30 inch) flight auger is common. The engineer or geologist is then lowered down the hole using a bosun's chair on a hand winch to inspect, profile and sample the soil. Depths of 15m (50ft) are common and occasionally holes are drilled and profiled to depths of 25m (80ft).

Where inspection of the in-situ material is required to depths in excess of 4m and chert gravel prohibits the use of auger drilling, hand dug shafts with lateral support comprising shoring and wire mesh are sometimes used. This method is however time consuming and expensive and is used only in special instances.

When profiling the trenches or holes described above, the engineer or geologist generally descends into the hole and examines the profile in-situ. Each layer in the profile is described in terms of its moisture content, colour, consistency, structure, soil type and origin (Jennings et al 1973). Gravels are described in terms of packing, shape, size, orientation, weathering and rock type and the matrix described as above. Disturbed and undisturbed samples may be cut for laboratory testing and in-situ tests such as lateral plate load tests may be performed by jacking across the hole.

Diamond drilling is seldom used on dolomites. Core recovery in residuum containing chert gravel is extremely difficult. As interpolation of results between holes is generally unreliable due to the variability of the residuum and bedrock, a large number of holes are required and the high cost of diamond drilling is seldom justified.

Occasionally penetration testing is used but is limited to areas where no chert gravel is encountered. Dynamic cone penetration testing using a 50mm (2 in) cone and standard SPT hammer is occasionally used for determining the depth of transported material in fault zones or overlying a "clean" rock surface. Some success has been achieved using a hand operated dynamic penetrometer to determine ballast thickness below railway lines to identify subsidence of the subballast.

TYPICAL APPROACH TO AN INVESTIGATION ON DOLOMITE

The overall approach to the investigation of dolomitic terrain adopted by the authors would be as follows. After the scope and nature of the proposed development have been determined, a photo interpretation and study of the thermal imagery is conducted. Tentative geological and soil type boundaries are annotated onto a photo-mosaic. This assessment of site conditions is confirmed and revised where necessary by a visit to site and field mapping and is then used to plan the detailed investigation. The positions of investigation points such as percussion holes or backactor trenches are then selected to prove the interpreted boundaries and to investigate any anomalies detected by photo interpretation or thermal scanning with due regard for the positioning of the proposed development. Drilling on a regular grid is seldom undertaken as this is considered both wasteful and ineffective. It is considered essential that the engineer or geologist who will be responsible for making foundation recommendations is intimately involved with the fieldwork.

Following the initial fieldwork, the results are evaluated and further detailed investigation conducted as required. This may include a gravity survey (if not already undertaken), delineation of dykes using magnetometers, further percussion drilling and the like.

When assessing the results of the investigation, all the available information is considered together rather than concentrating on one set of results in isolation. For example, it is not justified to restrict development along a gravity low trend and flanking gradients if percussion drilling or some other method of investigation shows the overburden and underlying bedrock are sound. It is the authors' policy that the report produced after completion of the investigation should be as practical as possible. The developer should be made aware of the risks involved in the development. However, he must also be provided with practical recommendations which will enable him to reduce the risk of subsidence by judicious layout of the development and the adoption of appropriate founding methods.

CASE STUDIES OF THE USE OF INVESTIGATION METHODS

The various methods mentioned above differ in terms of the type of site on which they may be used and the nature of the results obtained. In this section, two case studies are described. The first is an example of a general investigation of a large residential area while the second is a detailed investigation for bridge foundations (Day et al, 1981).

General Investigation of a Proposed Residential Area

The development comprising about 1 000 houses is situated on the Malmani dolomites in the South Western Transvaal. The site is approximately 120 hectares (300 acres) in extent. As such, a detailed investigation of the entire site would have been extremely time consuming and costly. The policy was therefore adopted to cover the site with aerial photo and thermal scanning interpretation and a gravity survey to identify the main features of the geology and to delineate problem areas. During this phase, a limited number of percussion holes were drilled to assist in establishing the regional gravity gradient.

Using the residual gravity map, areas of steep gravity gradients (sloping bedrock) and gravity low trends (zones of deep weathering possibly along faults) were delineated. The air photo mosaics were then annotated to show thermal scanning cold areas (possibly sandy infilling along faults) and photo interpretation boundaries corresponding to changes in gradient or vegetation. Figure 1 shows the extent of these areas and location of the photo interpretation boundaries. In general the correlation between thermal scanning and gravity low trends is reasonable.

A number of percussion holes were drilled to determine the nature of the material in the suspect areas. Backactor trenches were also excavated to assess the nature and consistency of the overburden. The correlation between the depth to bedrock predicted by the gravity survey and that measured during percussion drilling and backactor trenching was reasonable but exhibited a fair amount of scatter. This may be attributed partly to the irregular topography of the rock subsurface and the shortcomings of gravity surveys.

The end result of the investigation was that the area was divided into zones with the higher risk zones being set aside for parks, sportsfields and the like. In general, where thermal anomalies corresponded either with gravity low trends or rapid drill penetration rates, the area was regarded as potentially hazardous while areas of steep gravity gradients not accompanied by thermal anomalies or poor drilling results were released or regarded as marginal areas. In this way, the township plan was drawn up around the constraints imposed by the geology. The one thousand houses built in this area were founded on raft foundations on top of a 500mm thick mattress of compacted fill.

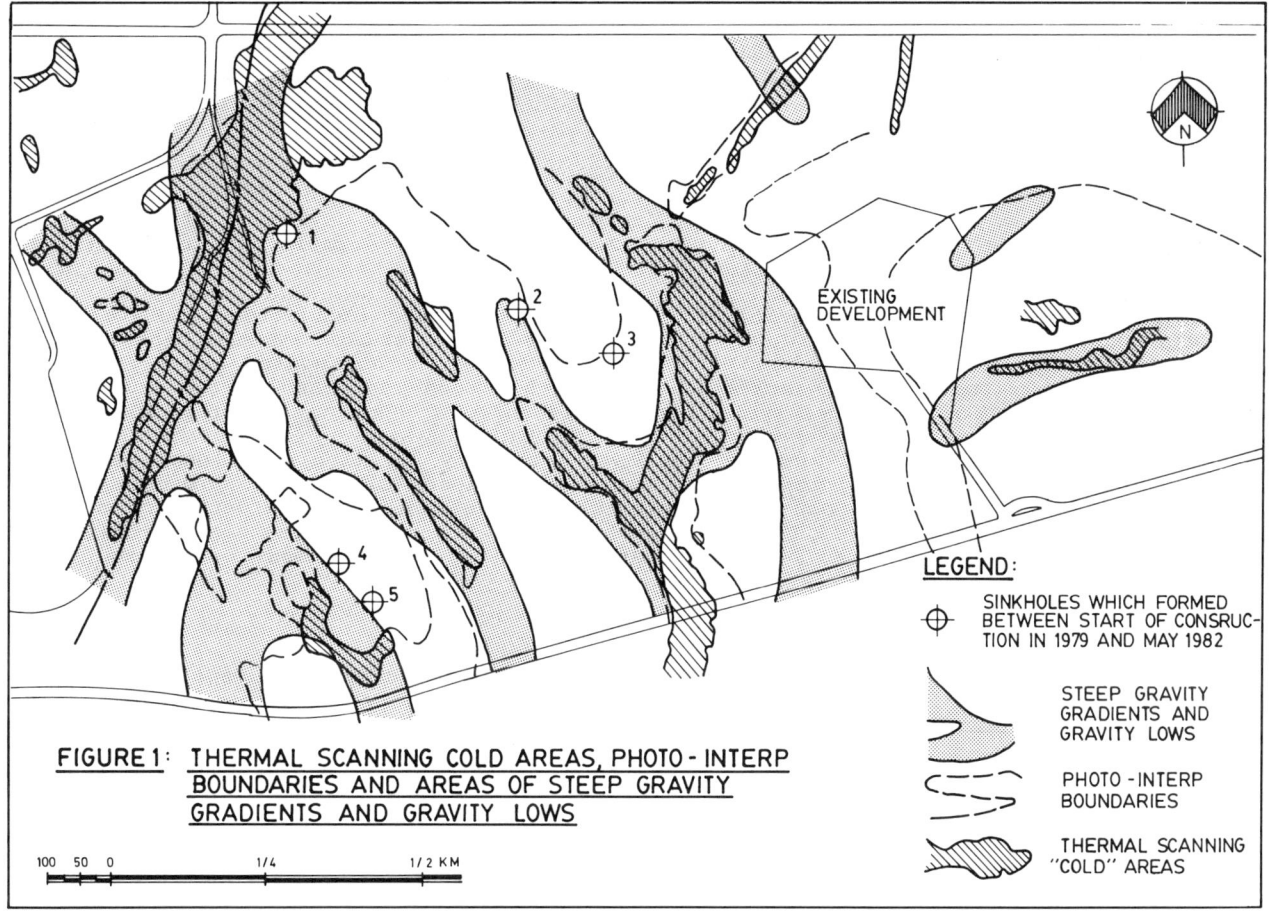

FIGURE 1: THERMAL SCANNING COLD AREAS, PHOTO-INTERP BOUNDARIES AND AREAS OF STEEP GRAVITY GRADIENTS AND GRAVITY LOWS

As shown in Figure 1 five small sinkholes have occurred in the area. None of these were within the areas released for development. Sinkhole 3 which occurred in a relatively stable area was the result of a burst slimes pipeline. All other sinkholes were associated with poor drainage during or after construction.

Detailed Investigation for Bridge over Vaal River

The detailed investigation for the bridge over the Vaal River near Orkney was preceded by a general investigation. The bridge comprised 7 spans of 21m each. The river depth at the crossing point was about 4m.

Once a site was chosen using photo interpretation, a seismic survey was conducted along the bridge centreline and along the two banks to gain an overall idea of the depth to bedrock. Due to the scouring action of the river, the dolomite surface was relatively level and free from pinnacles and the seismic survey yielded acceptable results. On the basis of this work and limited percussion drilling, the position of the bridge was finalised. A cost analysis and preliminary design were carried out and the number and length of spans were selected.

The detailed investigation which followed made use of percussion drilling at each pier position and the drilling of auger holes and excavation of backactor trenches into the alluvium on both banks. Eight percussion holes were drilled at each pier and abutment position on a 2,5m x 3,3m grid covering a total area of 2,5m x 10m. Drilling in the river was carried out using a percussion rig with a "down-the-hole" hammer mounted on a barge. Drilled depths were measured from water level which was controlled by a weir downstream. The drilling contractor was careful to note the depth to the mudline, top of solid rock and the existence of any detectable leached zones or fissures in the rock. All holes were cased and the casing was sealed into the rock to reduce water inflow directly from the river. The rate of water inflow through fissures in the rock was estimated by the driller. Chip samples were collected over each metre drilled and penetration rates recorded. On this job the drilling contractor's logs were invaluable as they provided much detailed information.

In order to assess the properties of the overburden, large diameter auger holes were drilled into the 10m thick layer of alluvium on the south bank and backactor trenches were excavated into the chert gravels on the north bank. The use of the auger gave an idea of what problems could be expected during the possible installation of piles on the south bank.

The investigation showed that the upper surface of the dolomite had been scoured by the river and was free of pinnacles. The further from the middle of the river, the greater the variation in depth to solid rock became. The standard deviation of rock depth over each pier position was computed and was found to vary from 0,2m in the middle of the river to 0,9m on the banks.

Large diameter caissons were selected as the most suitable foundation type for both the piers and abutments. Conventional concrete caissons were used on the banks. The caissons in the river were constructed from sections of Armco corrugated steel pipe and filled with concrete once the rock at founding level had been inspected in the dry. This founding method proved highly effective and was implemented without any major problems.

In all cases, the rock at founding level was proved by drilling additional holes 4m into the dolomite from founding level. At two piers, layers of leached dolomite up to 100mm thick were encountered in the fresh rock. These layers had escaped detection during the investigation stage. These layers may have been detected had limited diamond drilling been carried out in conjunction with the percussion drilling during the investigation.

CONCLUSIONS

The investigation of dolomitic terrain frequently requires the application of specialised techniques. No one technique is capable of accurately describing the site and a number of methods are frequently employed on the same site to obtain an overall picture. When interpreting the results of the investigation, the limitations of the various methods and variability of the material should be taken into account. In the final analysis, however, no amount of investigation can take the place of experienced judgement and practical founding recommendations.

REFERENCES

1. Brink A.B.A. (1979). Engineering Geology of Southern Africa, Volume 1. Building Publications, Pretoria. August, 1979.

2. Day P.W. and Wagener F. von M. (1981). A comparison and discussion of Investigation Techniques on Dolomites. Ground Profile, Geotech. Div. of SAICE No. 27, July 1981.

3. Jennings J.E. (1966). Building on Dolomites in the Transvaal. The Civil Engineer in S.A. February, 1966.

4. Jennings J.E., Brink A.B.A. and Williams A.A.B. (1973). Revised Guide to Soil Profiling for Civil Engineering Purposes in South Africa. The Civil Engineer in S.A. January, 1973.

5. Savage R. (1978). The Detection and Mapping of Sub-surface Cavities by Resonance Techniques. Applied Dynamics Ltd. Leatherhead, Surrey, England.

6. van Schalkwyk A. (1981) (in Afrikaans). Development Pattern and Risk Evaluation in Dolomite Areas. Eng. Geol. of Dolomite Areas. Dept. of Geol. Univ. of Pretoria. November, 1981.

7. Wagener F. von M. (1982). Engineering Construction on Dolomite. PhD Thesis. University of Natal, Durban. November, 1982.

New Jersey sinkholes: Distribution, formation, effects, geotechnical engineering

JOSEPH A. FISCHER *Geoscience Services, Millington, New Jersey, USA*
RICHARD W. GREENE *Storch Engineers, Florham Park, New Jersey, USA*

ABSTRACT

Any attempt to "engineer" structures in sinkhole prone areas is fraught with pitfalls unless one has a good understanding of the geological environment. The authors have selected New Jersey, in general, and several sites specifically, as case studies in this most interesting problem. A Geologic Map showing the distribution of the state's suspect formations has been prepared. Such a map provides the initial planning document for evaluating the possibility of sinkhole formation, and the suitability of engineering solutions. The chemical and physical properties of the limestones are important in the design of structures and the prediction of groundwater flow quality and quantity. Weathering has created soil like materials from the original rock as well as open and clay filled cavities, the geotechnical problem. Rock strengths are variable and well yields can range from near zero to more than 2000 gpm. From an engineering standpoint, the existence of voids in the residual soils above cavernous limestone is perhaps more critical than cavity formation or cavity enlargement in 500 million year old rock. Their formation is related to the underlying limestone cavities, joint and fault system orientation, thickness of residual soils, and groundwater flow. Their detection is limited to test drilling and aerial photos (after the collapse). In limestone areas a number of viable alternatives are available to the owner/planner/design team. The simplest is the "no build" alternative. The next, quite similar alternative "move the critical facilities of the site to the good" area. The third alternative is to provide engineering solutions to the site conditions. All are viable, but all depend upon: 1) geological awareness early in the site selection stage; 2) geotechnical understanding of the site conditions; 3) close liaison among the financial/planning/design and geotechnical team.

INTRODUCTION

The concerns connected with constructing in limestone areas are real. Unfortunately, these concerns are oftimes: 1) magnified; 2) ignored; or 3) misunderstood. A prospective site located in a limestone area should not automatically be discounted, however, the owner and designer should be realistically aware of the problems that they may face if they wish to support facilities in a particular location.

In organizing the presentation of the information in this paper the authors have attempted to follow a thought process similar to that followed in an investigation of a site:

1. What are the geologic concerns at the location of a particular site?
2. What are the physical properties of the subsurface soils, rock and groundwater at that site? and
3. What planning or engineering solutions are available?

We have chosen the State of New Jersey for our example, not only because of our personal background, but also because of the applicability of New Jersey limestone studies to areas to the north, west, and south. New Jersey limestones are found in the Valley and Ridge, and Highlands physiographic provinces (the easterly Appalachian ridges). The discussion is thus applicable to many of the eastern states (except for the geologically recent addition of Florida).

NEW JERSEY LIMESTONES

The limestones of New Jersey are in an area that was at the edge of a large continental land mass, some 500 million years ago. The area was folded and faulted during

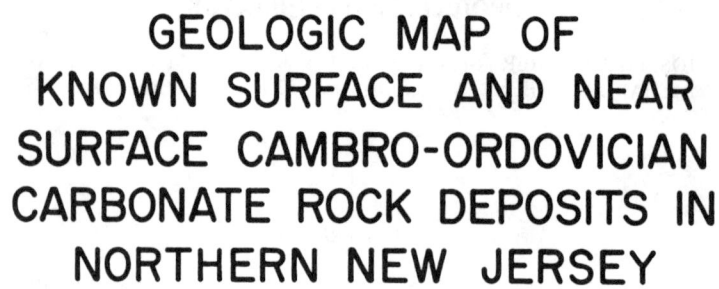

GEOLOGIC MAP OF KNOWN SURFACE AND NEAR SURFACE CAMBRO-ORDOVICIAN CARBONATE ROCK DEPOSITS IN NORTHERN NEW JERSEY

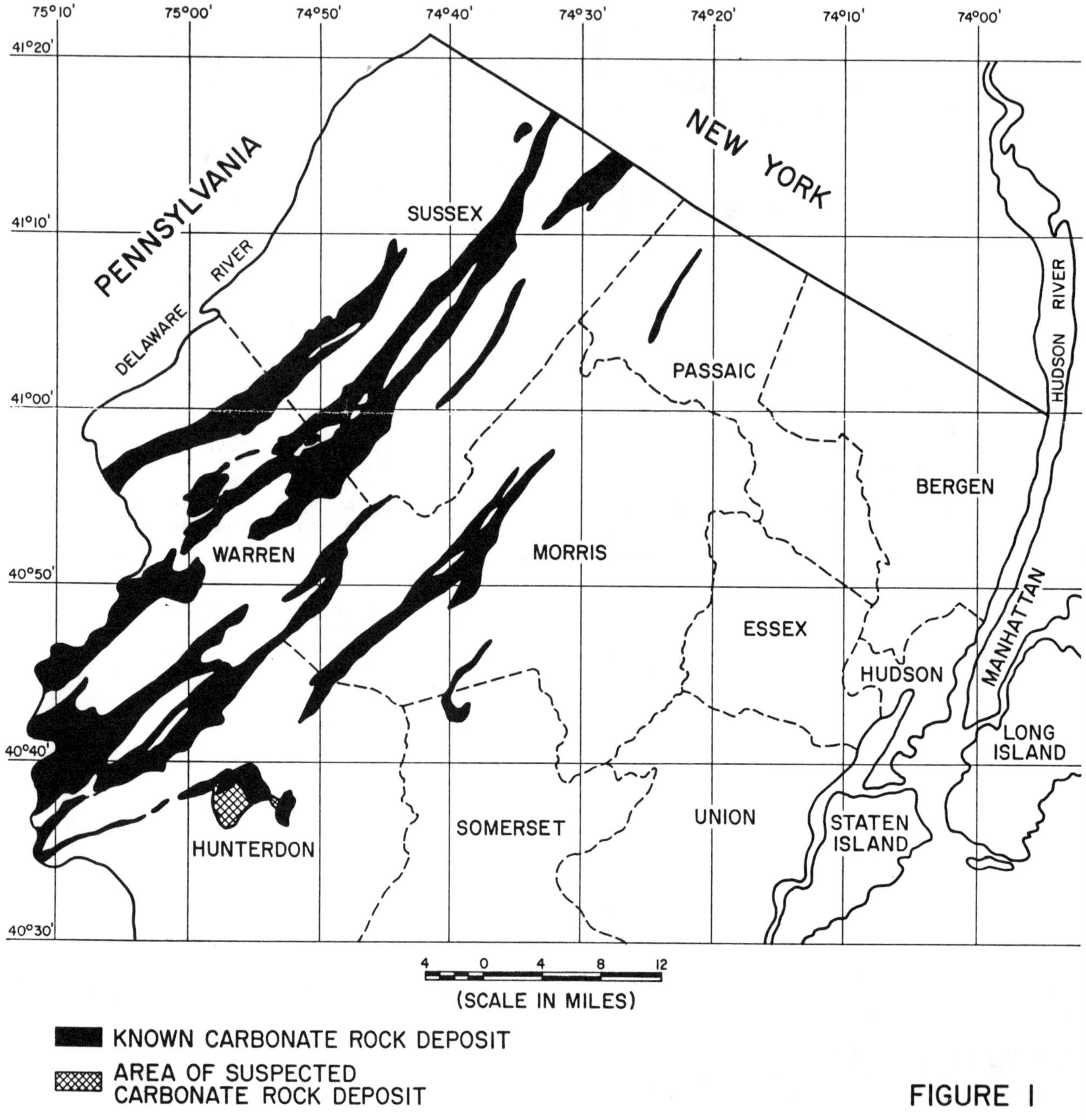

■ KNOWN CARBONATE ROCK DEPOSIT
▨ AREA OF SUSPECTED CARBONATE ROCK DEPOSIT

FIGURE 1

the closing of the proto-Atlantic Ocean and the limestones were subsequently deposited within the fold produced valleys and at the continental margin. Subsequent deformation along the edge of the continent resulted in faulted and broken rocks, and moved these limestones to their present positions. Thickness of New Jersey limestone formations vary from near zero to more than 1000 feet.

The Geologic Map (Figure 1) provides our best estimate of the cavity prone limestone areas. This map is essentially the first step in defining the problem that the owner/planner/design team faces. Most of the areas shown on the map were delineated by State of New Jersey Geology Survey and Department of Environmental Protection personnel, with only minor additions by the authors.

The State geologists have done a remarkable job, providing a great deal of useful information regarding the carbonate rocks in New Jersey. Unfortunately, many civil, municipal and design engineers, planners, developers and architects in New Jersey and surrounding areas are not aware of the existence of such data, or the engineering implications of the carbonate rock deposits in the state. Solution channelling, cavernous and sinkhole conditions can be anticipated in the carbonate rocks and residual soil of the Kittatinny group, the Jacksonburg Formation and the carbonate portions of the Jutland member of the Martinsburg Shale.

Table 1 provides a brief description of the rocks delineated on Figure 1, together with an evaluation of their potential for sinkhole development. Our estimate of sinkhole potential is based on grain size, known faulting, fracturing and/or jointing, groundwater hydrology and historical sinkhole or solution channel development.

The available generic information on well yields in these formations, combined with local well data, can provide a useful indicator of the potential for sinkhole or solution channel existence. In general, the larger the well yields in an area, the more likely cavity and channel implications to a project.

Groundwater flow in the New Jersey limestones travels in the joints, fractures and cavities in the rock. There is virtually no primary permeability or porosity. Groundwater is found under both semi-artesian and water table conditions. Well yields vary dramatically, from some ½ gallon per minute to several thousand gallons per minute. Obviously, wells penetrating larger fractures or cavities, in some instances near streams, have the largest specific capacity. Apparently solutioning is more abundant in valleys, depressions, and close to streams.

The water quality from wells is generally hard and low in iron. As would be expected, bicarbonates are high, as well as dissolved solids, calcium, and magnesium.

GEOTECHNICAL ENGINEERING

Introduction - The mere existence of cavities in limestones, is in itself, not necessarily of concern to manmade structures (unless water loss from the project is of significance). The cavities were formed over a long time period, in rock that is hundreds of millions of years old. Thus the enlargement of existing cavities, or the creation of new cavities in foundation rock is usually not a threat to engineered construction with conventional economic lifetimes of 50 to 100 years. Furthermore, a surprisingly thin layer of sound limestone overlying nominal sized cavities can support relatively high loads. Samples extracted from sound, unweathered strata have unconfined compression strengths of over 10,000 psf.

What is of concern, however, is the formation, or enlargement, of voids in the residual soils above the limestone. Water flow through these soil voids is related to cavities in the underlying rock. The water flow, apparently along the top of an impermeable rock layer, can erode these soils, forming sinkholes, well within the economic lifetime of conventional structures.

Planning - The first step in any site evaluation is to define the possibility and probable extent of the "sinkhole" problem. Is all, or some part of the site subject to sinkhole formation? Geologic data such as Figure 1, combined with site reconnaissance, and aerial photo inspection, are vital first steps in assessing the problem.

Thus, for a very small investment, a very significant assessment of the areal extent and degree of possible problems can be obtained. This information can be assessed and decisions made as to develop more definitive information, or abandon all, or a portion of the site.

Investigation - Many indirect procedures have been advanced for the detection of subsurface cavities (which have not yet become sinkholes). These include geophysical studies, such as seismic refraction, gravity and conductivity techniques. All can be useful in certain situations, however, one must almost know the nature of the subsurface before utilizing these forms of indirect sensing. Even air-percussion drilling, albeit quite fast, in most instances does not provide sufficient data to realistically interpret subsurface conditions encountered, even with the aid of widely spaced borings.

Perhaps the authors are of an overly suspicious nature, but carefully drilled test borings, qualified full time inspection, experienced and careful drillers, and large diameter double tube core barrels are the only way to develop definitive measures of the geotechnical concerns of a site. Other procedures can be useful in some instances to correlate between test holes, but not as the only tool to investigate a site. There is nothing as valuable as observing all of a 5 foot run; the clay seams, stained joints, noting the amount of water loss in a clayey residual soil, or watching the drill rods actually fall through a void; to aid in one's understanding of the subsurface.

On one job the authors observed the drilling of grout holes 10 feet apart. At times rock depths varied 20 feet or more in the 10 foot horizontal distance between borings. The amount of grout placed in the grout holes varied from 2 cubic feet of non-accelerated 1:1 grout to 140 cubic feet of non-accelerated and accelerated (as fast as 15 second set time) grout in 30 to 50 foot deep adjacent holes.

Thus we believe that a judicious program of subsurface exploration (with allowances for flexibility as a result of field obtained information) using experienced personnel, rotary wash boring drilling equipment, and double tube core barrels may represent the only positive way to identify the nature and extent of the problems that can be generated by limestones in the subsurface on an unexplored site.

Alternatives - Once the owner/planning/design team is aware of the nature and extent of the problems, they can 1) move away from the area of concern completely, 2) place non-critical facilities in sinkhole prone areas or 3) design solutions to account for the problem.

As the formation of a sinkhole in an open area is not likely to be life threatening, parking areas, ball fields, parks, golf courses and similar, can many times be economically located in limestone areas. From a planning standpoint the location of non critical facilities with an assumption of future maintenance costs included in project financial planning may be an extremely viable solution.

Detention basins to control runoff are mandated by law in most New Jersey communities. The use of such detention basins in limestone areas can only exacerbate the problem by increasing water flow and pressures in subsurface soils, thus increasing the possibility of soil movement and eventual development of sinkholes or open channels in the detention basins and adjacent areas. Unlined detention basins in sinkhole prone areas should be avoided. Conversely, it is possible to take advantage of potentially large well yields from solution prone limestone, or perhaps aesthetically utilize, limestone pinnacles, disappearing streams etc. keeping in mind the potential for migration of subsurface soils and ensuing problems associated with uncontrolled ground water movement.

Some of the more viable foundation solutions, if soil voids and cavities are found below planned construction sites, include: 1) Excavating to rock, filling any cavities large enough to be detrimental to the provision of adequate foundation support with concrete and returning to grade with a controlled structural fill; 2) Installing caissons or piles to sound rock strata. In this instance test probes should be drilled at each pile cap to assure that adequate support is available below the caisson or pile. Recommended bearing pressures should consider the expected or allowable thickness of rock that may be above a cavity. Typical design formulas are presented in Reference 3. The erratic nature of the subsurface topography must always be considered in both design and field installation. 3) Grouting is a well recognized procedure. Either cement, chemical or accelerated grouts have been used to fill cavities, or prevent seepage from reservoirs. Grouting may be used to a) fill cavities (sometimes extremely expensive) or b) merely to provide load bearing columns for lightly loaded areas (for example sanitary sewer lines, houses or similar) or c) a membrane type seal.

CASE HISTORIES -
Construction and post construction problems and expenses which affected projects in New Jersey limestone areas are due mainly to the project teams lack of knowledge of the existence, and or, implications of building in limestone areas. Examples of the bad and the good are summarized below.

High Income Housing Development (50 Units, $250-350,000/Unit) - Califon Member - Sanitary sewer line failure. The borough engineer questioned whether cavernous limestone was responsible. The resulting subsurface investigations indicated numerous areas of potential non-support of the sewer line. Columns of cement grout were installed below the entire line within the limestone area of the site. All subsequent construction in the limestone area of the development require foundation investigations and geotechnical recommendations. Currently investigation requirement appears to be informally applied to all new construction in the Borough.

Detention Basin - Califon Member - Sinkholes developed in the low flow channel, in one outlet area and in the embankment. During all but the heaviest storms no water exits the detention basin outlets but rather runs into the sinkholes, thus increasing the likelihood of additional sinkhole development in the area. The Borough has requested the County, who now owns the basin, to be responsible for remedial investigations and actions.

Bank Addition - Califon Member - The proposed addition, which contained the new vault, was to be constructed in a known sinkhole area. The existence of the cavities in the expansion area was verified by a foundation investigation which was conducted subsequent to a planning and preliminary design stage. This post planning awareness by the architects resulted in a partial redesign of the addition and the vault foundations from spread footings to pile and caisson foundations; new site plans and layout to account for the deletion of the proposed onsite detention basin and the acceptance of future maintenance costs in the parking area.

Insurance Company National Headquarters - Leithsville Formation - 10,000 feet of seismic refraction surveys, 15 borings, and a gravimetric survey on a 50 foot grid failed to detect cavernous conditions in either the residual soils or upper limestone formation at this large building site. Subsequently, 79 borings and 36 test pits encountered a "possible" soil filled void in but one boring. Foundation recommendations were for a mat supported on rock or undisturbed residual soils (except for any localized soft soil zone encountered during construction that might require special treatment). Monitor wells installed prior to excavation encountered cavities. After extensive additional investigation, the foundations were redesigned utilizing a combination of shallow footings and H piles using percussion drilling at each column location to select the appropriate foundation design.

350 Acre Residential, Recreational and Commercial Development - Jutland Member Leithsville and Allentown Formations - The prospective purchaser requested an opinion of the site conditions from their geologic consultant, despite a previous feasibility study by a national geotechnical firm for the original owner. The previous field explorations had consisted of 25 test borings, 10 test pits and 60 seismic refraction "tests," which apparently led to no suggestion of on-site limestones, nor mention of local sinkhole formation, or large local subsurface caverns (spelunker type). The pre-purchase study (12 test pits, a geologic reconnaissance and discussions with the State Geological Survey) indicated some 25% of the site is underlain with potentially cavernous limestones. Post purchase site explorations and design team liaison have resulted in outlining the extent of the problem, the relocation of proposed facilities, and plans for engineering solutions in areas where relocation is not possible. The economic penalty associated with development of the site is being minimized by early awareness, planning, and realistic cost projections for geotechnical solutions to building support and utility concerns.

Summary and Conclusion

The State of New Jersey provides virtually unlimited opportunities for those interested in sinkhole phenomena. Engineers and geologists with an understanding of the distribution and significance of the states limestones can provide an extremely useful resource for the owner/planner/design team, from initial stages of a project (pre-purchase) through field inspection of engineered construction.

It seems improbable that so may sites have not been effectively investigated, or the problems recognized, prior to starting construction. To most of those reading this paper the concerns are obvious. Unfortunately too few members of a design team will recognize that limestones exist in a particular area, nor understand the magnitude of the problem.

As discussed in this paper and throughout the conference, the geotechnical concerns of limestones in the subsurface are real. However, the problem is not insurmountable. The application of intelligent solutions allow the economic development of many cavity prone limestone areas. No radical procedures are embraced in this paper, however, the authors believe as Samuel Taylor Coleridge did, "Common sense in an uncommon degree is what the world calls wisdom."

References

Geologic Map of New Jersey 1911-1912, Lewis and Kummel, New Jersey Department of Environmental Protection (formerly Department of Conservation and Economic Development).

State of New Jersey County and Municipality Maps and Geologic Overlays No.'s 21, 22, 23, 24, 25, 26 and 27, 1956 to 1973, New Jersey Department of Environmental Protection (formerly Department of Conservation and Economic Development).

Fischer, Joseph A.; Szymanski, Jerzy S.; Fox, Robert H. "Foundation Design for a Cavernous Limestone Site" April 1983, Proceedings of the Twentieth Annual Engineering Geology and Soils Engineering Symposium. Boise, Idaho.

Geyer, A.R.; Wilshusen, J.P., 1982. Environmental Geology Report 1, Engineering Characteristics of the Rocks of Pennsylvania, Pennsylvania Department of Environmental Resources.

Markewicz, F.J. 1967, "Geology of the High Bridge Quadrangle, State of New Jersey, Department of Conservation and Economic Development (Now NJDEP)

Markewicz, F.J., Dalton, Richard, Canace, Robert, 1981. "Stratigraphy, Engineering and Geohydrologic Characteristics of the Lower Paleozoic Carbonate Formations of Northern New Jersey" Civil and Environmental Engineering Department, New Jersey Institute of Technology.

Subitzky, Seymore (editor), November 1969 "Geology of Selected Areas in New Jersey and Eastern Pennsylvania. The Geological Society of America and Associated Societies Annual Meeting - Atlantic City, NJ, Rutgers University Press.

Special thanks to Geri Murtha and Vicki Donnelly.

TABLE 1

NEW JERSEY CARBONATE ROCKS

	Formation	Sinkhole Potential
	The Kittatinny Group to date has been divided into five major formations (in decreasing age):	
1.0	**Leithsville Formation** - Lower Cambrian:	
1.1	Califon - medium to coarse grained dolomite.	High
1.2	Hamburg - shaly to siliceous dolomites, shales and sandstones.	Low
1.3.1	Walkill (Lower Unit) - fine to medium grained rubbly bedded, locally vuggy dolomite.	Moderate
1.3.2	Walkill (Upper Unit) - fine to medium grained, locally coarse, crystalline dolomite.	High
2.0	**Allentown Formation** - Upper Cambrian:	
2.1.1	Limeport (Lower Unit) - fine grained, cyclically bedded oolitic and cryptozan bearing dolomite.	Moderate
2.1.2	Limeport (Middle Unit) - medium to coarse grained cryptozan and oolitic dolomite containing silt and sand.	High
2.1.3	Limeport (Upper Unit) - fine grained oolitic dolomite.	High
2.2	Upper Allentown - fine to medium grained dense dolomite (the upper 100-200 feet becomes more siliceous).	High

TABLE 1 Continued

	Formation	Sinkhole Potential
3.0	**Rickenbach Formation** - Lower Ordovician:	
3.1	Lower Rickenbach - thin to medium bedded, fine to medium grained dolomite with local coarse grained beds.	Moderate
3.2	Hope - medium to coarse grained, massive bedded dolomite, interbedded with aphanitic to finely crystalline medium bedded dolomite.	High
3.3	Crooked Swamp Dolomite - fine to coarse grained (euhedral) dolomite.	Low
4.0	**Epler Formation** - Lower Ordovician:	
4.1.1	Branchville (Lower Unit) - variable sequence of very fine to coarse grained, medium to massively bedded dolomite.	Low
4.1.2	Branchville (Upper Unit) - fine grained, massively bedded finely laminated dolomite.	Low
4.2	Big Springs - fine to medium grained dolomite with bands or lenses of siliceous dolomite, quartzite or siliceous shale.	Low
4.3.1	Lafayette (Lower Unit) - fine to medium grained, massively bedded dolomite with shaly laminations.	Low
4.3.2	Lafayette (Upper Unit) - finely laminated massively bedded very fine to fine grained dolomite.	Low
5.0	**Ontelaunee Formation** - Lower Ordovician:	
5.1.1	Beaver Run (Lower Unit) - massive, medium to coarsely crystalline dolomite.	Moderate
5.1.2	Beaver Run (Middle Unit) - massive, medium to coarsely crystalline dolomite.	Moderate
5.1.3	Beaver Run (Upper Unit) - massive, fine to medium grained dolomite, with little chert.	Moderate
5.2	Harmonyvale - dense fine grained dolomite, in 1 to 4 foot beds.	Low
6.0	**Jacksonburg Formation** - Middle Ordovician:	
6.1	Lower Unit - "cement limestone facies" - calcium, medium to coarse grained largely well bedded.	Moderate
6.2	Upper Unit - "cement rock facies" - dark grey to black, fine grained, thin bedded argillaceous limestone.	Moderate
7.0	**Jutland Member/Martinsburg Formation** - Middle to Upper Ordovician: (Carbonate Facies only)	
7.1	Bottom Unit - Thin to moderately thickly bedded, fine to coarse well rounded frosted quartz, suspended in a carbonate matrix consisting of crystalline calcite.	Low
7.2	Middle Unit - chip conglomerate and breccia containing chips and irregular fragments of fine grained limestone and medium grained oolitic quartzoze in a quartz-carbonate matrix.	Low to Moderate
7.3	Upper Unit - very fine to medium grained, locally coarse crystalline, ribbony to platy to massive limestone, in bands from one-half to six inches thick.	Low to Moderate

Sinkhole risk analysis for a selected area in Warren County, New Jersey

D.RAGHU *New Jersey Institute of Technology, Newark, USA*
CHARLES TIEDEMAN *Paulus, Sokolowski and Sartor, Warren, New Jersey, USA*

ABSTRACT

For the design and construction of structures in and on carbonate formations, a sinkhole risk analysis provides valuable data. In this paper, a method of performing a sinkhole risk analysis for a selected area in Warren County, New Jersey is presented. This method is based on field reconnaissance and review of geologic maps to provide a probability base for additional site specific investigation.

Introduction

A considerable amount of construction activity is anticipated in Warren County, New Jersey. Some areas in the county are underlain by carbonate formations that either contain sinkholes and are prone to the formation of cavities. The supporting capacities of foundation of structures in and on these formations are significantly effected by the cavities in them. In order to design and construct new developments in this area, it therefore becomes necessary to perform a sinkhole risk analysis.

For the purpose of this study, an area located in Warren County, New Jersey was selected. Information regarding sinkholes in this area was obtained from various sources. Sinkhole probability distribution curves were generated from this data for the site. Also, a technique using percussion probes to obtain further information regarding sinkholes is outlined.

Area of Study

Topography - Warren County

The area of sinkhole studies in Warren County was located in a valley bounded by the Musconetcong Mountains to the south and the Scotts Mountains to the north. The valley ran northeast from the Philipsburg area to Washington Borough parallel to New Jersey, Route 57, and the Pohatcong Creek.

Geology

Bedrock

The area of study is located in an upland area in a region known as the Reading Prong. Age of the limestones are dated to the Cambrian and Ordovician geologic time periods which are about 520 and 580 million years old. The mountains bounding the Warren County study are of Precambrian age. The geologic formations, consisting primarily of carbonate rock in the Reading Prong, are as follows:

Formation	Symbol	Constituence
Leithsville	€l	Dark to light gray fine to coarse grain dolomite, light gray to tan phyillite, interbedded
Allentown	€a	Dark to light gray fine to medium grain dolomite
Rickenbach	Or	Interbedded limestone and dolomites
Epler	Oe	High calcium, medium to coarse grain limestone

The Leithsville, Allentown, and Rickenbach formations are in some areas mapped as the Kittatinny carbonate terrain. New Jersey State Geologist Frank Markewicz has performed field studies and determined that older formations, i.e., Leithsville belong to the Kittatinny formation[1]. According to reports issued by the United States Department of Agriculture, Soil Conservation Service, the soils were said to overlie bedrock at depths of three feet and greater in the area of study.

Collection of Field Data

To best acquire data for this project, available sources of information, such as initial geologic maps and areal photographs, were examined to determine the location and extent of sinkholes. Since this area is relatively undeveloped, very little information was available from other sources. Two field trips were made to the area in order that printed data could be verified and measurements could be made on observed sinkholes. Figure 1 shows the area visually examined for sinkholes. The general relief, location of sinkholes, and plots of geologic formations and highways in the area are presented on this map. For the purpose of this study, a specific area of approximately 2,600 acres in Lopatcong and Greenwich Townships was selected.

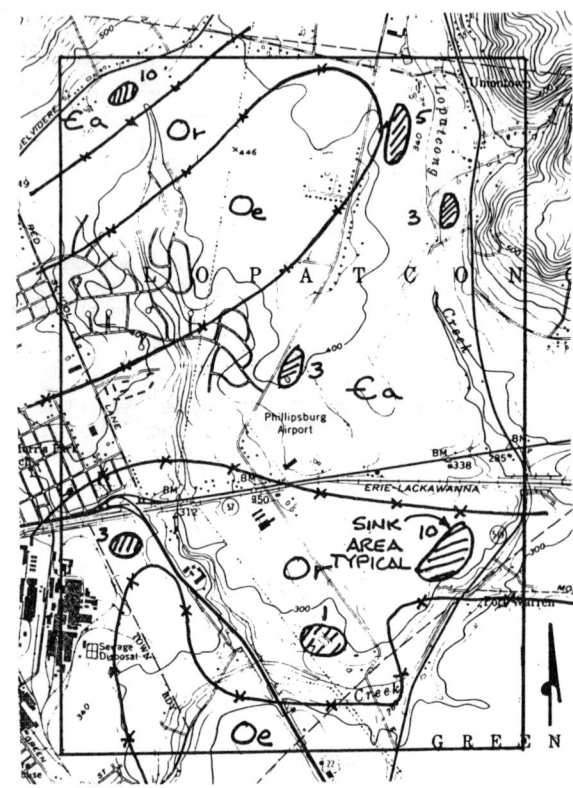

FIGURE 1
AREA OF STUDY
REFERENCE MAP -
U.S.G.S. EASTON PA QUADRANGLE
SCALE: 1" = 2,900'
STUDY AREA = 2,600 ACRE ±

Analysis of Data

With the available data, it is difficult to derive a probability distribution function. For this study, a Poisson's distribution function has been chosen to best model the data. This model is usually employed for representing natural occurring events in time and space[2,3].

Poisson's distribution is written as such:

$$P(n) = \frac{(V \times A)^n}{n!} e^{-VA}$$

where: V = mean rate of occurrence

A = area of interval

n = number of occurrences

$P(n)$ = probability of occurrence

This study indicates the average areal extent of each visible sinkhole occurrence was approximately 110 square feet. The percent area of coverage of active sinkholes is 0.25 percent. Therefore, the mean rate of occurrence can be calculated.

$$V = \frac{110 \text{ sq. ft.}}{43,560 \text{ sq. ft./AL}} \times \frac{1}{0.0025} = 1$$

From these observations, a probability mass function curve (PMF) were developed for sites two, five, ten, and twenty acres in area which are presented on Figure 2. These curves can be used to anticipate the number of events for a given area and develop additional field study programs.

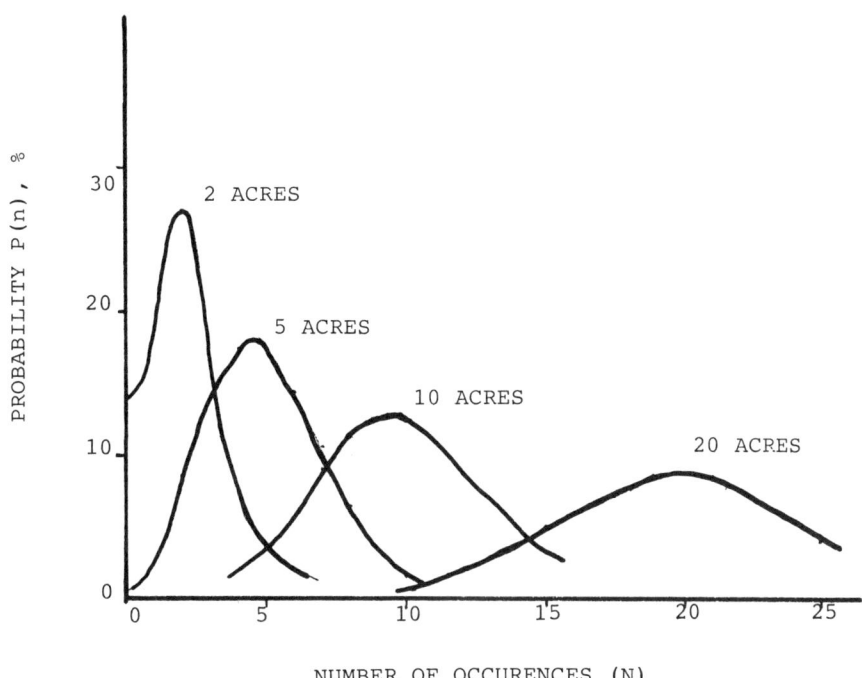

FIGURE 2 PMF CURVE SHOWING DISTRIBUTION WITH INCREASE IN AREA

Discussion

The data collected from the study could be used to determine the economic feasibility of the development of a site in the area. For example, for a site in this selected area containing ten acres, the largest probability of sinkhole occurrence corresponds to ten events, i.e., 12.5 percent. If for design purposes, a probability of sinkhole occurrence is five percent, then the corresponding number of events is 13. As stated earlier, the average area of visible sinkholes at the surface is found to be 110 square feet. This means foundations should be designed for a loss of bearing area of 110 square feet in this area.

The selection of percussion probes for foundation design, however, would be based on a number of factors, such as type of structure, structural loading, and the structure use. Once such a probability level is defined for a structure, the factors for design can be developed as stated earlier.

Little information regarding solution activity at depth can be obtained by performing a more detailed subsurface investigation. Percussion probes in conjunction with standard test borings and other conventional investigation techniques can be employed for this purpose. The design and construction of foundations using such techniques is in a companion paper[4].

References Cited

(1) Markewicz, Frank J. and Dalton, Richard, "Lower Paleozoic Carbonates: Great Valley", New Jersey Dept. of Environmental Protection, Division of Water Resources, Trenton, New Jersey.

(2) Lilly, W. W., 1976, "A Probability Study of Sinkhole Distribution".

(3) Ang, A. H-S and Tang, W. H., 1975, "Probability Concepts in Engineering Planning and Design".

(4) Raghu, D., Lifrieri, J. L., and Rhyner, F. C., 1984, "Use of Percussion Probes for Designing and Constructing Foundations in Carbonate Formations".

Use of percussion probes for the design and construction of foundations in and on carbonate formations

D.RAGHU *New Jersey Institute of Technology, Newark, USA*
J.J.LIFRIERI *Paulus, Sokolowski and Sartor, Warren, New Jersey, USA*
F.C.RHYNER *Storch Engineers, Florham Park, New Jersey, USA*

ABSTRACT

The supporting capacity of foundations in and on carbonate formations is considerably influenced by (1) the location and extent of cavities under them and (2) the quality and thickness of bridging material between the foundation and the sinkholes. In order to determine this, a subsurface exploration has to be performed under each foundation. Conventional techniques such as borings are too expensive and slow. Geophysical methods do not yield absolute subsurface information. Percussion probing is a fast and economical method of obtaining design data in such a case. In this paper, through case histories, a technique of designing foundation in karst regions using percussive probe penetration rates is presented.

Introduction

Limestones and dolomites, called Carbonate formations, often contain voids. Whenever the hydrological regime in these formations is altered either by manmade and/or natural causes, these voids tend to increase in size. The supporting capacity of foundations in and on these foundations depends on the size and location of the cavities and the dimension and quality of rock between the voids and the foundation. Several conventional subsurface exploration methods are available to determine the nature and extent of cavities. Borings provide useful information. But, they tend to be slow and expensive, if performed at each footing location. Geophysical methods do not give an absolute and accurate assessment of subsurface conditions under each footing. It is thus necessary to employ a fast and inexpensive means of exploring subsurface conditions in cavernous carbonate formations.

In this study, percussion probes were employed for determining the dimensions and extent of cavities and the quality of rock at each footing. This method was successfully employed for two sites, one in Morris County and the other in Hunterdon County in the State of New Jersey. Comparing percussion probe rates and data from borings, a site specific correlation was developed between penetration rate of probe and the quality of rock. These correlations provided the following data (1) type of foundation for each structural unit (footings or piles) and (2) depth of foundations and (3) verification of the soundness of bottom of pile.

Foundation Problems in Karst Regions

In karst terrain, of primary concern is the catostrophic loss of support beneath foundations. This can be a result of migration of soil from beneath footings into underlying cavities or voids, or of the sudden collapse of a cavity roof or ledge shear failure somewhere beneath the foundation. In the former case, soil is carried through fissures or solution channels in the rock by downward percolating water. The water may result from surface infiltration of precipitation, or from a leaking piping system. In the latter case, increased stresses caused by imposed foundation loadings might be sufficient to cause the roof of an underlying cavity to collapse or shear existing rock ledges, thereby undermining the support of the foundations.

Of secondary concern is the differential weathering of the bedrock into weathered incompetent zones that may manifest themselves as compressible layers. These layers may be found between more competent, relatively incompressible material thereby posing a potential settlement problem. These and other relevant aspects of design and construction are discussed in this paper.

Case History Number One

The first case history presented here pertains to a headquarters complex located on a 140 acre site in western central New Jersey. Ten three and four story buildings interconnected by a two level below-grade parking structure formed main part of the building. The individual building units were of steel-frame construction above ground and of reinforced concrete construction below ground. Column loads ranged from 400 to 1,000 kips. Wall loads did not exceed 24 kips per foot.

Geology

The site geology and subsurface conditions were explored prior to the start of the testing program. The area was determined to be underlain by at least two members of a cavernous limestone formation. The major member present is a fine grained and thinly bedded dolomitic limestone with interbedded residual clay layers. The other member is a massive crystalline dolomitic limestone which weathers in a pinnacled configuration. The site is intensely folded and faulted and contains voids and cavities of varying sizes. Some of these voids were found to have been filled with sediments. Groundwater levels at the side were determined to be below the final mass excavation level in the bedrock but occasional pockets of perched water were observed in the excavated overburden.

Philosophy of Foundation Design at Site

Due to the frequent occurrence of soft seams within the bedrock and the need for liners, caissons were conddered to be too costly. The cost of a mat foundation was also determined to be prohibitive. A combination of spread footings and piles was chosen as the most economical approach. In all cases, it was decided that the foundations would be founded on sound rock to preclude any possibility of loss of overburden soil from beneath the foundation and to minimize differential settlements. In order to ensure that all foundations were founded to sound rock, an exploration program at each column location was considered necessary due to the variability of the underlying rock conditions. This program would have to answer several design-related questions regarding foundation type. If shallow foundations could be utilized, it would be necessary to determine whether they could be founded at or close to the mass excavation level or if dropped footings would be required. For the dimensioning of the footings a bearing capacity, dependent upon the rock quality, would have to be determined. Wherever piling would be required, bearing levels would have to be determined. In addition, it would be necessary to ensure that a sufficient amount of competent rock existed beneath the founding level to distribute imposed stresses sufficiently so that neither a cavity roof collapse nor a punching shear failure would occur.

In order to determine the necessary design data discussed above it was decided to utilize and develop foundation bearing quality correlations using penetration rates from percussive type probings.

Literature Review

A detailed literature search was performed prior to the determination of the foundation design parameters. Papers published by the U.S. Bureau of Mines, the Colorado School of Mines and various mining and professional journals were reviewed.[2,4,10,11,12,17,18] The information provided by these papers dealt mostly with the drillability index and penetration rates of "air track" probes in specific rock types. These papers were generally focused on the expected rate of penetration for rock of varying type. Literature concerning foundation design in karst terrain was also searched as well as manufacturers' literature covering capabilities of equipment. Several papers were studied discussing the use and installation of H-piles as a foundation element in limestones and karst terrain.[1,4,6,7,8,12,14,15,16]

Test Program

Based upon review of the available literature, it was evident that the use of data obtained from probe holes as a foundation quality indicator has been limited to gross differentiations between rock and voids. Literature reviewed as part of this study does not contain any indication that a systematic quantitative approach to predict foundation bearing levels or rock quality has been developed. Therefore a testing program was designed and performed at the site to develop a relationship utilizing probe penetration rates. A Crawlair ECM-350, equipped with a VL-140 valveless drifter, was used. Of the various size drill bits tested, the 3-inch bit size was selected for use since it provided penetration rates that were easier to record but rapid enough to be economical. To provide the necessary compressed air, an Ingersoll Rand Spiro Flo DXL 750P Compressor, delivering 750 cubic feet of air per minute at 125 psi was used. To insure reliability of data, the equipment was checked and maintained and the drill bits changed when worn.

A test area was selected where borings were previously drilled. Probes were made adjacent to the borings and the penetration rate data obtained from probes were compared to the material recovered from the borings enabling preliminary correlations between the observed probe penetration rates and the recovered rock core quality to be made. These preliminary correlations had to be modified as the explorations progressed. It became evident that a visual inspection of the material penetrated was required before bearing quality parameters could be assigned to specific penetration rates. To accomplish this, a probe rig was set up at the top of slope of a near vertical excavation exposing rock and soil of various quality, and probing performed. The quality of the rock exposed in the excavation cut could then be visually inspected and compared to corresponding penetration rates observed during probe operations. Similarly, several test locations were made in each representative rock type to develop a reliable correlation. A comparison of penetration rates from different drill rigs at the same locations was also made. In all cases the maximum variation in penetration rates between different rigs was approximately ± 2 seconds per foot. More details of the program are presented in reference 9.

A comparison of probe penetration rates within rock of similar quality, but belonging to the different members of the formation present at the site, indicate significant differences. This is illustrated in the typical probe log shown on Plate 1. It was observed that the more massive coarse grained member exhibited much faster penetration rates, approximately 0.7 times the penetration rate for rock of equivalent quality within the finer grained, thinly bedded member.

TEST PROBE LOG PROBE NO. 18-3	
PROJECT	Proposed Corporate Headquarters
CLIENT	
CONTRACTOR	Rock Drilling, Inc.
GROUNDWATER	
EQUIPMENT DETAILS: BIT Carbide DIAMETER 3 in	
HAMMER: TYPE VL-140 RATING 100 PSI 750 CFM	
RIG TYPE: Crawlair ECM 350 RODS: Steel	
COMPRESSOR: TYPE IR DXL 750-P RATING: 125 PSI 750 CFM	
SHEET NO. 1 OF 1	
JOB NO. 10000G	
ELEVATION +232	
DATUM Mass Excavation	
DATE START 2/6/80	
DATE FINISHED 2/6/80	
DRILLER L. C.	
INSPECTOR —	

Depth Below Surface (ft)	Resist Seconds Per Foot	Length of Rods	Depth	Geologic Symbol	Identification of Strata	Remarks
0						
	9	12			RESIDUAL SOIL	
	30	11			HARD DOLOMITE	
	31	10				
	35	9				
5	2	8				
	2	7			VOID OR CAVITY	
	2	6				
	2	5				
	2	4				
10	2	3				
	35	2			HARD DOLOMITE	
	36	1				
	1	12				
	1	11				
15	1	10			VOID OR CAVITY	
	1	9				
	1	8				
	26	7				
	28	6			MEDIUM TO HARD DOLOMITE	
20	30	5				
	31	4				
	6	3			RESIDUAL SOIL	
	7	2				
	1	1				
25	1	12			VOID OR CAVITY	
	2	11				
	30	10				
	31	9				
	35	8			HARD DOLOMITE	
30	29	7				
	31	6				
	32	5				
	36	4				
	28	3				
35	31	2				
	43	1			BOTTOM OF PROBE 36'	

Figure 1

Determination of the Foundation Quality Index
The correlation between the penetration resistance and the bearing capacity of the rock, referred to as the "Foundation Quality Index" (FQI) was defined as follows:

$$\text{Foundation Quality Index (FQI) \%} = \frac{\text{Penetration resistance in seconds/foot}}{\text{Penetration resistance in seconds/foot for intact competent rock}} \times 100$$

The penetration resistance for intact competent rock is unique to each site, dependent upon the type of rock and the kind of equipment used. For this particular site the penetration resistance for intact competent rock was determined to be 40 seconds per foot within the finer grained, thinly bedded rock member. In order to account for difference in penetration rates observed between the differing rock units, the penetration rates obtained from the probes made within the massive coarse grained rock member were increased by multiplying each observed rate by 1.43. In this manner, these penetration rate values were normalized to those observed in the fine grained member upon which most of the foundations were established.

The allowable net bearing capacity of a rock exhibiting FQI values between 75 and 100 percent was observed to be 40 tons per square foot. This rock would be similar to the Class 2-65 rock as described by the New York City Building Code (1970). Similarly, rocks exhibiting lower FQI values could be assigned allowable bearing values dependent upon observable quality and could also be correlated with equivalent classes of rock as described in that code. Any rock zone exhibiting FQI values greater than 100 percent were arbitrarily set equal to 100 percent for calculation purposes. Table I has been prepared and shows the relationships between the FQI value, type of rock and allowable bearing capacity observed for the subject site. For purposes of comparison, the corresponding classes of rock as defined by the New York City Building Code and their nominal values for bearing are also presented in Table I.

FQI values less than 30 percent indicate soil conditions. Values less than 15 percent were indicative of soil filled cavities, very soft seams or voids in the bedrock. In this case no bearing value or type was assigned.

TABLE I

MATERIAL DESIGNATION TYPE	FQI %	ALLOWABLE BEARING VALUES (TSF)	EQUIVALENT N.Y.C. BUILDING CODE CLASS	ALLOWABLE BEARING VALUES (TSF)*
A	75-100	20-40	2-65	40
B	40-75	7-20	3-65	20
C	30-40	4-7	4-65	12
D	15-30	Appropriate Soil Values	N/A	N/A

* Values of class 2-65, 3-65 and 4-65 rock may be increased with embedment in accordance with N.Y.C. Building Code Requirements.

Design Application
In cavernous areas foundations must be supported on material which is capable of bridging over underlying cavities or soft zones. Expressions have been developed to determine the minimum desired distance between the base of the foundation and the top of the cavity or soft zone. One of these expressions was given and discussed by Kitlinski[8] and Dismuke[3]. The desired distance is dependent upon several factors including breadth of the cavity or soft zone, bedding planes or discontinuities in the rock and the strength of the rock. The strength of the rock is related to the quality of the rock and may be determined using the FQI values discussed previously.

To dermine the quality of the bridging material, a weighted allowable FQI value may be determined. Depending upon the FQI value obtained in allowable bearing capacity and allowable tensile stress may be assigned. The weighted allowable FQI value is determined by cummulatively adding the FQI values observed for each foot of bridge material and dividing the sum by the depth of bridge considered. The weighted FQI value may then be compared to the values shown in Table I and appropriate strength values selected. This method may be used providing material with FQI values less than 30 percent (Type D) are not present within the considered bridge zone and all FQI values in excess of 100 percent are reduced to 100 percent. The presence of Type D material within the weighted bridge zone invalidates this approach and a new bridging zone has to be defined.

Shallow foundations were designed for a rock bridge of at least Type B quality and of sufficient thickness as determined by the FQI values and previously mentioned expressions. In addition, it was necessary that the rock bridge be encountered within a specified depth below the slab subgrade. This cutoff depth was determined by the construction manager to be the depth to which the construction of shallow or dropped footings were more economical than driven piling. At columns where the depth to usable bridge zones exceeded the economical cutoff depth, pile foundations were utilized. By load test, a weighted FQI value of 40 percent was observed to perform adequately and was therefore selected as the minimum criterion whereby a pile bridged zone was judged to be satisfactory. If the analyses of the initial probe data indicated that a pile foundation was required, probes were made at each proposed pile location within a pile cap. The probes defined the anticipated depth to which the pile would be driven and determined the need for predrilling prior to pile installation. Predrilling was considered necessary when thin layers or zones of insufficient bridging material, too hard to penetrate without damage to the pile, were underlain by weaker materials exhibiting the FQI values less than 40 percent or by voids. By driving test piles it was determined that materials with FQI values less than 40 percent could be penetrated using the pile driving equipment available at the site. When the pile fetched up above or below anticipated depths, the adequacy of the bridged zone was established by probing beneath the pile tip alongside and within the pile flanges.

Case History Number Two

This pertains to the construction of research and office facilities covering a ground floor area of over 300,000 square feet. The three different buildings were a maximum of three stories and included partial basements.

The site was of gently undulating topography that masked the troublesome limestone conditions below. Geologic maps and published literature indicate that the side is underlain by residual silts and clays and by the Kittatinny limestone. Borings indicated that the limestone was actually a dolomite and included occasional to sometimes frequent cavities, as deduced by the dropping of rods during rock coring and poor rock core recovery.

The selected foundations were spread footings on sound rock, where such rock was shallow, or drilled piers socketed into sound rock. The columns loads were relatively light for such foundations and were typically several hundred kips to a maximum of approximately 1,500 kips. Footings were designed for bearing pressures not to exceed 16 tsf. Rock sockets were designed for skin friction along the shaft of 100 psi and end bearing at the base of 16 tsf. All drilled piers were 30 inches diameter to allow access by workers for cleaning and inspection.

Initially, probes were drilled after the drilled pier had been advanced to a proposed founding depth based upon interpolation between borings. The probes sometimes disclosed the location of cavities. Also, visual inspection sometimes disclosed that the rock was unsuitable for founding rock sockets. Accordingly, the drilled pier was carried deeper. Subsequently, the procedure was modified so that the probes were made in advance of the drilled pier to predetermine the founding depth. This reduced the need to manipulate the drill rig over each drilled pier a number of times. To achieve this use of the percussion probes, a correlation was developed between the rate of percussion probe penetration and the apparent quality of rock based upon visual examination. Probes for correlation purposes were drilled into different rock types exposed in the basement area of the building. They were also made adjacent to previously made drilled piers so that a fair representative sampling was obtained. Results of the correlation program showed that penetration rates could indicate three different categories. Firstly, penetration rates less than about 10 to 15 seconds per foot were generally in soil seams, open cavities, or highly weathered rock. Penetration rates of greater than about 45 to 60 seconds per foot were generally in relatively sound to sound rock capable of supporting rock sockets. With this information, it was possible to determine the depth of caissons and the adequacy of foundation to carry design loads.

Closing Remarks and Recommendations

In general, from the two case histories presented, it appears that the foundation design concept, using FQI values, appears to be successful. From the experience gained, the following recommendations are being made:
1) Ascertain and develop a thorough working knowledge of the geologic conditions at the site.
2) Develop a testing program for each rock type and normalize the penetration rates by using those of the predominant rock tupe as a base value.
3) Insist upon competency and continuity of personnel used to drill and inspect probing operations.
4) Ensure that accurate surveying controls are established at the site during probing, predrilling and piling operations.
5) Foundation bearing levels should be established as shallow as possible since accuracy of prediction and economics diminish with depth.
6) Foundation design plans should contain equivalent foundation tables for converting from shallow foundations to deep foundations and vice versa, should the need arise.

It would be remiss on the part of the author if the weaknesses and drawbacks of probing techniques are not pointed out.

The most significant limitation of a design approach using air track probes has been one of depth. It would appear from the results of the foundation construction program that the greater the depth, the less reliable the design predictions. For the site in question, it was determined that for depths greater than 40 feet the accuracy of the predicted pile lengths diminished rapidly. Its usefullness is further limited by changes in geologic conditions at depth, local soft seams, flexibility of drill rods and variable factors inherent with human involvement.

References

1. Ashton, W.D. and Schwartz, P.H., "H-Bearing Piles in Limestone and Clay Shales", Journal of the Geotechnical Engineering Division, A.S.C.E., July 1974.

2. Bruce W., and Selim, A.A., "Prediction of Penetration Rates for Percussive Drilling", Report No. 7396, Investigations, U.S. Bureau of Mines, June 1970.

3. Dismuke, T.D., "Structure Foundations in Limestone Regions", Proceedings, Symposium on Engineering, Construction and Maintenance Problems in Limestone Regions, Lehigh University, Bethlehem, Pennsylvania, August 1976.

4. Fang, H.Y., Dismuke, T.D. and Lim, H.P., "Pile Foundations in Pinnacled Limestone Regions", Analysis and Design of Foundations for Tall Buildings.

5. Foose, R.M., "Engineering Geology Karst Terrain", Bulletin of The Association of Engineering Ceologists, Vol. XVI, No. 3, 1979.

6. Kim, J.B. and Brungrober, R.J., "Full Scale Lateral Load Tests of Pile Groups", Journal of Geotechnical Engineering Division, A.S.C.E., January 1976.

7. Kim, J.B., et al., "Pile Cap Soil Interaction from Full Scale Lateral Load Tests", Journal of the Geotechnical Engineering Division, A.S.C.E., May 1979.

8. Kitlinski, F.T., "Soil and Foundation Studies for Milton S. Hershey Medical Center", Pennsylvania Professional Engineer, June 1969.

9. Lifrieri, J.J., "Foundation Analysis and Design in Karst Terrain Utilizing Percussion Probe Penetration Rates and the Foundation Quality Index", Project, Submitted towards Partial Fulfillment of Master of Science Degree in Civil Engineering, New Jersey Institute of Technology, Newark, New Jersey, May 1980.

10. Liljestrand, W.E., "Rotary Percussion Air Hammer Drilling" Colorado School of Mines Quarterly, Vol. 56, No. 1, January 1961.

11. Misra, B.C., "Penetration Rates in Rock Drilling" Journal of Mines, Metals and Fuels, September 1960.

12. Paone, J., Padison, D., and Bruce, W.E., "Drillability Studies-Laboratory Percussive Drilling", Report, U.S. Bureau of Mines, September 1969.

13. Peck, R.B., "Rock Foundations for Structures", Proceedings, Specialty Conference on Rock Engineering for Foundations and Slopes, A.S.C.E., University of Colorado, Volume II, August 1976.

14. Peck, Hanson and Thornburn, Foundation Engineering, II Edition, John Wiley and Sons, 1974.

15. Sowers, G.F., "Failure in Limestone in Humid Subtropics", Journal of the Geotechnical Engineering Division, A.S.C.E., August 1975.

16. Sowers, G.F., "Foundation Bearing Weathered Rock", Proceedings, Specialty Conference on Rock Engineering for Foundations and Slopes, A.S.C.E., University of Colorado, Volume II, August 1976.

17. Tandanand, S., and Unger, H.F., "Drillability Determination, A Drillability Index for Percussion Drills", Report, U.S. Bureau of Mines, 1979.

18. White, C.G., "Development of a Rock Drillability Index", Proceedings, 7th Symposium on Rock Mechanics, Volume I, June 1976.

19. White, C.G., "Rock Durability Index", Colorado School of Mines Quarterly, Volume 64, Volume 64, April 1969.

Methods for describing and predicting the occurrence of sinkholes

ALBERT E.OGDEN *Edwards Aquifer Research and Data Centre/SWTSU, San Marcos, Texas, USA*

ABSTRACT

This paper summarizes the various techniques that can be utilized to describe the morphology and spatial occurrence of sinkholes for the purpose of determining sinkhole origin and predicting future subsidence. Insight to the origin of sinkholes can be obtained from the width/depth ratio, depth/diameter ratio, utilization of discriminate and fourier series analysis, and an analysis of man's effects. To aid in predicting future doline occurrence, present dolines should be compared to factors such as geologic formation, cavern occurrence, topographic setting, relief, depth to the water table, and proximity to local base level. If lowering of the water table by man is a suspected cause, the change in depth to water should be compared as well. Probable direction of future doline occurrence or growth can be ascertained by comparing short and long doline orientation axes to the orientations of cave passages, joints, and photo-lineaments. If the types of dolines can be classified as to origin, then subsets of the data should be made prior to statistical treatment. All the data can then be further analyzed using multiple regression to determine which factors are most significant to doline occurrence. By combining two or more factors on a grid system, subsidence susceptibility maps can be made. In addition, by combining construction-restraint factors such as slope or soil permeability, additional maps can be made for land-use planning purposes.

Introduction

Sinkholes have been listed as an important hydrologic hazard and land-use issue in the recent National Water Summary-1983 of the United States Geological Survey (1984). The report details eight states that consider sinkhole collapse an important water issue. These states are Alabama, Florida, Georgia, Tennessee, Kentucky, Missouri, Pennsylvania, and Iowa. In addition to these states, this author has seen newly collapsed depressions in West Virginia, Virginia, Arkansas, New York, Indiana, Illinois, and Texas. Therefore, since the collapse of sinkholes occurs with frequency in at least 15 states, a need exists to predict their occurrence to help prevent loss of life and property. The purpose of this paper is to describe the processes forming the different types of sinkholes and to briefly review various research approaches that can be utilized in predicting future occurrence of sinkholes.

Types of Sinkholes

In order to predict sinkhole occurrence, it is first important to be able to distinguish the different types of sinkholes and to understand the processes which form them. The term sinkhole is defined by many geologists as a circular depression found primarily in carbonate rocks and generally thought to be a result of collapse. Most sinkholes are, in fact, primarily formed by slow, preferential dissolution of rock along fractures or by slow subsidence due to piping of a surface cover. Therefore, rapid collapse seldom occurs. For this fortunate reason, sinkholes are not usually considered an important hazard in most places. Many of the karst areas of the central and eastern United States have been utilized for farming due to their fertile soils and have thus been sparsely populated. As population pressures have increased, urban and industrial development has moved more and more onto the karst. Thus man is now beginning to be more impacted by natural collapse, and, in addition, to induced collapse and rapid subsidence caused by excessive ground-water pumping.

Since enclosed depressions in soluble rock form by a number of different processes, the term doline is used by most karst researchers so as to eliminate any genetic confusion caused by the term sinkhole. Figure 1 shows the four most common end members of doline types found in either covered or exposed karst terranes (Jennings, 1971). The type of end member to predominate in a specific karst region is related to a combination of numerous geologic, hydrologic, climatic, and topographic factors. As seen in Figure 1, collapse dolines can form either where there is an unconsolidated cover or where consolidated clastic deposits overlie the soluble rock. Collapse dolines involving overlying consolidated

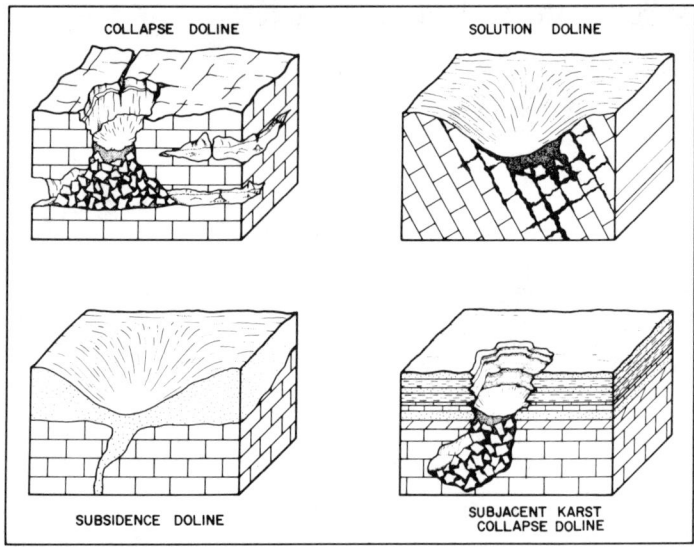

Figure 1. Block diagrams of collapse, solution subsidence, and subjacent karst collapse dolines. (adapted from Jennings, 1971.)

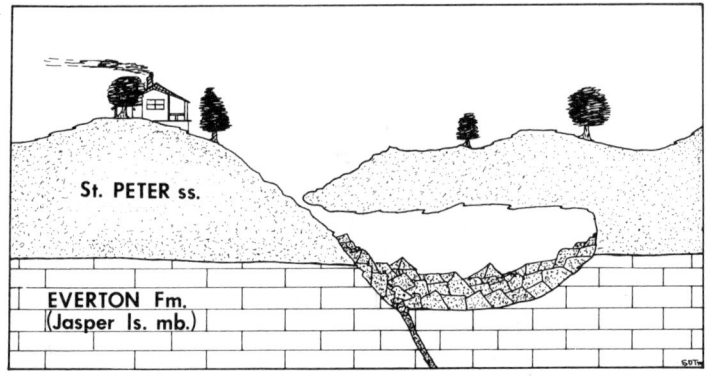

Figure 2: Diagrammatic representation of a cave formed by collapse into a solution cavity accompanied by piping of sand through solution-enlarged fractures. (Ogden, 1981.)

clastics have been termed subjacent karst collapse dolines. Throughout the central United States the Ordovician St. Peter Formation, a loosely cemented and friable sandstone, commonly collapses into underlying Ordovician carbonates (Ogden, 1981). Figure 2 diagrammatically shows a cave which formed first by limestone solution, followed by collapse of the overlying St. Peter sandstone. Subsequent weathering enlarges the doline as sand is piped downward through solution-enlarged fractures in the limestone.

Covered karst is the most common type of karst in the central and eastern United States and can be further classified as to the type of cover. The cover can be just a thin soil, or it can be a thick mantle of glacial, fluvial, or marine deposits. Sub-soil karst, occurs where only a relatively thin cover of soil overlies the rock. In these areas such as in Pennsylvania, West Virginia, and Virginia, solution and collapse dolines predominate. Collapse dolines usually are initially steep-walled and more deep than wide. A general width to depth ratio of 1:3.5 is sometimes used to distinguish between collapse and solution dolines (Jennings, 1971). A doline with a width to depth ratio of 1:20, for example, would most likely be collapse in origin, whereas a ratio of 50:1 would likely represent a doline of solution origin. A study of 500 dolines in West Virginia using the 1:3.5 width to depth ratio, suggests that 91 per cent of the dolines are primarily solution in origin (Ogden, 1982). Unfortunately, this ratio is only a general indicator of origin, since most dolines in sub-soil karst terranes form primarily by slow, preferential solution along fractures with some small-scale collapse and subsidence caused by slow, downward piping of residual material. Therefore, large, rapid collapses are very rare in the sub-soil karst terrances of the United States.

Collapse and rapid subsidence are a much more significant hydrologic hazard where the karst has been mantled by a cover of loosely consolidated material such as marine, glacial, or fluvial clastic deposits. Mantled karst occurs, for example, in parts of Iowa and Missouri (glacial and glacial-fluvial) and parts of Alabama, Georgia, and Florida (marine and fluvial). Subsidence dolines are the predominate type, being formed by piping of the cover through solution-enlarged fractures. Piping commonly does not occur uniformally throughout the cover, thus subterranean voids held together by a few "keystones" are formed. Sudden collapse can occur, but it is generally preceded by longer term, imperciptively slow subsidence.

Rapid lowering of the water table by man's activities, increases the rate of piping. This can lead to large-scale catastrophic collapse. Newton (1977) states that there are an estimated 4000 "induced" dolines or related fractures which have formed in Alabama since 1900. Most of these he attributes to water-table lowering associated with ground-water withdrawals and construction. He states that the lowering of the water table causes the downward migration of the doline cover or fill due to : (1) "loss of buoyant support, (2) increase in the velocity of movement of water, (3) water-level fluctuations at the base of the unconsolidated deposits, and (4) induced recharge (Newton, 1977)." Foose (1968) and Quinlan (1974) have described the formation of large subsidence dolines in a dolomitic karst in South Africa also due to excessive ground-water withdrawals associated with gold mining operations. Sinclair (1982) has likewise described the occurrence of subsidence dolines due to ground-water withdrawal from the Floridia aquifer around Tampa.

Loading of the land surface by construction of water impoundments can also cause subsidence and collapse. Newton (1976), Sinclair (1982), and others have shown that leaky impoundments can accelerate piping due to increased hydrostatic head. The weight increase alone can cause collapse into underlying voids. Many reservoirs throughout the world have drained suddenly due to bottom collapse into cavernous voids. Aley, et al. (1972) and Alexander (1980) have documented cases where waste lagoons have collapsed into karstic limestones and have contaminated the ground water. Reitz and Esklidge (1977) have reviewed construction methods that can be utilized to decrease the risk of doline collapse.

This review of doline types demonstrates the complexity and interactions of processes and physical conditions that form dolines. Through an analysis of these factors, tools can be developed that can aid in predicting their future occurrence.

Approaches for Predicting Doline Development

Numerous karst researchers have discussed the processes involved in doline development, and they have analyzed the causes of specific catastrophic collapse events. But few researchers have quantitatively described the factors affecting doline development for the purpose of predicting future occurrence. Cramer (1941) was the first to classify dolines by the dominant process forming them. Ford (1964) analyzed the occurrence of dolines in the Mendip Hills of England and related doline density to topographic setting. He found that eighty per cent of the dolines occur within narrow valley floors, thirteen per cent are found on steep side slopes, and seven per cent occur on the valley interfluves. Ford (1964) also showed that most dolines in the Mendip Hills have little relationship to the courses of known caves. Williams (1966) first applied the concept of morphometric analysis to doline development. As a result, he devised several new parameters to relate doline occurrence to topographic setting and depth to water, as well as for describing doline shapes. Palmquist (1977) performed a detailed analysis of the distribution and density of dolines in the glacial-mantled karst of Iowa. He was able to relate doline occurrence to the age of the glacial surface. In addition, he performed a multiple regression analysis relating doline occurrence to relief, gradient, elevation, proximity to a drainage way, and amount of soil cover.

Many times dolines are completely filled and give no surface expression to allow their detection. Instead of expensive drilling prior to construction, Kirk and Rauch (1977) have demonstrated in West Virginia and Pennsylvania that the tri-potential method of earth resistivity can be used to locate filled or buried dolines. Kirk and Snyder (1977) have successfully used shallow seismic reflection techniques as well to locate caves and to delineate bedrock-soil interfaces.

On site measurements and observations of doline growth have been utilized by several researchers. In order to help predict doline collapse caused by excessive ground-water withdrawal in South Africa, Quinlan (1974) has suggested using deeply-set strain gages, deeply-set telescopic bench marks, precise surface leveling, and close-spaced drilling. In areas of excessive ground water pumping on the covered Floridan aquifer, Sinclair (1982) lists the following physical features that may be precursors of doline collapse: (1) slumping, sagging, or cracking of man-made structures, (2) new areas of rainfall ponding, (3) areas of vegetative stress caused by man-induced water table lowering, and (4) turbidity in water wells. Some of these changes can be detected on aerial photographs. Newton (1976), for example, has used remote sensing in Alabama for the early detection and correction of doline problems. Through numerous borings and use of neutron-probe access tubes, Ruhe (1975) advanced the knowledge of doline "mechanics" by modeling downward flow in an "ideal" doline in the southern Indiana karst. He discovered that seasonal fluctuations in the water table and the intensity of precipitation events could be directly related to soil and alluvium piping and the associated growth of the doline.

Many dolines are believed to develop along lines of structural weakness. Researchers have used various techniques to demonstrate structural control. LaValle (1967) analyzed dolines in central Kentucky and used multiple regression to relate doline characteristics to other karst parameters. He found that twenty-five per cent of the dolines have a direction of elongation that could be related to structural and topographic factors. Kemmerly (1970) statistically related the long-axis orientations of dolines to joint orientations in Tennessee, and also found a relationship between the length and width of dolines. Day (1976) also found some structural control of the orientations of dolines in Jamaica, and utilized the depth/diameter ratio to help postulate depression origin. Both Ogden and Reger (1977) and Wigal (1978) were able to find some structural control on the orientations of dolines in the Greenbrier Limestone karst of West Virginia. They also found doline density to be related to lithology (geologic formation), with the greatest doline densities occurring on the more pure, cavernous, and thick-bedded limestones. To further analyze the distribution of dolines, Brook and Mitchelson (1981) used a two-dimensional, double Fourier Series transform on some Puerto Rico cockpit karst topographic data. Using low and medium frequency components of this transform, they were able to filter karst topographic variations and better define orientations that could be related to structural trends.

Ogden and Reger (1977) proposed a grid-format method of quantifying doline density, cave density, fracture density, and slope factors and combining these to produce a subsidence susceptibility map. Another type of subsidence susceptibility model was made by Brook and Allison (1983) based on doline and fracture distribution data for an area in Georgia. In their study area, they mapped 1011 subsidence dolines developed in thick surface residium over fracture-located cavities in the Eocene Ocala limestone. Using a grid-format data base, maps were made combining doline density, doline area, fracture density, fracture length, and fracture intersection density. They believe that most of the dolines formed due to lowering of the piezometric surface, but often times were elongated along lines of structural weakness. These last two described grid-format methods for quantifying doline occurrence may presently be the best tools for predicting subsidence susceptibility over a broad area from map data only. Other reseachers have described site-specific measures that might help predict subsidence in a given area. The need now exists to devise a method of combining map data with site-specific data so that subsidence and collapse can be better predicted.

Summary
Due to the complexities of geology, hydrology, and climate, no singular, accurate method of predicting the catastrophic occurrence of dolines has been developed. Therefore, to aid in the predictive process, the following set of suggested procedural steps has been devised for doline analysis:

1. Doline Delineation
   All dolines should be delineated from topographic maps and aerial photographs.
2. Size and Shape Analysis
   The depth, width, and elongation trends of the dolines should be determined. The width to depth ratio will help distinguish collapse dolines from solution dolines. If preferred trends of individual doline axes occur or dolines are found to align, these orientations should be statistically compared to joint, fracture, fault, and photo-lineament trends to determine the amount of structural control on doline growth.
3. Lithologic Controls on Doline Development
   The density of doline development should be compared to the various carbonate lithologies of a study area to determine which formations may be more prone to collapse. Subsets of the dolines as to type should be used in this analysis, if possible.
4. Comparison of Cavern Occurrence to Doline Development
   Many collapse dolines occur above limestone caves. Thus, it is important to understand the relationship between doline development and cavern occurrence. Maps of caves are seldom available to the general public. In many cases, they are available through cave exploring organizations. Over two hundred caving clubs comprising the National Speleological Society (NSS) exist in the United States. A library containing published maps can be found at the NSS headquarters in Huntsville, Alabama. With this information, the relationship between dolines and known caves can be better ascertained.
5. Relationships Between Doline Development and Topographic Factors
   Topography is known to affect the development of dolines. Therefore, in order to better predict the future occurrence of dolines, it is necessary to determine the distribution of dolines and doline types with regard to topographic features such as valleys, uplands, valley interfluves, etc.
6. Comparison of Doline Development to Depth to the Water Table
   The density of doline and their depth can be dependent on the depth to the water table. Dolines will sometimes develop more rapidly where the water table is near the surface and/or fluctuates substantially. Dissolution occurs most rapidly at the top of the water table due to mixing of waters of different carbon dioxide partial pressures and degrees of calcite saturation (Bogli, 1964; Davies, 1960; White, 1960, 1977). Also, piping of doline fills is enhanced where there is a significant change in the depth to the water table, be it natural or man-induced. Therefore, a water table (piezometric surface) map should be made for any area of

concern and the water levels in wells monitored.

7. <u>Search for Surface Signatures of Newly-Forming Dolines</u>
Fresh scarps on known dolines demonstrate recent subsidence or collapse. Likewise, deformation of man-made structures, ponding of rain water, and vegetation stress may all suggest the development of dolines. Aerial photographs taken repeatedly through time at a site of interest may show surface changes not noticeable from the ground.

8. <u>Utilization of Shallow Geophysical Techniques</u>
Once a specific site is under investigation, and the detection of future dolines is deemed of utmost importance, a shallow geophysical survey should be conducted. Earth resistivity and seismic reflection methods are the most affordable for delineating filled fractures and dolines. Such features can then be verified by shallow drilling.

9. <u>Determine the Effect of Man's Activities on Doline Development</u>
In most of the United States where carbonate rock is exposed at or very near the surface, the activities of man have little effect on doline collapse. But, where dolines have been buried or have formed beneath a loosely consolidated cover, excessive ground-water withdrawals can be related to catastrophic collapse. Mining, leaky impoundments, water diversions, and excessive ground-water pumping can cause dolines to develop. A site-specific analysis must be performed and predictions of future doline occurrence made through utilizing the previously described techniques.

10. <u>Model Development for Collapse Prediction</u>
Utilizing all the collected map and field data, a model should be developed in which each factor is weighted. Once the most important factors affecting collapse and subsidence are statistically determined, a series of maps combining two or more of these factors should be made to aid in predicting collapse and subsidence susceptibility.

These ten steps have no particular order in which to be performed. In many cases, not all of the steps will be necessary. Through the use of these proposed procedures, the factors that affect doline development can be determined and the loss of property and life hopefuly prevented.

References

Aley, T.J., J.H. Williams, and J.W. Massello, 1972, Ground water contamination and sinkhole collapse induced by leaky impoundments in soluble rock terrain: Missouri Geol. Survey and Water Resources, Engineering Geol. series no. 5, 32 p.

Alexander, Jr., E.C., 1980, Geology field trip guide: Proc. 1980 Nat. Speleol. Soc. Conv., Minneapolis, Minn., pp. 143-188.

Bogli, A., 1964, Mischungskorrosion-ein Beitrg zur Verkarstungs problem: Erdkunde, v. 18, pp. 83-92.

Brook, G.A. and T.L. Allison, 1983, Fracture mapping and ground subsidence susceptibility modeling in covered karst terrain: Dougherty County, Georgia: distribution data: in Environmental Karst, Geo Speleo Publ., Cinn., Ohio, pp. 91-1086.

Brook, G.A. and R.L. Mitchelson, 1981, Single and double fourier series analysis of cockpit karst in Puerto Rico: Proc. Eighth Int. Cong. of Speleol., v. 1, pp. 53-55.

Cramer, H., 1941, Die systematik der karstdolinen: Neues Jb. Miner. Geol. Palaont, v. 85, pp. 293-382.

Davies, W.E., 1960, Origin of caves in folded limestone: Bull. Nat. Speleol. Soc., v. 22, pp. 5-18.

Day, M., 1976, The morphology and hydrology of some Jamaican karst depressions: Earth Surface Processes, v. 1, pp. 111-129.

Foose, R.M., 1967, Sinkhole formation by ground-water withdrawal, Far West Rand, South Africia: Science, v. 157, n. 3792, pp. 1045-1048.

Ford, D.C., 1964, Origin of closed depressions in the central Mendip Hills, Somerset, England (abs.): Abstracts to papers presented at the 20th Int. Geog. Congress, pp. 105-106.

Jennings, J.N., 1971, Karst: An Introduction to Systematic Geomorphology: The M.I.T. Press, London, England, v. 7, 252 p.

Kemmerly, P.R., 1976, Definitive doline characteristics in the Clarksville quadrangle, Tennessee: Bull. Geol. Soc. Amer., v. 87, pp. 42-46.

Kirk, K.G. and E.R. Snyder, 1977, A preliminary investigation on seismic techniques used to locate cavities in karst terranes: in Hydrologic Problems in Karst Regions, Western Ky. Univ., Bowling Green, Ky., pp 79-91.

Kirk, K.G. and H. Rauch, 1977, The application of the tri-potential method of resistivity prospecting for ground-water exploration and land-use planning in karst terrains: in Karst Hydrogeology, Proc. 12th Int. Congress of Hydrogeology, Univ. of Alabama Press, Huntsville, Alabama, pp. 285-300.

La Valle, P., 1967, Some aspects of linear karst depression development in south central Kentucky: Ann. Assoc. Am. Geogr., v. 57, pp. 49-71.

Newton, J.G., 1976, Early detection and correction of sinkhole problems in Alabama, with a preliminary evolution of remote sensing applications: Alabama Highway Dept., HPR Report No. 76, 83p.

_____, 1977, Induced sinkholes--a continuing problem along Alabama highways: in Karst Hydrogeology, Proc. 12th Int. Congress of Hydrogeology, Univ. of Alabama Press, Huntsville, Alabama, pp. 303-304.

Ogden, A.E., and J.P. Reger, 1977, Morphometric analysis of dolines for predicting ground subsidence, Monroe Co., W.Va.: in Hydrologic Problems in Karst Regions, Western Ky. Univ., Bowling Green, Ky., pp. 130-139.

Ogden, A.E., 1981, Pseudo-karst caves of Arkansas: Proc. Eighth Int. Cong. of Speleol., v. 1, p. 766-768.

_____, 1982, A morphometric analysis of the Monroe Co., Karst, W.Va., with comparisons to Ky. Tn., and In. (abst.): Abstracts of the 1982 Nat. Speleol. Soc. convention, Bend, Oregon, p. 33.

Palmquist, R.C., 1977, Distribution and density of dolines in areas of mantled karst: in Hydrologic Problems in Karst Regions, Western Ky. Univ., Bowling Green, Ky., pp. 117-129.

Quinlan, J.F., (1974), Origin, distribution, and detection of development of two types of sinkholes in an anthropogenic karst, South Africa: Proc. 4th Conf. on Karst Geology and Hydrology, W. Va. Geol. and Econ. Surv., p. 161.

Reitz, H.M. and D.S. Eskridge, 1977, Construction methods which recognize the mechanics of sinkhole development: in Karst Hydrogeology, Proc. 12th Int. Congress of Hydrogeology, Univ. of Alabama Press, Huntsville, Alabama, pp. 432-438.

Ruhe, R.V., 1975, Geohydrology of karst terrain, Lost River Watershed, southern Indiana: Water Resoruces Research Center, OWRT, Indiana University, Bloomington, Indiana, 91 p.

Sinclair, W.C., 1982, Sinkhole development resulting from ground-water withdrawal in the Tampa area, Florida: U.S. Geol. Survey, Water-Resources Inv. 81-50, 24 p.

United States Geological Survey, 1984, National water summary-1983 hydrologic events and issues: U.S.G.S. Water Supply Paper 2250, 243 p.

White, W.B., 1960, Termination of passages in Appalachian caves as evidence for a shallow phreatic origin: Bull. Nat. Speleol. Soc., v. 22, pp. 43-53.

_____, 1977, Role of solution kinetics in the development of karst aquifers: in Karst Hydrogeology, Proc: 12th Int. Congress of Hydrogeology, Univ. of Alabama Press, Huntsville, Alabama, pp. 503-517.

Wigal, J.M., 1978, A study of the lower Greenbrier Group and some geologic controls on cavern development within Greenbrier Co., W. Va.: Unpubl. M.S. thesis, West Virginia University, 167 p.

Williams, P.W., 1966, Morphometric analysis of temperate karst landforms: Irish Speleol., v. 1, pp. 23-31.

Factors affecting the collapse of cavities

THOMAS F.BEGGS *Soil and Materials Engineers, Altamonte Springs, Florida, USA*
BYRON E.RUTH *University of Florida, Gainesville, USA*

ABSTRACT

Sinkhole occurrences along state highways in Florida have been documented by the Florida Department of Transportation since 1958. These records contain information on 533 sinkholes of which only 265 contained sufficient information for detailed analysis. These records were used to establish trends and relationships between those variables considered to have potential influence on the frequency of sinkhole occurrence.

A comparison of the number of recorded sinkholes according to year and month indicated the highest frequency occurred during April through August in 1973 to 1975. However, the variation in mean monthly values for 1969 to 1983 was small and conformed closely to the mean of 18.4 collapses per month. Exceptions to this were the low frequency periods in November and December.

Rainfall records were reviewed and the two year period of 1973 to 1974 was selected for analysis because it had the highest frequency of sinkhole occurrences. Actual rainfall 90 days prior to collapse was used in this selection process. Similar rainfall data were obtained for the 1976 to 1977 period, but the collapse frequency was very low. Comparison of the data for these two time periods indicated similar rainfall but substantially different mean monthly sinkhole occurrence (6.3 versus 0.92 collapses/month). Therefore, it was concluded the frequency of collapses cannot be based on precipitation records.

Geological information was used to identify the depth to the Floridan aquifer (limerock) for each sinkhole location. The frequency of sinkhole occurrence was found to be related to the depth to limerock. Similarly, the depth to the phreatic surface correlated well with the collapse frequency using a probability function.

These relationships indicate a greater frequency for sinkhole occurrence as the depth to the limerock and phreatic surface decreases. However, when both limerock and phreatic surface depths were analyzed, there appeared to be an equal potential for collapse when the phreatic surface is above or below the limerock.

Introduction

Florida geology is dominated by unique near-surface soluble limerock deposits which are overlain by recent clay and sand formations. In some areas of the state, the water-deposited sedimentary base is very soluble and prone to development of solution cavities or sinkholes. Cavity collapse can occur rapidly with potentially catastrophic results. Therefore, the study of geological and environmental factors associated with cavity development and collapse is of prime importance to geologists, engineers, and developers.

A study was conducted to analyze sinkhole data collected by the Florida Department of Transportation (FDOT). The FDOT had documented the location, date, size, and other visually observable factors for those sinkholes which have developed within state highway right-of-ways. This data was organized to fit a format established by the Florida Sinkhole Research Institute. The format was established to provide a computerized inventory of sinkhole data which can be easily accessed for site specific information.

Temperature and rainfall data, depth to Floridan aquifer and phreatic surface, and topographic information were compiled for each sinkhole location using available maps and other documentation. This data was used in various analyses in an attempt to develop relationships with the frequency of sinkhole occurrence. The results of this investigation (Beggs, 1984) indicate that a probability function can be used to define the frequency of sinkhole occurrence using either the depth to limerock or depth to the phreatic surface. Attempts to establish a relationship with rainfall was unsuccessful.

The Data Base
    Data collection was initiated by the FDOT in 1958. By 1983, data for a total of 533 sinkholes had been accumulated. However, detailed information needed for analysis was available for only 265 sinkholes. Each sinkhole was coded and its location was identified on county road maps. This provided for the determination of sinkhole locations according to 1) township, range, and section, and 2) latitude and longitude. United States Geological Survey (USGS) 7.5 minute quadrangle maps were used to obtain ground elevation, topography (flat, hillside, hilltop, valley, or closed depression), and comments relating to specific site characteristics. Air temperature (high and low) and precipitation data were extracted from the U.S. Weather Service reports. Total rainfall occurring one, six, and 90 days prior to the collapse of a cavity were documented. The elevation and depth to the Floridan aquifer and the phreatic surface were obtained from Florida Bureau of Geological maps. The exact water table level immediately prior to collapse could not be determined without a continuous monitoring system. This type of data would be valuable for site specific investigations.

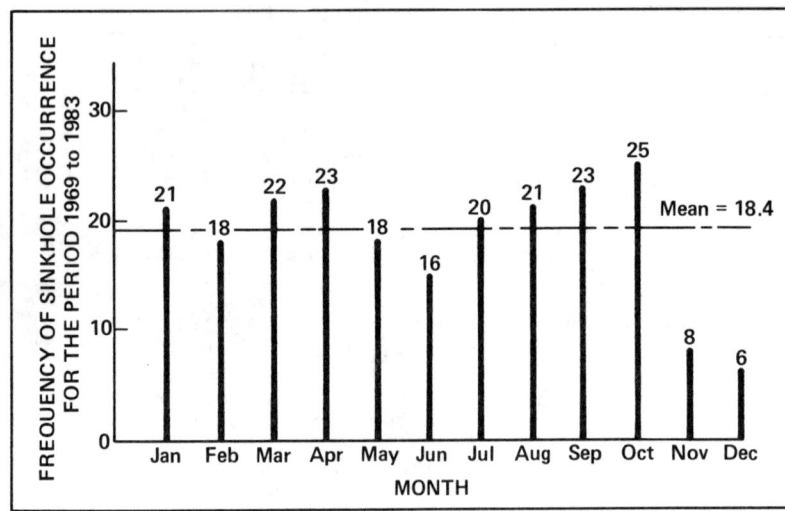

Figure 1   Sinkhole frequency vs month (1967-1983)

The collected data was organized into a format suitable for a computer data bank or inventory system. Some of this information was used in the analyses described in the subsequent sections of this paper.

Variation in Frequency with Time
    Histograms were prepared to identify trends in the frequency of sinkhole occurrence. The data for the time period prior to 1969 were not included because the month of occurrence was not identified in the records. Figure 1 indicates the greatest number of sinkholes (25) occurred during the month of October while the least (6) occurred in December. The months of March, April, August, and September each had between 21 to 23 occurrences. The number of occurrences in October was slightly greater than observed in these months.

    Analysis of the data provided a mean of 18.4 sinkholes per month with a standard deviation of 5.9. The October frequency was 1.1 standard deviation above the mean whereas the November and December frequencies were 1.8 and 2.1 standard deviations below the mean, respectively.

    Yearly variations in sinkhole occurrence for each month are depicted in Figure 2A through 2L. A preponderance of activity was observed between 1970 and 1974. Figure 3 clearly illustrates that frequency peaks occurred in 1970 and 1974. This may have resulted from natural causes or by increased emphasis on reporting within the FDOT. A memorandum to field personnel in the districts requesting increased diligence in the documentation and reporting of

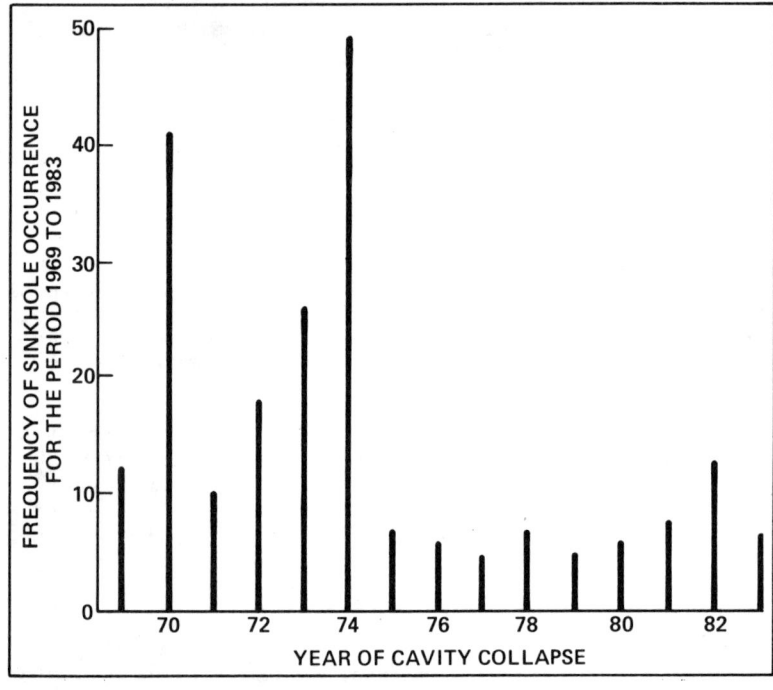

Figure 3   Sinkhole frequency according to year of occurrence

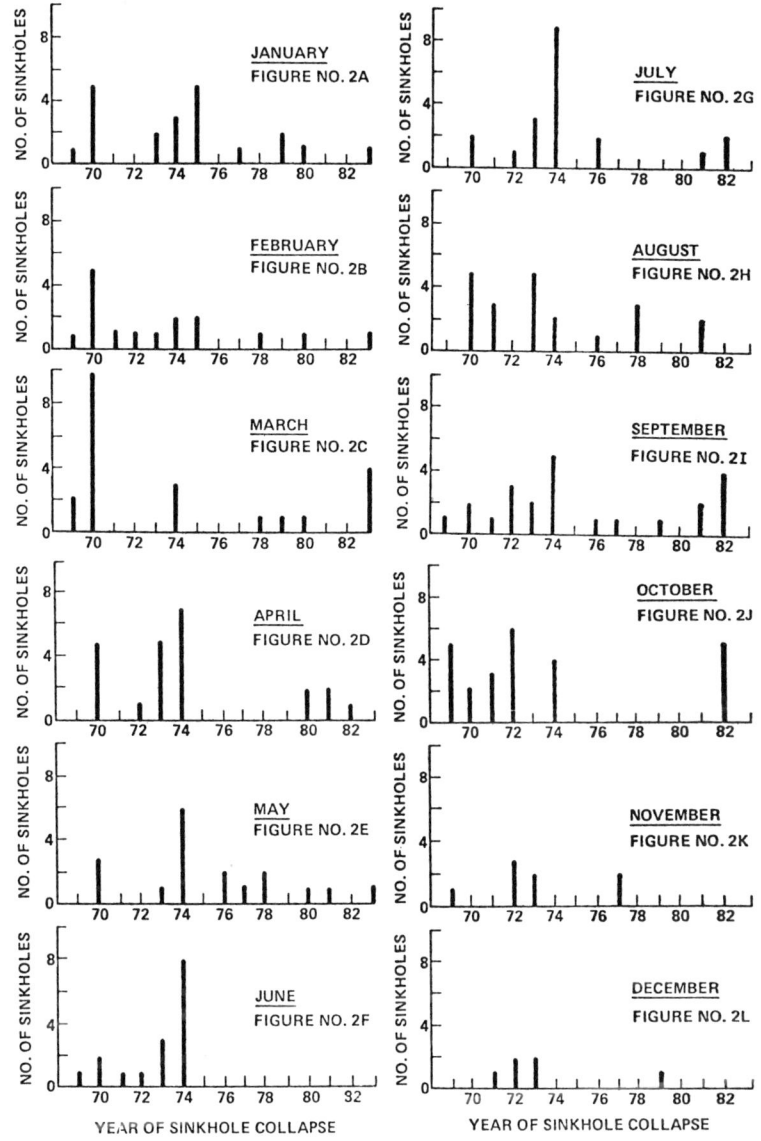

Figure 2 Yearly sinkhole frequency according to month of occurrence

sinkhole occurrences was sent some time during this period. Another possible explanation for these anomalies could be unusual weather (rainfall) effects.

Analyses to Evaluate the Effect of Rainfall

Rainfall records were reviewed to determine whether or not extreme variations in precipitation existed during the periods of increased sinkhole activity. The two year period of 1973 through 1974 was chosen to represent the highest recorded frequency of sinkhole ocurrence. A low frequency period (1976-1977) was selected for comparison. The actual rainfall during the 90-day period prior to collapse of each sinkhole was used to generate the mean 90-day rainfall for both time periods. Figure 4 illustrates both rainfall and sinkhole frequency according to month of occurrence for the two time periods. Comparison of the 90-day rainfall for the two periods indicated very little difference. This suggested that the high frequency of sinkholes during the 1973 through 1974 period could not be attributed to variations in precipitation. However, the indicated range in actual 90-day rainfall values varied between about 15 and 56 centimeters between April and September when the sinkhole frequency exceeded the mean. Therefore, additional analyses of rainfall seemed warranted.

Expected rainfall information was obtained for comparison to the collected actual rainfall data. Table 1 presents average monthly, daily, and tri-monthly expected rainfall values for north central Florida (Fernald, 1981). This data shows the summer months from June to September to have the highest expected rainfall. The mean ratio of the 1973-1974 actual rainfall to the expected rainfall was approximately 1.2. The ratio for the entire 1969-1983 period was almost exactly the same as for the 1973-1974 period.

Histograms of sinkhole frequency versus the rainfall ratio for daily and 90-day rainfall are illustrated in Figures 5 and 6 respectively. A ratio greater than one represents wetter than expected conditions. Conversely, a ratio less than one denotes dryer than expected conditions. If no rainfall was recorded on the day prior to collapse, the ratio was zero. Of 182 known data points, 90 or approximately 50 percent had a nearly zero daily rainfall ratio. Drier than expected conditions (Ratio<1.0) existed for 63 percent of the 182 sinkhole occurrences.

Long term effects of rainfall were evaluated using the 90-day rainfall ratio. Figure 6 shows a distribution which appears to have a tendency toward a normal distribution. Of 230 sinkholes, for which data was available, 92 sinkholes (40 %) had rainfall ratios less than one. About 57 percent of the sinkhole occurrences correspond to wetter than average long term rainfall conditions.

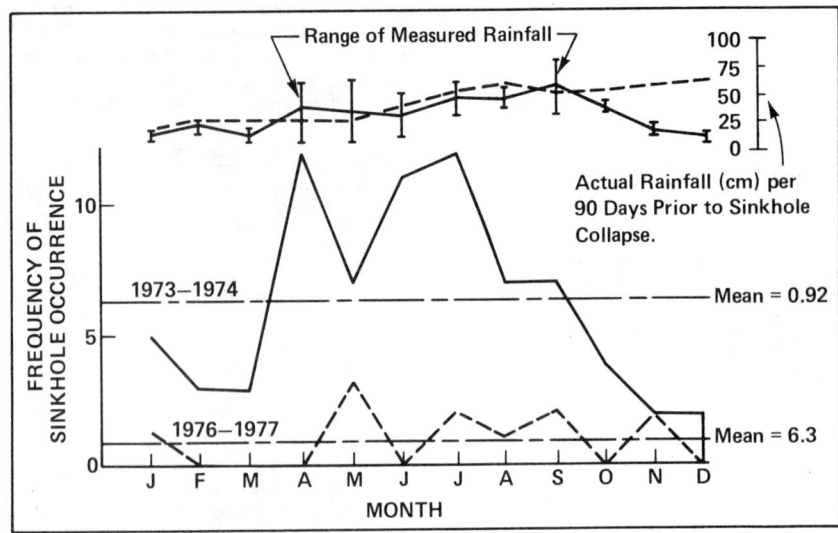

Figure 4 Comparison of sinkhole frequency and rainfall

Interpretation of the long term rainfall conditions suggest there is little difference in sinkhole frequency between drier and wetter than normal rainfall conditions. Similarly, the short term rainfall (dry preceding collapse) does not indicate any relation except an approximation of a rainfall intensity probability curve. Therefore, it was concluded that rainfall does not correlate with the frequency of sinkhole occurrence. This does not imply that rainfall contribution to shallow subsurface flow into existing cavities does not have an affect on the propogation, enlargement, and collapse of cavities.

Analysis of In-situ Conditions

The depth to limerock and to the phreatic surface were compiled and used to develop relationships with the frequency of sinkhole occurrence. Since the limerock depth values refer to the Floridan aquifer depth, only data for sinkholes situated in areas directly underlain by the aquifer were used in the analysis. Sinkholes occurring in the southern (south of Polk County) or extreme northwestern areas of the state were excluded from the data base.

The data plotted in Figure 7 illustrates an increased frequency of sinkhole occurrence with reduction in depth to limerock. The trend appeared to aproximate an exponential probability curve. Therefore, a regression analysis was performed which gave the following regression equation.

$$\omega = 95\ e^{-0.0985d}$$

where: ω = frequency of sinkhole occurrence, d = depth to limerock, meters, and $R^2 = 0.86$.

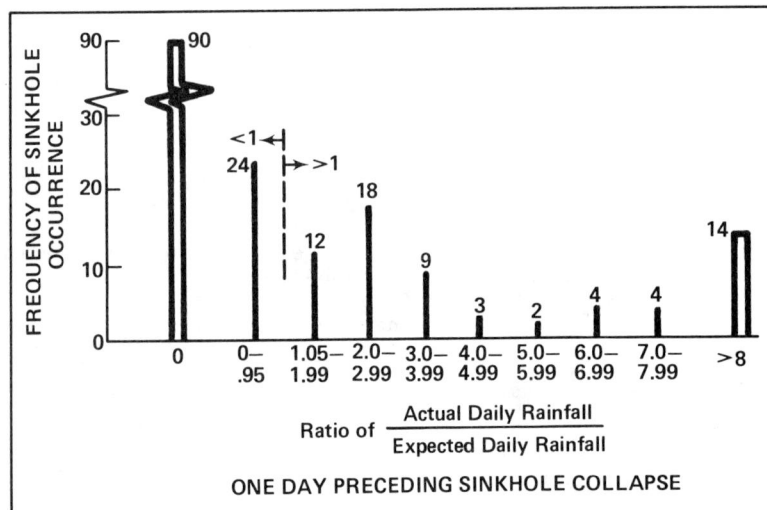

Figure 5 Sinkhole frequency vs one-day rainfall ratio

TABLE 1
EXPECTED AVERAGE RAINFALL FOR NORTH CENTRAL FLORIDA

		Jan	Feb	Mar	Apr	May	June	July	Aug	Sept	Oct	Nov	Dec
Per month	inches	4.0	4.0	3.5	3.0	4.0	6.0	8.0	8.5	6.0	2.0	1.5	3.0
	cm	10.2	10.2	8.9	7.6	10.2	15.2	20.3	21.6	15.2	5.1	3.8	7.6
Per day	inches	0.13	0.14	0.11	0.10	0.13	0.20	0.26	0.27	0.20	0.06	0.05	0.10
	(cm)	0.3	0.4	0.3	0.2	0.3	0.5	0.7	0.7	0.5	0.1	0.1	0.2
Per preceding 90 days	inches	6.5	8.5	11.0	11.5	10.5	10.5	13.0	18.0	22.5	22.5	16.5	9.5
	(cm)	(16.5)	21.6	27.9	29.2	26.7	26.7	33.0	45.7	57.1	57.1	41.9	24.1

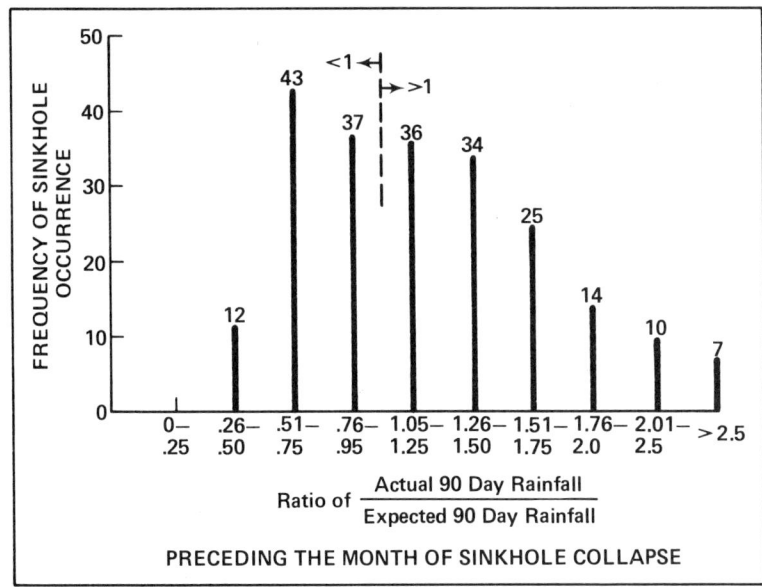

Figure 6 Sinkhole frequency vs 90-day rainfall ratio

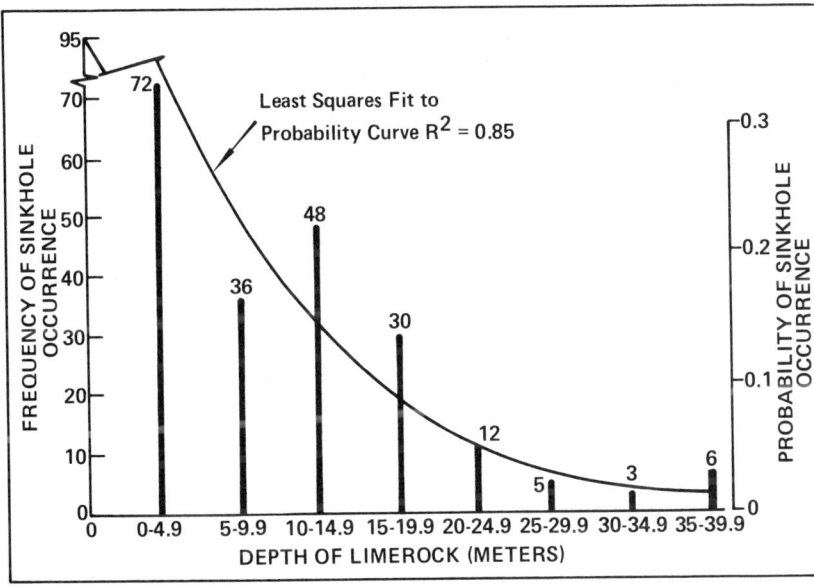

Figure 7 Sinkhole frequency vs limerock depth

Similarly, regression analysis of depth to the phreatic surface provided the equation

$$\omega = 145\ e^{-0.1313 D_p}$$

where: ω = frequency of sinkhole occurrence, D_p = depth to phreatic surface, meters, and R^2 = 0.92. Figure 8 shows the data trend and the curve depicting the regression equation for the frequency-depth to phreatic surface relationship.

The excellent correlation obtained for both limerock and phreatic surface depths indicates the desirability of extending this analysis to more complex geological situations. The results suggest that variations in phreatic surface are more prevalent at shallow depths which increases the frequency of collapse when limerock is near the surface. Logically, sinkhole size should be small in comparison to locations where phreatic surface and limerock depth are greater.

The interaction between phreatic surface and limerock depth was investigated using the ratio of these depths. The median value of phreatic surface depth to limerock depth ratio was 0.95 (almost equal depths) which indicated 50 percent of the collapses occurred when the limerock was submerged and the remaining 50 percent when water elevations were below the rock surface. Since no relationship was established, it is surmised that actual fluctuations in phreatic surface combined with inflow from rainfall create both high gradient inflows and accelerated piping which eventually contributes to cavity collapse. These effects are documented by both Sowers, 1976 and Casper, et al., 1981.

Conclusions

The results of this investigation indicate that it is possible to develop probabilistic methods to assess cavity collapse potential in Florida's karst terrain. This is a first effort based on limited and incomplete data. The developed relationships should be considered as preliminary and only for use in assisting future investigators in achieving a thorough understanding of the factors affecting cavity growth and collapse mechanisms.

Specific conclusions derived from the study are:
1. Over a 15 year recording period, the frequency of sinkhole occurrence for each year varied as much as ten times the lowest recorded yearly value.
2. Monthly totals for 15 years provided fairly uniform sinkhole occurrence frequencies except for the months of November and December which exhibited very low frequencies.
3. No correlation could be found between rainfall and frequency of sinkhole occurrences.

4. The frequency of sinkhole occurrence increases as the depth to underlying limerock (Floridan aquifer) decreases according to a probability function.
5. The frequency of sinkhole occurrence increases as the depth to the phreatic surface decreases following a probability function.
6. There appears to be an equal probability of sinkhole occurrence when the phreatic surface is either above or below the surface of the limerock.

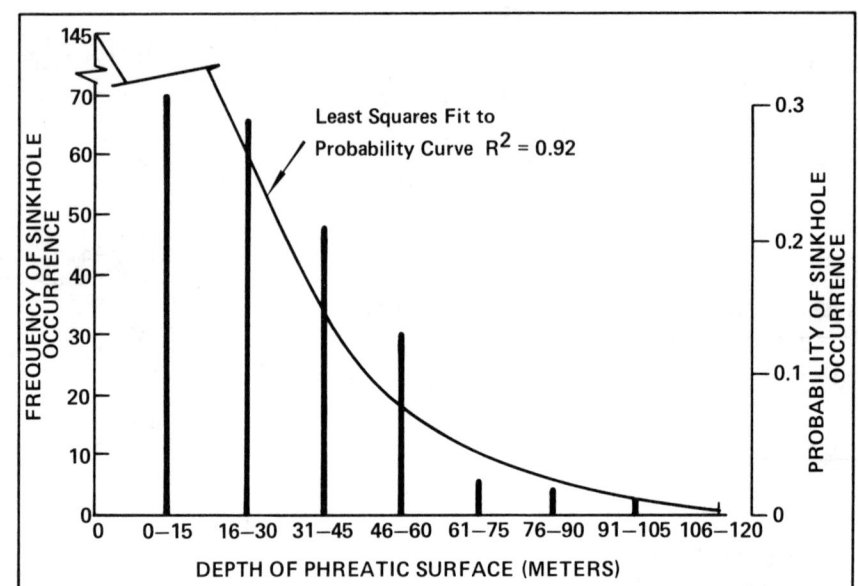

Figure 8 Sinkhole frequency vs phreatic surface depth

Acknowledgements
    The financial support provided by the Florida Sinkhole Research Institute is gratefully acknowledged. Special thanks goes to Dr. Barry F. Beck for his technical advice and assistance.

References

Casper, J. Ruth, B.E., and Degner, J., A remote sensing evaluation of the potential for sinkhole occurrence: Remote Sensing Applications Laboratory, Department of Civil Engineering, University of Florida.

Beggs, T.F., 1984, Environmental and geological factors influencing sinkhole development in Floridian karst topography: Unpublished Master of Engineering report, Department of Civil Engineering, University of Florida.

Fernald, Edward A., 1981, Atlas of Florida: Published by the Florida State University Foundation.

Sowers, G.F., 1976, Mechanisms of subsidence due to underground openings: Transportation Research Record No. 612.

Relationship of modern sinkhole development to large scale-photolinear features

J.R.LITTLEFIELD, M.A.CULBRETH, S.B.UPCHURCH & M.T.STEWART *University of South Florida, Tampa, USA*

ABSTRACT

An inventory of reported sinkhole occurrences for the past 10 years in west-central Florida reveals several surprising results. First, sinkhole development commonly occurs in linear patterns on a regional scale. These linear patterns do not coincide with linear cultural features, such as roads. Second, while sinkhole development is more frequent in areas of high water use, such as agricultural, mining, or urban areas, there is a lack of obvious concentrations of modern sinkhole in the immediate vicinity of well fields or other major groundwater withdrawl areas.

The linear alignments of modern sinkholes are large scale and some extend for lengths in excess of 30 km. Many modern sinkhole occurrences have been reported along each feature. While linear features, such as joints, fracture zones, or faults, are widespread throughout the area and can be detected at all scales of observation by tonal variations, surface drainage patterns, and the presence of ancient sinks, the linear features along which modern sinkholes are reported are usually the largest. These regional photolinear features, or lineaments, can best be differentiated at the scale of high altitude and/or satellite imagery.

Therefore, it appears that, while sinkhole development over geologic time can occur along linear features and conduits of any scale, short-term sinkhole development is most probable along the largest photolinear features. If this pattern holds for other areas, a predictive model for recognition of high sinkhole probability can be easily developed.

Introduction

A standard practice in investigations of foundation stability, sinkhole probability, and ground-water availability involves delineation of photolinear features that may predispose affected areas to instability and high transmissivities (Ray, 1960; Lattman and Parizek, 1964; Parizek, 1976; Holz, 1985). This approach has been used successfully in west-central Florida by many investigators (e.g., Vernon, 1951; Coker, 1969; Miller, 1977) and is commonly required by local regulatory agencies prior to land-use permitting.

This paper is a product of two seperate investigations: (1) a compilation of data for sinkhole occurences reported in west-central Florida over the last 10 years (Beck, 1984) and (2) a geophysical study of the causes of major linements in south Florida. Juxtaposition of the results of these two studies has produced a surprising result: the modern sinkhole occurences align with regional linements as opposed to the more widespread, but shorter, local features.

Study Area

Reports of sinkhole development are suspect because they are generally only of interest when they impact people. Thus, there is a probability that sinkhole occurrence reports are biased towards highway corridors; urban, densely-populated areas; or major water-withdrawal areas.

The study area is located in west-central Florida (Fig. 1). This area includes the major urban region of Tampa-St. Petersburg and much rural, undeveloped land. In order to negate the possibility of biased data, this paper concentrates on the relationship of sinkhole occurrence to photolinear features in the urban area of Hillsborough County, Florida, which should provide the most reliable census of sinkhole development. Two 7.5 minute quadrangle areas are emphasized (Fig. 1). These are the Sulphur Springs Quadrangle, which is characterized by densely-populated, urban development of Tampa, and the Thonotosassa Quadrangle, which is suburban. While the sinkhole occurrence data are best in these areas, recognition of photolinear features is most difficult owing to disruption of drainage, soils, and sinks by man.

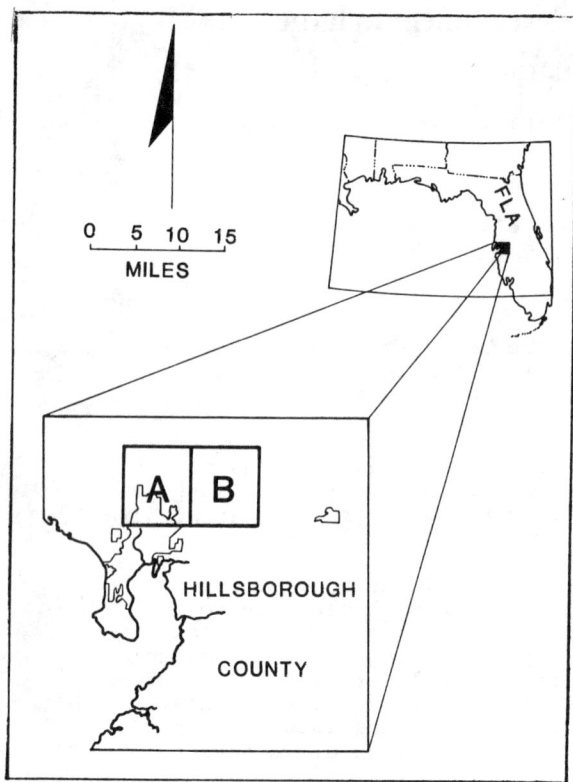

Figure 1: Location of the study area. A is the location of the Sulphur Springs Quadrangle and B is the Thonotosassa Quadrangle.

Methods

The sinkhole occurrence data were obtained from a number of diverse sources, such as governmental agencies and the press. All reports were cross checked with other reports and many were field checked as well. While it is certain that some occurrences were not reported, the data obtained herein constitute the most complete set yet obtained (Beck, 1984). In all 138 sinkholes were reported to have developed in the 12 quadrangle area of Hillsborough County over the last 10 years. Forty five of those occured in the Sulfur Springs area and 10 in the Thonotosassa area.

Remote imagery used for the study include U.S.G.S. quadrangle maps for topography and identification of closed depressions, 1:20,000 aerial photographs for location of fracture traces, and 1:500,000 LANDSAT images for recognition of lineaments.

Regional Geology

Hillsborough County is underlain by a thick carbonate sequence ranging in age from early Tertiary to Miocene. This sequence comprises the Floridan aquifer (Stringfield, 1966). The Eocene Avon Park and Lake City Limestones, Oligocene Suwannee Limestone, and Miocene Tampa Formation are the most productive strata in the aquifer (Menke et al., 1961). The upper Eocene Ocala Group varies in transmissivity and serves as a weak aquitard. Depth to the top of the Floridan aquifer ranges from 0 to 75 m. The Floridan is unconfined and its upper zone is exposed in the two quadrangles of interest.

Clay beds in the Tampa Formation and the Miocene Hawthorn Formation confine the Floridan. The eastern part of Hillsborough County is underlain by thick Hawthorn. The only significant Hawthorn present in the two quadrangles of emphasis is a thin clay residuum throughout the area and a modest section of intact Hawthorn to the northeast. The thickness of the confining beds ranges from 0 to 3 m, and in the study area the residuum is usually less than 1 m.

Undifferentiated sands and clays of Pliocene and Holocene age mantle the area. These deposits include marine terrace clayey sands, shell beds, and aeolian sands, which may accumulate into dunes 1 to 2 m in height. The total thickness of the Pliocene and Holocene ranges from 0 to 45 m.

Both the Floridan and the Hawthorn thicken and deepen toward the southwest and southeast. Thus, the sedimentation pattern indicates some differentiation of the Tertiary section into small basins. Menke et al. (1961) suggested that the Cenozoic sedimentation was associated with downwarping into these small basins, which are associated with the Ocala Arch. They argued that downwarping was associated with stresses that were relieved by jointing and much "faulting". Some of the "faults", which had been proposed by Vernon (1951), were thought to have displacements of about 60 m. While opinion on the relationship of these "faults" or fractures to the regional photolinear features is reserved until the geophysical study is completed, there is a possibility that they are related.

Sinkholes and Photolinear Features

Lattman and Parizek (1964) have shown that sinkholes and associated highly transmissive zones are related to differential solution along linear fractures, faults, and/or joints in limestones. In fact, these linear features are recognized by the common alignment of sags, sinks, sinkhole lakes, drainage, and soil tonal zones. The scale of these linear features has not been a factor considered for prediction of sinkholes or transmissive zones.

A lineament is defined by O'Leary et al. (1976) as a mappable, simple or composite linear feature, whose parts are aligned in a rectilinear or slightly curvilinear relationship which differs distinctly from the patterns of adjacent features and presumably reflect a subsurface phenomenon. Lineaments are, therefore, large-scale features that are composed of smaller scale, fracture zones or joints. At the scale of low altitude photoimagery, joints and fracture traces can be recognized. Where they occur in great density, a lineament may be present. High altitude imagery, such as obtained from u-2 aircraft and satellites, filters recognition of the smaller-scale linear features and the regional lineaments become the dominant photolinear features seen. Parizek (1976) assumed a 1 km width for structurally-controlled lineaments and 1 to 10 m widths for joint or fracture-trace features.

Withington (1973) has shown that lineaments in the Atlantic Coastal Plain near Washington, D.C., represent fault traces. The geophysical study of Florida lineaments suggests a similar relationship, although the lineaments appear to reflect translation of tidally-generated joints and fractures upward from basement lithologic contrasts and faults, rather than faults in the Tertiary strata.

Results

A total of 2,303 ancient sinkholes have been identified in the 12 quadrangle areas that include Hillsborough County. This compares with a total of 138 reported in the last ten years. Most of the ancient sinks are concentrated in the karst belt in the Sulphur Springs area of Tampa. The Sulphur Springs Quadrangle contains 537 identified sinks, which is 23% of the total. The Thonotosassa Quadrangle contains 181 ancient sinks, or 8% of the total identified. For comparison, of the 138 sinks reported in the last 10 years, 45 (33%) were in the Sulphur Springs Quadrangle and 10 (7%) were in the Thonotosassa Quadrangle. There seems, therefore, to be a slightly higher incidence of sinkhole reports in the urban area than is supported by the density of ancient sinks. This may reflect the importance of reporting these sinks in an urban area or the changes in hydrology that have taken place owing to urbanization. The Thonotosassa data are consistent, however, and there is no indication that significant numbers of sinks go unreported.

Figure 2 illustrates the relationship of sinkhole occurrences reported within the last ten years to lineaments. The large-scale alignment of modern sinks was the first indication

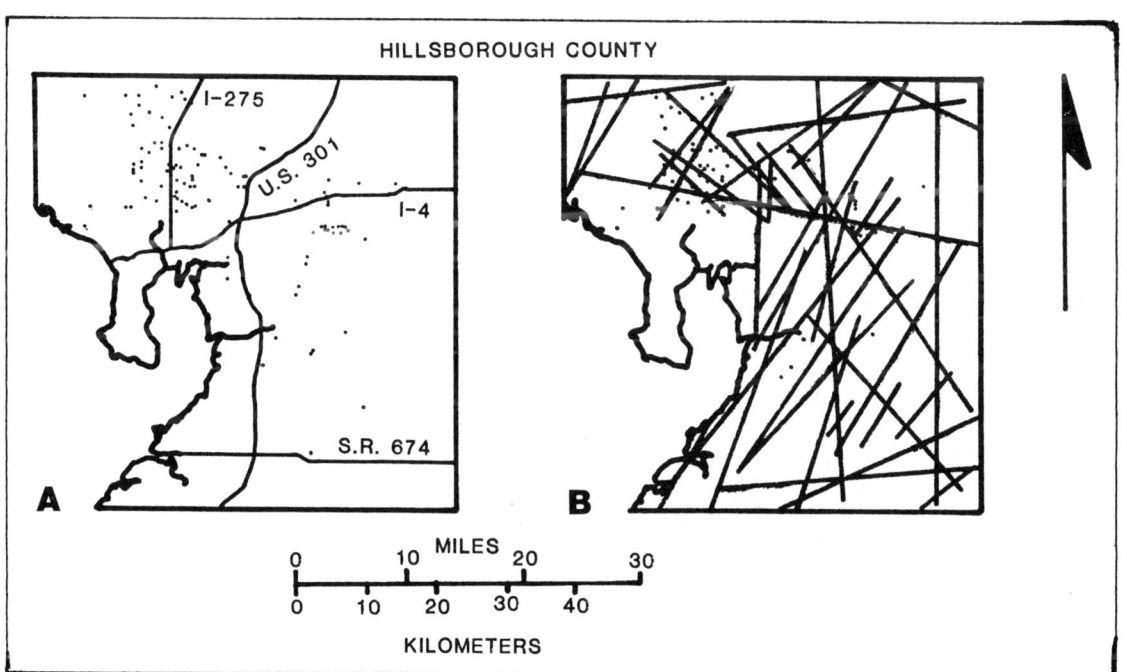

Figure 2: Comparison of sinkhole occurrences reported in the last 10 years in Hillsborough County to major lineaments observed in LANDSAT imagery.

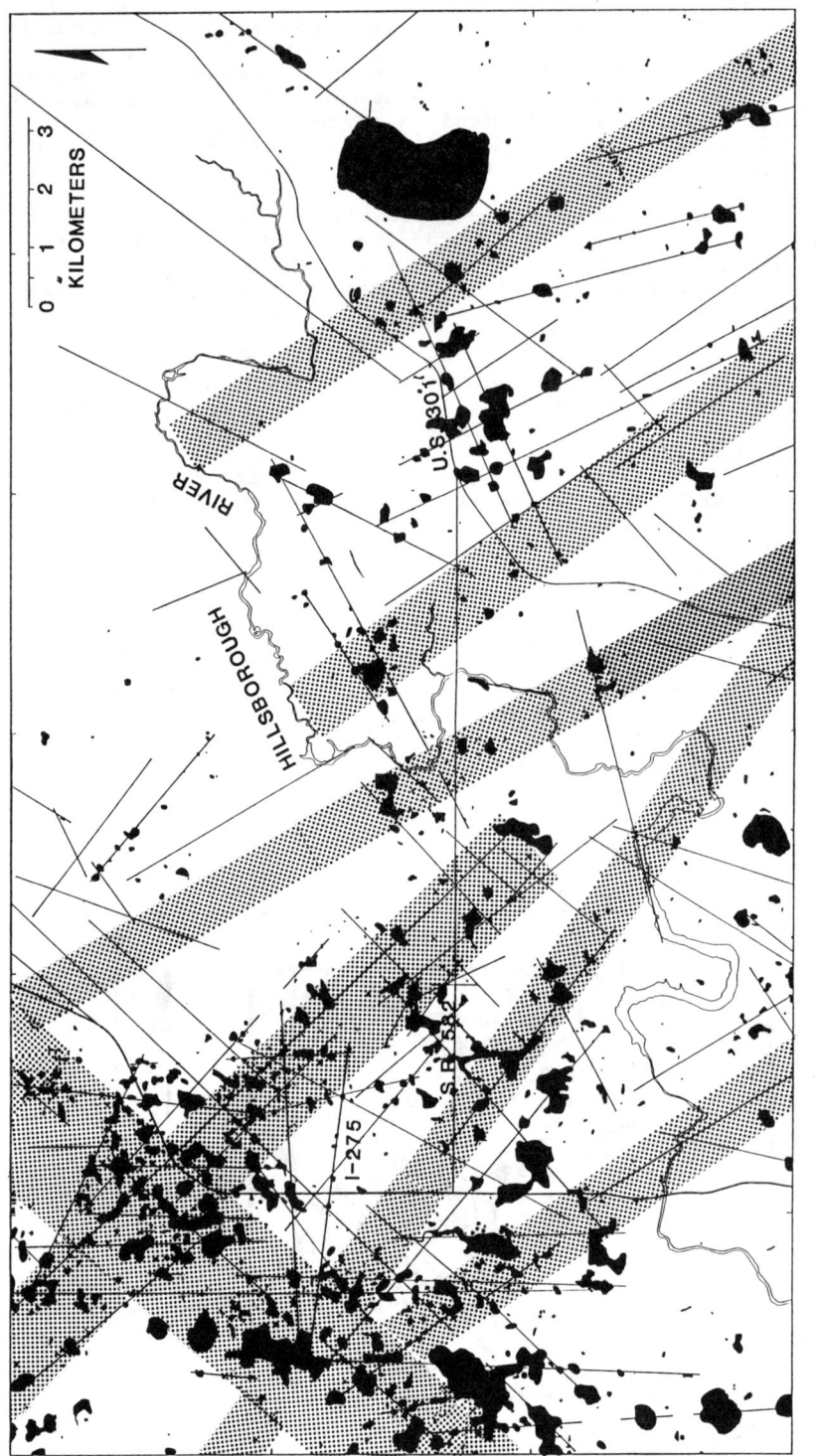

Figure 3A: Relationship of closed depressions (sinkholes and sinkhole lakes) to fracture traces and lineaments in the Sulfur Springs and Thonotosassa Quadrangles, Florida. Black areas are closed depressions identified on the topographic maps. Lines are fracture traces from 1:20,000 aerial photographs and topographs. Dotted regions are linements seen in 1:500,000 LANDSAT imagery.

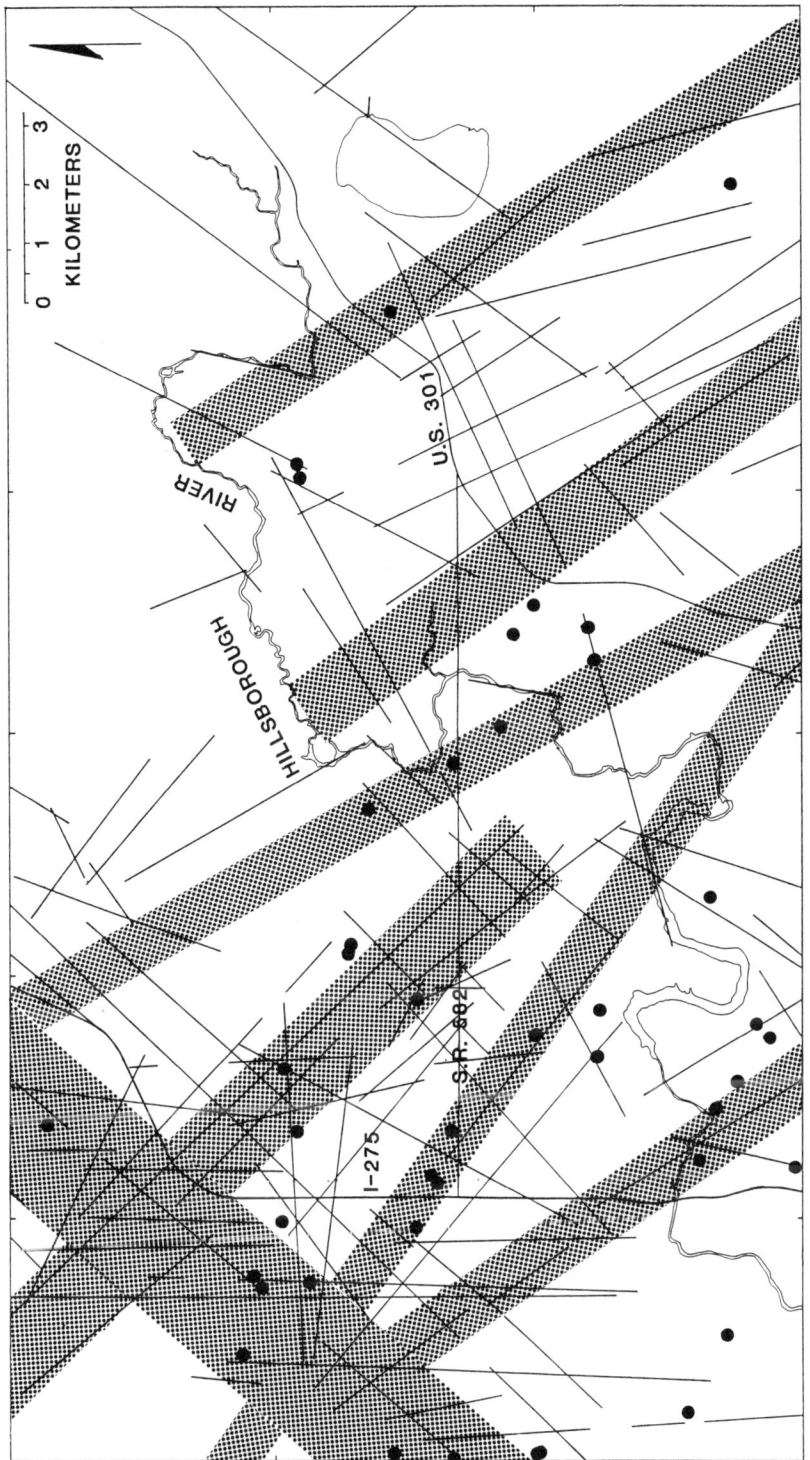

Figure 3B: Relationship of modern sinkhole occurences to fracture traces and linements on the Sulphur Springs and Thonotosassa Quadrangles, Florida. See Figure 3A for explanations of lines and patterns. Dots are sinkholes reported within the last 10 years.

that there is a relationship with lineaments. Note that Figure 2A shows several trains of sinks that extend for long distances. Figure 2B shows the major lineaments superimposed on the sinkhole occurrence data. Note that few of the reported sinks fall outside of the lineaments.

In order to evaluate the relationship of the lineaments with small-scale fracture traces and ancient sinks the Sulphur Springs and Thonotosassa Quadrangles were intensively studied (Fig. 3A). Note that many of the fracture traces coincide with lineaments and that lineaments are the loci of major sinkhole development through time. However, there are many ancient sinks that do not fall on lineaments and many fracture traces that show no correlation with sinkhole or lineament development.

Figure 3B compares modern sinkhole occurrences with the fracture trace and lineament data. Even in the urban areas of the Sulphur Springs Quadrangle, where fracture trace and lineament recognition is partly obscured by cultural features, there remains a correlation between modern sink occurrences and lineaments.

Discussion and Conclusions

The data presented in this paper are not unexpected. Sinkholes occur wherever vertical permeability makes them possible. The greater the permeability, the more likely sinks are to develop. Since large-scale lineaments include high density swarms of fracture traces, sinks should be most probable therein. Fracture-trace intersections (Lattman and Parizek, 1964) produce highly probable sinks with the probability increasing with magnitude of the fracture-trace system. Over geologic time, sinks from all probability situations accumulate throughout a karst terrain. Lineament-related sinks occur in wide swarms (Fig. 3A), while less probable sinks fall elsewhere.

Sinkhole development per unit time, therefore, should show a preference for lineament-dominated areas. The data presented in this paper support this conclusion and emphasize the necessity of lineament analysis as part of sinkhole probability determination.

These data also show that urban areas do have a greater incidence of sinkhole reports. It cannot be determined if this is a function of population awareness or altered hydrology. Both are probable. The Thonotosassa Quadrangle data, however, suggest that reports from lower population-density areas are not inconsistent with the actual distribution of ancient sinks. This suggests, but does not prove, that sinkhole development is accelerated in urban areas.

Swarms of sinkhole occurrences have been noted near wellfields (Sinclair, 1982) and regions of intense ground-water withdrawal for irrigation (Hall and Metcalfe, 1981) in the study area. The areas in which these sinkhole swarms occurred are within lineaments and lineament intersections. These swarms are highly localized and regional patterns do not show isolation of sinks near water-withdrawal centers. Thus, sink swarms associated with localized withdrawal seem to be associated with large-scale, lineament-related sinkhole patterns and their probability is related to the overall probability of sinks on lineaments.

If this pattern of short-term sinkhole development being associated with large-scale lineaments holds for additional areas, a predictive model for sinkhole probability will result. At present, these data strongly suggest that highest sinkhole probability areas are in zones of major lineament development. Lineament intersections seem especially probable areas. An hierarchy of sinkhole probability in a karst terrain results:

    MOST PROBABLE   (1) sinkholes at major lineament intersections,
                           (2) lineament-related sinks,
                           (3) off-lineament, fracture-trace intersection sinks,
                           (4) off-lineament, fracture-trace related sinks, and
    LEAST PROBABLE (5) sinks in unfractured rock.

Acknowledgements

This study was based on data partially obtained under the direction of Barry F. Beck and supported by the Florida Sinkhole Research Institute, and was also supported by a grant from the Standard Oil Corporation of Ohio (SOHIO) to S.B. Upchurch.

References

Beck, Barry F., 1984, A computer based inventory of recorded, recent sinkholes in Florida: Florida Sinkhole Research Inst. (U. of Central Fl.), Report 84-1, 12 p.

Coker, A.E., 1969, Application of remote sensing to occurrence of collapse sinkholes in the Alafia and Peace River basins, Florida: Nat. Aeronautic and Space Admin., Earth Resour. Aircraft Prog., Status Review Vol. III, Hydrol., Oceanogr., and Sensor Studies, pp. 22A-0 - 22A-14.

Hall, L.E., and Metcalfe, S.J., 1981 (in press), Sinkhole collapse due to groundwater pumpage for freeze protection irrigation near Dover, Florida, January, 1977: in Karst Hydrology, Int. Assoc. of Hydrologists.

Holz, R.K., 1985, The Surveillant Science: Remote Sensing of the Environment: New York, John Wiley & Sons, 2nd Ed., 413 p.

Lattman, L.H., and Parizek, R.R., 1964, Relationship between fracture traces and the occurrence of ground water in carbonate rocks: Jour. of Hydrol., 2:73-91.

Menke, C.G., Meredith, E.W., and Wetterhall, W.S., 1961, Water Resources of Hillsborough County, Florida: Fla. Geol. Surv., Rept. of Invest. No. 25, 101 p.

Miller, J.C., 1977, Fracture trace analysis for well siting in carbonate karst terrane, Crossbar Ranch Wellfield, Pasco County, Florida: unpubl. rept. to West Coast Regional Water Supply Authority, Florida, 12 p.

O'Leary, D.W., Friedman, J.D., and Pohn. H.A., 1976. Lineament, linear, lineation: Some proposed new standards for old terms: Bull. Geol. Soc. Amer.: 87:1463-1469.

Parizek, R.R., 1976, On the nature and significance of fracture traces and lineaments in carbonate and other terranes: In V. Yevjevich (ed.), Karst Hydrology and Water Resources, Vol. 1, Karst Hydrology, Ft. Collins, CO, Water Resour. Publ., pp. 3-1 - 3-108.

Ray, R.G., 1960, Aerial Photographs in Geologic Interpretation and Mapping: U.S. Geol. Surv., Prof. Pap. 373, 230 p.

Sinclair, W.C., 1982, Sinkhole Development Resulting from Ground-Water Withdrawal in the Tampa Area, Florida: U.S. Geol. Surv., Water-Resour. Invest. 81-50, 24 p.

Stringfield, V.T., 1966, Artesian Water in Tertiary Limestone in the Southeastern States: U.S. Geol. Surv., Prof. Pap. 517, 226 p.

Vernon, R.O., 1951, Geology of Citrus and Levy Counties, Florida: Fla. Geol. Surv., Bull. No. 33, 256 p.

Withington, C.F., 1973, Lineaments in Coastal Plain sediments as seen in ERTS imagery: In Symposium on Significant Results Obtained from the ERTS-1, Vol. 1, Tech. Present., Sect. A, Nat. Aeronautic and Space Admin., SP-327, pp. 517-521.

Application of double Fourier series analysis to ground subsidence susceptibility mapping in covered karst terrain

MARCUS J.W.THORP & GEORGE A.BROOK *University of Georgia, Athens, USA*

ABSTRACT

Grid point elevations were interpolated from 1:24,000 scale topographic maps of a 9 x 12 km area in suburban Orlando, central Florida to produce 3,780 data points with X, Y, and Z coordinates. The data were analyzed by the double Fourier series technique and the dominant waves making up the landscape were isolated. Wave troughs were found to parallel the major fracture directions in the area indicating significant bedrock structural control of topography. Results indicate that double Fourier series analysis may be useful in isolating significant fracture zones in bedrock beneath overburden in covered karst regions. Troughs in the dominant waves isolated by Fourier analysis were found to correspond well with the locations of 23 subsidence sinkholes that have developed since 1962, suggesting that this approach may be useful in defining areas of increased susceptibility to ground subsidence.

Introduction

Accelerated development of subsidence sinkholes due to man's activities is a widespread problem. For example, Newton (1977) estimates that 4,000 man-induced sinkholes have formed in Alabama since 1900 compared to perhaps 50 natural collapses. The induced sinkholes occur through the subsidence of residual or other unconsolidated deposits into openings in the underlying carbonate bedrock. Newton (1977) argues that induced sinkholes result from a decline in the water table due to ground water withdrawals or from construction. Decline in the water table results in (i) loss of buoyant support, (ii) increase in the velocity of movement of ground water, (iii) water-table fluctuations at the base of unconsolidated deposits, and (iv) induced recharge. Most sinkholes resulting from construction are due to the diversion of drainage over openings in bedrock.

Man-induced sinkhole development can be costly. On the Far West Rand of South Africa a sinkhole which developed in December, 1962, swallowed the West Driefontein Mine crushing plant taking 29 lives; a second sinkhole engulfed a domestic house at Blyvooruitzig Mine in August, 1964, taking 5 lives (Jennings 1966). In Florida, the Winter Park sinkhole of May, 1981, caused more than $2 million in damage. Not surprisingly, there have been numerous attempts to devise a methodology for identifying subsurface cavities in limestone that might eventually induce subsidence of overlying unconsolidated material. Two principal methods have been tested--visual observation of the ground surface, and mapping of areas of potential sinkhole collapse using remotely sensed information. Evidence of potential subsidence from visual observations includes circular or linear soil cracks, subsidence of structures and cracks, and vegetation stress. Remote sensing techniques include the study of black and white, infra-red, microwave and side-looking airborne radar (SLAR) imagery, and utilization of geophysical gravity, resistivity, and seismic data. All of the studies of subsidence susceptibility involving remotely sensed data have had only limited success. Ogden and Reger (1977) and Brook and Allison (1983), have integrated the physical variables that control subsidence to produce predictive models that differentiate areas of varying susceptibility to sinkhole subsidences. However, data are not presently available to test the validity of the models produced.

Central and northern Florida is a covered karst terrain dominated by sinkholes. Sinkhole development continues so that ground subsidences present serious problems to urban and rural development. Subsidences are particularly problematical in urban areas such as Orlando where ground subsidence resulting from natural and man-induced causes has resulted in damage to roads, buildings and public facilities. This paper tests a new approach to explaining the spatial characteristics of sinkhole subsidences in a sample area of suburban Orlando. The method is based upon the double Fourier series analysis of 1:24,000 scale topographic quadrangle elevation data. The method could be useful in developing maps of subsidence sinkhole susceptibility in covered or mantled karst terrains.

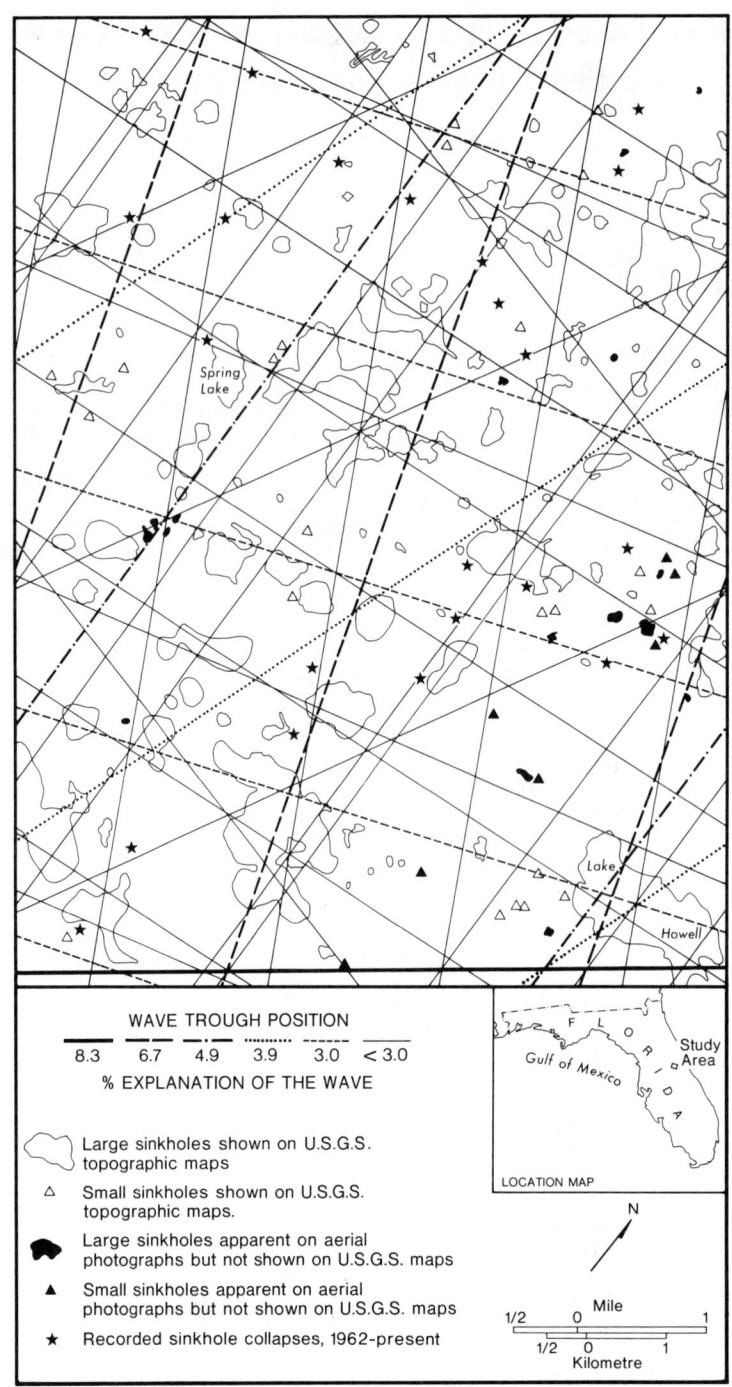

Figure 1. Sinkhole Locations and Troughs of the Dominant Waves Isolated by Double Fourier Series Analysis, Suburban Orlando, Florida.

Theoretical Considerations

Ground water flow and solutional denudation in carbonate rocks are concentrated along the major lines of secondary permeability--namely the major faults and joints. In areas of bare karst water migrates laterally at the surface and at shallow depth towards these major permeability zones. At the surface, fracture zones ultimately become depressions while the resistant areas of more massive rock between them become areas of positive relief. Application of double Fourier series analysis to the study of doline karst in Kentucky and cockpit karst in Puerto Rico revealed that complex karst terrains can be decomposed into a few, simple waves that explain a high percentage of the topographic variation (Liang et al. 1983). Furthermore, studies of Jamaican doline and cockpit karst have shown that the troughs of the most important waves making up a karst landscape parallel faults and joints that are visible in black and white aerial photography of the areas (Hanson and Brook, Department of Geography, University of Georgia, unpublished data). Double Fourier series analysis of covered and mantled karsts should also reveal the presence of bedrock fractures where surface topography mirrors the form of the underlying bedrock-overburden contact. Fractures mapped in this way should define linear zones of high subsidence susceptibility as it is within these zones that large subsurface cavities in carbonates are to be expected.

Study Area and Fourier Analysis

A landscape can be viewed as a complex spectrum made up of a series of sine and cosine waves. Spectral analysis is the "process of calculating and interpreting a spectrum" (Rayner 1971). Through Fourier transformation, the frequencies and amplitudes of a complex spectrum can be decomposed into a series of simple constituent waves. The frequencies, directions, and amplitudes of these constituent waves provide a useful description of the basic characteristics of the original spectrum or landscape. To be suitable for spectral analysis a landscape should exhibit periodicity of relief. Karst topographies, which are controlled by sets of fractures of different orientation with each set consisting of approximately evenly-spaced fractures, are ideal landscapes for spectral analysis.

A 9 x 12 km area in northern suburban Orlando, central Florida was selected for study by the double Fourier series technique. At least 23 sinkholes have developed in the area since 1962 (Fig. 1). The area is therefore ideal to test Fourier prediction of sinkhole susceptibility against actual sinkhole occurrences. Elevations range from 6-36.5 m a.s.l. Geologically the area consists of Avon Park limestones which are overlain by cavernous Ocala limestones of Eocene age. The upper surface of the Ocala lies at approximately 50 m below the ground surface. Above the Ocala are less permeable sediments of the Miocene Hawthorn Formation and approximately 20 m of well-drained Pleistocene sands and recent deposits. Most sinkholes in the area are caused by subsidences of Miocene to recent sediments into cavities in the Ocala limestone.

Topographic data used in the Fourier algorithm (Liang et al. 1983) were obtained by placing a grid of 54 x 70 units over the 1:24,000 scale U.S. Geological Survey topographic maps covering the study area. X, Y, and Z coordinates were digitized at each grid intersection producing 3,780 data points. Elevations or Z-values were interpolated from contours, the contour interval on three maps covering the area is 5 ft (1.52 m) and on the fourth map it is 10 ft (3.05 m). Prior to Fourier analysis the data set was detrended using the General Linear Model (GLM) procedure of SAS. Residual values from the regression were inputed to the Fourier algorithm.

Using the 3,780 topographic data points a Fourier model was generated which explained 91% of the karst topography in the Orlando area ($R^2 = 0.91$). Twelve pairs of dominant frequencies explained 40% of the topographic variation, the most significant wave explained 8.3% of the topographic data (Table 1). The R^2 value is comparable with values of 0.9 for two 2 x 2 km doline areas in Jamaica, and higher than values of 0.7 and 0.6 for two 2 x 2 km cockpit areas (Hanson and Brook, unpublished data). The major difference between the Jamaica and Florida results lies in the higher explanation of the dominant waves in the first area. In the Jamaican doline areas the 10 dominant waves explain 74% and 76% of the topographic variation, in the cockpit areas the 10 dominant waves explain 58% and 61% of the variation. All figures are substantially higher than the 37% of the topographic variation explained by the 10 dominant waves isolated in the Orlando study area. This difference almost certainly reflects the fact that the Orlando study area is 26 times larger than the Jamaican doline and cockpit study areas.

Comparison of Fourier Wave Troughs with Sinkholes and Fractures

Individual wave trough directions were calculated for the 12 dominant waves from the paired frequency data. Three dominant trough directions are apparent. Waves with troughs oriented at 0-26° (NNE), 45-63° (NE), and 329-341° (NNW) explain 14.6%, 13.3% and 10.5% of the variation in the Orlando topographic data respectively. Sinkhole long axes and linear portions of sinkholes show generally similar preferred orientations. Length-weighted data indicate that 34% of long axes and linear sinkhole elements are aligned at 0-10°, 52% at 30-70°, and 14% at 280-290°. Fractures in central Florida, mapped by Vernon (1951) from aerial photography, display predominant NW-SE and NE-SW orientations which approximately parallel two of the three dominant Fourier wave directions. Comparisons suggest that fractures and sinkholes in the Orlando area have generally similar preferred orientations and that these orientations approximately parallel the main trough directions of the 12 dominant waves making up the topography of the area. It is possible therefore that the dominant wave troughs do mark the positions of fracture zones in the underlying Ocala limestone.

Wave Rank	Frequency (k_1, k_2)	Power	Amplitude (m)	Wavelength λ (m)	Trough Direction (Degrees)	Contribution (%)	Total Contribution (%)
1	1,0	33.4	1.76	12000	45	8.3	8.3
2	-1,2	27.0	1.58	4336	341	6.7	15.0
3	-1,1	19.6	1.35	7351	0	4.9	19.9
4	-2,1	15.9	1.22	5042	18	3.9	23.8
5	4,1	12.2	1.06	2855	59	3.0	26.8
6	-1,4	9.8	0.95	2282	329	2.4	29.2
7	-3,1	8.5	0.89	3674	26	2.2	31.4
8	-3,3	7.6	0.84	2450	0	2.0	33.4
9	3,1	7.8	0.85	3674	63	2.0	35.4
10	1,1	6.3	0.76	7351	90	1.7	37.1
11	-2,2	5.9	0.74	3675	0	1.6	38.7
12	6,3	5.1	0.69	1681	333	1.4	40.1

Table 1. Dominant Waves in the Orlando Area of Florida Isolated by Double Fourier Series Analysis of Topographic Data.

Figure 1 shows trough positions for the 12 major waves isolated by Fourier analysis. These are compared with the locations of sinkholes shown on U.S. Geological Survey 1:24,000 scale topographic maps, with sinkholes that are not shown on these maps but which are visible on 1:24,000 scale black and white aerial photographs, and with the locations of 23 sinkholes that are known to have formed since 1962. Approximately 90% of the sinkholes shown on the topographic maps are aligned along wave troughs and the larger sinks are located where two or more wave troughs intersect. 35% of the sinkholes that are only shown on the aerial photographs are bisected by wave troughs. Fourteen of the 23 recorded sinkhole subsidences (61%) are located within 110 m of a wave trough and 20 (87%) within 220 m of a wave trough. Four of these sinkholes are located on a wave trough. Only three sinkholes (at 264, 312, and 384m from the nearest trough) are at a considerable distance from a wave trough.

Conclusions

Results show that double Fourier series analysis of topographic data can reduce a complex landscape into a small number of simple component waves. Twelve such waves explain 40% of the topographic variation in a 9 x 12 km area of suburban Orlando, Florida. If the troughs of these 12 waves represent fracture zones approximately 400 m wide, they provide a good explanation of the locations of 87% of the 23 sinkholes that are known to have formed in the area since 1962.

References

Brook, G. A., and Allison, T. L., 1983, Fracture mapping and ground subsidence susceptibility modelling in covered karst terrain: Dougherty County, Georgia: In: Environmental Karst (ed. P. H. Dougherty), GeoSpeleo Publications, Cincinnati, Ohio, 91-108.

Jennings, J. E., 1966, Building on dolomites in the Transvaal. Transactions of the South African Institution of Civil Engineers: 8(2), 41-62.

Liang, C-l., Mitchelson, R. L., and Brook, G. A., 1983, A computer program for two dimensional spectral analysis: Abstracts of the 38th Annual Meeting, Southeastern Division, Association of American Geographers, Orlando, Florida, November, 1983, p. 22.

Newton, J. G., 1977, Induced sinkholes--a continuing problem along Alabama highways: In: Karst Hydrogeology (eds. J. S. Tolson and F. L. Doyle), University of Alabama Press, Huntsville, 303-304.

Ogden, A. E., and Reger, J. P., 1977, Morphometric analysis of dolines for predicting ground subsidence, Monroe County, West Virginia: In: Hydrologic Problems in Karst Regions (eds. R. R. Dilamarter and S. C. Csallany), Western Kentucky University Press, 130-139.

Rayner, J. N., 1971, An Introduction to Spectral Analysis: Pion Limited, London.

Vernon, R. O., 1951, Geology of Citrus and Levy counties, Florida. Florida Geological Survey Bulletin, 33.

Evaluation of subsidence or collapse potential due to subsurface cavities

RICHARD C. BENSON & LESTER J. LA FOUNTAIN *Technos, Inc., Miami, Florida, USA*

ABSTRACT

Though the methodology to provide accurate location and assessment of subsurface cavities exists, the knowledge to properly implement the appropriate methodologies is fragmented.

Three key methods that may be used in subsurface investigations are:

- Direct sampling methods such as drilling and observation
- Indirect methods such as remote sensing and geophysics
- Statistical methods

It is critical to recognize that limited direct sampling (e.g., borings) will affect the accuracy of a site investigation. It is also important to understand how the indirect and statistical methods may be employed to improve the accuracy of an investigation by providing additional data in a cost effective manner. Methodology selection is dependent upon the area of investigation, the size, depth, and stability of the cavity system being investigated.

The above concepts and methods need to be incorporated into an integrated systems approach along with a working knowledge of geology, hydrology, geomorphology, geostatistics, geochemistry, soil mechanics, and rock mechanics as they apply to karst problems.

Selecting the appropriate methodology to accomplish these goals depends to a high degree on site-specific conditions. By selecting the most suitable methods and utilizing the synergistic benefits of an integrated systems approach, high levels of technical accuracy and cost effectiveness can be achieved.

Background

Subsidence or collapse due to the presence of subsurface cavities is a common problem in many areas of the continental United States. W.E. Davies of the United States Geological Survey estimates that 15% of the United States is composed of limestone or other soluble rock at the surface, and that 50 to 75% of the continental United States may be susceptible to solution and subsidence problems if deep, soluble rocks and pseudo-karst effects are included.

Subsurface cavities range in size from the small pore spaces between soil or rock particles to large, cavernous rooms within solid rock. Small cavities or pore spaces are important in that they can contribute to subsidence similar to that found in California's San Joaquin Valley. There, up to 8 meters of subsidence has occurred as a result of water withdrawal for irrigation purposes. Small cavities in rock with a characteristic diameter of approximately one meter often occur in abundance. Uncovered, this rock resembles swiss cheese. Large cavities of up to 100 meters or more in diameter can also occur. The ultimate collapse of these large cavities is responsible for many of the sinkhole lakes found throughout the State of Florida (see Figure 1).

Most large cavity systems can be described in terms of regular shapes such as vertical or horizontal planes associated with fractures or bedding planes, vertical or horizontal cylindrical conduits, and large rooms of approximately spherical shape. These cavities are the result of long term solution of the cavity walls at a rate of a few centimeters per 1000 years. Although small cavities can contribute to serious problems, only large cavities will be considered in this paper to simplify the discussion. The same philosophy of investigation and methodology can be scaled down to address any size cavity, even the pore space between soil.

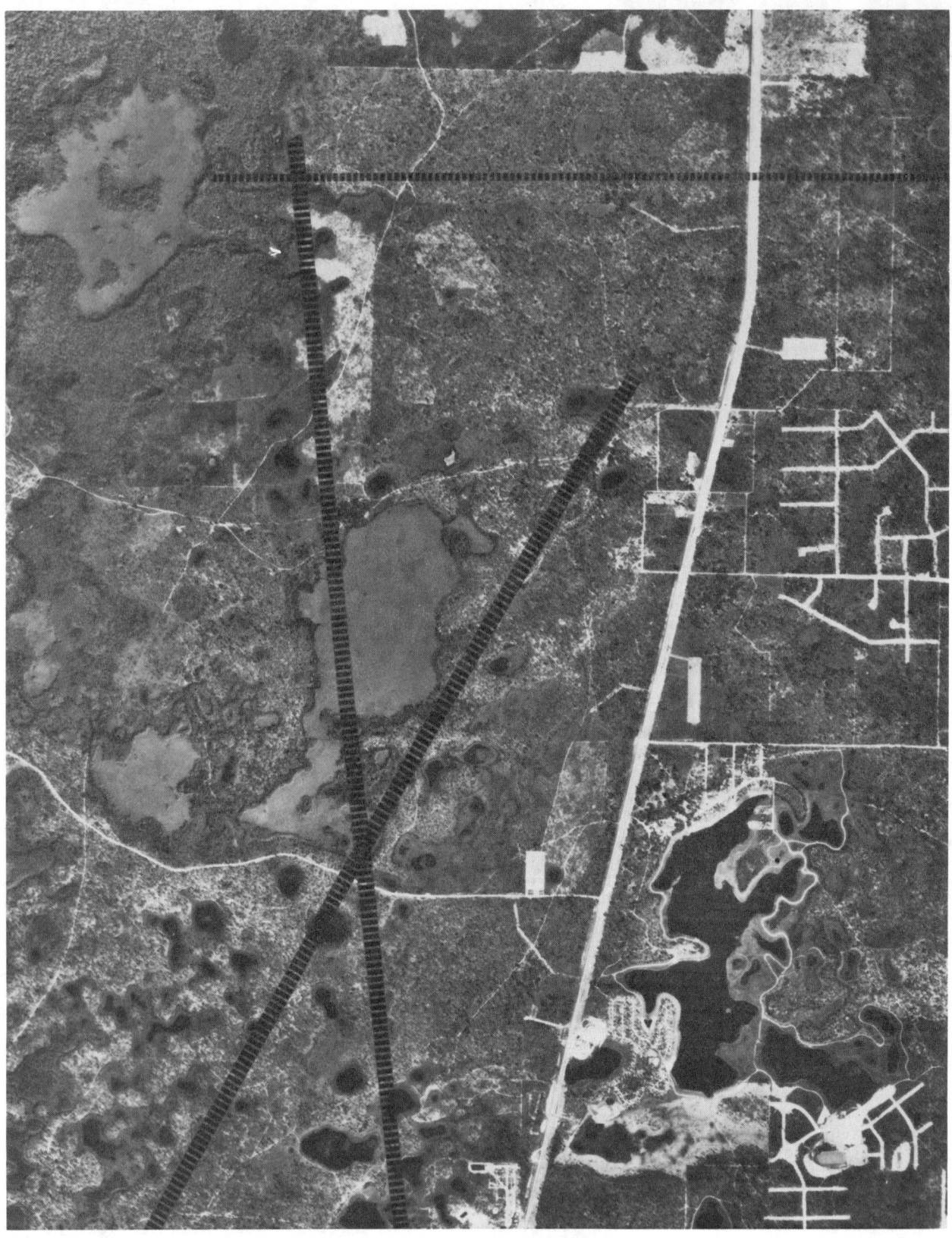

Figure 1: Aerial photo from the west coast of Florida showing numerous sinkhole lakes. Dashed lines indicate linear trends formed by the sinkhole lakes. Large cavities of up to 100 meters in diameter are found in this area where major collapse has occurred.

Causes of Collapse and Triggering Mechanisms

The cause and effect relationships of subsidence and collapse due to the presence of large cavities within rock are numerous. As limestone is dissolved by slightly acidic ground water and eroded, voids form. When voids enlarge to the point that the overhead supporting structure fails, surface collapse occurs. Collapse of the overhead rock and soil is accelerated by loading which may result from the static weight of the overburden, man-caused changes to the environment, rainfall, or a combination of factors, all of which represent increased static and/or dynamic loads to the overhead structure. Although it is safe to say that long term geologic conditions such as the natural solution and erosion of bedrock set the stage for the occurrence of subsidence and collapse, variation in rainfall and man-caused changes to the environment over the short term are by far the most significant factors that impact man's construction.

Changes in surface water runoff and ground water levels as a result of variations in rainfall are major factors in developing and triggering collapse. A lack of rainfall, for instance, results in lowered ground water levels causing a loss of buoyancy that leads to general soil stress, and ultimately, collapse. An abundance of surface water from increased rainfall, on the other hand, can accelerate vertical seepage, increase piping activity, and trigger collapse.

The effects of man-caused changes on the natural environment are the most important factor in developing and triggering collapse. Two of the most common collapse-precipitating activities are the withdrawal of ground water for residential and industrial use and the concentration of surface runoff or change in surface runoff patterns resulting from the construction of major roads, paved parking lots, or airport runways. Though many variables contribute to the ultimate cause of collapse, a singular event usually acts as the final triggering mechanism.

The following data obtained from the Florida Department of Transportation summarizes the causes of collapse. The majority of these statistics represent roadway-related collapses. Included in the data are 96 cases of collapse recorded over a 5 year period.

Blasting	5%	Construction	11%
Drilling	5%	Other or Unknown	11%
Low Water Table	8%	Heavy Rainfall	58%

It is not surprising that the figures show the dominant cause of collapse to be associated with heavy rainfall since excess surface water is concentrated by roadway drainage. Although limited, other nonroadway-related data compiled in 1977 shows many of Florida's large collapses to be associated with low ground water levels occurring predominately during April and May -- the last two months of Florida's dry season.

The key point made from these data are that within the lifetime of a manmade structure, 100 years or less, the solution of rock and even the mechanical erosion of rock have little to do with the final cause of collapse; they merely set the stage for the event at some time in the future. Furthermore, the factors contributing to collapse are not necessarily singular. In most cases, they appear to be cummulative and from many different causes.

Surface subsidence or collapse generally manifests itself within a limited area over or near a ruptured cavity and may take the form of a single, centralized collapse or a large collapse with numerous satellite sinkholes and fractures around the perimeter. One example of a single, centralized collapse is the Winter Park Sinkhole in Orange County, Florida. Examples of a major collapse with numerous satellite sinkholes and fractures around the perimeter are the December Giant in Shelby County, Alabama and the collapse that occurred in Hernando County, Florida during a water management district's well drilling attempt.

One common misconception is that a cavity is a singular occurrence. In general, this is not true, and in particular, it is not generally true for large cavities. Each "cavity" is a member of a large system of enlarged fractures, bedding planes, vertical pipes, horizontal conduits, and large rooms similar to those observed in caves throughout the world. In Florida, most cave systems are water-filled. Treating a cavity as a single entity for assessment or remedial purposes can only result in errors that may have significant impact in the future. Understanding the numerous cause and effect relationships of subsidence and collapse as a result of subsurface cavities is important. It will certainly lead to better forecasts about the behavior of cavities and the impact that environmental and man-caused factors have on them.

Methodologies Available for the Evaluation of Subsurface Cavities

Many cavities cannot be analyzed using a single methodology such as aerial photography or surface observations. Narrow, vertical fractures, and small cavities, for instance, may be virtually impossible to detect through a normal drilling program. Features such as piping over large, deep cavities can also go undetected in a normal field investigation. Missing such features can result in serious construction problems and subsequent, catastrophic failures. The technology and methodology to completely define the existence of both large and small cavities at any depth does exist. Though drilling is the most commonly used investigation tool, other approaches are necessary. The remote sensing geophysical and <u>in situ</u> methods are listed below. In addition, there are other tools that may be employed such as geomorphology and statistics.

POSSIBLE METHODOLOGIES FOR CAVITY DETECTION AND EVALUATION

Airborne or Satellite Spatial Methods

- Black and White Photography
- Color Photography
- Infra-Red Photography
- Thermal Imagery
- Radar Imagery
- Satellite Imagery
- Multispectral Satellite Imagery

Surface Methods

- Thermal Imagery
- Seismic Techniques (Various)
- Resistivity (Various)
- Electromagnetics (EM)
- Ground Penetrating Radar (GPR)
- Micro Gravity
- Magnetics

Downhole Methods

- Camera/Television
- Acoustic Scanning (Sonar)
- Dyes/Tracers
- Conventional Logging Tools
- Seismic Techniques
- Electromagnetics
- Ground Penetrating Radar
- Gravity
- Magnetics
- Geochemical
- Nuclear

In Situ Sensors

- Piezometers
- Pressure Sensors
- Thermal Sensors
- Acoustic Emission Sensors
- Displacement Sensors
- Precision Leveling

In addition to defining the presence of cavities at any depth, cavity stability can be measured, and, to a reasonable degree, cavity behavior can be predicted. Though the tools to do both detection and evaluation exist, they are seldom applied because of:

- o limited budgets
- o not knowing that the methods exist
- o lack of knowledge about the methods and how to apply them
- o lack of a single person or firm with the expertise to utilize them

Numerous conferences and papers have attempted to address the problems associated with subsurface investigation, subsidence, and cavity detection (see Bibliography). Most of these documents focus on one methodology to solve a problem. Since each methodology has advantages and disadvantages, and since improperly utilized methodology does not produce positive results, it follows that any single method can fail under a given set of field circumstances. Therefore, reliance on a single approach usually results in failure. This paper, in contrast, focuses on a broad, systematic approach that incorporates a range of skills and technology, then selectively applies them to bring about an economical and technically optimum solution. Every investigation requires a tailored, site-specific systems approach that takes into consideration the available budget and the required level of accuracy. In keeping with these concepts, cavity detection and evaluation methodology can be broadly grouped into the following four categories:

1. Direct measurement methods such as drilling or direct observation
2. Indirect measurement methods such as aerial photography or geophysical methods
3. Statistical methods such as those used to characterize direction, size, and spacing of cavities
4. Use of an effective systems approach.

Direct Measurement Methods

Direct measurement methods reveal the presence of subsurface voids through direct contact with the cavity. For example, a loss of fluid or a drill stem drop during drilling constitute direct measurement of the presence of a cavity. Visual observation of a cavity using a borescope or television camera also constitute direct measurement. Direct cavity hits by drilling are unusual for most subsurface investigations because the number of borings must be limited in order to be cost effective and the probability of hitting a cavity is low.

The number of borings required to provide an acceptable probability level for cavity detection can be estimated. By dividing the area of the site by the estimated area of the smallest cavity the investigator wishes to detect a site to cavity ratio is established. Then, statistical tables can be used to determine the number of borings for a given level of confidence. A simple example is shown in Figure 2. The larger the site to cavity size ratio, the greater the number of borings necessary to provide an acceptable level of confidence for cavity assessment at a given site.

A 10:1 site to cavity ratio involves a rather large cavity. For example, a one-acre site with a 10:1 ratio implies that a cavity of about 23 meters in diameter exists. It is not unusual for ratios of 100:1, 1000:1, or greater to occur. On a one-acre site, a 100:1 ratio implies that a cavity of about 7 meters in diameter exists, and a 1000:1 ratio implies that a cavity of about 2.3 meters in diameter exists. Even a cavity 2.3 meters in diameter can be significant on a one-acre site. The following example using a one-acre site and a 90% detection probability level shows the number of borings necessary to provide an acceptable level of confidence from direct detection drilling programs.

ONE-ACRE SITE WITH A 90% DETECTION PROBABILITY LEVEL

Cavity size of 23 meters (As/At=10):   Requires approximately 10 borings
Cavity size of 7 meters (As/At=100):   Requires approximately 100 borings
Cavity size of 2.3 meters (As/At=1000): Requires approximately 1000 borings

This example assumes that uniform grid spacing is used to locate borings. If drilling locations are randomly selected, the number of borings required increases significantly. Although the use of this procedure assures a given level of confidence for cavity detection, the boundaries of the anomaly must still be defined. Defining them requires additional drilling. Furthermore, if the smallest cavity size estimated is too large, significant error will be induced into the program.

It is obvious, therefore, that the achievement of an adequate evaluation of complex subsurface conditions by borings alone is not generally practical. Neither is it cost effective. To provide such an evaluation would necessitate the installation of an excessive number of borings. While critical projects such as dams, tunnels, and nuclear plants may justify high density drilling and subsequent grouting, most investigations do not.

Based on the example given, it should be clear that most subsurface investigations do not begin to approach 100% accuracy. In fact, many investigations are probably less than 10 to 20% accurate. Yet, many professionals and their clients continue to think of subsurface investigations in terms of high accuracy. It is obvious that alternatives to direct measurement methods must be used if realistic cavity investigation programs are to be implemented.

Indirect Methods

A drill stem drop during drilling indicates the presence of a cavity even though it has not been seen. A downhole television camera gives visual proof of a cavity's presence even though it may not be touched. Although these methods of direct measurement provide a high level of confidence in the subsurface information obtained, the information is localized and must be interpolated between sample points or extrapolated beyond them. At the sample points, a high level of confidence exists. Beyond each sample point, guesses must be made.

In order to fill in the low levels of confidence between sample points, various indirect measurement methods can be employed such as remote sensing or geophysical techniques. Where a drill stem drop allows a cavity to be detected and a downhole television camera allows it to be seen, indirect methods measure the physical, chemical, or electric anomalies associated with the cavity or the disturbed zone surrounding the cavity.

Using this approach, continuity between direct sample points can be provided to eliminate or at least minimize errors associated with interpolating and extrapolating information from direct sampling points. Better yet, boring locations can be selected based upon prior knowledge, thereby increasing the validity of data from a given number of borings. Just as a surgeon uses X-rays and CAT scans to locate a tumor before surgery, indirect methods can be used to indicate the presence of a cavity before a direct drilling program begins.

A large number of indirect measurement methods can be used to evaluate the presence of a cavity. Figure 3 shows the general application of indirect methods. The first two methods

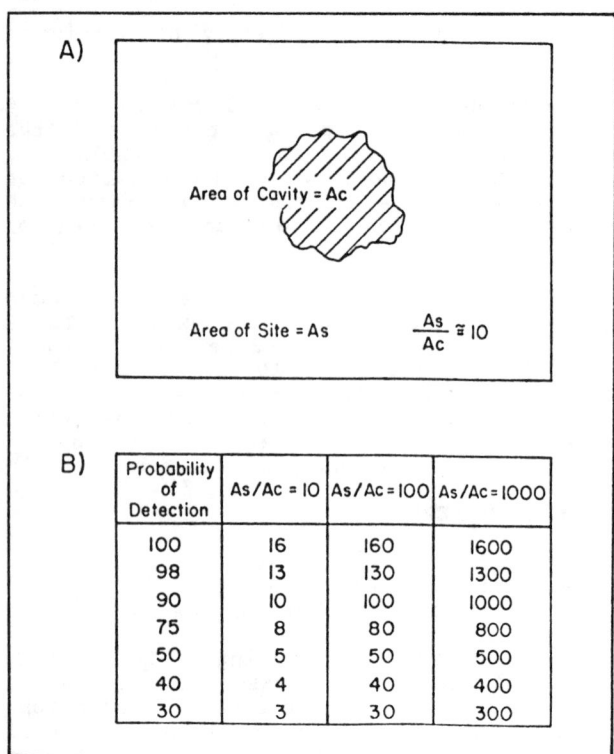

Figure 2: Figure A shows a site to cavity area ratio of approximately 10. Table B shows various site to cavity ratios and the probability of detecting a cavity with a given number of borings.

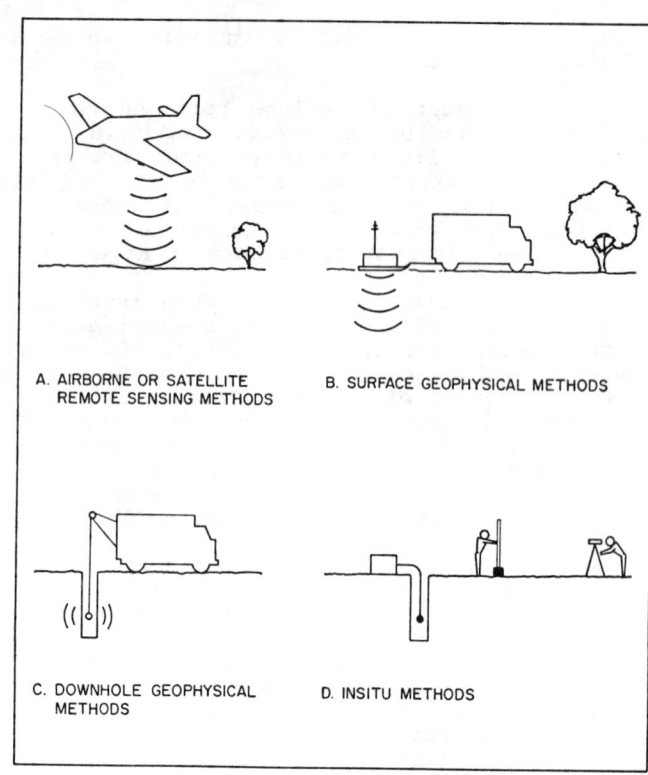

Figure 3: Four indirect methodologies for detection and evaluation of subsurface cavities.

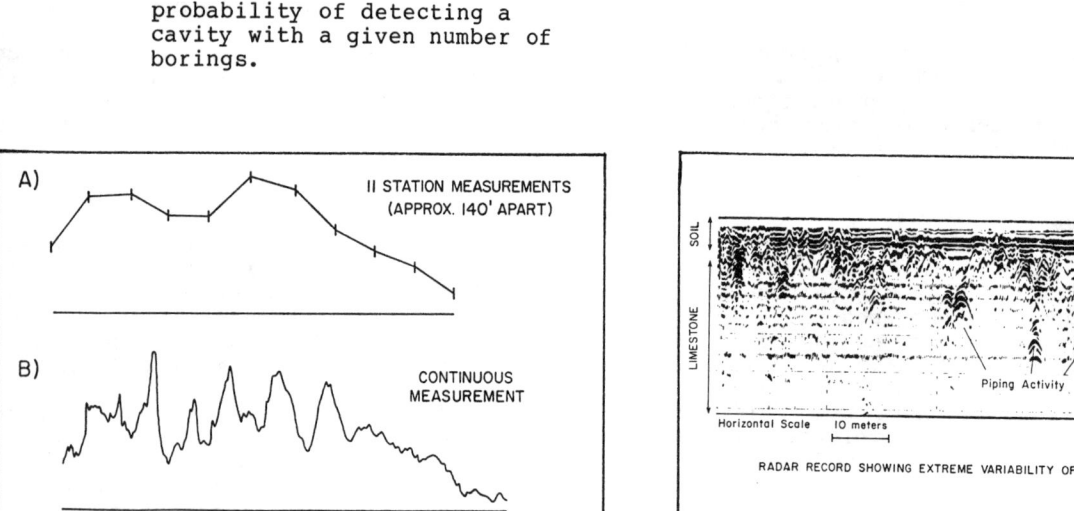

Figure 4: Comparison of station and continuous electromagnetic conductivity measurements along the same traverse. Continuous data shows fractures in rock based upon moisture content.

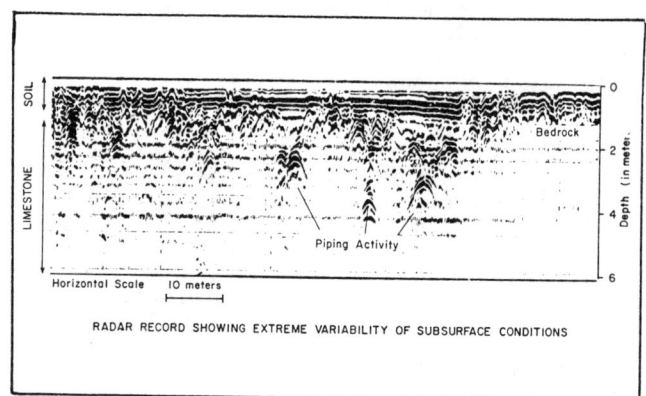

Figure 5: Ground penetrating radar record showing piping. This example illustrates the use of near surface indicators to locate and evaluate the activity of deep cavities. In this case, shallow piping activity indicates the presence of a major cavity system at a depth of 30 to 45 meters.

illustrated in Figure 3 are the airborne and surface geophysical methods. They provide the benefits of _in situ_, nondestructive measurements.

Airborne remote sensing is beneficial in terms of spatial coverage per unit time and cost; however, subsurface data can only be obtained through interpretation (see Figure 1). Surface geophysical methods, on the other hand, yield less spatial coverage per unit time and cost than airborne methods, but they significantly improve depth resolution while they provide subsurface information. A three-dimensional subsurface picture can often be generated using special measurement and imagery techniques. Surface geophysical methods are quite cost effective for shallow investigations, but resolution and the ability to define details decreases with increasing depth.

Downhole measurement methods also improve the resolution of local details. Furthermore, resolution does not decrease with depth as it does with surface geophysical methods. The volume of soil or rock sampled by downhole methods is usually much less than that attained by surface geophysical methods; however, it is much more than that achieved by drilling alone. The major benefit of downhole measurement methods is that detailed, continuous information may be acquired at significant depths. The cost per unit area of coverage is high, but existing boreholes can often be used to reduce the cost.

In situ sensors are another indirect measurement method. They can be implanted at a site and sampled periodically to detect changes in subsurface conditions. Sampling with _in situ_ sensors can be done manually or electronically depending on the specific method employed. Generally, airborne, surface, and downhole methods provide a number of measurements at one point in time. These measurements are known as spatial measurements. _In situ_ measurements provide a number of measurements in one place over a period of time. These measurements are known as temporal measurements. Though airborne, surface, and downhole measurements can be repeated periodically to yield a series of quasi-temporal measurements, and _in situ_ measurements can be made at a number of locations to provide quasi-spatial measurements, there are limits to the compromises that can be made.

Continuous Surface Geophysical Techniques:
Two contemporary geophysical measurement techniques known as ground penetrating radar (GPR) and electromagnetic conductivity (EM) provide unique cavity detection capabilities in that they provide a means to obtain continuous subsurface information at rapid traverse speeds. For these reasons, they are effective for both reconnaissance and detailed site investigations.

The benefits of continuous subsurface sampling can be seen by comparing the two sets of data in Figure 4 which were taken from a dam site leakage investigation. The upper set of data in Figure 4 is comprised of discrete measurements taken at 11 points along a traverse line. These points are joined by a line to produce a data profile. The lower set of data is the result of continuous measurements taken along the same traverse line. Comparing the two data sets, it is obvious that continuous measurements are the most effective for sampling complex subsurface site conditions because they provide more detail. The peaks in the electromagnetic conductivity data shown in Figure 4B indicate the presence of fractures within the underlying rock. The benefits of rapid traverse speeds are lower cost and more detailed site coverage. In many cases, 100% site coverage can be economically obtained. Most detailed surveys are run at slow speeds of about 3 kilometers per hour, however, high speeds for less detailed, reconnaissance surveys are possible.

GPR is a reflection technique using high frequency electromagnetic radiation. GPR surveys produce graphic profiles of subsurface conditions that resemble the side walls of trench cuts. Figure 5 shows the radar record of a thin veneer of soil over limestone. Considerable piping can be seen in the data indicating the presence of a deep, active cavity. The reflections shown on the radar record are produced as a result of contrasts in the complex dielectric constant of individual, subsurface materials. This method provides the highest resolution of all surface geophysical methods. Depths of one to fifteen meters or more may be obtained; however, the depth of penetration is quite site-specific and depends upon soil conditions. In some cases, penetration depth is limited to 1 meter or less.

The EM conductivity technique permits rapid measurements of the bulk electrical conductivity of the subsurface to be made. EM conductivity values are a function of the site's porosity, permeability, saturation, natural subsurface materials, and the specific conductance of pore fluids. This measurement is similar to that made by the more familiar resistivity method, but is accomplished without ground/electrode contact. The EM method permits high lateral resolution profiling measurements to be made which are particularly effective for locating lateral anomalous conditions. Figure 6 shows the data resulting from an EM survey over fractured limestone. The high EM conductivity values indicate a fracture

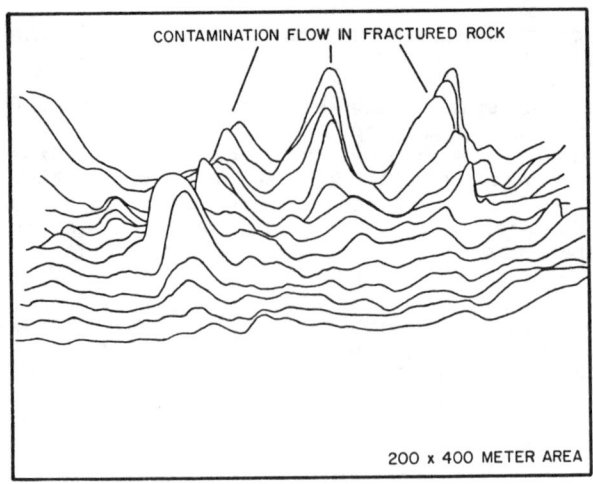

Figure 6: Parallel electromagnetic conductivity profiles showing migration of salt water in fractured limestone.

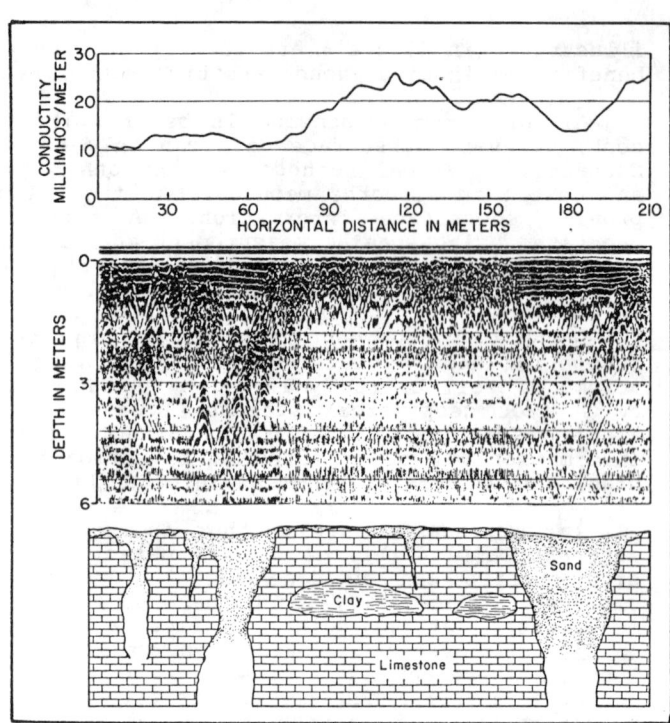

Figure 7: Electromagnetic conductivity (top) and ground penetrating radar (middle) profiles over karst terrain with a geologic cross section (bottom). Note the correlation between electromagnetic conductivity and ground penetrating radar data where paleo karst features occur.

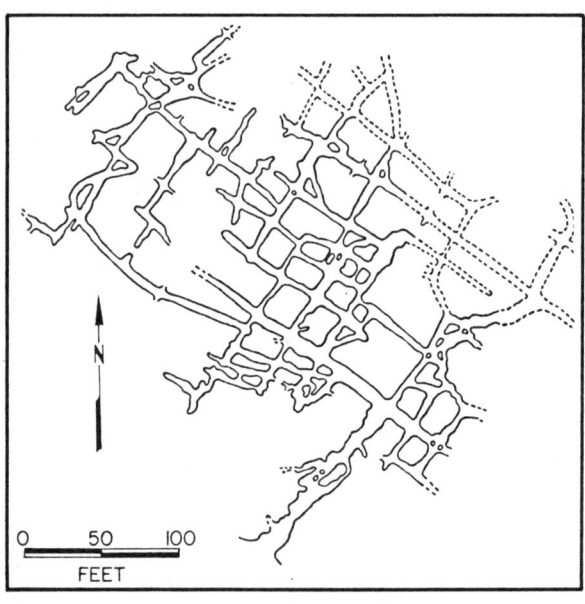

Figure 8: Plan view of cave system (W.E. Davies). Note the repeatable pattern that lends itself to statistical analysis.

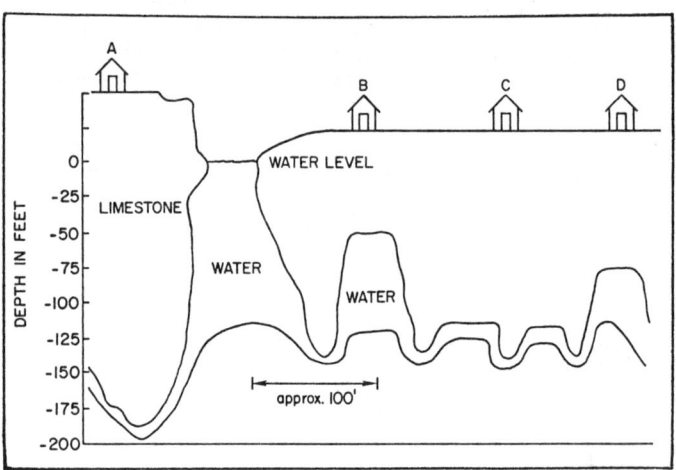

Figure 9: Schematic cross section of sinkhole cave system mapped by cave divers. Note the periodicity associated with cavity growth. Risk to structures built at points A and C may be low; however, structures built at points B and D have a distinctly higher risk of damage.

zone. The linear trends observed in the data are related to fluids moving within the fracture system. Locating these vertical fractures by drilling would be economically prohibitive.

While GPR and continuous EM conductivity techniques are typically limited to depths of 15 meters or less, considerable insight into problems occurring at deeper levels can be acquired through the use of near surface indicators.

Using Near Surface Indicators:
Long before subsidence or collapse occur, indicators at or near the surface generally appear. In other words, deep-seated cavities and fractures often show signs of their presence in the near surface before actual collapse occurs. For example, lineaments are commonly identified on aerial photos as evidence of deep-seated fractures or cavity systems (see Figure 1). Often, these fractures manifest themselves at the surface in subtle ways, such as by disturbed vegetation patterns. In such cases, the fracture or cavity is not observed directly, but its presence is implied by observing vegetation patterns -- a near surface indicator (NSI). Local piping of soil due to downward flow of surface water into fractures or cavities can often be detected by means of surface geophysical methods and the use of NSI. Identification of NSI provides a rapid and cost-effective means of locating deep cavities. In many cases, the use of NSI has been found extremely effective when used in conjunction with continuous sampling surface geophysical methods.

Synergism:
A synergistic increase in the certainty of interpretation occurs when many methods are combined into a systems approach. For example, geophysical methods such as GPR and EM conductivity may be combined to yield synergistic results. The EM conductivity values in Figure 7 are high over limestone due to interbedded clays and clay-filled pockets. Over paleosinks (old sinkhole collapses filled by the natural deposition of sands) filled with quartz sand, EM conductivity values are substantially lower. GPR data, located in the middle of Figure 7, shows a continuous cross section of the site to a depth of approximately 6 meters. A distinctive paleosink can be seen to the right side of the radar data. This sink is greater than 30 meters across. Smaller paleosinks and piping activity can be seen to the left. The combined results of the EM conductivity and GPR geophysical surveys using NSI and geologic knowledge about the local area were used to draw the interpretative section shown on the bottom of Figure 7. These data were used to accurately locate drilling locations. Consequently, "smart holes" were drilled instead of proceeding with a blind drilling program. Three borings along the 200-meter traverse confirmed the major collapse and active piping zones with a certainty well above 80%.

Statistical Methods
The approach for evaluating large areas is different in that they simply cannot be investigated at the same level of detail as localized areas. Other approaches, therefore, must be used. Assessing regional problems to maintain reasonable levels of accuracy in an investigation or mapping program depends heavily upon the integration of information from many sources to provide an overview of conditions that can be thought of as a statistical data base. For example, a lineament map can be developed from regional aerial photography (see Figure 1) or satellite imagery and used to characterize the extent and direction of fractures or karst activity in the region as well as to illustrate trends through a specific area of interest. Using regional data such as geologic and hydrologic information, aerial photo interpretation, and records of recent collapse, regional probability maps can be generated to show areas susceptible to collapse.

A few kilometers of continuous geophysical data obtained along easily accessible roads and fields can also provide a valuable statistical base from which to work. Based upon the presence, absence, or number of NSI encountered, a reasonable statistical assessment can be made. In addition, potential problem areas can also be identified for subsequent, detailed studies.

Cave explorers are an important source of critical information that can be used to evaluate local trends. The cave map in Figure 8 shows the orderly periodic nature of fractures and subsequent solution of limestone. This information is invaluable for planning a site investigation or predicting potential problems. The profile of a water-filled cave in Figure 9, mapped by cave divers, shows the potential of sinkhole collapse as roof sections spall and grow toward the surface, eventually resulting in failure. Here, the periodicity of the potential sinkhole collapse areas are clearly illustrated. Both the map and profile examples provide significant statistical information that can be used to evaluate the presence of a cavity system and the potential for local subsidence, piping, or collapse.

The presence of existing cavities is often confused with the activity of subsidence or

SITE AREA / APPROACH	LOCAL (.1 sq. mi.)	INTERMEDIATE (1 sq. mi.)	REGIONAL (10 sq. mi. or more)
DIRECT MEASUREMENT	Primary	Secondary	Secondary
INDIRECT MEASUREMENT	Primary	Primary	Secondary
STATISTICAL MEASUREMENT	Primary (limited)	Primary	Primary

Figure 10: Applicability of direct, indirect, and statistical approaches to cavity investigations versus the scale of investigation. "Primary" indicates the cost-effective approach. "Secondary" indicates a support approach. Note: Areas are provided for relative comparison only.

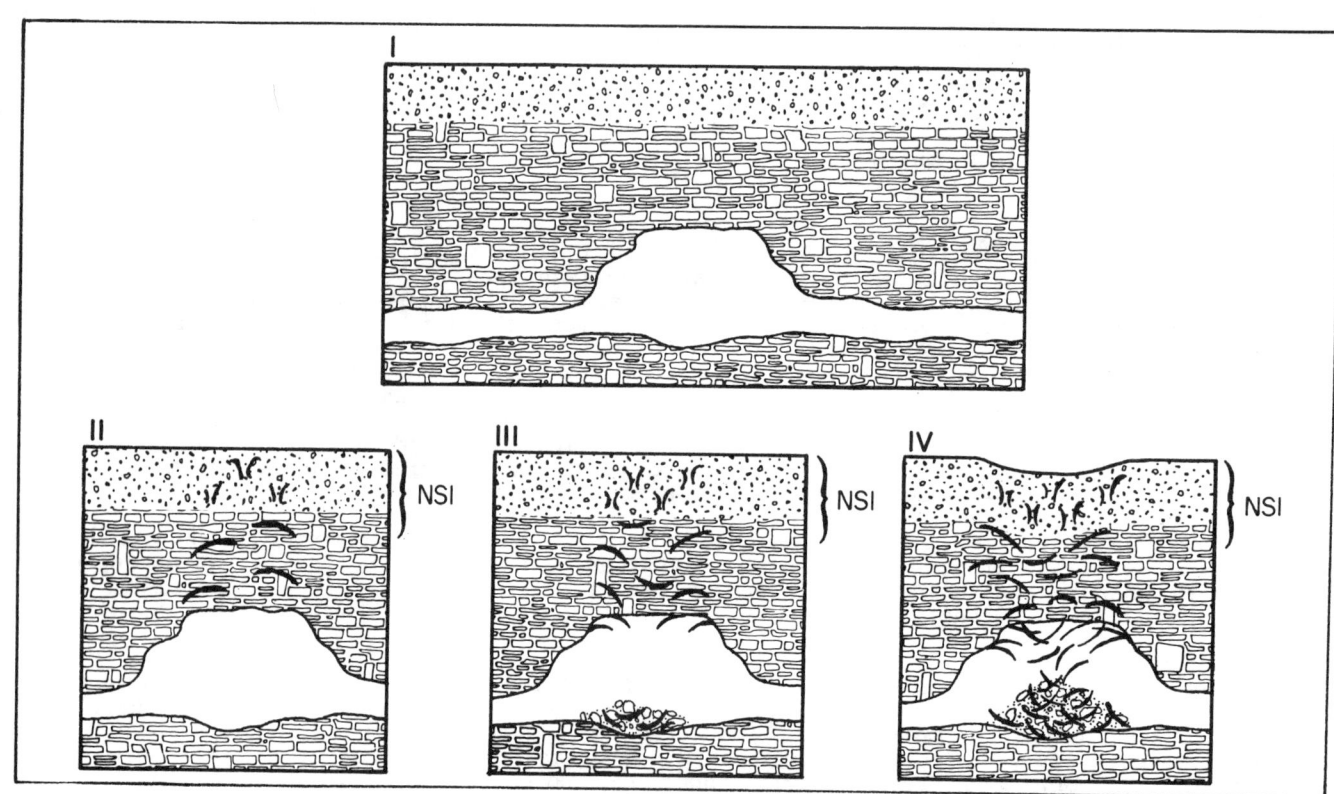

Figure 11: Characterization of cavity system stability. I - Totally stable cavity system and overburden. II - Stable cavity system with some overburden instability. III - Moderate cavity system and overburden instability. IV - Considerable cavity system instability resulting in gross overburden instability and small surface displacement. Note the presence of near surface indicators (NSI) in Stages II, III, and IV.

collapse. For example, Central Florida is clearly an area of active subsidence, whereas South Florida has very little active subsidence. The lack of active subsidence in South Florida does not imply that cavities do not exist, however. In fact, they do. Even though surface subsidence would be rare in South Florida due to the high water table, cavities exist and still present a problem for major structure construction or deep well injection.

An Effective Systems Approach

A wide scope of techniques are available for subsurface cavity detection and assessment. Yet, many practitioners continue to investigate for subsurface cavities with a limited number of borings.

Because no single method or approach can solve every problem, it is imperative that the practitioner understand the problem, the tools available, and how to produce the desired results. All methods have advantages and disadvantages; they all produce useful results when they are properly applied and fail when they are improperly applied. The selection of methods and the approach used should only be made by persons thoroughly familiar with the problems associated with cavity detection as well as the tools at his disposal. In addition to the methods available, the practitioner's professional training and years of in-field survey experience are essential to produce meaningful results. Tools are not an end-all answer, they are merely an aid to the experienced professional.

A number of key factors must be considered in order to construct an optimal systems approach. Four key factors are presented here. They are:

1. The area to be investigated
2. The size of the cavity
3. The stability of the cavity
4. The site perspective

The approach to be implemented is dependent upon the relative scale of the site investigation. Figure 10 illustrates how direct and indirect sampling methods, together with statistical approaches, can be used most effectively taking into consideration the size of the area being investigated. Drilling, for example, is a primary method employed for localized site investigations; however, as the area investigated increases, the sample density decreases, due to the cost and time involved, and accuracy is sacrificed. At that point, drilling becomes a secondary tool and the use of indirect and statistical methods must be employed to maintain an acceptable level of accuracy. Indirect sampling methods can be cost effectively applied to both small and intermediate-sized areas to fill in the information gaps between direct sample points. Here, the indirect methods become a primary tool. Over very large areas, they can only be applied on a statistical sampling basis and become of secondary importance. Various statistical approaches can be used effectively for regional and intermediate-sized investigations. While most statistical approaches may not yield site-specific results, there are a limited number of cases in which statistical data can be used effectively in site-specific local surveys.

Cavity size clearly impacts the approach as well. Assuming that all other factors and conditions are properly met and that the survey is well-designed, most measurement methods must still pass over or reasonably near the cavity in order to get a response. It is much like locating an object in the dark with a flashlight -- the light must shine on the object before it can be seen. The cavity must also be big enough to be seen. For example, a cavity 1 meter in diameter located at a depth of 100 meters cannot be detected from the surface. However, a cavity with a 10-meter diameter located at a depth of 10 meters can be detected from the surface. The size to depth ratio must be large enough and other system noise sufficiently low to permit detection. If the minimum size of the cavity of interest can be defined and the maximum depth of interest can be estimated, the optimum approach can be selected. If it cannot, it will at least be obvious where a given approach is deficient.

The stability of a cavity plays an important role in choosing an approach. Figure 11 shows four stages of cavity stability. They are summarized as follows:

Stage I: Those in which the cavity and the overburden are totally stable
Stage II: Those where some instability in the overburden has occurred
Stage III: Those with moderate instability in the cavity and overburden
Stage IV: Those with significant instability in the cavity and the overburden, yielding displacement and small surface subsidence.

Stage I: Stage I cavities are the most difficult to detect. Detection is primarily dependent upon the ability of the method to directly detect the cavity's presence since no NSI exist. The lack of piping in these types of cavities indicates a level of stability; therefore, they may not present a short term problem.

Although stage I cavities usually cannot be detected by airborne methods, surface geophysical methods have and can be used successfully to detect them. Generally, surface geophysical methods are dependent on the depth and relative size of the cavity. Typically, ratios less than 10, and as small as 1 may be required for reliable cavity detection. Statistical methods can also be applied to Stage I cavities to characterize the area and may sometimes be used to support local site investigations.

<u>Stages II and III</u>: Stages II and III cavities are more readily detected than Stage I cavities because of subtle changes that occur in the shallow overburden. By observing these changes, indirect detection of deep cavities is often possible. Both airborne and surface geophysical methods become quite effective for detecting and assessing these types of cavity conditions due to the presence of NSI. The NSI may include such manifestations as vegetation stress, temperature differentials, soil piping, and electrical properties of soil. Instability associated with these cavity types indicate that they are a potential hazard over the short term. Furthermore, construction activity over or near a Stage II or III site can trigger collapse.

Since NSI are shallow, these anomalies are more readily detected by airborne and surface geophysical methods. When airborne methods can be used, they are highly cost effective, particularly over large areas. Surface geophysical methods have a clear cost advantage and provide an improvement in resolution for site-specific investigations. Statistical methods can also be applied to Stage II and III cavities to characterize the area and may sometimes be used to support local site investigations.

<u>Stage IV</u>: Although surface subsidence is already underway in Stage IV cavities, it may go undetected by the naked eye due to little displacement and slow rates of occurrence. The instability associated with these types of cavities indicate that they are clearly a potential hazard over the short term. Furthermore, nearby construction or drilling can easily trigger collapse.

Stage IV cavities are even more readily detected using airborne and surface geophysical methods to detect NSI. As more subsidence and cracks occur in the near surface, indirect sensing methods are more easily applied because increased activity tends to emphasize the parameter or parameters being monitored.

When airborne methods can be used they are highly cost effective, particularly over large areas. Surface geophysical methods have a clear cost advantage and provide an improvement in resolution for site-specific investigations. Statistical methods can also be applied to Stage IV cavities to characterize the area and may sometimes be used to support local site investigations.

The Need for a Perspective:
Localized field investigations generally focus on the immediate area of concern and ignore the regional setting. Omitting the regional perspective as it relates to the local site can result in critical gaps in understanding the site. While the specific site of interest may only be one acre in size, knowledge of the regional setting is still important because the regional setting reveals information about geomorphology. For example, regional fracture trends may be observed in aerial photos and may extend to the local site, whereas knowledge of only the local site might not provide adequate insight into these trends. Information from a localized drilling program provide considerable detail, but they must be put into perspective by considering the regional setting. On the other hand, interpreting aerial photos on a regional basis without detailed results of local drilling to support a cause/effect interpretation can also be misleading.

Risk Assessment:
A risk assessment can be made for any site. The important question to ask is how site-specific and how accurate need the risk assessment be? A fairly accurate regional collapse probability map can be generated by considering geologic and hydrologic data as well as past level of activity. Such an assessment, however, is not applicable to site-specific problems within the region.

A reliable local approach would be to evaluate the presence of NSI at and around the site. NSI can be obtained from reconnaissance data using aerial methods or surface geophysical methods. An even more reliable approach would be based on a site-specific drilling program designed as a result of previous regional and geophysical knowledge obtained from the site.

It is important to recognize the inherent limitations of any investigation and balance them against realistic project objectives and constraints. Smaller sites of one acre can be assessed to high levels of confidence with total coverage in a reasonable time and economic framework. Larger sites must utilize geomorphology and statistical data to minimize the guessing involved in cavity detection. Total coverage is unrealistic over large areas due to time and cost restrictions. Something less than 100% coverage, therefore, must be acceptable, yet the confidence level must be maintained as high as possible. High levels of confidence with limited coverage can only be accomplished through considerable insight gained from experience.

Here, statistics, geology, geomorphology, and geometric patterns and trends become of great importance. If patterns can be established with some level of confidence, the location of high-probability hazard areas can be predicted. For example, Figure 9 shows that construction may be reasonably safe at points A and C with only a limited site investigation, but not at points B and D. Points B and D require a detailed site-specific stability analysis because they are in a high-risk area. Having established the location of the high and low-risk areas through the qualitative data of Figure 9, the site's construction suitability can be evaluated. More detailed investigations can be carried out until an acceptable level of confidence is achieved. Such an approach allows problem areas to be defined without 100% surface coverage. In many cases, effective detection of cavities or delineation of problem zones and site assessment stability can be accomplished with high levels of confidence at minimal cost before problems occur.

Four Levels of Site Investigation:
A site can be evaluated in various detail to yield different levels of data accuracy and assessment confidence. Each level of evaluation improves upon the previous information, coverage, and level of confidence. Many times only a preliminary, first order approximation is needed to determine whether a project is in a highly sensitive area or an area that is relatively safe. On the other hand, a project may require detailed information necessitating a much higher level of confidence; hence, a second, third, or fourth level of assessment. Unfortunately, the problem is all too often glossed over or ignored, and a first level assessment is sometimes all that is done.

Four levels of site investigation can be applied to cavity detection methods. They are:

1. <u>Review of Existing Data</u>: Aerial photos, geologic maps, general geologic/hydrologic literature, and any specific statistics or data that are readily available should be reviewed and analyzed to provide preliminary information on a site. The results of such assessments are only preliminary, however, and <u>must be used with caution</u>.

2. <u>Site Visit</u>: Site visits include a geologic and environmental visual inspection. Interviews with local land owners, drillers, contractors, quarry operators, county agents, and state and federal personnel can provide numerous unpublished details.

3. <u>On-Site Reconnaissance Measurements</u>: On-site reconnaissannce measurements may include aerial techniques or surface geophysics. If no drilling data is available from the local site, selected borings or "smart holes" whose locations are based upon previous reconnaissance work should be included. The methods selected should be effective reconnaissance tools and should be used as such.

4. <u>Detailed Site Assessment</u>: A detailed site assessment can be used to prove the existence of cavities in areas thought to be high risk or to prove the nonexistence of cavities in areas assigned as low risk. On small sites, the entire site may be examined by detailed methods to provide coverage approaching 100%. On larger sites, however, statistics and geomorphology must be used to locate areas of high and low risk. Sufficient measurements must be taken to achieve the selected level of program confidence.

The various levels of site investigation must be interactive, for, as local data is obtained, greater insight and resolution of details about the site is gained. After information is gained from Level III, it may be advisable to return to Level I and review any new possibilities. It is essential to have a flexible program with in-field analysis and feedback to optimize field activity throughout the overall program. Although remedial action and monitoring may follow a detailed site investigation, they are not included as part of this discussion.

The level of site assessment undertaken should be a function of:

- o   The known susceptibility of the site to subsidence
- o   The critical nature of construction
- o   The level of probability or confidence desired by the investigation
- o   The overall project economics

A complete systems approach should include all of the following:

1. The statistical spatial sampling requirements for an effective drilling and remote sensing program
2. The need for regional and local perspective
3. The use of indirect sampling with contemporary methods
4. Understanding the benefits of continuous data and making use of both airborne and surface sensors wherever appropriate
5. The use of near surface indicators (NSI)
6. The benefits of a well-planned and executed direct sampling drilling program
7. Application of various statistical approaches that may be applied to regional and local problems
8. Application of various measurement methods depending upon the size of the area
9. Having a working knowledge of the principles of geology, hydrology, geomorphology, geostatistics, geochemistry, soil mechanics, and rock mechanics as they apply to karst problems
10. Understanding the cost versus accuracy tradeoffs of site investigations
11. A blending of experience and judgement
12. On-site presence of key professional project personnel

Summary

An accurate evaluation of subsidence or collapse potential due to subsurface cavities requires an accurate definition of the problem area. While the methodology to solve the problem already exists, knowledge of its use and thorough understanding of the problem is fragmented. Furthermore, most programs are restricted by cost and schedule limitations. One of the major problems of subsurface evaluation continues to be the errors developed through a lack of perception and adequate sampling. In many cases, a balance between high-density spatial sampling requirements and cost-effective drilling programs can be achieved by combining the contemporary and traditional approaches discussed in this paper.

It is important to remember that no single method or approach will solve all site investigation problems. Although the methods referred to in this paper are founded on solid scientific principles, they can fail if they are improperly implemented or applied to the wrong problem. The process of proper implementation requires trained, experienced personnel. By selecting the most suitable methods and utilizing the synergistic benefits of an integrated systems approach, high levels of accuracy and cost-effectiveness can be achieved and the project can be done right the first time.

The technical methods and systems approach discussed in this paper have been successfully applied to a number of site investigation problems including reconnaissance and detailed surveys for the location of cavities, fractures, and differential soil conditions. Location and evaluation of rock fracture, subsurface cavities, and collapse potential have been evaluated using a model based upon these general principles. Both the techniques and the model have been tested in a number of locations in and out of the continental United States for nearly two decades. They have been proven effective for providing improved confidence levels, accuracy, cost-effectiveness, and for predicting hazardous geologic and man-induced conditions.

BIBLIOGRAPHY

American Society of Civil Engineers Convention and Exposition, 1979, Geophysical Methods in Geotechnical Engineering: American Society of Civil Engineers, October 23-25, 1979, Preprint 3794.

Fountain, L.S., Herzig, F.X., Owen, T.E., 1975, Detection of Subsurface Cavities by Surface Remote Sensing Techniques: Department of Transportation, Federal Highway Administration Offices of Research and Development, Washington, D.C., Report No. FHWA-RD-75-80.

Franklin, A.G., Patrick, D.M., Butler, D.K., Strohm, W.E., Jr., Hynes-Griffing, M.E., 1981, Foundation Considerations in Siting of Nuclear Facilities in Karst Terrains and Other Areas Susceptible to Ground Collapse: Geotechnical Laboratory, U.S. Army Engineer Waterways Experiment Station, NUREG/CR-2062 R6, RA, CA, CG.

Kirk, K.G., and Werner, Eberhard, 1981, Handbook of Geophysical Cavity-Locating Techniques U.S. Department of Transportation, Federal Highway Administration, Implementation Package FHWA-IP-81-3.

Newton, J.G., 1976, <u>Early Detection and Correction of Sinkhole Problems in Alabama with a Preliminary Evaluation of Remote Sensing Applications</u>: U.S. Geological Survey and Alabama Highway Department and U.S. Department of Transportation Federal Highway Administration, Bureau of Public Roads, HPR Report No. 76.

U.S. Army Engineer Waterways Experiment Station <u>Symposium on Detection of Subsurface Cavities, 1977</u>: Office of U.S. Army Engineer Waterways Experiment Station Soils and Pavements Laboratory, 12-15 July, 1977.

Engineering Foundation, <u>International Conference on Evaluation and Prediction of Subsidence, 1978</u>: American Society of Civil Engineers.

All figures and photographs are provided through the courtesy of Technos Inc. unless otherwise specified.

Examination of sinkholes by seismic reflection

DON W.STEEPLES, RALPH W.KNAPP & RICHARD D.MILLER *University of Kansas, Lawrence, USA*

ABSTRACT

Salt dissolution sinkholes have developed at more than a dozen localities in Kansas during the past 25 years. Most of the sinkholes subside gradually over a period of years, although catastrophic collapse has occurred in some cases. We have performed high resolution seismic reflection surveys across more than a half dozen of these sinkholes. It is possible to discern considerable geologic detail at depths of 10 to 500 meters within the sinkholes by seismic reflection methods. At one site astride Interstate Highway 70 we obtained acoustic images of grabens within the sinkhole that showed about 40 to 50 meters of vertical down-drop at a depth of 400 meters in an area where surface displacement was less than 5 meters. At another site we detected two paleo-sinkholes adjacent to a presently active sink. The paleo-sinks are filled with alluvial material of probable Pleistocene age, and one of them shows indications of two different geologic ages of active sinking. While many of the new sinkholes that have formed appear to be related to oil-field brine disposal or salt-solution mining activities, the detection of the paleo-sinks by seismic reflection methods establishes the natural occurrence of some salt-dissolution sinkholes in Kansas prior to the encroachment of civilization.

Introduction

Seismic reflection surveys offer a powerful method of acoustically imaging portions of the subsurface of the earth in the vicinity of some sinkholes. The successful use of the technique depends upon the existence of velocity and/or density contrasts in the geological materials within and surrounding the sink. In the cases discussed in this paper, geologic structures related to or defining the sink were detected and mapped in the subsurface. Under favorable conditions, it is likely that cavities can be detected directly in the future as seismic instrumentation and computer processing techniques continue to improve. Direct detection of cavities is likely to depend upon recognition and delineation of diffracted seismic waves (rather than classic reflections).

Seismic reflection techniques have been used in petroleum exploration for nearly 60 years. This paper, however, is an example of use of seismic reflection surveys in Kansas oil fields to evaluate subsurface salt dissolution and subsequent surface sinking around salt water disposal wells. We present results of seismic reflection studies in the depth range of 30 to 500 meters performed across actively developing sinkholes located astride Interstate Highway 70 (I-70) in Russell County, Kansas and astride a rural road in Reno County, Kansas. The results shown in this paper indicate that high-resolution seismic reflection surveys can be useful in subsurface investigation of some sinkholes. We believe that the seismic reflection method is potentially very useful in engineering studies of sinkholes and Karst features. We note that the seismic data were obtained in the presence of a few dozen vehicles per minute of highway traffic using the MiniSOSIE recording technique. (MiniSOSIE is a registered trademark of SNEA(P), France.)

Geologic Background - Russell County Location

While sinkholes were known to have formed in Kansas prior to the onset of modern oil, gas, and mineral exploration (Smith, 1940), there seems to be little question that the activities of man have either increased their number or accelerated their development (or both). The following discussion is pertinent to almost all of the salt dissolution sinkholes that have been geophysically and geologically investigated in Kansas. For a more extensive discussion of geological investigations of these sinkholes, see the excellent work of Walters (1977).

During the past 25 years progressive development of at least three sinkholes has occurred beneath I-70 in Russell County, Kansas. The sinkholes are all centered at the sites of salt water disposal wells associated with an oil field. Sinking at the ground surface has occurred at the rate of roughly 15 to 30 centimeters per year since the late 1950's. I-70 has twice been rebuilt to grade and the gradual sinking has been monitored with caution but without great alarm. Concern for public safety mounted in 1978 when another sinkhole developed catastrophically and without apparent warning at a salt water disposal well about 25 kilometers to the northwest.

The whole area in question is underlain by the 90-meter thick Hutchinson Salt Member of the Permian Wellington Formation at a depth of about 400 to 500 meters (Figures 1-3). The formation of the sinkholes is consistent with dissolution of the Hutchinson salt by undersaturated brines introduced into leaky disposal wells or by fresh to brackish waters leaking downward along the outsides of the well casings. Either cause involves leaky pipes or insufficient cement sealing of the salt beds. Rock units above the salt dissolution cavity progressively cave into the water-filled void, eventually reaching the surface as a gentle depression or as a catastrophic collapse.

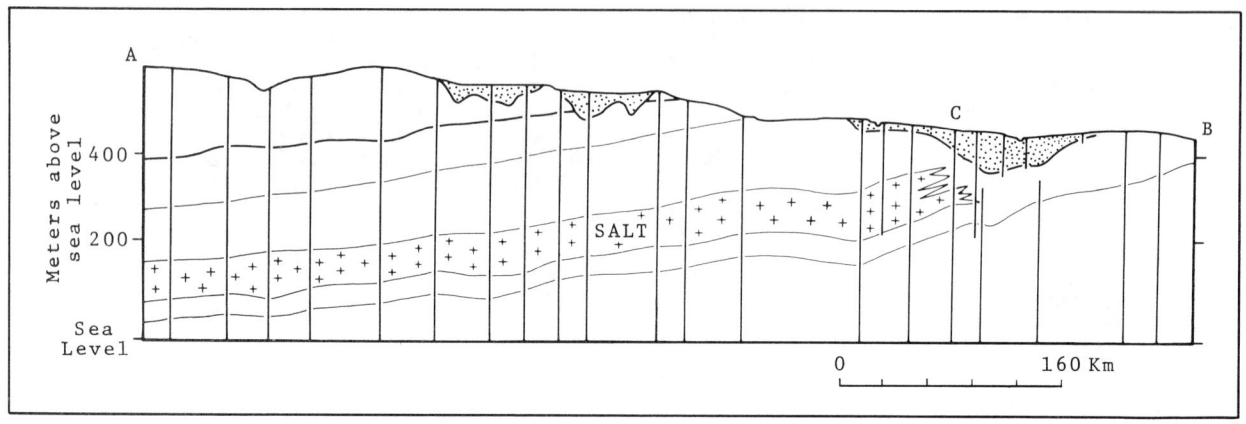

Figure 1: Geologic section along ACB of Fig. 2. Vertical lines show locations of well control in Walters (1977). Locations A and C depict the vicinity of the Russell and Reno County locations, respectively.

Figure 2: Lateral and vertical extent of Hutchinson Salt. Russell County (I-70) location is near A and Reno County location is near C. Fig. 1 shows a geologic section of line ACB. Modified from Walters (1977).

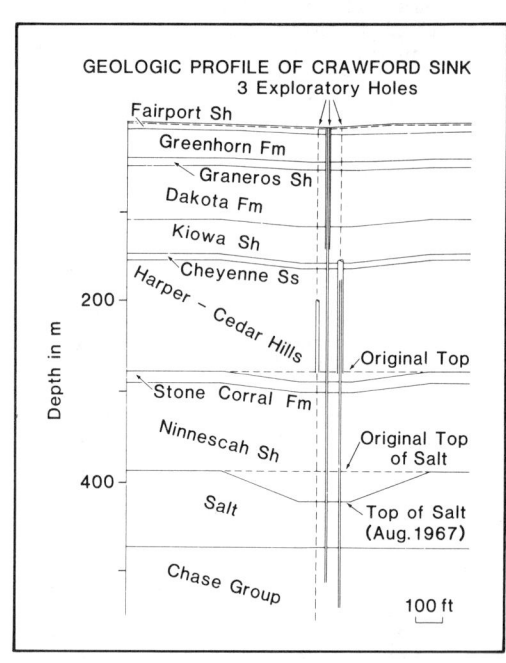

Figure 3: Geologic section based on geophysical logs and drilling by Kansas Department of Transportation.

The Crawford sinkhole on I-70 was drilled and geophysically logged by the Kansas Department of Transportation in 1967 and the results are summarized in Figure 3. Note that at that time the tops of the Stone Corral anhydrite and the Hutchinson salt were about 12 meters and 35 meters below normal elevation, respectively. Seismic reflection data presented in this paper indicate that the Stone Corral in particular has dropped almost 30 meters since 1967.

In order to mitigate public concern about the safety of I-70 in Russell County after catastrophic formation of the sinkhole 25 kilometers to the northwest, the Kansas Geological Survey performed a seismic reflection survey along I-70. We obtained about 4 kilometers of 6-fold common depth point (CDP) seismic reflection data by the MiniSOSIE method (technique discussed later). The Stone Corral anhydrite at a depth of about 300 meters is an excellent marker bed for seismic reflections. In addition to the Stone Corral, we were able to image both the top and the bottom of the salt at critical points in the vicinity of the largest sinkhole.

The structure in the sinkholes is that of a classic graben when imaged in cross-section. On the basis of seismic evidence discussed later in this paper, the Stone Corral had been down-dropped by as much as 45 meters by 1980 in the largest sinkhole. The minimum remaining salt thickness is less than half of the original. At least five normal faults symmetrically bound the graben blocks that are down-dropped into the sinkhole. The seismic records indicate that collapse has apparently occurred gradually with progressive salt dissolution and that substantial bridging of voids in the sinkhole vicinity was not present in 1980. We believe that catastrophic collapse of known sinkholes along I-70 is unlikely because of the brecciated nature of the material between the earth's surface and the top of the still-dissolving salt.

Finally, it should be noted that this seismic reflection survey was performed along the shoulder of I-70 in the presence of normal traffic including heavy trucks. The MiniSOSIE method is an excellent recording technique in a noisy environment. The technique provides substantial energy in the frequency range of 80 to 120 Hz, providing relatively high resolution.

Geologic Background - Reno County Location

As can be noted on at location "C" of Figure 1 and 2, the Reno County site is in a similar geologic situation, except that the salt is shallower and the Stone Corral anhydrite is missing. At this site the salt is between 15 meters and 60 meters thick at a depth of 120 to 200 meters. Examination of drillers logs at nearby wells indicates a high degree of variability in the occurrence of the evaporite sequence. Since the salt dissolution front (Figure 1) subcrops just a few kilometers to the east of the site, the variation is not surprising. Since original construction of the county road at this site, approximately 3 meters of downward vertical displacement has occurred.

Common Depth Point (CDP) Seismic Reflection

For targets in the 50 meter to 600 meter depth range, the MiniSOSIE seismic reflection technique is useful (described later in this paper). Earth compactors known as Wackers rapidly thump the ground to provide seismic source energy. Geophones are placed at one meter intervals on the ground to sense the ground motion resulting from the Wacker footplate impact. At any given shotpoint, there are 10 geophones providing data to the seismograph for each recording channel. The seismograph has 24 recording channels and is mounted on a recording truck. Twelve live data channels are recorded both ahead of and behind the shotpoint. The live channels are between 20 and 130 meters from the shotpoint. Shotpoints are located at 10-meter intervals along the line of survey. Normal rate of progress is 20 to 30 shotpoints per hour, depending upon terrain, weather, and how many Wacker impacts are summed at each shotpoint.

In a 10-hour working day, about 150 shotpoints can be occupied. The resulting data have 12-fold common depth point redundancy. This means that using ray theory, each subsurface reflection point is replicated 12 times on the digital tape. The data are processed to sum the 12 replicated reflection records in an enhancement manner and unwanted signals (noise) tend to cancel out. The resulting data are then displayed in a seismic record section format.

A seismic record section (such as Figures 4 and 5) may be thought of as a "pseudo-road cut." In other words, there are places along most highways where excavation through hills has exposed layers of rock. It is possible to look at the rock units in cross-section at such road-cut localities. A seismic record section is a display of the acoustic properties of the geologic cross-section in much the same manner as a road-cut displays the rock layers. In the case of this paper, the "pseudo road-cut" displays are about a kilometer long and a few hundred meters deep.

MiniSOSIE Recording Technique
After examination of environmental conditions and determination of the target zone, we chose the MiniSOSIE technique of data acquisition. MiniSOSIE is the land version of the SOSIE technique originally developed as a marine seismic source (Barbier and Viallix, 1973).

Physically in the field, recording is done by summing signals from about 10 to 40 impacts per second from one or more civil engineering earth compactors known as Wackers. Typically, signals from 1,000 to 2,000 impacts are stacked at each shotpoint. The impacts are usually made along the seismic line over a linear segment equal to geophone group interval (i.e., a source array) rather than at a single point and one to four Wackers are run simultaneously. Each Wacker has a transducer attached to its base place and the transducer sends a time-break pulse by radio or wireline to the recording truck each time the Wacker base plate strikes the ground.

The mystery about MiniSOSIE arises from the fact that typical seismic records are the order of one second in duration, while the time between successive Wacker impacts is of the order of a tenth of a second. Intuition tells us that the signals from successive impacts should interfere in an unpredictable and possibly noisy, if not destructive manner. The key to the MiniSOSIE technique is overcoming this intuitive difficulty by performing a simple processing step in the truck during recording.

Real-time processing is done in the recording truck according to the following scheme:

$$\text{Recorded signal} = (\text{source}) * (\text{earth function}) * (\text{ACF time series}) \quad (1)$$

where (source) is pulses of energy transmitted into the earth, (earth function) varies with geology, (ACF time series) is the auto-correlation function of the time series of impulses from the Wacker base plates, and * is the convolution operator. This compares with conventional techniques (i.e., dynamite) where:

$$\text{Recorded signal} = (\text{source}) * (\text{earth function}) \quad (2)$$

Note that if ACF time series in (eqn. 1) is a spike (i.e., an impulse or Dirac delta function), the recorded signals (eqn. 1) and (eqn. 2) will be the same. MiniSOSIE takes advantage of the fact that the auto-correlation function of a random time series is a spike and that convolution with a spike is essentially multiplication by unity. In essence, this is why MiniSOSIE works.

The random time series is generated by randomly varying the engine speed (and, hence, the impact rate) of the Wackers. Real-time processing in the recording truck is done by a 20-bit microprocessor to produce MiniSOSIE field data (eqn. 1) that look very much like dynamite field data (eqn. 2). Except for the unique energy source and the auto-correlation processing in the recording truck, MiniSOSIE seismic recording is identical to conventional dynamite recording.

Experience has shown MiniSOSIE surveys to provide good high-resolution results at depths between 50 and 600 meters in most localities. It is an especially good technique in areas of high ambient random noise (such as automobile traffic) because random noise tends to cancel during the tens of seconds required to stack coherent signal from 1,000 or more Wacker impacts.

Discussion of Seismic Section at Crawford Sink (Figure 4)
The Crawford sink begins in the subsurface at CDP 131, about 60 meters west of an overpass, and extends to near CDP 98, about 260 meters east of the overpass - its total east-west extent along I-70 slightly exceeds 300 meters. The sink is marked at depth by at least 5 faults in the Stone Corral which is down-dropped in graben-like fashion into the central area of the sinkhole. The total vertical movement of the Stone Corral has been about 45 meters (30 msec two-way travel time) at the deepest part of the sinkhole along I-70. The highway does not pass exactly over the center of the sink, so it is not possible to evaluate conditions at the center of the sink from our data.

Approximately 50 percent of the thickness of the salt has been dissolved beneath the highway at the point of maximum drop in the Stone Corral. This compares favorably with a 40 percent dissolution estimate obtained by drilling done by the Kansas Department of Transportation in 1967 (Figure 4), suggesting that dissolution may have slowed since the mid-1960's, assuming the drilling was on or near the highway right-of-way.

In addition to the data shown in Figure 4, we ran a north-south profile which supports the Crawford sink interpretation discussed above for the east-west I-70 profile. The profile runs along the road that crosses over the top of the overpass and intersects the east-west line at CDP 128. These data are not shown in this paper, but are available from the authors and are shown in Steeples and Knapp (1982).

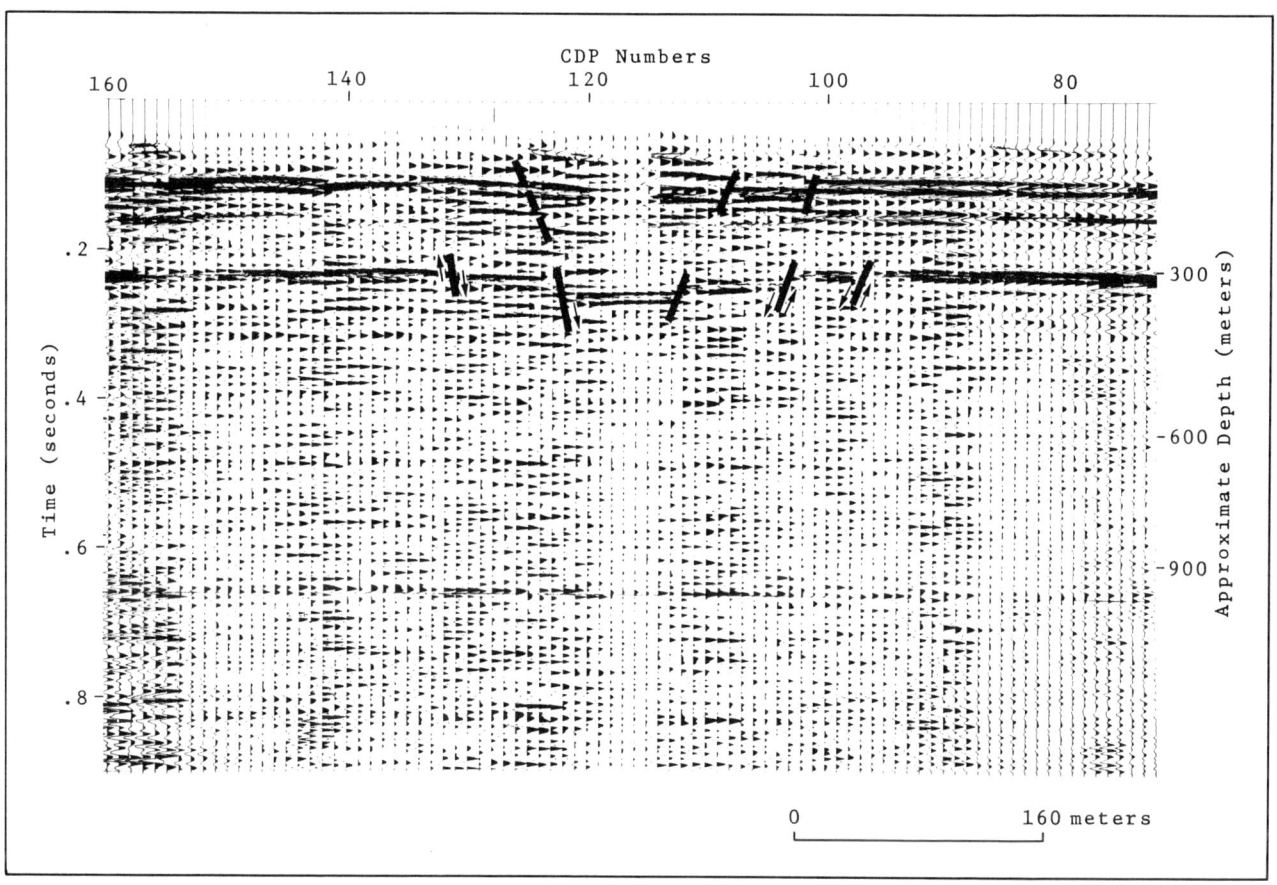

Figure 4: Seismic section along median of I-70 near borehole of Fig. 3 at Crawford sink in Russell County. Graben structure is exhibited near center of section, particularly by Stone Corral anhydrite at depth of about 300 meters. Note Stone Corral is down-dropped about 30 msec at center of sink near CDP numbers 115-121.

Discussion of Seismic Section in Reno County

The seismic section obtained at the previously mentioned sinkhole in Reno County is shown in Figure 5. The geologic interpretation of the seismic section is shown in Figure 6. Since drillers logs in the area are old and not very reliable, and seismic velocity is not constrained by sonic logs or uphole tests, we do not assign lithologic types to the geologic units. There are important observations that can be made from Figure 6 without knowing the exact lithology of the layers.

Rock units below 200 milliseconds are flat, indicating lack of structural deformation below the Hutchinson salt. Rock units above 200 msec have been structurally deformed by dissolution of the Hutchinson salt and collapse of the overburden. Perhaps the most important observation that can be made from Figures 5 and 6 is that at least two episodes of sinkhole development have occurred. The most recent episode is still going on. The earlier episode, however, predates the deposition of the bottom of the upper geologic layer. The upper layer is of Pleistocene or older age. This shows that the hydrologic conditions necessary for dissolution of salt were present before any petroleum exploration drilling occurred. The occurrence of a paleo sinkhole and a modern active sinkhole adjacent to each other signifies that salt dissolution sinkholes in Kansas can occur either naturally or as a result of inducement by drilling or mining.

General Conclusions
    1. The reflections from rock units below the Hutchinson salt are essentially flat at

both sites, indicating a lack of geological structure beneath the sinks.

   2. The Stone Corral was used as a marker bed in our report to the Kansas Department of Health and Environment dated 3 April 1980. In that report, we were unable to trace the Stone Corral into the sinkholes. CDP methods and computer processing allowed us to trace the Stone Corral across the sinkholes, delineating structures with detail that was previously impossible. For the first time we were able to delineate the top and bottom of the Hutchinson salt along part of the profile beneath I-70 (Steeples et al., 1984, in press).

   3. As noted in the discussion of Figures 5 and 6, there is excellent evidence that at least two episodes of sinkhole development have occurred, and that a paleo-sinkhole exists beside a presently active sink beneath the seismic line in Reno County.

   4. Seismic reflection methods are capable of delineating geologic structure within sinkholes at some localities. Expected developments in high resolution seismic reflection methods will probably allow the direct detection of voids in some sinkholes in the near future.

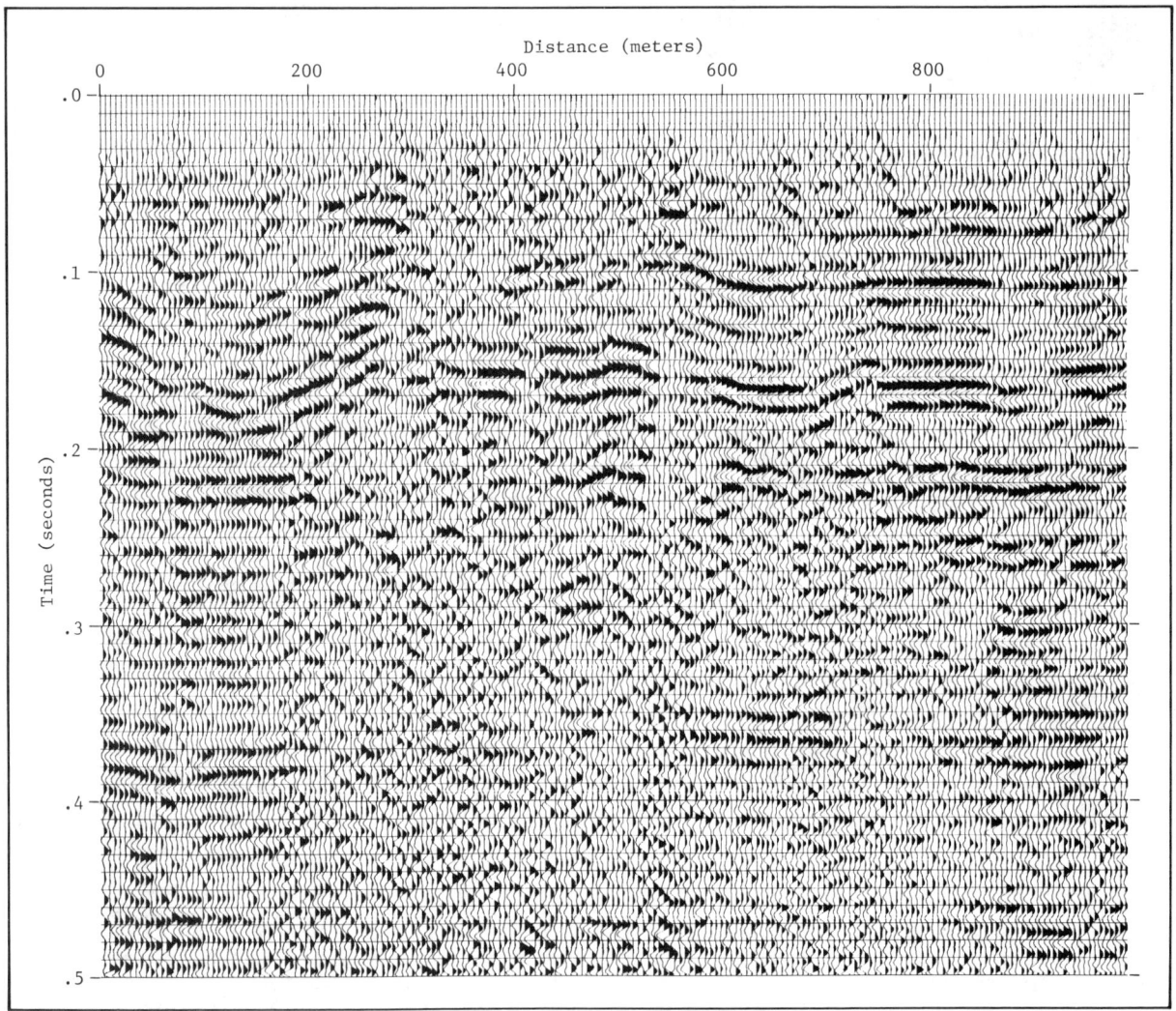

Figure 5: Seismic section at sinkhole in Reno County. Fig. 6 shows geologic interpretation and horizontal distance scale can be used to correlate between the two figures. Units below salt are flat. A paleo-sinkhole near left end of figure between distance 0 and 200 meters is filled with Pleistocene or older sediments.

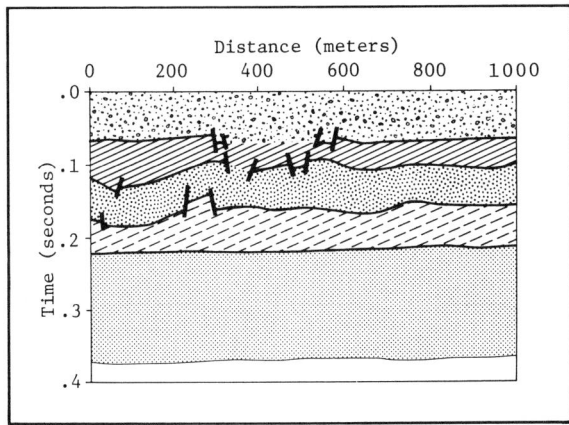

Figure 6: Geologic interpretation of seismic section of Fig. 5. Note paleo-sinkhole near left end of figure and apparent faulting near center of figure which is beneath site of present surface sinking.

Acknowledgements

The support of the Kansas Department of Health and Environment and the Kansas Department of Transportation is gratefully acknowledged. Free access to seismic data processing facilities was kindly provided by Jon Weigand.

References

Barbier, M.G., 1976, MiniSOSIE for land seismology: Geophysical Prospecting, v. 24, p. 518.

Barbier, M.G. and Viallix, J.R., 1973, SOSIE: a new tool for marine seismology: Geophysics, v. 38, p. 673-683.

Mayne, W.H., 1962, Common reflection point horizontal data stacking techniques: Geophysics, v. 27, p. 927.

Smith, H.T.U., 1940, Geological studies of southwestern Kansas: Kansas Geological Survey Bulletin 34, 212 p.

Steeples, D.W., Knapp, R.W., and McElwee, C.D., 1984, Seismic reflections for other than oil exploration in the petroleum industry: Oil and Gas Journal, annual geophysical activity issue, in press.

Steeples, D.W. and Knapp, R.W., 1982, Seismic investigations of sinkholes in Russell and Ellis counties, Kansas: Final report to Kansas Department of Health and Environment, Topeka, Kansas.

Walters, R.F., 1977, Land subsidence in central Kansas related to salt dissolution: Kansas Geological Survey Bulletin 214, 82 p.

Geophysical characteristics of fracture traces in the carbonate Floridan Aquifer

MARK STEWART & JOHN WOOD *University of South Florida, Tampa, USA*

ABSTRACT

Several geophysical methods were used to investigate the geophysical and geologic character of fracture traces at two sites in west-central Florida. Both sites are underlain by the carbonate Floridan Aquifer. The limestone surface is 10-20 m below ground surface at both sites, and is covered by a sequence of unconsolidated sands, silts, and clays. The geophysical methods used were horizontal electrical profiles, vertical electrical soundings, tri-potential profiles, and microgravity and triple-track gravity profiles. The geophysical results reveal two different subsurface expressions of the fracture trace. At the Cross Bar Ranch site the fracture zone is marked by a V-shaped depression in the limestone 30 m deep and 60-90 m wide. At the Crystal River quarry site the fracture trace is underlain by a bedrock ridge 10 m high by 60 m wide, with a central depression 10 m deep by 10 m wide. The feature at Cross Bar Ranch seems to have formed through the solution of the limestone along the fracture zone. At the Crystal River quarry the limestone along the bedrock ridge under the fracture trace has higher resistivities than the surrounding limestone and may represent preferential cementation and/or recrystallization along the trace.

Introduction

Photolinears in carbonate terranes often represent the surface expression of vertical, subsurface zones of higher fracture density in the underlying limestone (Lattman and Parizek, 1964; Parizek, 1976). These fracture zones are usually sites of increased solution activity, and can influence the development of karst features such as sinkholes. It has been suggested that modern sinkhole activity in west-central Florida has been principally along fracture traces (Littlefield and others, 1984, this volume). However, sinkholes do not develop along all fracture traces or along the entire length of a particular trace.

In order to better understand the geologic character of fracture traces a study was initiated to investigate the geologic and geophysical expression of fracture traces at two sites in west-central Florida. The study has two objectives. The first is to determine the the response of several common geophysical methods to the fracture traces, and second, to determine the stratigraphy and lithologic character of the fracture zones by direct sampling. The geophysical surveys have been completed and are the subject of this paper.

Procedure

Two sites in west-central Florida were chosen for this study (Figure 1). The first site is at the Cross Bar Ranch, a major municipal wellfield in Pasco County. The second site is at a limerock quarry near Crystal River in Citrus County. The fracture trace at Cross Bar Ranch is over 1 km long and 100-300 m wide. Its surface expression is a series of shallow depressions with a relief of 1-2 m. Depth to the limestone is 10-20 m, and the bedrock is overlain by an upward-coarsening sequence of clay, silty-clayey sand, and sand. The Crystal River site fracture trace is marked by a distinct linear swale with relief of several meters. The bedrock is 10-15 m below landsurface and is covered by a sequence of sediments similar to those at Cross Bar Ranch. Figures 2 and 3 show the locations of the traces at the sites.

Several geophysical methods were used in this investigation, including three types of resistivity surveys and two gravity methods. The resistivity methods are horizontal profiling, vertical soundings, and tri-potential profiles. Tri-potential profiles use three arrangements of the Wenner array (Carpenter and Habberjam, 1956). Over laterally homogeneous ground all three give the same apparent resistivity. However, over vertical discontinuities the three apparent resistivities differ (Ogden, 1984).

Microgravity profiles were completed across the fracture traces. Microgravity surveys measure gravity anomalies of 10^{-1} milligals, as compared to anomalies of 10^0-10^1 milligals in standard gravity surveys. At several locations, closely-spaced gravity readings were taken along three parallel tracks. As the gravity field follows the Laplace equation, this triple-track procedure allowed the calculation of the first and second vertical gravity derivitives from the horizontal gradients. Vertical gradients tend to accentuate shallow anomalies at the expense of deeper ones.

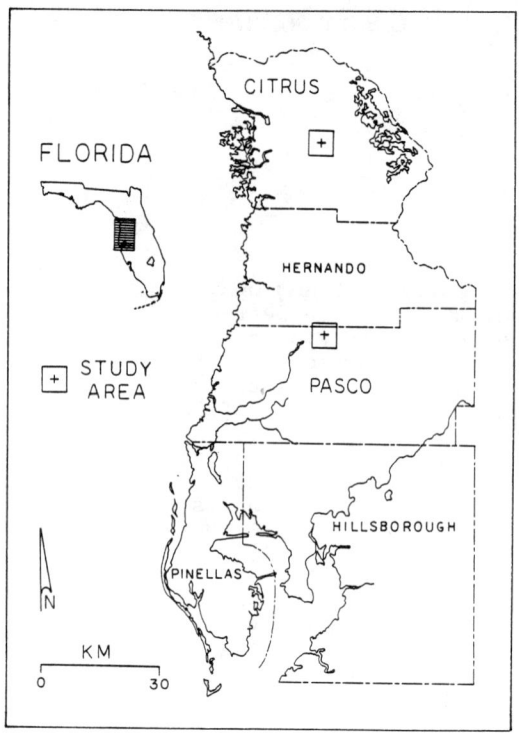

Figure 1: Location of study sites.

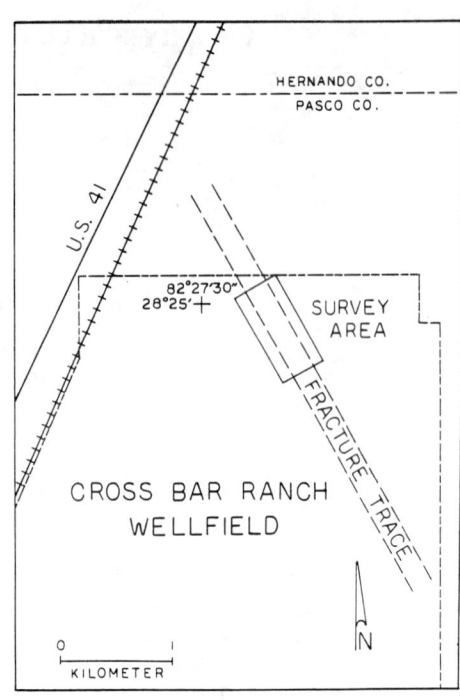

Figure 2: Orientation of fracture trace and location of geophysical profiles at Cross Bar Ranch.

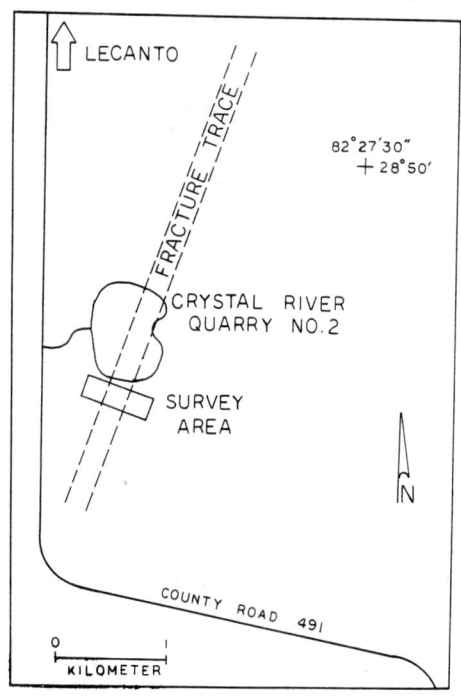

Figure 3: Orientation of fracture trace and location of geophysical profiles at Crystal River quarry.

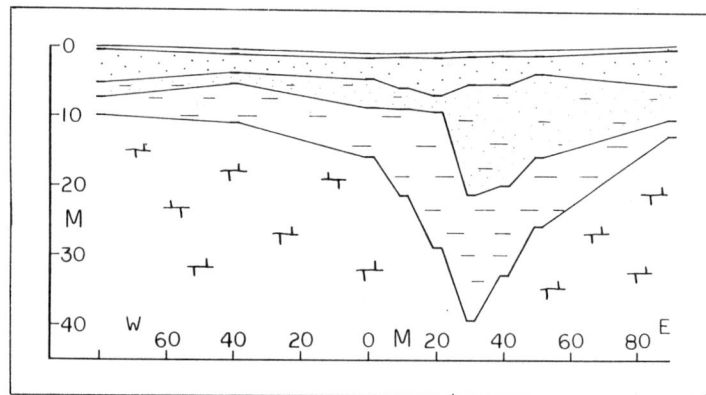

Figure 4: Geoelectric section at Cross Bar Ranch. Short horizontal lines indicate location of soundings. Vertical exaggeration 2X.

Results
	The vertical geoelectric section obtained at the Cross Bar site is shown in Figure 4. The section reveals a deep, V-shaped depression in the limestone 30 m deep and 60-90 m wide. The depression is filled with sandy clay, silty sand, and sand, coarsening upward, based on the resistivities of the units. The greatest depth to the top of rock is not directly under the topographic low which forms the fracture trace at the surface, but is displaced about 30 m to the east. The Crystal River geoelectric section is shown in Figure 5. Here, the fracture trace is associated with a bedrock ridge 10 m high and 60 m wide, with a V-shaped depression 10 m by 10 m in the center of the ridge. The ridge is superimposed on a sloping bedrock surface. The central depression underlies the topographic low at the surface. The resistivities of the limestone on the ridge are higher than the resistivities of the limestone on either side. Along the ridge resistivities are 200-500 ohm-m, while they are 100-200 ohm-m on the sides.

	The results of a tripotential survey completed at the Cross Bar site along the same line as Figure 4 are shown in Figure 6. The CPPC array is the standard Wenner configuration. The AB/3 distance is 15 m and the station spacing is 5 m. The CPPC response is equivalent to a horizontal electrical profile. It shows a broad, strong resistivity low, with the lowest values at about 17 m east of the trace center, near the position of the V-shaped feature of Figure 4. The CPCP and CCPP arrays show a sharp divergence at the same position that the CPPC resistivity reaches its minimum value, indicating a vertical fracture or dicontinuity (Ogden, 1984).

	The microgravity profile corresponding to the electrical profiles of Figures 4 and 6 is shown in Figure 7. The gravity values show a sharp, strong low east of the trace center, corresponding to the location of the V-shaped depression of Figure 4. Figure 8 is a plot of the second vertical derivative of gravity derived from a triple-track profile. It has the same general pattern as the microgravity plot of Figure 7, but is considerably noisier. Both profiles used the same stations spaced 5 m apart, starting at the trace center.

Discussion
	Of the geophysical methods used closely-spaced vertical electrical soundings yield the most information on the geometry and the stratigraphy of the fracture trace. Horizontal electrical profiles can reveal the general area of the fracture zone. By measuring tri-potential resistivities a better estimate of the location of the fracture zone can be made. Tri-potential surveys simply require switching electrode wires at the receiver to vary the arrays. This can be accomplished with a simple switch box. The electrical profiling methods are quite sensitive to the AB/3 separation used. Too small or too large a spacing can diminish the resistivity contrasts. Electrical soundings should be used to determine the optimum spacing.

	The microgravity profile locates the bedrock low delineated by the vertical electrical soundings. However, the second derivative or triple-track survey did not give any better resolution of the anomaly. Because the second derivative method also enhances instrument and survey inaccuracies, great care must be taken with elevations and meter readings to reduce the noise level in the data. This survey is too noisy to allow enhancement of the anomaly by the second derivative method.

	The geometry of the fracture zone at the Cross Bar Ranch is what would be expected for a vertical fracture zone in a soluble limestone. The limestone surface has a V-shaped depression created by ground water moving downward through the fractures. The depression is not sand filled, but contains a thick clay unit. The total relief is about 30 m over a horizontal distance of 60-90 m. The lowest point on the bedrock surface is not directly below the center of the photolinear at the surface, but is displaced about 30 m to one side, suggesting that the narrow fracture zone in the bedrock is not always centered within the fracture trace.

	The limestone surface under the fracture trace at the Crystal River site has an unexpected geometry. Instead of a deep, V-shaped depression as at Cross Bar Ranch, the limestone surface under the trace exibits a ridge with 10-15 m of relief and a narrow V-shaped depression in the center. A possible explanation for this geometry is that the limestone in the fracture zone is recrystallized or cemented, forming a more resistant unit. Three observations support this suggestion. First, the limestone on the ridge has higher resistivities than the adjacent rock, suggesting that it has lower porosities. Second, the fracture trace passes through the quarry, and in two of the three exposures of the trace zone in the quarry walls the limestone is harder, fractured, and recrystallized. Third, vertical fractures observed in the quarry walls had a central filling of clay, while their walls had a thick rim of recrystallized limestone.

Conclusions
	Electrical methods can yield significant information about the geologic character of

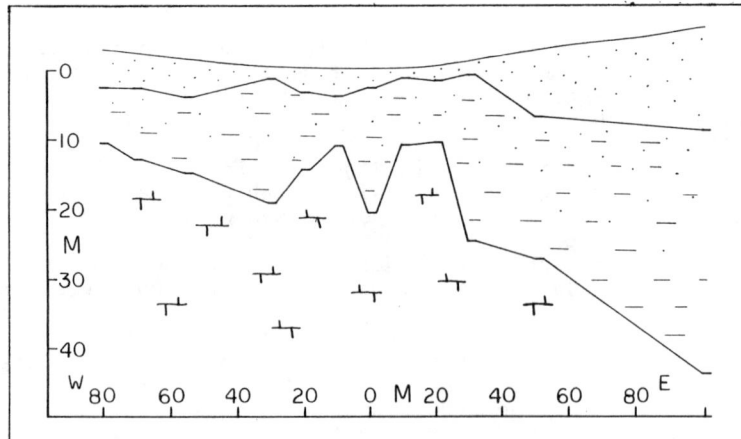

Figure 5: Geoelectric section at Crystal River quarry. Short horizontal lines indicate locations of soundings. Vertical exaggeration is 2X.

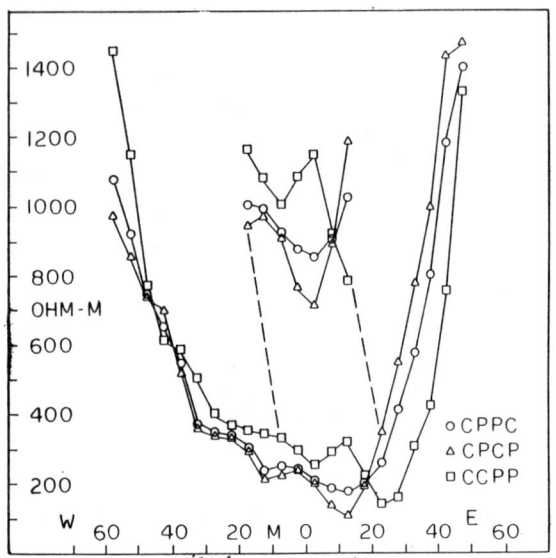

Figure 6: Tri-potential profile at Cross Bar Ranch. AB/3 is 15 m, station spacing is 5 m.

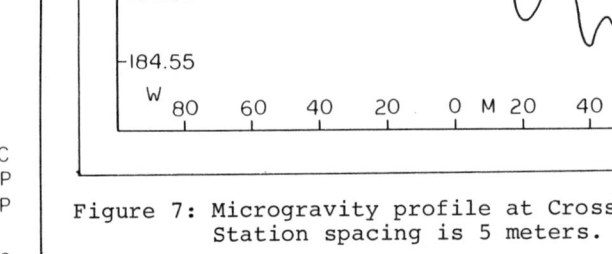

Figure 7: Microgravity profile at Cross Bar Ranch. Station spacing is 5 meters.

Figure 8: Second vertical derivative of gravity, Cross Bar Ranch, for same profile as Figure 7. Station spacing is 5 m.

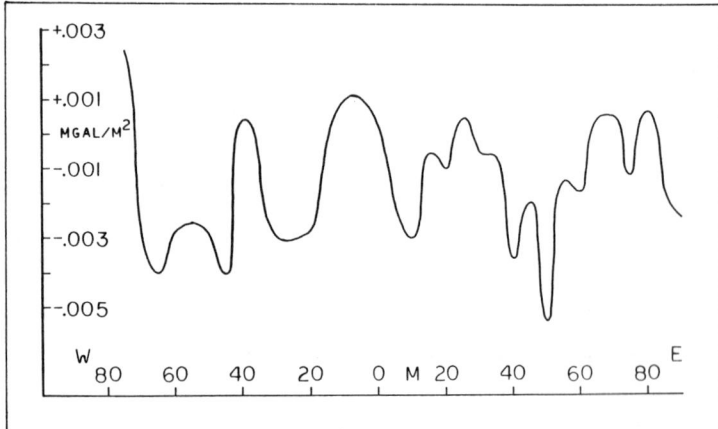

fracture traces. Closely-spaced vertical soundings give the most detailed geoelectric section. However, profiling methods can effectively locate the center of the fracture zone if care is taken to use the appropriate electrode spacing. Tri-potential surveys provide additional information while requiring only a modest increase in survey effort. Microgravity profiles can be used where resistivity methods are not effective, such as in very dry, sandy soils. However, they require considerably more survey effort and care than resistivity methods.

While both fracture traces have similar surface expressions, their subsurface geometries are very different. The Cross Bar Ranch feature is a V-shaped depression in the limestone surface with considerable relief. The Crystal River feature is a linear ridge of moderate relief with a central depression. The difference between the two may be related to the effect the fractures have had on the limestone. At Cross Bar solution has been the dominant process, producing a deep cleft in the rock. The development of the feature at the Crystal River site has been more complex. Movement of ground water through the fractures has created a harder, less porous limestone within the trace, which then forms a ridge through differential weathering. The smaller V-shaped depression at the center probably represents the actual fracture zone, and has formed by solution of the limestone.

References

Carpenter, E.W., and G.M. Habberjam, 1956. A tri-potential method of resistivity prospecting: Geophysics, v. 21, p. 388-402.

Lattman, L.H., and R.R. Parizek, 1964. Relationship between fracture traces and the occurrence of grond water in carbonate rocks: Journal of Hydrology, v. 2, p. 73-91.

Littlefield, J., M. Culbreth, S.B. Upchurch, M. Stewart, 1984. Relationship of modern sinkhole development to large-scale photolinear traces: Proceedings, First Multidisciplinary Conference on Sinkholes, Orlando, Sinkhole Research Institute (this volume).

Ogden, A., 1984. The use of tri-potential resistivity to locate fractures, faults, and caves for siting high-yield water wells: Proceedings, Surface and Borehole Geophysical Methods in Ground Water Investigations, NWWA Confrence, San Antonio, February.

Parizek, R.R., 1976. On the significance of fracture traces and lineaments in carbonate and other terrranes: In; Karst Hyrology and Water Resources, Proc. Dubrovnik Symposium, June, 1975, Water resources Publications, Ft. Collins, CO, p. 47-108.

Sinkhole prediction – Review of electrical resistivity methods

EBERHARD WERNER *GeoAnalysis, Morgantown, West Virginia, USA*

ABSTRACT

The cost of drilling programs for the detection of sinkhole potential has encouraged the investigation of geophysical methods. All these methods have certain advantages and disadvantages. The electrical resistivity methods have the advantage of low cost and ease of operation, but they cannot be used where there are buried pipes and cables or where high-current or high-voltage power lines exist in the vicinity. The problem of the detection of sinkhole potential resolves to two aspects of local geology: the presence of a subterranean cavity and the presence of weakness, usually represented by fractures, in the overlying materials. Neither condition is easily detected directly by the resistivity method unless the features are very large; however, increased weathering along zones of weakness and changes in ground-water drainage caused by cavities provide volumes of rock material with electrical resistivity different than that of the country rock, and these differences can be easily detected. Instruments are available for contact or inductive coupling to the earth and using frequencies from direct current to radio frequencies which permit a choice of horizontal and vertical resolution and depth penetration and allow trade-offs of these against cost and speed of operation. Horizontal and vertical electrical resistivity profiling surveys can usually provide sufficient information to interpret for the production of sinkhole-potential hazard maps.

Introduction

The unexpected appearance of sinkholes has plagued construction projects in karst areas on many occasions. Although often no provisions are made in the planning phase of the project for the detection of a potential problem, sinkholes sometimes develop unexpectedly during a project even when such provisions are made. Most commonly, such unexpected occurrences are found where limestone with widely-spaced cavity development is overlain by relatively impermeable rocks which are removed during excavation for the construction project. Once the limestone is exposed, the roofs of these cavities collapse, either immediately or in response to the next heavy rain. Pre-excavation investigations by conventional boring programs may have missed the cavities entirely or not accurately determined their frequency or extent.

Although this type of problem is wide-spread and fairly well-known, two examples might be instructive. One example occurred during the construction of a shopping center just outside Harrisburg, Pennsylvania, in the early 1960's. In this instance, limestone containing well-developed cave passage is overlain by a shale. The cave development extends to the top of the limestone. Excavation plans called for removal of all or most of the shale to provide a level area for the buildings and parking lot. Test borings were made and no significant cavities were intersected. Excavation proceeded and some concrete foundation blocks were poured, when collapses occurred beneath the excavation machinery into the two largest caves. Excavation and construction were halted, and by the following spring, as many as one-fourth of the already poured concrete blocks had shifted because of additional sinkhole development.

A second example of a sinkhole problem occurred in the early 1970's when construction of Appalachian Corridor H (US Route 33) was done across the limestone terrain east of Elkins, West Virginia. This is an area of high relief with a limestone sequence about 100 meters thick between sandstones, siltstones and shales. The road building involved a hillside cut below, through, and above the limestone. The problems in this cut were encountered principally where the cut passed through the rocks overlying the limestones. Although the conventional testing was done, construction proceeded without knowledge or consideration of possible karst-related problems. Excavation machinery caused collapses into several cavities as the level of the limestone was approached, and several cavities were also dug into. Construction was halted for a period of time while investigations were conducted into the extent of the problem, and for formulating a solution which would not further impact the groundwater of the area.

In both of the above cases, the open holes were eventually bridged or filled, before construction could continue. The delay in the construction was quite expensive, and knowledge of the conditions prior to excavation would have saved significant amounts of time and money in both projects.

Definition of the Problem

Sinkholes often form at the most inopportune times and places. The potential for sinkholes is generally provided by the pre-existence of cavities coupled with some mechanical weakness of the material which forms their roofs. In a typical karst area, the cavities are natural caves and the weakness is one or more natural fractures (joints) developed in the roof. Sinkholes form in several ways in this kind of terrain. One of these ways is when the rock material loses cohesion across the fractures and the roof collapses. In this case, it is rock which falls in. In other cases, a cavity may have developed along a fracture, and weathered rock material (that is, soil) may have lodged in a higher part of the fracture, say above a constriction. With changing groundwater/surface water conditions, this loose material may fall into the underlying cavity (ravelling failure, for example). In relatively rare cases, the rock roof may collapse through loading which exceeds its beam strength. Of course in this case, no pre-existing fractures are involved.

Cavities often seem to meander inexplicably and can escape detection even though high-density boring programs are carried out. The cavity development is an extremely complex process involving many geological factors. Likewise, the location of fracture systems which contribute to the instability of terrains underlain by cavities is not always easy. There may be no outcrops present, nor other signs of such fractures. Areas with such sinkhole potential may be perfectly stable until certain of man's activities interfere with the balance. Such was the case in both of the examples described above. If a relatively simple procedure had been applied to either area before excavation had begun, it is likely that the location of potential sinkholes could have been delineated.

Cavity Detection Methods

Several geophysical techniques exist which are capable of providing information on the existence of both the cavities and the accompanying fractures that contribute to most of the sinkholes developed in certain karst terrains, particularly those in the Central Appalachians, but also elsewhere. The techniques discussed in this report generally work best in such areas. The characteristics of the Central Appalachian karst areas can be generalized as areas of relatively thin limestone sequences with thin soil cover and fairly high hydraulic gradients. This produces the majority of cavity development along vertical or near vertical fracture systems. Although bedding-plane cavities do develop, these rarely contribute to the sinkhole problem in such areas.

Given the combination of conditions, engineering geophysical techniques are well suited to a solution to the problem of sinkhole prediction. Probably the best suited, especially from the viewpoint of cost and time requirements are the electrical or electromagnetic techniques. Micro-gravity and high-resolution seismic methods have been used with some success, but require more complex equipment and are time- and labor-intensive and therefore expensive.

Of the numerous electrical and electromagnetic methods, telluric and streaming potential methods have relatively poor resolution, and are somewhat undependable except under specialized conditions. Ground-probing radar can be good at times, but suffers from poor depth penetration and fairly high cost. The remaining methods are based on some measure of ground resistivity.

Electromagnetic methods require no actual contact with the ground. Therefore they can be done very rapidly. The technique is based on the induction of electrical currents in buried conductors. Fractures with lower resistivity serve as these buried conductors. If a fracture is filled with clay or weathered, permeable rock material that is wet, its conductivity will be significantly higher than the low-permeability limestone on either side. Two different types of apparatus are commonly used for such surveys. One of these is a receiver which is tuned to one of the several very low frequency (in the 20 kHz range) communications transmitters that are located in many parts of the world. The fracture zone to be detected should be parallel to the line of wave propagation or nearly so. The obvious disadvantages of this method are two: fractures not properly aligned with the station cannot be detected, and the method cannot be used when the stations are not on the air. A second electromagnetic method uses its own low power transmitter, and therefore is not subject to the limitations of the passive system, but the equipment is significantly more expensive and more awkward to handle. A problem common to electromagnetic methods is that they cannot be used near powerlines, fences, or where conductors of any kind are buried.

The electrical systems are all quite similar in that contact is made to the ground through some type of electrode, generally stakes driven into the ground. One pair of electrodes serves to introduce current into the ground, and a second pair measures the potential created between them. In essence, the apparatus is a Wheatstone Bridge which measures the resistance of some volume of soil and rock. Problems with powerlines, fences, and buried conductors are much less severe than with electromagnetic methods; only high-voltage or high-current powerlines interfere, and conductors must be in electrical contact with the ground and parallel to and near the survey line to be serious problems.

Electrical Resistivity Methods

Electrical resistivity methods are the most popular among the engineering geophysical techniques because the equipment is readily available and relatively inexpensive. The different procedures which have been developed have a long history of searching for and delineating shallow geological features (Kirk and Werner, 1981).

Although numerous arrangements of electrodes are possible for the electrical resistivity survey, the colinear, equal-spaced array is popular for fracture detection because of the relative ease of operation and interpretation. Variations in arrangement and positioning of the electrodes provide the ability to define horizontal or vertical resistivity profiles of the area under consideration. These can then be visually interpreted.

Typically, the electrical resistivity methods provide information of the electrical conductivity of a certain volume of rock located below ground level. If this volume contains an air- or water-filled cavity, then a horizontal profiling survey across it would show either a rise or drop in resistivity, respectively, when traversing the cavity. There are limitations to the size of the cavity that may be detected; theoretical considerations show that the standard Wenner array (colinear, semi-symmetric) cannot detect a cavity buried deeper than its radius (Van Nostrand, 1953). However, in practice, the weathered rock around a cavity will significantly enhance its detectability. Similarly, an expanding-array vertical profiling survey provides information on the resistivity of a progressively larger hemisphere of rock and soil. If no significant vertical discontinuities or variations exist within that same volume (these would have been detected by horizontal profiling), then the vertical position of a cavity or geologic discontinuity (usually bedding planes) may be detected (Manley and Garton, 1977). Combination of the two procedures gives the location and depth to the cavity.

Other electrical resistivity procedures, particularly with the Logn configuration, have been successfully used for locating cavities (Logn, 1954; Bristow, 1966; Bates, 1973). This method uses a relatively long wire layout which may present problems in heavily vegetated terrains.

A second feature of electrical resistivity surveys is that, given the proper electrode configurations, geological discontinuities may be detected. The tri-potential technique (see Kirk, 1976; Kirk and Rauch, 1977) is routinely being used to detect vertical fractures. A horizontal profiling survey using this technique produces a characteristic signature when traversing across a fracture.

A Procedure for Detecting Sinkhole Potential

A modification of the procedure being routinely used for the detection of fracture zones by the oil and gas industry (Werner, Hempel, and Garton, 1983) can serve to provide information on potential sinkhole hazard. This procedure works best in the relatively unusual Central Appalachian karst terrains, but should be applicable elsewhere as well.

An initial indication of potential fractures and especially their orientations is determined by conventional photogeologic procedures such as originally outlined by Lattman (1958). The orientations are particularly important so that the grid for the resistivity surveys may be laid out as nearly at right angles to the fractures as possible. The horizontal profiling resistivity surveys may then be done and the actual location of the fractures determined and plotted. To determine the locations of cavities, vertical profiling (sounding) surveys would be run. Anomalies on the plotted curve would then indicate the existence of and depth to cavities. Air-filled cavities would show high-resistivity anomalies; water-filled cavities would show low-resistivity anomalies.

Combination of the information gathered could provide the surface locations of a potential sinkhole hazard to be considered for any construction project. The electrical resistivity survey gives the capability of looking through insoluble rocks which may overlie the limestone, and provides more continuous information than a boring program would (although boring programs would still have to be done for rock sampling).

Conclusion

Potential sinkhole hazards in many types of terrains can be delineated by use of a combination of electrical resistivity methods. These determinations are usually less costly and more informative of such hazards than conventional boring programs. Variations of the methods exist, and could be applied in various combinations to provide redundancy to improve the accuracy of the information derived from them.

References

Bates, E. R. (1973) Detection of subsurface cavities: U. S. Army Waterways Experiment Station Miscellaneous Paper 5-73-40: 63p.

Bristow, C. (1966) A new graphical resistivity technique for detecting air-filled cavities: Studies in Speleology 1(4):204-227.

Kirk, Keith G. (1976) Evaluation of the tri-potential resistivity technique in locating cavities, fracture zones, and aquifers: Masters thesis, West Virginia University Department of Geology and Geography: 90p.

Kirk, Keith G. and Rauch, Henry (1977) The application of the tri-potential method of resistivity prospecting for groundwater exploration and land use planning in karst terrains: Memoirs, 12th Congress, International Association of Hydrogeologists, University of Alabama Press, Huntsville, Alabama.

Kirk, Keith G. and Werner, Eberhard (1981) Handbook of geophysical cavity-locating techniques with emphasis on electrical resistivity: U. S. Department of Transportation, Federal Highway Administration Publication FHWA-IP-81-3: 175p.

Lattman, L. H. (1958) Technique of mapping geologic fracture traces and lineaments on aerial photographs: Photogrammetric Engineering 24:568-576.

Logn, O. (1954) Mapping vertical discontinuities by earth resistivities: Geophysics 19:739-760.

Manley, Thomas R. and Garton, E. Ray Jr. (1977) The A, B, C's of finding and delineating caves with apparent resistivity measurements: Hydrologic Problems in Karst Regions, Western Kentucky University: 92-95.

Van Nostrand, R. G. (1953) Limitations of resistivity methods as inferred from the buried sphere problem: Geophysics 18(2):423-433.

Werner, Eberhard, Hempel, John C., and Garton, E. Ray (1983) Well site location by combined data analysis, photolineament mapping, and earth resistivity profiling surveys (abstract): West Virginia Geological and Economic Survey Circular C-31:73.

3. Sinkhole-like features (subsidence pits)

Hydrocompaction sinkholes in the San Joaquin Valley, California

NIKOLA P. PROKOPOVICH *US Bureau of Reclamation, Sacramento, California, USA*

ABSTRACT

Karst processes are the main, but not sole, source of sinkhole development. Sinkholes also may be developed by melting of buried ice blocks, mining, geothermal activity, subsurface nuclear explosions, etc. One of the least known, and fortunately not widespread, type of sinkholes is that created by hydrocompaction. Hydrocompaction is a peculiar property of some dry unconsolidated sediments to lose their high dry strength and to collapse spontaneously after wetting, creating sinks, cracks, and slumps. The process is common in some loess deposits in the USA, Europe and Asia.

On the west side of the California San Joaquin Valley, hydrocompaction occurs locally in clayey piedmont alluvium derived from the Coast Ranges. Irrigation of prime agricultural flat land in some areas of the valley resulted in up to 4.6 m of spectacular settling, including creation of numerous, frequently round, 100-200 m diameter, 1-3 m deep sinkholes. In the past, such sinkholes were particularly typical around leaky irrigation wells. The recent change from the use of well water to canal water and from row irrigation to more efficient sprinklers insured more uniform water spreading, but did not completely eliminate minor accumulations of water on irrigated fields. Such accumulations resulted in local development of numerous scattered sinkholes. Surface irrigational runoff accumulates in some sinkholes in fields and causes migration and local accumulation of salts, which results in the degradation of good agricultural soils.

Introduction

Most of the sinkholes on the earth are created by collapses of overburden into preexisting karst cavities present in limestone and dolomite. Karst processes, however, are not the sole cause of sinkhole development. For example, some sinkholes present in the Holocene and Pleistocene periglacial zone were developed by melting of buried ice. Other sinkholes may be created by collapses of old underground mines, by geothermal activity, or even by subsurface nuclear explosions, etc. Probably one of the least known and fortunately not widely spread type of sinkhole is created by hydrocompaction. Description of such sinkholes in this paper is based mostly on some 25 years of observations in the west-central portion of the California San Joaquin Valley (Figure 1). Both field and office studies were made in connection with the design and construction of a major water convenience system known as the San Luis Unit of the Central Valley Project of the Federal Bureau of Reclamation (Anonymous, 1981, p. 165-232). The ideas expressed in this paper, however, are those of the author and may not represent the official view of the Bureau.

Hydrocompaction - General Data

Hydrocompaction is probably one of the most rapid and hazardous forms of subsidence, and is restricted to dry, semiarid and arid areas. It usually occurs as a direct result of human activity involving some form of water application such as irrigation, construction of dams and canals, urbanization, disposal of industrial wastewater, etc. The term identifies a peculiar property of certain dry, unconsolidated, porus, arid and semiarid deposits to loose their dry strength, and to develop, after wetting, spontaneous settling, slumpage, and cracking; causing damage to dams, canals, housing, roads, railroads, bridges, embankments, fields, pipelines and other structures.

The phenomenon has been well-known both in the United States and elsewhere. It was described under various names such as shallow subsidence, near-surface subsidence, hydroconsolidation, soil settling, soil settling by wetting, etc. Some of these names are misleading. For example, in the San Joaquin Valley and elsewhere, the process frequently is not restricted to near-surface deposits and can occur at a depth of 50 meters or more. The near-surface deposits may be stable, while deposits occurring at a depth of several meters may be susceptible to hydrocompaction, etc. The more proper term hydroconsolidation can be associated with some form of cementation, etc. The term hydrocompaction, properly reflects both the nature and the character of changes which occur in sediments and, at the present time, seems to be generally accepted in the literature.

Susceptibility to hydrocompaction is controlled by (1) the development of porus deposits with a low moisture content, but with high dry strength, and (2) aridity of climates preventing "natural" hydrocompaction of such deposits. Consequently, hydrocompaction is restricted to semiarid and arid regions and no hydrocompaction can be expected either below the ground-water table or in previously well wetted sediment. The thickness of deposits susceptible to hydrocompaction ranges from 1 to 2 meters to 30-50 meters or more. The process frequently may have a "spotty" occurrence and visually similar deposits in neighboring areas may be either susceptible or nonsusceptibile to hydrocompaction.

Rapid initial water application, such as row irrigation of highly susceptible deposits, results in rapid hydrocompaction, while minor, prolonged water application (such as sprinkling) causes less damaging but longer lasting hydrocompaction, which continues for years with subsequent water applications. After completion of the process, however, originally hazardous deposits become stable and not susceptible to hydrocompaction. Such stabilization, however, may occur several years after the initial wetting.

Hydrocompaction usually occurs in alluvial deposits and is particularly common in wind-blown loess. Hydrocompaction and/or damages caused by hydrocompaction were reported in several areas in the United States, including Wyoming, Washington, Montana, Arizona, Nebraska, Kansas, Utah, Colorado, New Mexico and elsewhere (Lofgren, 1969; and others). Large, loess-covered areas of India, China and Russia are susceptible to hydrocompaction (Anonymous, 1963; Drashevska, 1962; Lin and Liang, 1980; and others).

Location, Regional Geology and Hydrocompaction

The study area is located in the west-central portion of the deeply alluviated California San Joaquin Valley (Figure 1B). The valley is a giant structural trough surrounded on the east by the Sierra Nevada Mountains and on the west by the Coast Ranges. The near-surface deposits in the study area are represented by mostly clayey piedmont alluvium derived from the Coast Ranges. The general landscape is extremely flat and undissected. The piedmont alluvium can be subdivided into: (1) mostly fluvial fan deposits of major ephemeral creeks, such as Little Panoche, Cantua and Los Gatos Creeks, and (2) "interfan alluvium" deposited frequently as mud or debris flows produced by minor arroyos. Alluvium of some interfans has been known to be susceptible to hydrocompaction while no hydrocompaction was noted on several other interfans in the vicinity (Figure 1A). No hydrocompaction was noted on major alluvial fans.

The climate of the area is semiarid, of Mediterranean type, with dry, hot summers and foggy, relatively cool, rainy winters (Karhl, 1979). The soils are fertile, but modern agriculture is possible only with irrigation. Irrigation in the area started at the end

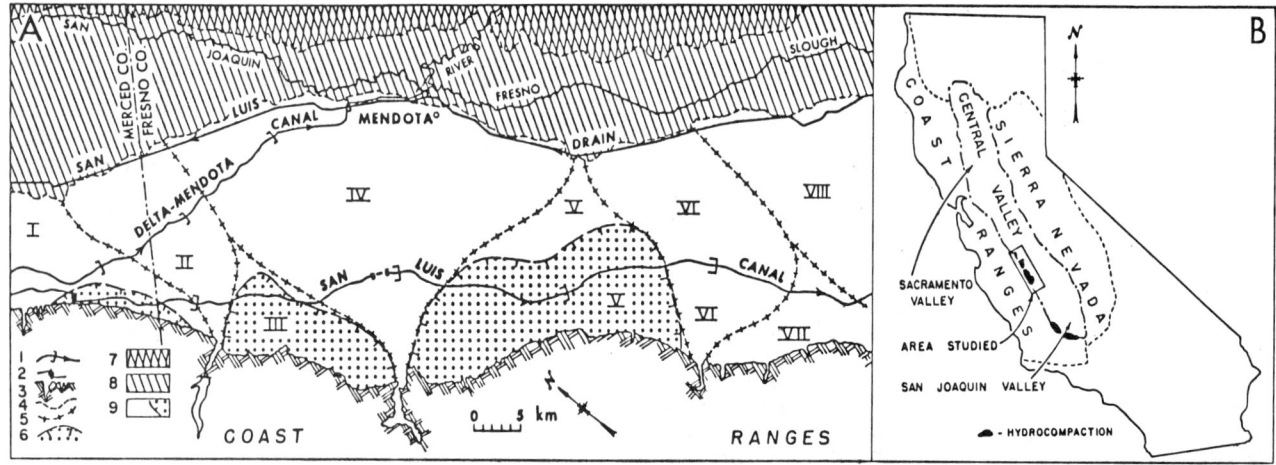

Figure 1: A. Location of the study area and distribution of hydrocompaction in the San Joaquin Valley, CA. B. Geologic map of a portion of San Luis Service Area showing: 1 - Major USBR canal; 2 - pumping plant; 3 - toe of Coast Ranges; 4 - geologic contact; 5 - boundary between alluvial fans and interfans; 6 - area affected by hydrocompaction; 7 - east side piedmont alluvium; 8 - basin alluvium; 9 - west side (Coast Range) piedmont alluvium; I - Little Panoche-Los Banos interfan; II - Little Panoche fan; III - Little Panoche-Panoche interfan; IV - Panoche fan; V - Panoche-Cantua interfan; VI - Cantua fan; VII - Cantua-Los Gatos interfan; VIII - Los Gatos fan.

of the XIX century with minor diversion of local surface water and was followed by irrigation based mostly on ground-water pumpage from a deeply seated confined aquifer system. (So-called "shallow" unconfined aquifer system occurs usually at a depth of 5 - 60 m, but is too salty for agricultural usage.) At the present time, after construction of the San Luis Unit, surface canal water replaced most of the ground-water pumpage. Original irrigation was accomplished with open ditches and furrows. Sprinkling irrigation is common at the present time. Historically, irrigation gradually spread in a westward upslope direction from the axial portion of the valley. Early irrigation did not reach areas susceptible to hydrocompaction and development of sinkholes probably started in the 1940's.

Over 95 percent of the hydrocompaction sinkholes in the study area have been developed through human activity-irrigation. Natural precipitation (always rain) in the area is small, some 15-25 cm/yr (Karhl, 1979). Moreover, pure rainwater deflocculates clay colloids in the uppermost 3-5cm of topsoil and virtually seals underlying alluvium, which remains "bone dry" after strong occasional desert downpours. Runoff of occasional storms in the foothills has a somewhat larger amount of electrolites and a somewhat better infiltration capability. Runoff of such storms sometimes creates spectacular flooding in the valley. Both the runoff and flooding are, however, restricted to terrains of alluvial fans of major creeks. These fans are not susceptible to hydrocompaction. Watersheds of small arroyos and washes which provide sediments for interfan alluvium locally susceptible to hydrocompaction are much smaller than watersheds of major fans. No major flooding comes from such small watersheds and the main mode of deposition, here, is by rapidly drying "mudflows." During some 25 years of studies in only two or three cases has hydrocompaction cracking and setting in never irrigated areas been caused by "natural" runoff. In all of these cases, the natural runoff was artificially blocked by man-made dikes or road embankments causing local flooding.

Hydrocompaction in the area was first noted in 1915 after construction of the Chaney Pumping Station in an undeveloped terrain some 65 km SW of the city of Fresno. Hydrocompaction here was kept well under control for decades by keeping the station (now abandoned) perimeter dry through diversion of industrial and domestic waste water and by restriction of irrigation. With a general increase of irrigated terrain in the valley, particularly after world wars I and II, local hydrocompaction became a well recognized geologic hazard damaging roads, fields, buildings, and particularly wells and ditches. Incomplete historical data seem to indicate that ultimate amounts of hydrocompaction appear to be smaller on the Little Panoche-Panoche interfan than on the larger Panoche-Cantua interfan, where ultimate amounts of hydrocompaction reach 4.6 m. The ultimate amounts of hydrocompaction seem to increase toward the foothills and decrease toward the valley trough. Particularly severe hydrocompaction was encountered southeast of the study area during State Water Project studies in the southwestern end of the San Joaquin Valley (Figure 1B). Since early 1960 hydrocompaction became particularly important because of scheduled construction of the San Luis Canal and its service area and associated construction of the State Water Project in the southern part of the Valley (Anonymous, 1974).

The origin of hydrocompaction, or more specifically the origin of excessive porosity of sediments susceptible to hydrocompaction is not fully known (Bull, 1961, 1964). By all probability this porosity is polygenetic, i.e., can be developed by several processes. In California's Central Valley several controversial and polygenetic features such as contorted bedding, loam wedges, hummocky topography, and rock polygons seem to indicate the existence here of severe Pleistocene periglacial paleoclimates (Prokopovich, 1983). The existence of such climate may explain the origin of excessive porosity of sediments susceptible to hydrocompaction by the freeze-drying of water-saturated Pleistocene summer mudflows expanded volumetrically by freezing during cold winter seasons and dried by westerly dry wind passing over the Coast Ranges. Such sublimation resulted in preservation of newly developed postdepositional porosity which was maintained during the Holecene sermiarid climate. Variations in microclimate and mode of deposition could be responsible for local occurrences of sediments susceptible to hydrocompaction. It is possible that hydrocompaction, widespread in loess deposits in the USA and elsewhere, has a similar origin.

Hydrocompaction Sinkholes
In general, the size, shape and severity of hydrocompaction sinkholes in previously nonirrigated alluvium depend upon the amounts of irrigation water used, rate and mode of application, and shape of wetted bodies. During the period of irrigation by ground water numerous rounded 2 -3 m deep sinks developed around irrigation wells as a result of leakage (Figure 2A). These wells were usually located at section line roads. The hydrocompaction resulted in extensive damage to wells (collapsing of casings, "telescoping," lifting of pumps above their foundations, etc.). Numerous 1 - 2 m deep

trough-like sinks, causing damage to roads, fields, and irrigation systems, also developed along major lined and unlined irrigation ditches paralleling section line roads. For example, a 1.5 km long and over 3m deep sink was developed in an area along Nebraska Avenue some 23 km SW of the town of Mendota after an initial row irrigation in 1945. After three irrigation cycles, no farming was possible here from 1950-55 when sprinkler irrigation was introduced.

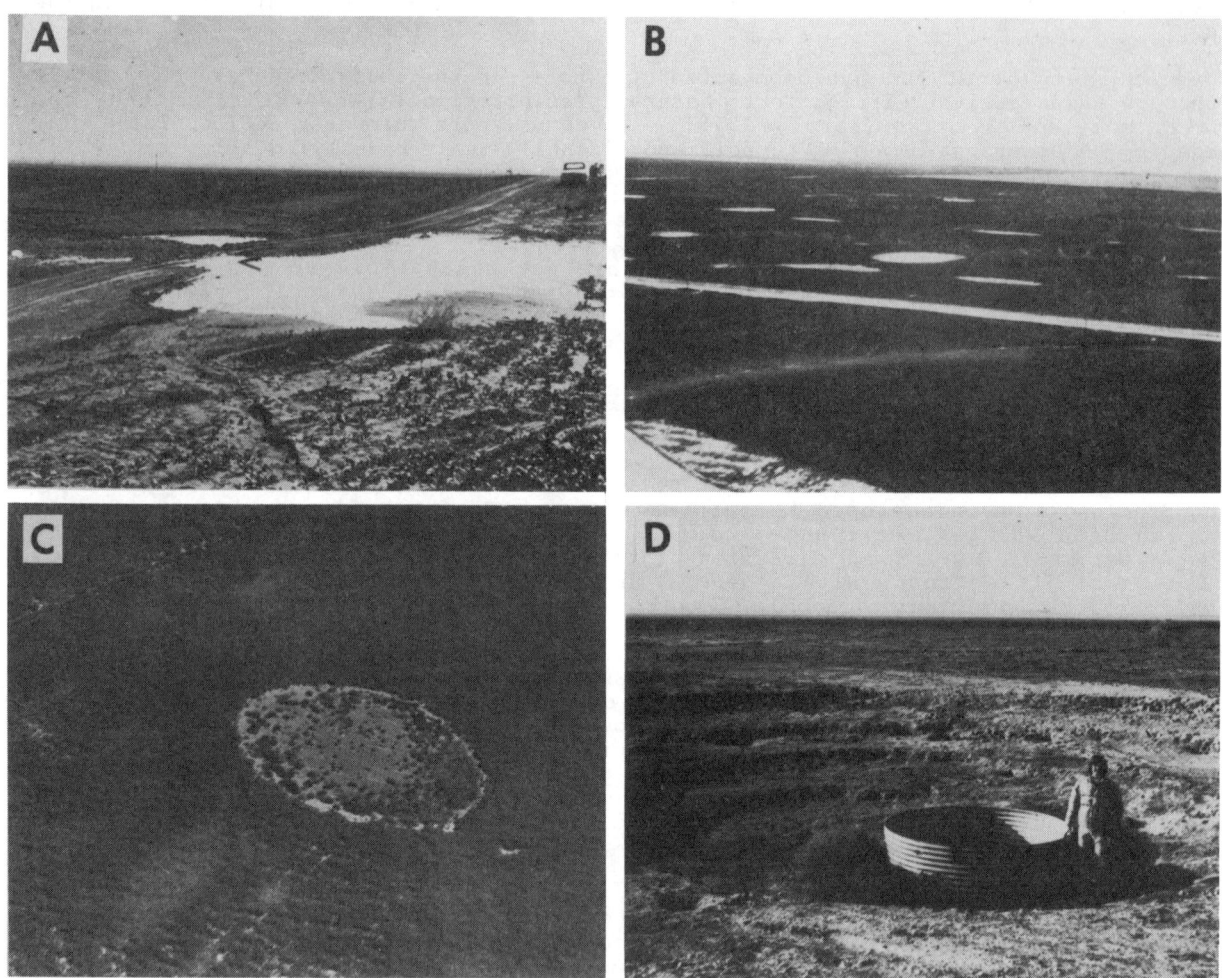

Figure 2: Typical examples of hydrocompaction sinkholes on the Panoche-Cantua interfan: A. Sinkhole on a section line road partially flooded by irrigation runoff; B. Aerial view of several sinkholes in an irrigated field; C. Salt accumulation in a sinkhole prevents normal crop growth; D. So-called "small test plot" experiment operated by the State Department of Water Resources. A sinkhole developed around a flooded shallow infiltraion well.

With the change to sprinkler irrigation, ditches were replaced by pipelines. However, numerous sinks began developing in irrigated fields around pipe leaks and minor depressions which collect irrigation water (Figure 2B). Such sinks are usually 2 - 3 m deep and 100 - 200 m in diameter. Irrigation surface runoff collecting in sinks may produce local accumulations of soluble salt resulting in a degradation of soil (Figure 2C). Gradual leaching of such polluted spots will occur for several years after the completion of hydrocompaction has restored the original flat relief.

References

Anonymous, 1963, Shallow subsidence (hydrocompaction): Annotated bibliography of foreign literature prepared for the United States Department of the Interior and National Science Foundation, Israel Program for Scientific Translation, Israel, 46 p.

Anonymous, 1974, California state water project; bulletin no. 200: State of California Department of Water Resources, Vol. I: History, planning, and early progress, 173 p.; Vol. II: Conveyance facilities, 349 p.; Sacramento, CA.

Anonymous, 1981, Project data, A water resources technical publication: United States Water and Power Resources Service (now Bureau of Reclamation), U.S. Government Printing Office, Denver, Colorado, 1,463 p.

Bull, W. B. 1961, Causes and mechanics of near-surface subsidence in western Fresno County, California: United States Geological Survey Professional Paper 424-B, U.S. Government Printing Office, Washington, DC, p. B-187 - B-189.

Bull, W. B., 1964, Alluvial fans and near-surface subsidence in western Fresno County, California: United States Geological Survey Professional Paper 437-A, U.S. Government Printing Office, Washington, DC, 70 p.

Drashevska, L., 1962, Review of recent USSR publications in selected fields of engineering soil science: In Reviews in Engineering Geology, Volume I, Geological Society of America, New York, N.Y., pp. 197-256.

Kahrl, W. L. (Editor), 1979, The California water atlas: Governor's Office of Planning and Research, Sacramento, CA, 113 p.

Lin, Z., and Liang, W., 1980, Distribution and engineering properties of loess and loesslike soils in China. Schematic map of engineering geological zoning: Bulletin of the International Association of Engineering Geology, No. 21, Krefeld, West Germany, pp. 112-117.

Lofgren, B. E., 1969, Land subsidence due to the application of water: Reviews in Engineering Geology II, Geological Society of America, Vol. 2, Boulder, CO, p. 271-303.

Prokopovich, N. P., 1983, Paleoperiglacial climate and hydrocompaction in California, USA: abstracts and program, 4th international conference on permafrost, University of Alaska, Fairbanks, Alaska, 1983, p. 204-205.

Sinkholes in southeastern North Carolina – A geologic phenomenon and related engineering problems

HENNING F. KOCH *North Carolina Department of Transportation, Raleigh, USA*

ABSTRACT

Sinkholes in southeastern North Carolina generally are found in an area underlain by shell-limestone deposits of Upper Cretaceous and Tertiary Age. The composition of these formations varies from sandy shell-limestone to silicified limestone with calcareous sand facies. The cavernous, sponge-like skeleton and interlayered structure of parts of these formations indicates a reef-like depositional environment. The hydrological characteristics are similar. The potential for sinkhole development was built-in as these units were formed. The occurrence of sinkholes is closely tied into the development of the natural drainage pattern of this area. As the rivers cut further back into the land the newly imposed relief induced a gradient on the water table. The resulting turbulent flow flushed out the sandy, silty or clayey infilling of the cavernous limestones. The lower ground water table increased the effective weight of the overburden. This then led to the eventual collapse of the sponge-like structure, causing land subsidence. Fossil collapse features are exposed in several localities in southeastern North Carolina, evidencing that this process has been an integral part of the geologic history of this region probably as early as the Oligocene.

For both the natural and man-induced sinkhole development the mechanism is basically similar.

Engineering problems related to sinkhole occurrence have increased in this area substantially in the past 20 years. But so has the number of detection methods and engineering solutions dealing with these problems. Some methods, primarily resistivity combined with conventional drilling, coring and sounding have produced satisfactory results. Others like ground penetrating radar and seismic methods have been tried with little success so far. Engineering solutions presently include structural bridging, excavation and backfilling, deep compaction, and grouting. To date the application of slurry walls or use of explosives has not been attempted in this area. The single most important factor for preventing or greatly reducing the undesirable occurrence of sinkhole development and subsidence is directed areal management and maintenance of the existing ground water table.

Introduction

Until about two decades ago little attention had been paid to sinkholes in North Carolina. The geologic literature almost ignores this phenomenon. Jasper Stuckey's, North Carolina: Its Geology and Mineral Resources (1965) does not index the term sinkhole. The Explanatory Text to the Geologic Map of North Carolina (1958) does not mention sinkholes. Ground-Water Bulletin No. 1 (1960): "Geology and Ground-Water Resources of the Wilmington - New Bern Area" however states that numerous sinks occur around Chinquapin and also mentions the "Natural Well" approximately 3 km (2 mi.) west of Magnolia, Duplin County. The North Carolina Department of Conservation and Development (now NRCD) Bulletin No. 54, "Marls and Limestones of Eastern North Carolina," by E.W. Berry also briefly acknowledges the "Natural Well" west of Magnolia and the presence of several sinks. Berry also describes another area, located 1.5 km (1 mi.) east of Pollocksville, Jones County: "The overburden of the marl is up to 5 ft. (1.5 m) thick and the top of the marl is very irregular and contains many tubular sinks full of sand and clay." However, the lack of sinkhole reference does not mean they were not known. Most of them were located in remote wooded areas and did not present problems. At best they were used as dumps or regarded as local curiosities.

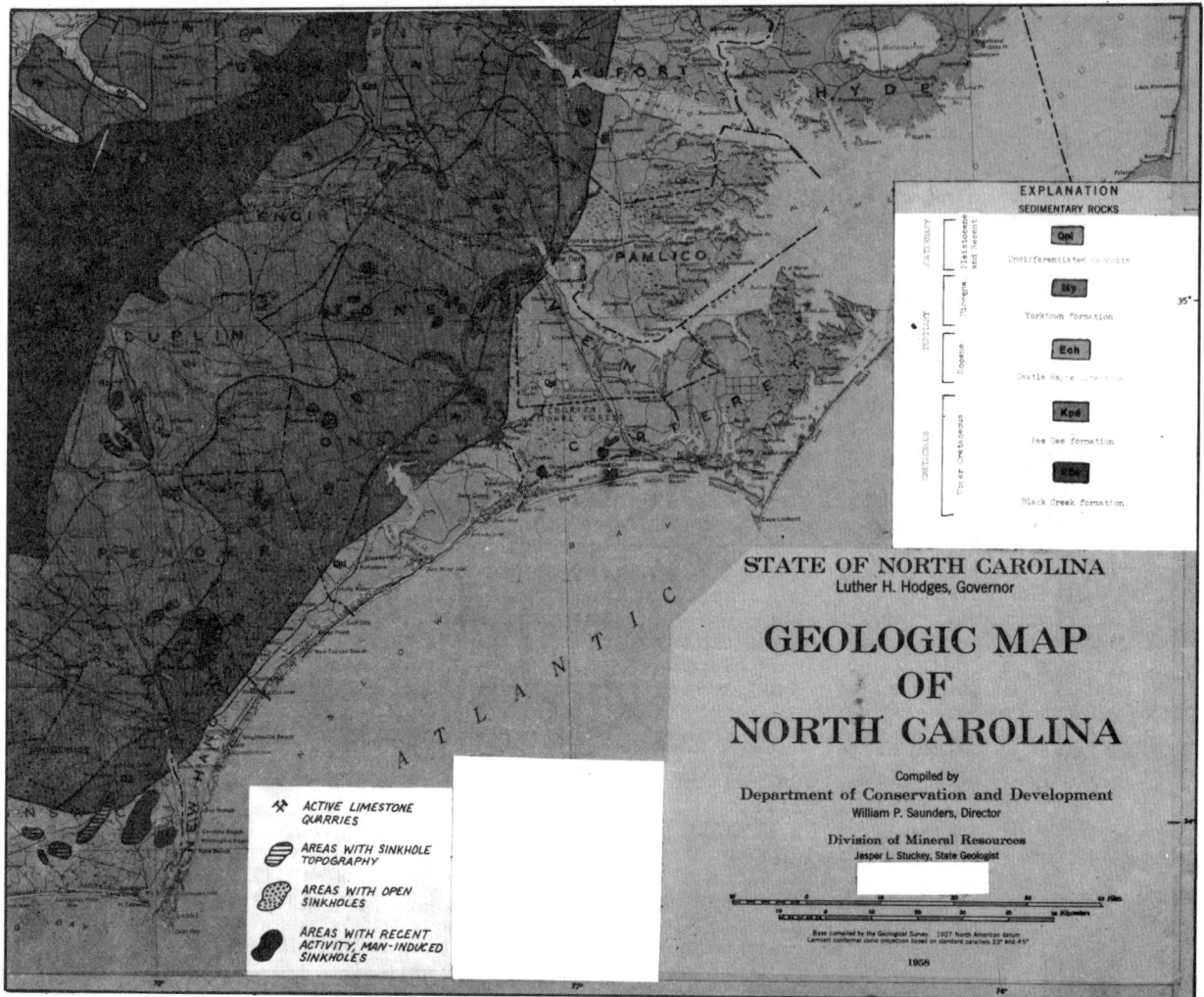

Figure 1: Sinkhole Areas of North Carolina

Geologic Setting and Mechanism

The Geologic Map of North Carolina of 1958 is used as a base map here. However, the Geologic Survey of N.C. is presently revising this map which should be available late 1985. Extensive revisions are anticipated, therefore, some of the stratigraphic statements in this text may not match the map units as shown. Sinkholes in southeastern North Carolina generally are found in areas underlain by shell-limestone deposits of Upper Cretaceous Age. In the regional stratigraphy of southeastern North Carolina this would include the Pee Dee Formation (Upper Cretaceous), Castle Hayne limestone (Eocene) and calcareous marine deposits of Oligocene, Miocene and Pliocene Age. Recent stratigraphic work has resulted in much greater differentiation and distinction of these deposits. The composition of these formations varies from sandy shell-limestone to silicified limestone with calcareous sand facies. The cavernous sponge-like skeleton and interlayered structure of parts of these formations indicates a reef-like depositional environment. The hydrological characteristics are similar. The potential for sinkhole development was built in as these units were formed.

The majority of sinkholes in North Carolina are concentrated in six southeastern counties: Brunswick, New Hanover, Pender, Jones and Craven. Type locations for open natural sinkholes are: Fort Barnwell in Craven County; Acme in Brunswick County and the sinkholes south of SR 1003, west of Magnolia in Pender County. Areas with pronounced sinkhole topography are in the vicinity of Boiling Springs, Brunswick County; northeast of Castle Hayne, Pender-New Hanover Counties and along the Lower Trent River between Trenton and New Bern, Craven County. From the topographic position of these areas it is quite obvious that the occurrence of natural sinkholes is closely tied to the development of the natural drainage pattern. As rivers cut further back into the land the newly imposed relief induced a gradient on the ground water table. The resulting turbulent flow flushed out the sandy silty or clayey infilling of the cavernous limestones. The lowering of the ground water table increased the effective weight of the overburden. This then led to the eventual collapse of the sponge-like interval structure causing land subsidence. Field mapping in the past few years has found evidence of fossil sinks. A type location is the Martin Marietta limestone quarry near New Bern, Craven County. Eocene and Oligocene shell-limestone collapse structures with Pliocene shell hash infilling have been exposed to a depth of 9.1m (30 ft.). A similar fossil sinkhole was found several years ago in the Fussel Quarry near Rose Hill, Duplin County. These fossil collapse structures bear evidence that formation of sinkholes and related land subsidence has been an integral part of the geologic history of this region probably dating back as early as the Oligocene.

The collapse mechanism as deduced from recent observations requires certain prerequisites: 1) A limestone facies that represents environment with the randomly arranged structure of a reef, including sand or clay filled calcareous tubes, pipes and crevasses, 2) A soil overburden, 3) A point lowering of the ground water table to create a gradient sufficient to create an erosional flow velocity and 4) An outlet for the water and the flushings. If these conditions exist, subterraneous erosion will take place resulting in land subsidence.

The effect of solution of the limestone by the water is considered minimal, simply because of the time required for this process. Man-induced sinks in southeastern North Carolina occurred shortly after quarry pumping lowered the ground water table and created a steep ground water gradient. There was hardly any time for chemical action.

Man-induced Sinkholes

During the past 20 years sinkhole related problems have increased in southeastern North Carolina significantly. At present they are mostly effecting transportation facilities. The area of Boiling Springs and Sunny Point Military Ocean Terminal is one region plagued by sinkhole activities. The other is located north of Wilmington where the eastern terminal segment of I-40 is under construction between Castle Hayne and Rocky Point.

I-40 East of Castle Hayne

Shallow sinkholes have been known to exist in this area which borders the Northeast Cape Fear River. But there are no reports of sinkhole activities until about 1981, when grading for the Highway project began. At the same time a nearby Martin Marietta limestone quarry began to extend operations in the direction of the project. Heavy ground water drawdown occurred and soon collapses within the construction limits were observed. A geotechnical investigation was initiated. Electric resistivity surveys combined with exploratory drilling and sounding programs determined potential problem areas. This investigation also revealed the triggering impact of ground water draw-down on the development of sinkholes, as illustrated in Figure 2.

Treatment

To minimize construction delays the decision was made to use the following construction procedure at sinkholes:
1) Sinkholes were identified in the field as to their configuration and boundaries. A rectangular stabilization area and their organic content was determined.
2) Organic material, if present, was removed.
3) Select granular material was used to backfill the cleaned depression to approximate natural ground level.
4) Then deep compaction was applied. This treatment was designed to densify the granular material placed in each sinkhole, locate underground voids and possibly retard future sinkhole expansion. Areas were treated by dropping a flat bottomed concrete weight with a minimum area of $2.32m^2$ (25 ft.^{2}) of 8 tons (10t) from a height of 12.2m (40 ft.). A drop grid pattern of 6.1m (20 ft.) center to center was used. One to five drops were made at each grid location. After releveling the site a second pass of 8 tons (10t) from a height of 3.05m (10 ft.) on a grid pattern of 3.05m (10 ft.) center to center was made. Both passes extended to 6.1m (20 ft.) outside the limits of each sink area. The impacted area was brought back to natural ground elevation by the addition of granular material.
5) Following soil stabilizing fabric of PVC, plastic lining was installed over the previously determined sinkhole area.
6) Then a minimum of 0.9m (3 ft.) of compacted earth embankment material was placed over the area covered by the fabric or plastic lining.
7) From this point on, the conventional construction of the roadway embankment proceded.

New sinkholes have appeared nearby, but treated areas are performing well to this date.

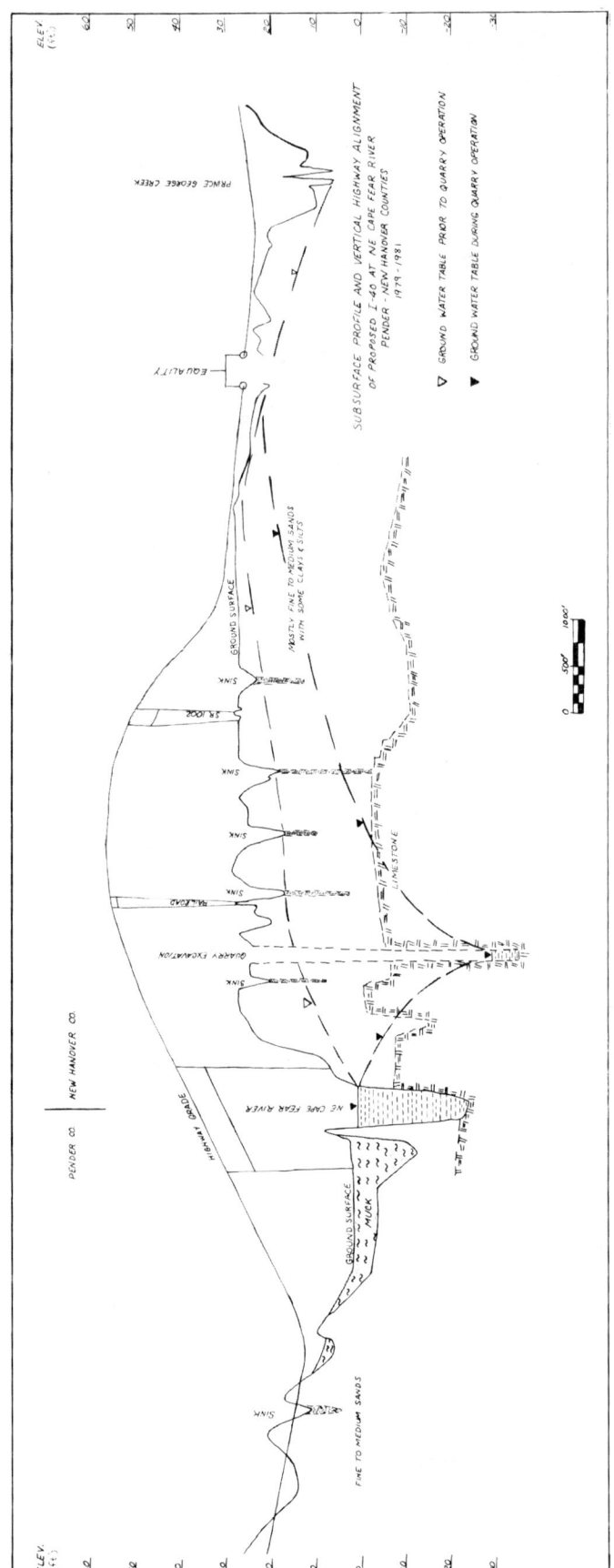

I-40 Southeast of Rocky Point, Pender Co., North Carolina

This region is underlain by shell-limestone. The lithology of these deposits varies from earthy to porous to massive. The presence of old sinkholes had been noticed. But there were reports or signs of activities.

As the grading for the portion of the I-40 between the Northeast Cape Fear River and NC 210, northeast of Rocky Point, was in progress a Martin Marietta limestone quarry began mining operations in the summer of 1983 adjacent to the highway project.

Water levels in the quarry were lowered to a depth of 10m (30 ft.) below normal levels. On August 29, 1983 the first signs of subsidence were discovered. After heavy rains a 7.5m (25 ft.) diameter depression, partly outlined by cracks, had formed in the southbound lane. Maximum vertical displacement was about 18cm (1.5 ft.). In the eastern segment of the circle a cavity of about 1.5m^3 (1.9 yd.^{3}) was observed. A series of auger borings indicated rather uniform soil conditions. Approximately 3.35m (11 ft.) of sand, including 1.8m (6 ft.) of roadway embankment, were underlain by 3m (10 ft.) of clay soil which rests on limestone. In one of the borings a loosely filled cavity 3m (10 ft.) thick was encountered at a depth of 7.6m (25 ft.). This was probably the cavity to which the overburden sands flowed during the collapse.

On another occasion a sinkhole was discovered in the right ditch of the southbound lane. The sinkhole pipe was about 0.6m (2 ft.) in diameter with the adjacent area eroded to a diameter of approximately 1.5 (5 ft.). There were indications that ditch drainage had been flowing into the sinkpipe, thus constituting a "swallow hole". Within the quarry area five sinkholes had occurred during the time of the investigation. The smallest was about 1.5m (5 ft.) in diameter and 0.3m (1 ft.) deep, and the largest was 3.66m (12 ft.) in diameter and 2.1m (7 ft.) deep. The absence of a large collapse in this particular area is explained by the slightly cohesive soils overburden and the shallow depth, ± 2m (6 ft.).

Resistivity Survey

An electrical resistivity survey has been made along the project between Sta. 700 and Sta. 820. This section covers more than a mile both north and south of the quarry site. The purpose

---

Figure 2: Subsurface Conditions (1979-1981) at I-40 Alignment Over NE Cape Fear River.

of this survey was to determine whether anomalous areas exist which could be related to sinkholes or subsurface cavernous zones.

Typically five lines of resistivity profile were run along the project - one in the highway median, one in each shoulder, and one near each ditch line or toe of slope. In areas of significant anomalies, additional lines were run along the right-of-way line or in the adjacent service road.

The surveys were made using the Wenner configuration (equal spacing of electrodes). A spacing interval of 10.2m (33.3 ft.) was utilized because: 1) the void encountered near the sinkhole at Sta. 796+80 lies at a depth between 7.9m and 10.7m (26 and 35 ft.), 2) it gives apparent resistivity values at reasonably close intervals, and 3) it facilitated the locating of reading sites by regular coincidence with project stations.

Apparent resistivity for each point was plotted on a plan view and then contoured to give a two-dimensional view of resistivity values.

Piezometers
Eleven standpipe piezometers were installed by Martin Marietta to monitor piezometric water levels in the rock. These piezometers were installed to depths ranging from 4m to 10.7m (13 to 35ft.) below natural ground. Six of the piezometers were placed adjacent to the right-of-way at about half mile intervals. Four were placed between the quarry and the project, and one was placed east of the quarry near a group of houses.

Ground Penetrating Radar
In the spring of 1984 a section of the effected highway segment near the quarry was surveyed using ground penetrating radar. The results were poor and uninterpretable. The survey program was discontinued.

Test Grouting
Evaluating the field data obtained during the geotechnical investigation north and south of the Martin Marietta quarry site at Rocky Point, five areas were identified as having significant resistivity anomalies. In two areas sinkholes had already occurred. When test borings were made in these two areas the loss of drilling fluid was common in the vicinity of the caprock. The loss of drill fluid is a significant indication of voids or highly permeable materials. In June, 1984, five test borings were made in the remaining three areas. In some of these borings some soft or loose layers were encountered below the hard caprock; however, loss of drill fluid did not occur and these areas were no longer considered potential treatment areas.

Consequently it was decided to test grout one portion of the affected roadway on a limited scale. Approximately 135m^3 (175 c.y.) of cement grout was pumped into voids and soft subsoils. For easier identification Tamms Red Mortar Coloring was added to the grout mix. Initially Rhodamine was used as a dye but proved to be too costly, not readily available and difficult to handle. It is estimated that about 90% of the grout pumped into the ground resulted in filling voids and densifying soft or loose cavity fillings. To determine the effectiveness of the grouting, nine core holes were drilled. In four of these core borings grout was recovered in thicknesses of up to 12 cm to 13 cm (5 in.). One meter core borings where grout was recovered lie 1.5 to 4.6 (5 ft. to 15 ft.) away from the grouting location. In general it is believed that grouting could be an effective means of preventing further sinkhole development within the affected roadway sections. Additional methods such as partial excavation, deep compaction, use of engineering fabric and others could be utilized to increase the factor of safety. Experience gained while test grouting this portion of the highway can be summarized as follows: 1) Avoid grouting at high pressure. Approximately 3.8 kg/cm^2 (50 psi) may be a reasonable maximum pressure when grouting under these conditions and in this type of rock formation. 2) The grout hole pattern should consist of primary spacings between 6m and 12m (20 ft. to 40 ft.) with secondary grout holes half the distance between each primary hole. 3) Grade elevations should be checked frequently to monitor any effects of jacking the surface as a result of the grouting operation. Other corrective measures, including the installation of slurry walls to control the ground water or the use of explosives to induce controled collapse have been considered. However, because of the lack of experience in these little tried methods, the limited time frame of the construction schedule and economic considerations the decision was made to apply conventional grouting.

Conclusion
During the recent 20 years sinkhole activity in southeastern North Carolina has increased considerably. It was found that this activity was associated with changes in the ground water table and in particular its flow gradient. It was noted that new subsidence occurred only in areas where existing depressions indicate earlier, natural sinkhole activities. The formation of sinkholes is mostly confined to areas underlain by a cavernous, sponge-like structured shell-limestone deposit (reef-facies). Due to an increase of the

ground water gradient the sand-silt infillings are flushed out. Simultaneous increase of effective overburden weight and effective stress in the cavernous limestone lead to collapse. Because of the short time span of only a few months between initial pumping and the first observed subsidence, the chemical action of dissolving the limestone thus causing "solution cavities" is nearly nil. <u>Corrective</u> methods that have been tried in highway construction in N.C. include: 1) Cleaning the sinkhole, backfilling and application of the deep compaction method and 2) Cement grouting. <u>Preventive</u> measures in areas with the potential for sinkhole activity are: 1) Recognition and identification, 2) Maintenance of ground water table. No change in ground water gradient or flow, and 3) Effective Regulatory measures.

An area with the potential for sinkhole activities represents a geologic hazard zone. Land values are affected. Engineering projects are costlier and are more likely to become a permanent maintenance liability. Certain projects (hazardous waste sites) should not even be permitted. Early recognition of these problems will reduce or avoid many of these calamities.

Acknowledgements

I would like to thank Mr. W.D. Bingham, Head of Geotechnical Unit, North Carolina Department of Transportation for his support and advice and review. Also, Mr. G.L. Bunch for input from observations, survey data and suggestions. Last but not least Mrs. V.K. Conyers and Ms. C.L. Baity for typographical and editorial work.

References

Berry, E.W., 1947, Marls and Limestones of Eastern North Carolina: N.C. Dept. of Conservation and Development, Bulletin No. 54.

Fountain, Lewis S., 1976, Subsurface Cavity Detection: Field Evaluation of Radar, Gravity, and Earth Resistivity Methods: Transportation Research Board, Transportation Research Record 612.

Goughnour, Roger D., Anderson, Robert, and DiMaggio, Jerome A., 1979, Limestone Related Problems on Transportation Facilities: Highway Focus (U.S. Dept. of Transportation), vol. 11, No.2.

Hannon, Joseph B., and McGee, Barry E., 1976, Ground Subsidence Associated With Dewatering of a Depressed Highway Section: Transportation Research Board, Transportation Research Record 612.

Legrand, Harry E., 1960, Geology and Ground-Water Resources of Wilmington-New Bern Area: N.C. Dept. of Water Resources, Ground-Water Bulletin No. 1.

Legrand, Harry E., 1977, Ground-Water Hydrology of the Boiling Springs Lake Area (As Related to the development of Active Sinkholes): Technical Report on the Sinkhole Development along the Sunny Point Access Railroad at Boiling Springs, U.S. Army Engineer District, Savannah Corps of Engineers.

Newton, J.G.(U.S.G.S.), 1976, Early Detection and Correction of Sinkhole Problems in Alabama, With a Preliminary Evaluation of Remote Sensing Applications: Alabama Highway Research, HPR Report No. 76, Research Project 930-070.

Newton, J.G., 1976, Induced and Natural Sinkholes in Alabama - A Continuing Problem Along Highway Corridors: Transportation Research Board, Transportation Research Record 612.

N.C. Dept. of Transportation internal report, 1984, Sinkhole Investigation Near Martin Marietta Quarry on Priority Route From North of SR 1509 Near Burgaw to the NE Cape Fear River Bridge: Raleigh, N.C., Project No. 8.1223332 X-3DW.

Ruth, Byron E., and Degner, Janet D., 1984, Characteristics of Sinkhole Development and Their Implications on Potential for Cavity Collapse: Department of Civil Engineering, Univ. of Florida, Gainesville, Florida, January 1984. (Publication by the Transportation Research Board.)

Sowers, George F., 1976, Mechanisms of Subsidence Due to Underground Openings: Transportation Research Board, Transportation Research Record 612.

Stuckey, Jasper L., and Conrad, Stephen G., 1958, Explanatory Text for Geologic Map of N.C.: N.C. Dept. of Conservation and Development, Bulletin No. 71.

Stuckey, Jasper L., 1965, North Carolina: Its Geology and Mineral Resources: North Carolina State University Print Shop, North Carolina, 550p.

Soil cavities formed by piping

RALPH J.HODEK, ALLAN M.JOHNSON *Michigan Technological University, Houghton, USA*
DEAN B.SANDRI *Giles Engineering Associates, Waukesha, Wisconsin, USA*

ABSTRACT

Subsidence pits are a common surface feature on the western portion of the Menominee Iron Range in the upper peninsula of Michigan. This paper reports both case histories of occurrences in the region and a combined field and laboratory investigation into the problem. The laboratory analysis models the field conditions and demonstrates the mechanism whereby a significant subsurface cavity in the soil can develop prior to any indication of noticeable distress at the surface. The results of this study indicate that a cavity and subsequent subsidence can occur when flowing groundwater, granular soil, and manmade or naturally occurring cavernous bedrock are present. The paper demonstrates that collapse of the bedrock cavern is not necessary for the occurrence of subsidence, nor is a competent cap necessary above the granular soil for a temporarily stable cavity to develop.

Introduction

The extreme western area of the Menominee Iron Range in the upper peninsula of Michigan has a long history of the development of subsidence pits in overburden above mine workings. In 1978 the last active mine in the area, the Sherwood, was closed. This closing presented the opportunity to monitor surface and subsurface conditions during the four-year period that it would take for the mine to flood and the groundwater levels to return to their premining states.

A typical mining method in this area was sub-level stoping in the steeply dipping ore body, which produces a grid-like pattern of large open stopes. At the Sherwood the stopes extended from a depth of 503 m (1650 ft) to within one hundred meters of the bedrock surface. Here, the bedrock is overlain by glacial outwash sands and gravels covered by a clayey till cap. The thickness of soil is commonly 30 to 60 m (100 to 200 ft), and the till cap, where present, is 6 to 9 m (20 to 30 ft) thick. The premining water table was well above the bedrock surface. In an effort to minimize pumping costs, most mines employed deep wells to bedrock to intercept groundwater before it seeped into the mine, and mine pumps were used to remove water which did enter. The Sherwood, during operation, pumped about 0.06 m^3/s (1000 gpm) from the mine and 0.32 m^3/s (5000 gpm) from dewatering wells.

Three systems were monitored for this study. They were the hydrologic regime, overburden soil conditions, and rock mass regime. Soil borings, piezometers, stream gauging stations, borehole extensometers, surface subsidence monuments, an acoustic emission array, and various geophysical procedures were utilized. Detailed descriptions and study results are given elsewhere (4, 6).

In addition to the most obvious causes of subsidence such as collapse of the intramine support system or rock mass failure, the physical setting on the Menominee Iron Range is susceptible to subsidence which is caused by groundwater transport (1, 2). With this in mind a simplified laboratory model simulating these conditions was successfully developed and tested.

Case Histories

Unless a void in the overburden can be located before the surface expression becomes obvious, it is difficult to say with certainty what the direct cause of that surface expression is. However indirect evidence can be helpful. It was generally acknowledged by miners and mine superintendents in this district that considerable water entered the mines from the saturated overburden above bedrock and that it was not unusual for this water to be transporting sand.

Prior to 1978, a number of subsidence pits had developed above mined areas on the West Menominee Range. The circular to elliptical-shaped pits varied in size and depth from 4.6 m (15 ft) to more than 122 m (400 ft) in diameter and from 3 m (10 ft) to more than 30 m (100 ft) deep. According to observers many had developed quite rapidly. Some reports verified that subsidence could occur in a matter of a few minutes. In one instance a section of highway disappeared within minutes of the last successful transit. In another incident a

Figure 1:   Subsidence in West Pit on August 31, 1979

Figure 2:   Additional Subsidence in West Pit on October 19, 1979

subsidence pit developed in overburden immediately following a nearby blast. At the time, all of these events were attributed to massive failures of supporting rock in upper stopes.

Three separate subsidence events occurred at the Sherwood Mine after the 1978 cessation of operations and while groundwater was flooding the mine. The first of these events was initially detected by an increase in acoustic emissions. The subsidence occurred over an area of 10 ha (25 acres) and required almost two years to achieve 0.4 m (1.3 ft) of settlement. Borehole extensometer data from the 305 m (1000 ft) and 366 m (1200 ft) levels proved this subsidence to be due to rock movement along a pre-existing shear zone. The other two events did not appear to be directly related to bedrock movement and warrant further description.

On August 31, 1979 subsidence occurred in an area known as the West Pit. The West Pit depression had formed in 1956 due to mine pillar failure and was subsequently partially refilled by inwashed sands, however the typical cap of till was not present. The August 31 event resulted in a circular pit 9 m (30 ft) in diameter and 3 m (10 ft) deep. The pit had a flat bottom and steeply sloping sides ringed with tension fractures. On October 12, acoustic emission levels increased greatly and continued for about 30 hours at which time they ceased. This quiet period lasted for 15 hours, being interrupted by short bursts of emissions until 10:45 P.M. on October 14 when a moderate increase was recorded. At this time renewed surface subsidence enlarged the pit to 30 m (100 ft) in diameter and 18 m (60 ft) deep. Acoustic emission activity increased again beginning at 3:00 A.M. on October 18. Emissions were high for more than 20 hours and then decreased about an order or magnitude until 4:00 P.M. on October 19 when the surface caved again causing the pit to achieve a diameter of 61 m (200 ft) and a depth of 30 m (100 ft). Figures 1 and 2 show the subsidence features after the August 31, 1979 and October 19, 1979 movements, respectively.

During the late evening of April 1, 1981 when the Sherwood was within 30 m (100 ft) of being flooded, an unusual sequence of events took place. During a brief time period that night, three pre-existing subsidence pits were greatly enlarged, and a new subsidence pit was formed. Concurrent with the subsidence was the breaching and disruption of concrete seals and mine shaft caps by surges of mine water under high pressure. Surface erosion channels indicated that large quantities of water were expelled through these openings. The acoustic emission data indicate that subsurface activity began several days earlier on March 26 and that the event occurred at 11:24 P.M. on April 1. This was confirmed by a nearby resident who heard a sound "similar to the crusher starting up" at that time. It is interpreted that the surface subsidence was essentially completed in less than 30 minutes.

On or about July 15, 1983 a subsidence pit suddenly developed in a pasture adjacent to known mining property in the same district. The ground caved-in under the weight of a grazing horse. The resulting hole was 1.2 m (4 ft) by 1.8 m (6 ft) wide at the surface and 12 m (40 ft) to 15 m (50 ft) deep with a diameter of 6 m (20 ft) to 8 m (25 ft) (7). There was no previous history of subsidence at this particular location.

Laboratory Studies

The general conditions necessary for the creation of a subsurface cavity in soil by piping were identified by Allen (1) as 1) groundwater with a hydraulic gradient sufficient to transport sediment and having access to an erodible bed, 2) the presence of an outlet or reservoir to allow the sediment to be transported away, and 3) the presence of a layer of sufficient competency over the erodible bed so that the piping process can form a void.

It appeared that conditions at the Sherwood Mine were such that a void could develop in the overburden without having an immediate collapse of the ground surface. Groundwater had been moving into the mine before shutdown and would continue until the mine was completely flooded. As the sand and gravelly sand just above the bedrock was erodible, the same channelways which let water enter the mine were probably large enough to pass sand-size soil. The mine itself provided a very large reservoir or sink into which the sand could be disposed of, and a cap of till able to resist collapse was present above the sand over much of the site.

In order to better understand and quantify the processes which were anticipated to occur as the Sherwood Mine flooded, a laboratory model was constructed to simulate the conditions at the Sherwood Mine and to observe the creation of an internal void in the soil. The model consisted of 770 kg (1700 lb) of uniform sand in a plexiglass container 1.2 m (4 ft) long by 1.2 m (4 ft) high and 0.6 m (2 ft) deep. In the base of the container at the front wall was a small port which simulated a passageway in the bedrock leading to a mine stope. After placement of the dry sand, the water table was established at a predetermined depth in the sand, the basal port was opened and steady-state seepage was allowed to develop. Tests were run with and without a clay cap above the sand. A complete description of the procedure and results is provided elsewhere (8).

In all model runs an internal cavity developed in the sand mass as granular soil was carried away through the basal port under the action of the flowing groundwater. The cavity developed to equilibrium or breakthrough to the surface by a combination of vertical falls and hydraulic transport through the basal port. The entire process could be visually observed through the forward plexiglass wall of the container. A typical cavity development is shown in Figure 3.

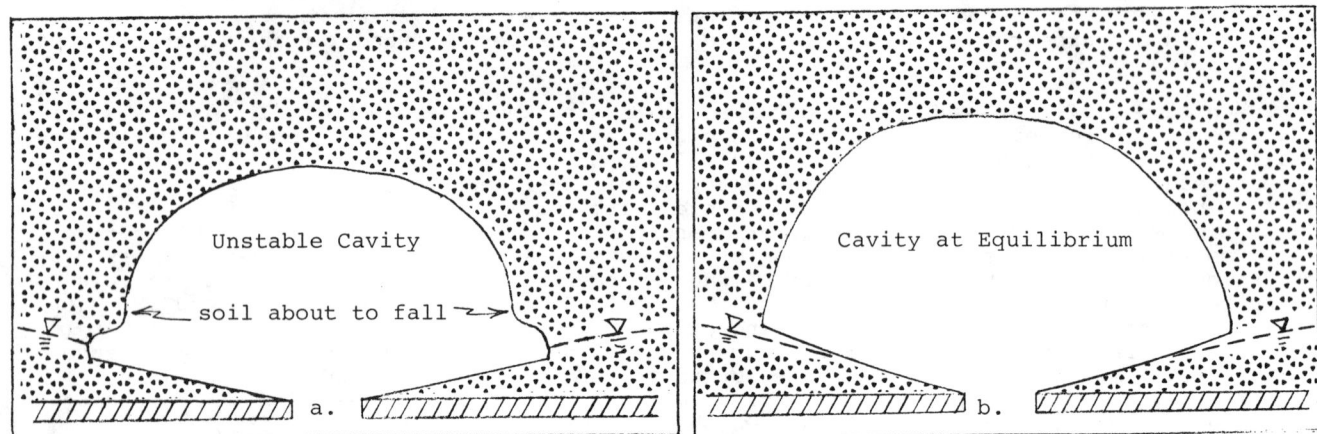

Figure 3: Idealized Representation of Cavity Development

The results of this small-scale testing program indicated the following:

1) A soil-free void can develop in granular soil.

2) In the absence of anisotropy in the soil character or groundwater flow pattern, an approximately hemispherical cavity will develop.

3) A strong cap above the granular soil is not essential to the presence of a soil-free cavity in the granular soil.

4) The dimensions of the cavity are linearly related to the groundwater position outside the zone of drawdown; cavity size and shape are reproducible, and the dimensions can be predicted.

Referring to Figure 3a, the condition here is unstable. Seepage water is entering the cavity above the intersection of the basal slope and the dome. The result is undercutting of the dome, a mass fall, the transport of the fallen roof material by viscous flow and saltation into the mine void, and a repetition of the process. Eventually complete roof collapse occurs or the situation depicted in Figure 3b develops.

In Figure 3b the groundwater's free surface develops entirely along the basal slope and the combination of seepage out of the gentle slope and sheet flow along the slope are insufficient to transport further granular material into the mine void. An unpublished finite element analysis of this situation has been performed by Bedenis (3). He showed that reasonable strength values for the angle of friction and apparent cohesion were sufficient to allow the development of a relatively large, stable hemispherical cavity. For example, soil with an angle of friction of 35° and a cohesion intercept of 4.8 kN/m^2 (1000 psf) would support a hemispherical void with a 5.2 m (16.9 ft) radius within a total overburden thickness of 7.6 m (25 ft). The source of apparent cohesion for this laboratory model was negative pore water pressure. In the field environment natural cementation would be required to provide the necessary cohesion.

The laboratory experiment under discussion was limited to a maximum phreatic surface of 0.13 m (5 in) in the sand, developed a maximum cavity diameter of 0.56 m (22 in), and utilized only one soil. Within these limits the following relation was determined:

$$x = 3.76y - 0.162$$

where x is the cavity's maximum height in meters and y is the groundwater height outside of the cone of depression in meters.

Extrapolating the laboratory results, and assuming that sufficient soil strength exists, yields a stable cavity 3.6 meters high for a background water table 1 meter above the bedrock.

Proposed Subsidence Mechanism

Based on field observations, reports, and the laboratory study it is concluded that a temporarily-stable, soil-free cavity can develop in predominantly granular soil whenever the necessary requirements are met. These requirements are:

1) a channelway and sink for the internal disposal of groundwater and solids (diamond drill holes or fracture zones could satisfy this requirement in underground mines)

2) groundwater table above the elevation of the sink

3) sufficient cementation or other source of cohesion in the granular soil

4) a sufficient thickness of granular soil in relation to the height of groundwater above.

Flowing groundwater carries soil particles into the sink as a more or less hemispherical cavity develops in the soil. If the soil becomes effectively dewatered and the groundwater flow ceases or the cavity grows to such a size that the groundwater can enter the cavity through the gently sloping soil floor, no further piping-induced volume increase will result.

The same result will occur if the sink becomes filled or clogged and groundwater loss is prevented. However the cavity is only temporarily stable and eventually a rising water table, leaching of the cementing agents, an additional boundary load or some other event will occur and the cavity's roof will fail.

It is not necessary that all of the granular material in the overburden be small enough to pass into the sink. Depending on the gradation a considerable volume reduction can occur in gravelly sand. An analysis of representative soil samples from the Sherwood Mine property showed that a volume reduction of 17 to 98% was possible if no soil particles larger than sand-size could be lost from the overburden. It can be imagined that the developed cavity in this case would contain a coarse debris pile centered about the bedrock fracture.

An understanding of the creation of internal soil cavities formed by piping suggests that upon the cessation of mining operations and abandonment of a property the further development of cavities may be mitigated. If a combination of overburden dewatering and in-mine pumping is being used during the active life of that mine, the overburden water table has been artificially depressed and should be expected to return to a higher level when overburden dewatering is discontinued. Consideration should be given to a sequence of removing the mine pumps, filling the mine with water from the overburden dewatering system, and finally allowing the overburden to recharge after the mine void is water filled so that post-operational subsidence events can be minimized.

Acknowledgment

This work was performed under a contract with the U.S. Bureau of Mines, Department of the Interior, and with the cooperation of the Inland Steel Mining Company.

References

1. Allen, A. S., "Geologic Settings of Subsidence," Reviews in Engineering Geology II, Geological Society of America, 1969, pp. 305-342.

2. Allen, A. S., personal communication, April 1975.

3. Bedenis, T. H., An Analytical Study of Localized Mine Subsidence, MSCE Thesis, Michigan Technological University, 1982, 116 pp.

4. Frantti, G. E., and Johnson, A. M., "Acoustic Emission Related to Mine Subsidence," Proceedings of the Third Conference on Acoustic Emissions/Microseismic Activity in Geologic Structures and Materials, Trans Tech Publications, 1984, pp. 321-337.

5. Johnson, A. M., Hodek, R. J., and Frantti, G. E., "Piping Induced Subsidence Over an Underground Mine," Proceedings of the Workshop on Surface Subsidence Due to Underground Mining, West Virginia University, 1981, pp. 268-273.

6. Johnson, A. M., Hodek, R. J., and Frantti, G. E., A Case Study of the Sherwood Mine Closing - Final Report, Institute of Mineral Research, Michigan Technological University, 1984.

7. Korach, P., personal communication, July 1983.

8. Sandri, D. B., An Experimental Study of the Creation of Voids Within a Cohesionless Soil Medium by the Piping Mechanism, MSCE Thesis, Michigan Technological University, 1981, 117 pp.

Sinkholes at Tarbela Dam project

IZHARUL HAQ & CH. ALTAF-UR-REHMAN *WAPDA, Lahore, Pakistan*

ABSTRACT

Tarbela Dam is located on a 125 to 215 m deep permeable alluvium layer containing extensive open work gravels. In order to control seepage and hydraulic gradients, an impermeable blanket (an extension of the dam core) was laid on the upstream side of the dam. During the first drawdown, the formation of a number of sinkholes was observed in the blanket. The maximum sinkhole diameter observed was 10 m and the maximum sinkhole depth was 2.5 m. The migration of sand below the blanket into the open work gravels was considered responsible for the majority of the sinkholes. These sands are thought to originate from river gravel deposits or intermittent sand lenses. The sinkholes were filled and covered by a thickness of well-graded blanket material equal to the diameter of the sinkhole and extending to a distance of four times the sinkhole diameter. The blanket was then monitored by a side scanning sonar survey. During subsequent reservoir filling, some sinkholes reappeared and were filled by barge dumping of specially prepared blanket material. This proved to be a successful method of dealing with the sinkholes and is used in dealing with the few sinkholes that appear every year.

Introduction

Tarbela dam, 150 m high and consisting of 180 million cu.m of material is one of the largest earth and rock fill dams in the world. The construction of the dam started in 1968 and water was first stored, to an elevation of 1463 m in 1974. Since the dam is founded on deep alluvium, an upstream impermeable blanket was provided. During August, 1974 the failure of the upstream part of diversion tunnel No. 2 occured and the reservoir was depleted. This revealed that a number of sinkhole had formed in the blanket. This paper describes the formation of these sinkholes, their causes, treatment, and subsequent monitoring.

Design Concept of Tarbela Dam

The Tarbela dam is located on a 125 to 215 m thick layer of Indus river alluvium consisting of sands, gravels, cobbles, and boulders which overlie consolidated rock. The installation of a vertical cut-off was considered not feasible due to the large thickness of the alluvium and high construction cost. A partial cut-off would not be effective because

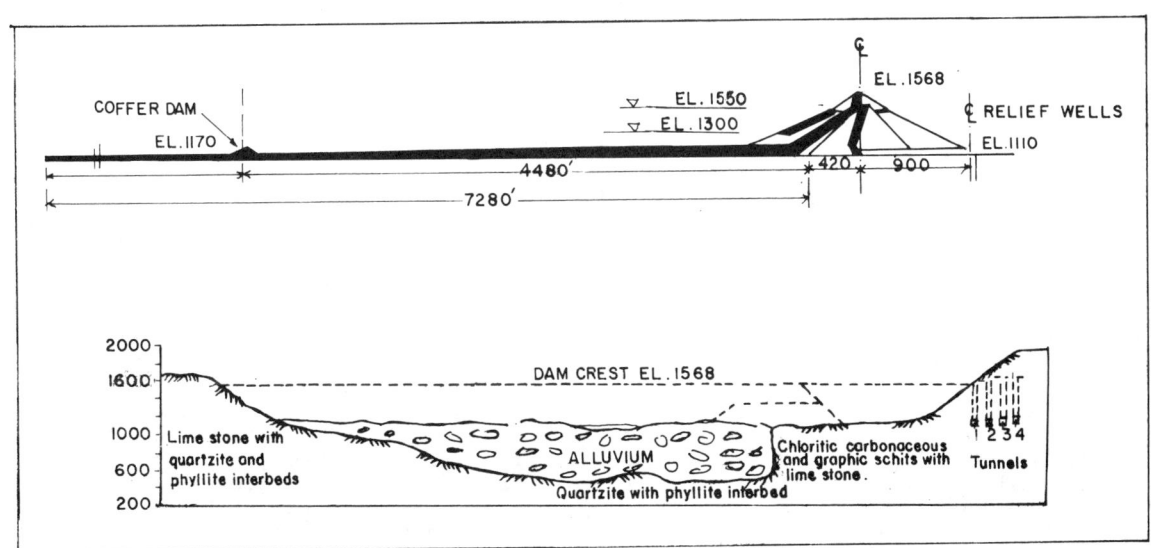

Fig. 1. Typical cross-section and long-section of Tarbela Dam.

the permeability does not decrease markedly with depth. Therefore, in order to control the foundation seepage and gradients, an impermeable blanket continuation of the dam core was laid on the upstream side. The initial length of the blanket was 1700 m and was later extended to 2260 m. The initial thickness near the toe of the dam was 8 m and was increased to 12.5 m which then tapers down to 1.5 m at the upstream end. The typical cross section and longitudinal section of the dam is shown in Figure 1.

The maximum head created by the Tarbela reservoir is 136 m and the corresponding ratio of length of the blanket to head is about 17:1. The blanket consists of the same material as that of the core and is well graded finer than 300 mm with fines (< No. 200 sieve) ranging from 20% to 50%. The blanket was so designed as to be well graded, resistant to erosion/migration, and self-healing. In order to control the pressure of seepage water and to avoid piping conditions relief wells were constructed at the downstream toe of the dam. These wells were spaced 15 m c/c and are 38 to 76 m deep so as to cross the known open work gravels.

First Observation of Sinkholes

The first impounding started in 1974 and water was stored up to an elevation of 1463m. In August, 1974 a mishap occurred in the upstream part of diversion tunnel No.2 which depleted the reservoir. This exposed the blanket and revealed a number of sinkholes and cracks. The sinkholes varied at the surface from a few cms. in size to up to 10 m. The

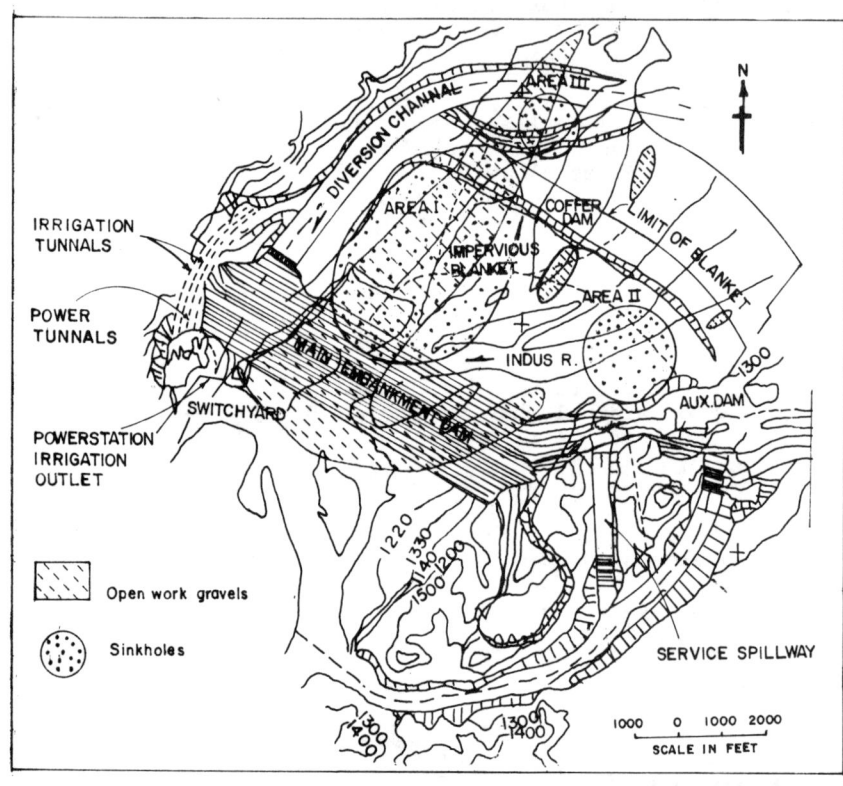

Fig. 2. Location of sinkholes and open work gravels.

sinkholes were concentrated mainly in three areas (Figure 2). The largest of these areas was on the right side of the main blanket extension between the main dam stations 15 to 68 and between the main dam and auxiliary diversion channel. The second area was on the left side between the left bank tunnel, the intake structure, and the coffer dam. The third area which was relatively smaller, was in the blanket extension between the main diversion channel and the auxiliary diversion channel. The sinkholes were of circular shape with almost vertical walls although there were several sinkholes which were the truncated cone type having the diameter decrease with depth and occurred mainly in the open work gravels. Investigations carried out revealed that none of the sinkholes penetrated to the full thickness of the blanket although some of the vertical cracks were found to go down to the foundation material. In the extended blanket between the diversion channel and the auxiliary channel, a number of sinkholes and subsidence features were deep enough to penetrate through the small thickness of the blanket in this area. There were over 200 sinkholes in the area between the main dam and the auxiliary channel. The largest sinkhole being 6 m in diameter and the deepest one being 4 m deep. In the left bank, there were over 50 sinkholes of 1.75 m in depth. There were less cracks found in this area, however substantial migration of fines from the blanket into the foundation materials was noticed. In the blanket extension at the right bank, the number of sinkholes was over 50 and a number of these holes penetrated through the full depth of the blanket with the maximum diameter being 6 m and the maximum depth being 2.5 m.

Causes of Sinkholes

Causes of the development of the sinkholes were discussed with Dr. Casagrande and other members of the panel of experts. The following causes were considered:

- i) Differential settlement and shear fracture due to the load of the reservoir.
- ii) Unfilled bore holes or open wells.
- iii) Migration of sand under the blanket.
- iv) Due to instrument cables.

i) Differential Settlement: Being that the alluvium is over 200 m thick in the valley and there is an uneven rock profile underlying the alluvium, the possibility of differential settlement exists. It is quite likely that some of the cracks formed due to this settlement. A few sinkholes formed in the cracks.

ii) Unfilled bore holes: Unfilled bore holes and improperly filled open wells or pits in the valley could have in some cases encouraged erosion in a vertical direction. This explanation could not be applied to all the sinkholes because these are conspiciously oriented along the cracks and fissures. However, some of the large sinkholes on the left bank may be due to open wells which existed in this area.

iii) Migration under the blanket: According to Dr. Casagrande, the most likely cause of the sinkholes requires that sand lying below the blanket migrate into open work gravels. These sands originate from river gravel deposits or intermittent sand lenses. Most of the sinkholes were located on the known position of the open work gravels, however some sinkholes were located over other positions. Figure 3 shows schematically the migration of sand from the

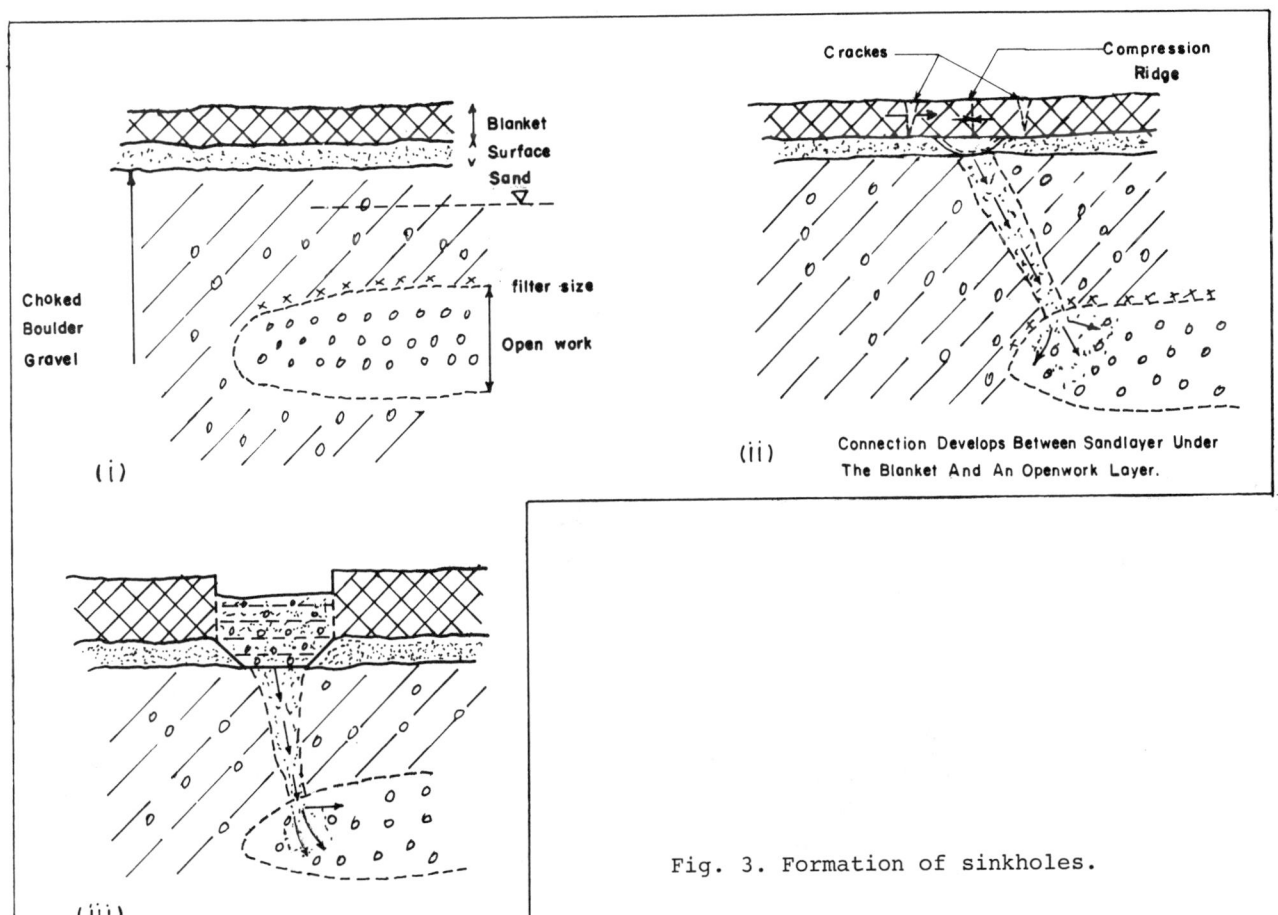

Fig. 3. Formation of sinkholes.

top sand layer and intermittent sand lense into the open work gravels. If a sinkhole starts to develope close to a piezometer, then the piezometer definitely responds in the form of recording higher pressure. However, a decrease in pressure recorded by the instrument does not mean an improvement in the area but that drainage has occurred at this location due to movement of sand. Areas showing a rise in piezometric levels have generally lead to detection of sinkholes or are due to enlargement of prexisting sinkholes.

iv) Instrument cables: In a few cases sinkholes were formed due to settlement of the fill around the instrument cables. The location of some of the sinkholes was positioned exactly above the vertical riser of the instrumentation cables. The risers installed in the embankment were 6 m in diameter and were shaped like a Christmas tree. These risers had been compacted by hand tools instead of heavy machinery. Due to the under compaction of this material it is liable to consolidation resulting in settlement and also to leaching of fine particles if migration towards the downstream was possible. Both these phenomena were observed.

Remedial Measures

In 1974, it was decided that the sinkholes be filled and then overlain by a protective layer of well-graded blanket material having a thickness equal to the diameter of the sinkhole and an extension of at least four times the sinkhole's diameter. The blanket thickness was increased by 1.5 m to 3 m in the areas underlain by the open work gravels. This additional material consisted of uncompacted filter material grading from silt to cobble size and was laid over the sinkholes and cracks. An additional line of relief wells was installed at the downstream toe of the dam. After the dam was repaired in 1975, it was steadily filled. Last year some sinkholes reformed in vulnerable areas and were treated by dumping through the use of special barges. A mix consisting of silt to cobble size particles with some percentage of bentonite was prepared, compacted at a moisture content level above optimum and dumped from the barge at the location of the sinkholes as determined by side scanning sonar. The sinkholes near the toe and in the main blanket area were treated on priority. The sinkholes upstream of the blanket were treated by dumping river sand over them. After a few years of sinkhole formation and treatment, the foundation stabilized and few sinkholes were formed. Marked effects were seen due to sedimentation.

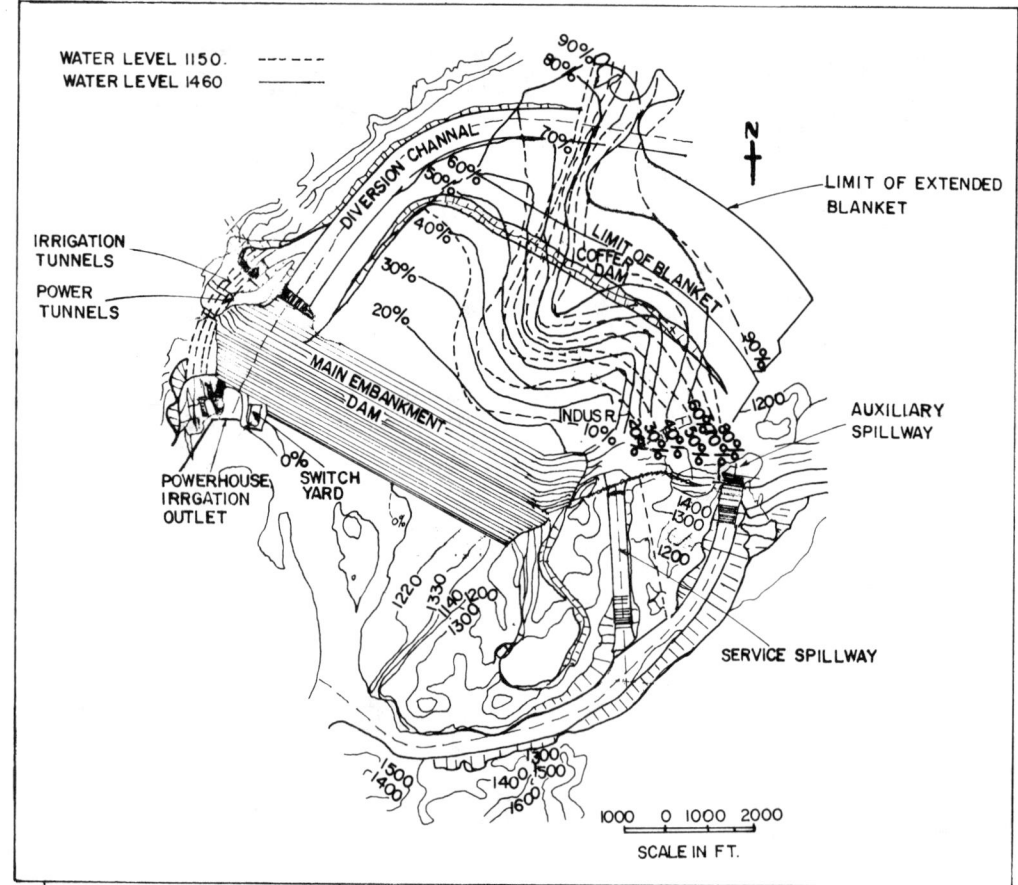

Fig. 4. Equipotential contours under the blanket.

Monitoring

For a large dam like Tarbela, it is extremely important to monitor the behavior of the dam's various structures in order to ensure proper functioning and safety. A large number of piezometers were installed under the blanket and dam foundation in order to observe the pore pressures and potentials under the blanket. The equipotential contours under the blanket for various levels of the reservoir are depicted in figure 4. During the dam's filling in 1974, the shift of the equipotential surface under the blanket for different reservoir levels was quite significant indicating that adjustments were taking place within the foundations and the blanket.

Side scanning sonar surveys were performed in 1975, during the second filling of the reservoir and during each filling thereafter. Subsequent surveys in 1975 indicated that the number of sinkholes increased with reservoir level. The initial sonar survey acted as a base for direct comparison with later surveys. Up to 1979, a total of 489 sinkholes had been detected within the main blanket. Sinkhole depths were determined with Ocean Research equipment. The treatment by barge dumping and sonar survey continued and the number of sinkholes decreased due to stabilization of the blanket. Eleven sinkholes formed in 1980, 2 in 1981, none in 1982, and 1 in 1983 and 1984. The sinkholes formed in 1983 and 1984 were above an instrument lead on the upstream slope. The seepage at the downstream toe in 1975 was 10 cms (elevation 1530 m) and has since decreased to 2 cms in 1984. The piezometric pressures have also decreased considerably. There is a possibility that with the sedimentation there would be a drop in foundation pressures leading to high gradients across the blankets. This may lead to new sinkhole formation but the system of monitoring and treatment will deal with these.

Conclusions

The sinkholes found at Tarbela dam were in most cases due to the migration of sand lying below the blanket into open work gravels. The treatment of the sinkholes, filling them with well-graded blanket material, proved successful. The side scanning sonar survey proved to be an efficient tool in the location of the sinkholes. After a few years of sinkhole formation and treatment, the foundation stabilized. The performance of the blanket and main embankment is now satisfactory. Detection and treatment in time will prevent further development.

References

TAMS Tarbela Dam Project Design Report.

Minutes of meetings of special consultants on Tarbela Dam Project (1974).

TAMS Yearly Progress and Monitoring Reports.

DMO (1979) Report No. 41. History of Sinkhole Development in Main Embankment Dam and Blanket.

Self-healing sinkholes in an earth dam foundation

NEIL H.WADE & LLOYD R.COURAGE *Monenco Consultants Ltd, Calgary, Alberta, Canada*

ABSTRACT

This paper describes and attempts to explain the phenomena of "self-healing" sinkholes which were discovered at the upstream toe of a 21 m high earthfill dam founded on 60 m of glacio-fluvio deposits. Constructed in 1951, the dam forms part of a hydroelectric development located in the continental ranges of the Rocky Mountains in Western Alberta. Sinkholes, which occurred on the lower reaches of the right abutment and in the intake channel for the buried penstock at the base of the dam fill, were first observed in 1974 when the reservoir was drawn down in preparation for downstream remedial work.

Additional sinkholes, observed in subsequent years during low reservoir levels, varied in size from small depressions characterized by an accumulation of twigs deposited as in a vortex to one measuring 5 m in diameter by 1.5 m deep. Although the sinkholes were backfilled upon discovery with impervious till, new sinkholes were later discovered, some active but the majority dormant. Sinkhole formation in the right abutment was attributed to heave of the intermixed glacial, alluvial and talus deposits due to artesian pressures generated by spring run-off at times when the reservoir was drawn down. Subsequent loading of these loosened deposits by the rising reservoir and concurrent artesian pressure dissipation resulted in migration of fines in the mixed grained formation sufficient to self heal most of the sinkholes shortly after they developed.

Introduction

THREE SISTERS Project forms the upstream part of the Spray Hydroelectric Power Development located in the Rocky Mountains of Western Alberta. The project consists of the Three Sisters Dam, a 3 MW remotely controlled Powerhouse, the Canyon Dam and a Spillway structure. Three Sisters Dam and Canyon Dam are constructed in a north-south trending valley and contain Spray Lakes Reservoir, a 20 km (12 mi) long lake about 12 km (7 mi) south of the town of Canmore. Completed in 1950, Three Sisters Dam is about 670 m (2200 ft) in length by 21 m (70 ft) high and is founded on interbedded deposits of alluvium, open gravels, glacial till and talus with a combined thickness over 60 m (200 ft). The embankment was constructed as a homogeneous dam with an upstream impervious zone to provide the main water barrier. A 3 m (10 foot) diameter steel penstock extends through the dam foundation from a stoplog intake structure near the upstream toe to the powerhouse located at the downstream toe. The general arrangement is shown on Figure 1.

Because of excessive seepage and numerous small boils discovered around the powerhouse and along the downstream toe upon first filling of the reservoir to full supply level (FSL), a row of steel sheet piling was installed close to and around the periphery of the powerhouse. The piling extends to about 6 m (20 ft) below the raft foundation to reduce the risk of erosion of fines in the foundation soils. The annular space between the powerhouse wall and the sheet piling was capped with concrete and grout pipes installed at 3 m (10 foot) centres to allow grout injection should the powerhouse experience differential settlements. In addition, the lower portion of the downstream slope of the dam was flattened to about 4 horizontal to 1 vertical (4H:IV), a 0.6 m (2 ft) thick blanket of gravel ballast was placed on the flat ground downstream of the dam, and an inverted filter blanket, graded from sand-size to boulders, was constructed in the tailrace channel. Upon completion of this work, it was concluded that, although flow in the tailrace channel from seepage and natural drainage of the mountain slopes was in the order of 2.8 m^3/sec. (100 cfs) measured 180 m (600 ft) downstream of the powerhouse, the dam was stable and would remain secure over the long term[3]. However, during a site inspection in the Spring of 1981, a sinkhole was discovered on the downstream slope behind the powerhouse. This occurrence, along with recurring sinkhole activity experienced in the intake channel during previous years, precipitated additional remedial measures in an attempt to reduce foundation seepage and minimize sinkhole activity.

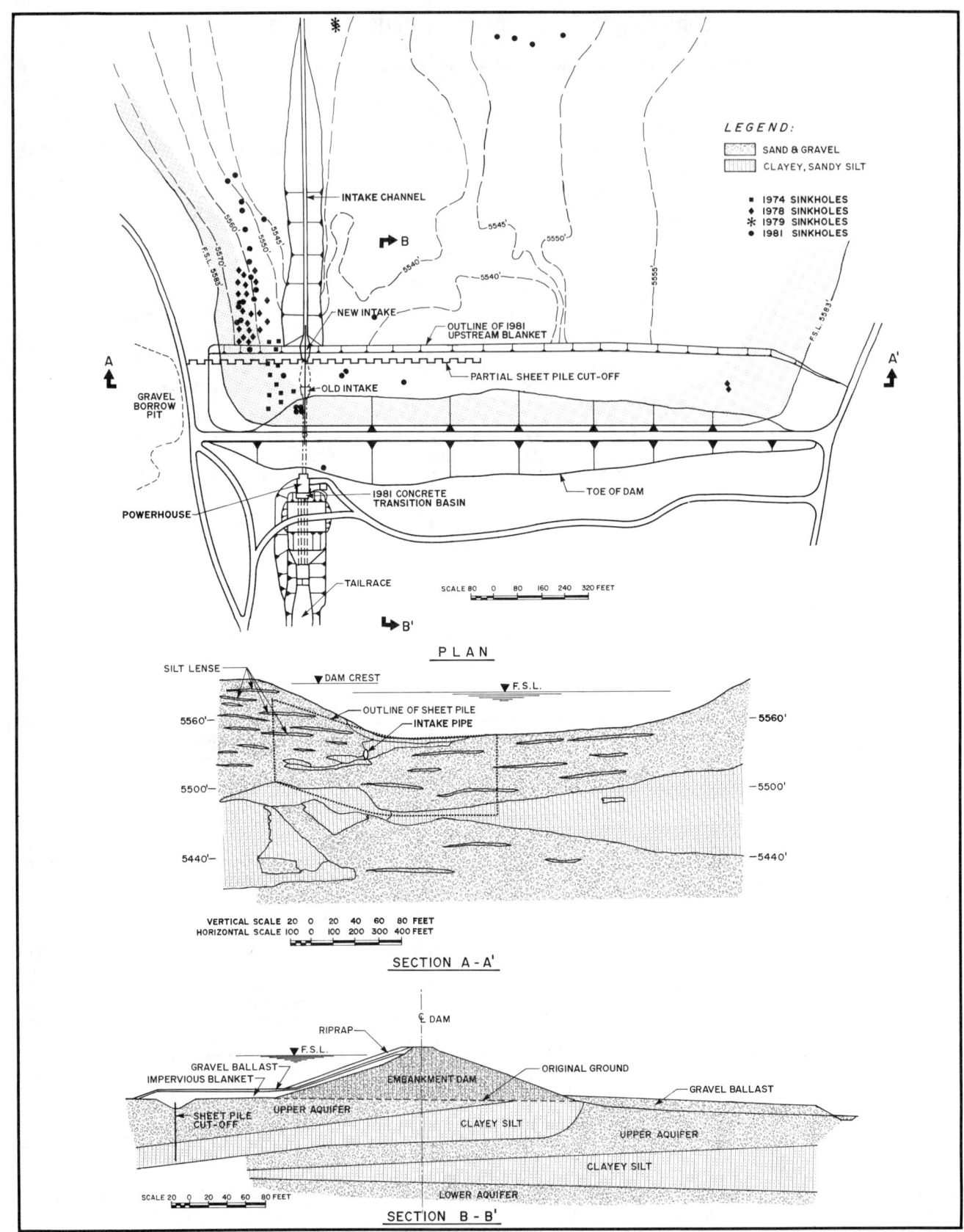

FIGURE 1. THREE SISTERS DAM - PLAN & SECTIONS

Site Geology

The Spray Lakes reservoir and associated structures lie within the Rocky Mountains physiographic region - an area of high relief displaying a marked correlation between geological structure and local topography. The major valleys and mountains trend parallel or sub-parallel from northwest to southeast corresponding to the strike of the strata. The local relief between valley bottom and mountain summits varies from 600 m (2000 ft) to 900 m (3000 ft). The valley at the dam site exhibits a U-shaped cross-section typical of glaciated highlands. The mountains to the northeast and southwest rise steeply from the relatively flat valley floor. The valley had been infilled by colluvial talus materials, glaciofluvial outwash and glacial moraine. The surface is dissected by a network of small intermittent streams. The deposits on which the dam is founded consist in general of discontinuous horizons of sand and gravels interbedded with talus fan materials, silts and clayey silt of variable extent.

Although bedrock does not outcrop in the valley floor, available geological maps indicate bedrock to be comprised of southwesterly dipping limestones and dolomites with minor shales, siltstones and cherts of the Mississippian Rundle Group. A southwesterly dipping thrust fault occurs about 180 m (600 ft) southwest of the Three Sisters Dam. This fault has a near-vertical surface expression and does not directly affect the structure.

Foundation Materials

The native soils comprising the embankment foundation are illustrated on Figure 1 and consist of the following:

(i) <u>Sand and Gravel Strata</u> - The widely graded sand and gravel formations occur generally at two different levels in the geologic profile - an upper layer forming the top third of the deposit and a lower stratum comprising the bottom half. The two layers are separated by a sandy clayey silt deposit varying in thickness up to about 18 m (60 ft). The upper sand and gravel sequence is slightly less dense and contains more fine sand interbeds than the very dense lower formation. Both the upper and lower sand and gravel strata, especially at the right abutment, contain zones of open-work and poorly graded gravelly talus apparently derived from the mountain fan. The fines content in the upper strata varies from zero to a high of 70%, the latter representing interbedded silts. The lower sand and gravel formation has fines contents up to about 35%. In-situ permeability tests carried out in drill holes indicated permeability values ranging from 10^{-2} to 10^{-4} cm/sec for the upper strata and 10^{-3} to 10^{-4} cm/sec for the lower. The results of dye injection tests carried out in the more pervious zones encountered in drill holes indicated apparent seepage velocities up to 15 cm/sec in the upper sand and gravel of the right abutment. Drillholes in which the dye test was run were located as far away as 210 m (690 ft) from the tailrace where the dye exited.

(ii) <u>Sandy Clayey Silt Strata</u> - These materials vary from soft to very stiff consistency and have plasticity index values from 0 to about 7. The liquid limit values were generally below 30% giving a classification of CL-ML to SC-SM. Fines content varied from a low of 45% to a maximum of 98%. At some locations the clayey silt strata is cemented with calcite, forming a sporatic, sometimes continuous over a large area, layer of hardpan. This cemented layer forms a significant interbed in the upper sand and gravel strata.

Embankment Construction

With the exception of the intake area, the dam was built as a homogeneous section with care being taken to place the finer portions of the fill material in the upstream half and the coarser in the downstream. The portion of the dam surrounding the intake structure for the buried penstock was constructed with an impervious blanket on the upstream face.

It was intended during the final design stage to construct an impervious blanket upstream from the toe to cover the gravel floor of the valley for a distance ranging from 60 m (200 ft) to 300 m (1000 ft). However, further examination of the area indicated impervious deposits below the valley bottom especially in the area of the intake canal excavation and, because the water in the reservoir was rapidly approaching the toe of the dam, it was decided that the extent of the upstream blanket should be considerably reduced. Accordingly no blanket was built along the western third of the upstream toe - it extended only about 60 m (200 ft) upstream of the toe over the remaining two-thirds of the embankment length.

Sinkhole Occurrence

Sinkholes in the area of the upstream blanket on the right abutment were first discovered in the spring of 1974 when the reservoir was drawn down in preparation for rehabilitation work in the canal downstream of the dam[1]. This reportedly was the first time the reservoir was lowered sufficiently to expose the blanket since the plant was commissioned in 1951. It is probable therefore that these sinkholes had been in existence long before they were first discovered.

The dozen or so sinkholes observed in 1974 were contained in a band extending from the right bank crest of the intake channel to a line 30 m (100 ft) upslope and from the toe of the dam for a distance of 90 m (300 ft) upstream (see Figure 1). Most of the sinkholes were in the form of shallow depressions up to 38 cm (15 in) in diameter and 10 cm (4 in) deep. They were distinguishable by an accumulation of twigs which appeared to have been deposited in a vortex. Only two or three were open holes extending for several centimetres into the slope. With the ready availability of construction equipment which had been mobilized for the canal work downstream, the affected area was stripped to a depth of 1 m (3 ft) and backfilled with compacted glacial till. The impervious till blanket was in turn covered with the granular stripped material to protect it against erosion.

Additional sinkholes discovered in late May of 1978 were similar to those found in 1974 except that they were much more numerous. Because the reservoir was being refilled at the time of the discovery and was within 9 m (30 ft) of FSL, most of the area which had been repaired in 1974 was inundated and could not be examined. The new sinkholes appeared to be contained in a 30 m (100 ft) wide strip, immediately upslope (east) and upstream of the area which had been blanketed in 1974. In view of the rising reservoir, each sinkhole was excavated to a depth of 1 m (3 ft) over an area of 2 m (6.5 ft) square and backfilled with impervious till. The till was then mounded over with the granular materials from the excavations to protect the till from erosion and to serve as identification markers should the sinkholes reappear.

During the inspection of 1978, there was no evidence of active sinkholes at the left (west) end of the upstream toe. Two depressions that were found appeared to have plugged themselves and were dormant.

In May 1979, an eight drum raft powered by an outboard motor was used as an observation platform and the reservoir bottom was viewed through a glass-bottom box fitted into the floor of the raft. Two areas considered to be sinkholes were found about 365 m (1200 ft) upstream of the intake structure. The larger was 5 m (16 ft) in diameter by 1.5 m (5 ft) deep with near-vertical walls. The bottom of the hole was free of organic material except for some logs. A pike pole was used to stir up fines but no ingress of water was observed. The smaller crater was located 6 m (20 ft) downstream of the large depression. Both craters, which were under about 2.5 m (8 ft) of water, had a new look about them as if they were freshly eroded or disturbed.

In the spring of 1981, thirty-two sinkholes were observed along the right abutment, on the upstream slope and in the reservoir floor. These sinkholes varied in size up to 1 m (3 ft) in diameter and 30 cm (1 ft) deep. At the right abutment, the sinkholes were concentrated in an area from 90 m (300 ft) to 150 m (500 ft) upstream of the intake structure and at an elevation approximately 5 m (16 ft) below reservoir FSL. After removal of rip rap on the upstream slope, numerous small sinkholes, more aptly termed pervious zones, were observed in the bottom third of the slope along a 180 m (600 ft) stretch extending west from the penstock alignment. Field permeability tests carried out in shallow test pits dug at a number of these zones indicated that the pervious materials penetrated into the homogeneous dam fill for considerable distance judging by constant water takes over a period of several days. The sinkholes observed in the reservoir floor were located generally in the impervious blanket and in areas where previous sinkholes had been backfilled.

In addition to the upstream sinkholes, a conical depression 1.2 m (4 ft) in diameter and 1 m (3 ft) deep was observed during the 1981 inspection on the downstream slope to the left of the powerhouse. It was later learned from conversations with plant operating staff that depressions had occurred in the same location over the previous two years and these had been backfilled during routine site maintenance work. A test pit excavated at the location encountered loose sand and gravel to a depth of 2.4 m (8 ft). The pit continued through medium dense sand to a depth of 5.5 m (18 ft) and near the bottom a hole about 13 cm (5 in) by 4 cm (1.6 in) led off in the upstream direction. Subsequent filling of the test pit with water and dry lime as a marker indicated that the hole or otherwise pervious material was in hydraulic connection with the tailrace since lime-water was observed entering the tailrace from the left side of the powerhouse.

Mechanisms of Sinkhole Formation

The mechanism causing sinkholes to develop apparently differs in the various areas depending on soil conditions as discussed below:

a) Right Abutment

Prior to the annual snowmelt, which generally occurs in late spring and early summer, the reservoir level is drawn down to allow runoff water impoundment for subsequent power generation later in the year. It is thus at low reservoir levels when peak runoff conditions exist and when excessive artesian pressures build-up in shallow pervious layers within the right abutment below FSL. These artesian pressures tend to heave and weaken any overlying impervious layers which, during and after impoundment, allows ingress of reservoir water and sinkholes develop. The relatively small size of most of the sinkholes, however, indicates that once initiated migration of fines occurs for only a limited period and is arrested within a relatively short distance. The sands and gravels which sandwich the thin silt layers appear to collapse and heal the rupture zone.

This mechanism was more clearly demonstrated in the Spray Canal which carries outflow from the Three Sisters Plant to Spray Powerhouse 5km (3mi) downstream. The canal, which runs along lower mountain slopes at the side of the valley, was formed by excavating into the slope and constructing an earth dyke on the downslope side of the excavation. The bottom of the excavation and water side of the dyke was blanketed with impervious till in those areas where the canal traversed pervious fan and talus deposits. At a number of locations after construction was completed, the impervious liner was ruptured during spring runoff by internal piping due to high ground water flows or by bottom heave from excessive artesian pressures.

(b) Dam Fill

Pervious zones within the dam fill probably existed since the time of construction and resulted from segregation of fill materials during placement. There was no evidence of excessive settlement of the dam crest or upstream slope to suggest significant loss of material in the upstream portion of the dam, although field pumping tests determined these zones to be permeable.

The sinkhole which occurred on the downstream slope was in all probability not activated by groundwater from the alluvial fan but by seepage from the reservoir, possibly through the foundation and along the buried penstock as extensive leakage has been observed exiting from both sides of the powerhouse since construction. During repairs carried out in 1981 when the dam was completely unwatered by upstream and downstream cofferdams, leakage past the powerhouse into the tailrace was estimated from pump capacity to be 1.1 m^3/s (40 cfs). At FSL, leakage flows measured in the tailrace 975 m (3200 ft) downstream of the powerhouse are as much as 4 m^3/s (140 cfs). Leakage through pervious zones within the dam fill could also be a contributing factor to formation of this sinkhole as a concentration of pervious zones was observed on the upstream face of the dam directly upstream of the sinkhole. Nevertheless, the sinkhole which occurred took about 30 years to manifest itself. Particle migration was obviously a slow process and, considering the large volumes of seepage flows, the presence of fines would be undetectable in the tailrace. It is concluded therefore that this particular sinkhole was caused by the slow loss of fines associated with longterm leakage conditions through the foundation.

(c) Reservoir Floor

The occurrence of isolated, generally small sinkholes in the reservoir floor within about 150 m (500 ft) upstream of the toe of the dam are apparently not related to mountain run-off. Relief of artesian pressure within the top 6 m (20 ft) of the floor deposits is provided by the intake canal which acts as an interceptor ditch between the talus fan materials and the glacio-fluvial sands and gravels in the reservoir floor.

These sinkholes are attributed to percolation of reservoir water and associated particle migration into localized zones of gap graded gravels within the floor deposits. Because of the mixed-grained nature of the foundation materials, the sinkholes develop slowly and tend to self heal with the ingress of sand and gravel particles into voids along the seepage path.

Remedial Measures

The development, frequency and locations of sinkholes gave rise to concern about the integrity of the dam and resulted in major modifications being implemented to the structure in the autumn of 1981 and summer of 1982. The remedial work, shown on Figure 1, consisted of driving a partial sheetpile cut-off, placing a new upstream impervious blanket and installing a heavily reinforced concrete transition section, backfilled with coarse gravel, downstream of the powerhouse. The sheetpile cut-off was located about 25 m (80 ft) upstream of the original intake structure and extended 300 m (1000 ft) from the right abutment along the upstream toe to about the midpoint of the embankment length. The sheet piling was driven to a maximum depth of 21 m (70 ft), approximately 1/3 of the depth of the foundation alluvium, and terminated in the silty clay stratum. In some areas, however, pile refusal occurred in the dense gravel and talus overlying the silty clay layer. The new impervious blanket was tied to the cut-off wall and the penstock was extended upstream to a new intake structure located over the cut-off wall.

Since construction of the remedial works, seepage flows in the tailrace at the measuring station 180 m (600 ft) downstream of the powerhouse were reduced by about 50% to 1.5 m^3/s (53 cfs) with the reservoir at FSL. However, at the measuring station 975 m (3200 ft) downstream of the powerhouse, the flow in the tailrace decreased only from 4.0 m^3/s (140 cfs) to 3.3 m^3/s (116 cfs) as a result of the remedial work[4]. This small reduction is attributed, in part, to increased inflows from mountain slope run-off due to a larger catchment area. It is also possible that substantial seepage volumes enter the tailrace channel some distance downstream of the powerhouse from the deep aquifer not penetrated by the cut-off wall and from around the ends of the cut-off in the upper aquifer.

The average piezometer head reduction in the upper aquifer after installation of the sheet pile cut-off and impervious blanket, as determined from 19 Casagrande-type standpipe piezometers collared at various locations downstream of the cut-off, was about 0.9 m (3 ft) with head drops ranging from negligible to 2.4 m (7.9 ft). However, two other piezometers located on the right abutment showed increased piezometric heads of 1 m (3.3 ft) and 5.5 m (18 ft) respectively, which indicates increased seepage flows around the right end of the cut-off or which may be due to a backwater effect caused by installation of the new concrete transition basin and associated backfill immediately downstream of the powerhouse. A smaller average piezometric head reduction (0.5 m) was observed in the lower aquifer, although one piezometer embedded deep in the right abutment showed a 3 m (10 ft) increase in head subsequent to cut-off installation.

Conclusions

It was concluded that the remedial measures had been moderately effective in reducing seepage flows through the alluvium foundation, especially in the upper aquifer, and that the effectiveness of the cut-off in reducing particle migration through the foundation should increase with time as a result of siltation upstream of the sheet piling. In the three years following completion of the remedial work, there has been a significant reduction in the formation of sinkholes upstream of the dam. It was recommended to the Owner that monitoring of the piezometers and seepage flow be continued on a routine basis, that a site inspection be carried out monthly to check for signs of distress, and that an updated safety evaluation report for this dam be prepared annually[2].

Acknowledgements

The authors would like to thank TransAlta Utilities Corporation, Owner of the project, for permission to publish this paper.

References

1. Chrumka, S.J. - "Spray Hydroelectric Plant Inspection Report", Inspection Appendix. Section D, 1978.

2. "Three Sisters Development Dam Safety Evaluation Report", prepared for TransAlta Utilities Corporation by Monenco Consultants Ltd., May 1982.

3. Eckenfelder, G.W. - "Spray Hydroelectric Power Development", The Engineering Journal, Vol. 35, No. 2, 1952.

4. "Report on Three Sisters Dam Seepage Monitoring", prepared for TransAlta Utilities Corporation by Monenco Limited, July, 1983.

5. "Dam Safety Evaluation for Spray Development", prepared for TransAlta Utilities Corporation by Monenco Consultants Ltd., September 1983.

Sinkhole problem related to dam engineering

ROBERT C. LO *Klohn Leonoff Ltd, Richmond, British Columbia, Canada*

ABSTRACT

A potential for sinkhole development within a dam or its foundation and/or reservoir base, may pose a serious threat to the safety and integrity of the dam and the reservoir it retains. The formation of sinkholes in a dam-reservoir environment can be attributed to one or more of the following factors: (1) presence of natural and/or man-made surficial and sub-surface voids, (2) presence of incompatible gradations at boundaries of different materials, (3) presence of cracks, (4) presence of soluble, dispersive or collapsible materials, and (5) increase of hydraulic gradients across critical zones associated with the impounding reservoir.

The first four factors set the stage for the action of the fifth, the impounding reservoir water. Thus, preventive and remedial measures are closely related to the interplay of the above outlined factors. These measures include: avoidance of difficult sites; filling of voids with concrete or grout; use of filters to prevent migration of materials; prevention of cracks and mitigation of their effects; reduction of hydraulic gradients across critical zones using positive cutoffs or upstream impervious blankets; use of drainage to allow safe exit of seepage; and use of instrumentation to monitor pore pressures and seepage rates. Representative case histories of sinkhole problems involving earth and rockfill, water-retention dams and tailings dams are discussed for illustrative purpose.

Introduction

The formation of sinkholes is closely related to the presence and/or creation of voids under the ground surface and the subsequent collapse of the surficial materials above, by a stoping process, into the underground voids. The existing voids are usually associated with particular rock types such as volcanic rocks, soluble limestones and dolomitic limestones; or open joints in rocks and in folded and faulted zones; or mined-out areas in underground mining. Upon reservoir impoundment, new voids can be initiated as tension or shear cracks, or can be created by such seepage related processes as: hydraulic fracturing; solution; dispersion; and internal migration of materials. Regardless of their origins, if not controlled, the continual growth of underground voids under seepage condition will worsen the leakage problem. If the growth of the voids tends to reduce the seepage path from the reservoir to the downstream area, the process can accelerate and eventually lead to dam failure by piping. Therefore, the presence and development of sinkholes within the dam, or its foundation, or the reservoir area to be inundated, pose a serious threat to the integrity and safety of the dam as well as to the watertightness of the reservoir. The catastrophic impact of a dam failure and the ensuing sudden release of reservoir water on loss of life and property damage downstream underscores the importance of the sinkhole problem in dam engineering. Even if failure does not occur, substantial leakage from a reservoir may curtail significantly and sometimes eliminate completely, the use and benefit of the reservoir. In the case of leakage from a tailings pond, potential pollution downstream is an additional concern. Unfortunately, these incidents do occur from time to time (Refs. 2, 3 & 10) and in order to prevent them from reoccurring, it is important to understand both the causes of sinkhole formation, and the appropriate measures used in dam engineering to control the problem.

Causes and Control of Sinkhole Problems

The development of sinkholes in a dam-reservoir environment is a complicated process. It involves the interplay between seepage from the reservoir and various existing or potential unfavourable conditions present in the dam, its foundation, or the floor and walls of the reservoir. It is, therefore, important to recognize the possibility of sinkhole development during initial geological, geotechnical, and geophysical investigations of the site. Early recognition of a potential problem could lead to the avoidance of a troublesome site, or to a more thorough study which would define the problem in sufficient detail to allow the application of appropriate design measures to prevent or control the problem. This paper briefly outlines the major factors that contribute to the formation of sinkholes, and the preventive and/or remedial measures that are used to control the problem.

(1) <u>Existing Surficial and Sub-Surface Voids</u>: Natural surficial and sub-surface voids are likely to be more prevalent in the following geological settings:
Soluble limestones and dolomitic limestones having openings, cavities, caverns, and

channels along joints, bedding planes, faults and folding fractures;
Extrusive volcanic rocks with open joints and scoriaceous contact zones;
Significant fissures and joints in rocks (caused by folding, faulting, stress-relief, instability, etc.);
Openwork gravels, cobbles and boulders in alluvial deposits;
Eskers, pervious terraces, divides and buried valleys with openwork gravels and boulders, kettles, and potholes in glacial terrains.

To detect sub-surface voids, foundation investigation methods include drilling, test shafts and adits, geological mapping, air photography, geophysical exploration and, in special situations, large-diameter calyx drilling in rocks or in cohesionless overburden after freezing treatment. It is important to understand the nature of existing voids and any infilled materials, and the potential changes which may be brought about by reservoir seepage. Once voids are discovered, they should be filled with concrete, or cement grout or other suitable materials by proper surface and subsurface treatment methods (Refs. 1, 17, 23 & 24). Where existing voids are infilled with loose and erodible materials, these infillings should be completely removed and backfilled with appropriate materials. In limestones and related carbonate rocks, the location and treatment of solution cavities and tunnels to produce a water-tight dam and reservoir can be a costly and time consuming process (Ref. 5). In some instances, the cost of preventive or remedial measures can exceed the total cost of a project. Therefore, it is extremely important to conduct thorough investigations at the stage of site selection (Ref. 12).

Man-made, sub-surface voids are largely due to underground mining. Thus, a potential conflict, in land-use planning, exists between the development of reservoirs and underground mining. Reference 6 gives an account of the current practice adopted by the New South Wales Dams Safety Committee in Australia to manage this conflict between the safety of dams and reservoirs and the development of underground coal mines. The regulations allow bord and pillar first workings under the reservoir storage area and total extraction only outside a perimeter zone, which is determined as a plan distance equivalent to an angle of 35 degrees measured from the vertical at full reservoir supply level to the appropriate coal seam. This controlled mining under the reservoir area was carried out on an experimental basis, which included stringent monitoring. To date, except for an incident of large water inflow into Wongawilli colliery, the majority of mining adjacent to dams or reservoirs has taken place without detrimental effect.

(2) <u>Incompatible Gradations</u>: At the boundaries of various materials within the foundations, or within the zones of earth and rockfill dams, the presence of incompatible gradations in the direction of the projected reservoir seepage will provide the opportunities for creating seepage-induced cavities, pipes, and channels, through the process of internal erosion (piping) in an initially intact dam and foundation.

Therefore, it is critically important to check the gradations of materials, using accepted filter criteria, across various zone boundaries along the direction of seepage, to ensure that each zone serves as a protective filter for the zone immediate upstream. Filter criteria established in 1940 (Ref. 4) have proved over the years to be entirely satisfactory (Ref. 19). However, it is equally important to remember Terzaghi's admonition in 1929 (Ref. 21) concerning the influence of minor geological details on the safety of dams, and to design for the most unfavourable possibilities which could be expected under the existing geologic conditions. This point was brought home once more by the tragic failure of Teton Dam (Ref. 11) in 1976 (see Fig. 1).

(3) <u>Cracks</u>: Cracks within an embankment dam, its foundation or embedded conduits, can be induced by a variety of causes, such as: differential settlement, horizontal movement, arching, hydraulic fracturing, faulting, and natural or reservoir-induced earthquake (Ref. 9). The potential for the development of cracks and the tendency of their subsequent growth or healing vary with the characteristics of the materials involved, such as plasticity and gradation. If the cracks are sustained or enlarged, they would tend to reduce the gradation compatibility and increase the hydraulic gradient in the vicinity of the cracks, thus initiating the process of internal erosion (Ref. 7).

There are two aspects to controlling cracking: First, is prevention of initial cracking, and second is mitigating the undesirable effects of cracks once they have formed. Preventive measures include: (1) minimizing differential settlement by trimming rock overhangs and abrupt slopes in dam foundations; (2) minimizing horizontal movement by using flat dam slopes; (3) reducing arching by maintaining compatible compression characteristics between the impervious core and the pervious shells; (4) avoiding hydraulic fracturing by controlling fluid pressures in both grouting and drilling operations, increasing the earth pressure within the dam by aligning the dam axis convex upstream, and controlling the rate of filling of the reservoir; and (5) improving the crack resistance of core materials by selecting more plastic materials from borrow and by controlling the placement water content above that corresponding to the optimum compaction. On the other hand, to mitigate the effects of cracking, it is essential to promote self-healing of any cracks that may form. This is accomplished by providing lightly-compacted, cohesionless materials in wide transition zones flanking both the upstream and downstream sides of the crack-susceptible core. The downstream transition zone, which is a filter, serves as crack-stopper, while the upstream zone provides material for crack-filling. Similarly, the use of well-graded, impervious materials in a wide core can also promote self-healing of cracks.

(4) <u>Soluble, Dispersive or Collapsible Materials</u>: Soil or rock materials that go into solution such as salt, gypsum, limestone, and related carbonate rocks; disperse into colloidals suspended in water such as dispersive soils; or collapse upon saturation such as löess; are potentially problematic, even if quite stable in their natural condition when not attacked by water. The impoundment of reservoir water will raise the regional water table and change the groundwater regime, and could result in the development of undesirable voids, tunnels and channels in these susceptible materials. Figure 2 shows an example of sinkhole formation in a dam constructed of dispersive soils (Ref. 18).

Fig. 1: View of sinkhole at Teton Dam moments before dam breach (from Ref. 11)

Fig. 2: Sinkhole formation in a dam constructed of dispersive soils (from Ref. 18)

For soluble rocks, the rate at which the solution process might take place depends on the chemistry, temperature and gradient of groundwater. This solution rate should be assessed and its impact on the proposed dam and reservoir evaluated. Grouting of joints, fractures, and other openings could retard the solution process. Monitoring of pore pressures and seepage flows is essential for early detection of any defects in dams and reservoirs already in service so that any required corrective measures can be carried out in an orderly and timely fashion. For dispersive soils, proper sand filters can protect them from the attack of internal erosion, and chemical treatment like lime-stabilization can render the treated soils non-dispersive. For collapsible soils, prewetting by ponding could induce the collapse of the unstable soil structure, and improve its characteristics to serve as a dam foundation or an embankment fill.

(5) <u>Hydraulic Gradients</u>: As the hydraulic gradients increase with increasing levels of the impounding reservoir, so do the seepage velocities of the percolating water. The increased seepage velocities associated with the increased flow volume tend to enhance the erosive power of the seepage flow. Thus, the potential weaknesses within a dam, its foundation, and reservoir base as outlined in preceding paragraphs will be subjected to increasingly more severe attacks by the seepage water. Consequently, seepage control measures should be in place before the reservoir is raised to high levels.

Seepage through embankment dams is controlled by impervious cores while foundation seepage is controlled by positive cutoffs, by measures to lengthen the seepage path, or by drainage measures to allow the safe exit of seepage without internal erosion. In fact, designers should adopt the principle of multiple lines of defence advocated by Casagrande (Ref. 8). The approach makes use of several independent design measures that compliment each other and enhance the safety of the dam and reservoir. Instrumentation, which continuously monitors seepage conditions, including pore pressures and flow rates, forms an integral part of the design.

Case Histories

Case histories involving sinkholes are quite numerous in the dam engineering literature. In most cases, the problem was corrected successfully. However, in some unfortunate cases, dam or reservoir failures did occur. Space limitations restrict this paper to a few representative examples which are discussed following for illustrative purpose. Their details are contained in the references cited; only those features related to sinkholes are highlighted here. These examples include dam and reservoir failures, successful sinkhole repairs, and sinkhole related construction problems involving tailings dams.

(1) Dam Failure: Teton dam (Ref. 11) was situated in a steep-walled canyon cut by Teton River in Idaho. It was a zoned, earthfill dam about 91 m high and 945 m long, founded on highly-jointed, welded, ash-flow tuff (rhyolite) at the abutments, and on alluvium in valley sections. The core and key trench backfill was the eolian silt abundant on the site. A grout curtain was extended along the full length of the dam to minimize seepage through the rock foundation. No instrumentation other than settlement monuments was installed. The failure was initiated by leakage occurring through the key trench on the right abutment. The highly erosive trench backfill was in direct contact with the jointed rocks with no protective filters placed between. The leakage ultimately caused further erosion along the downstream contact of the core and the abutment rock forming internal tunnels or pipes. From the initial detection of small springs which totaled around 6 l/sec about 460 m downstream from the dam to the final breach of the dam, it took only two days. Moreover, at the final phase the erosion progressed at such an alarming rate, with flows in the order of 3,200 l/sec accompanied by an upstream whirlpool and downstream sinkholes, it was not possible to save the dam. Thus, although the observation of sinkholes usually serves as a warning signal that internal erosion has already occurred, and remedial measures are required to rectify the problem, for this case, the signal came too late.

(2) Reservoir Leakages: Reservoir leakages associated with the formation of sinkholes in the reservoir floor and walls caused the final abandonment of Lone Pine reservoir (31 m high dam) in Arizona and severe reduction of available storage at May dam reservoir (28 m high dam) in Turkey (Ref. 10). In the former incident, reservoir seepage flushed out soil infilling that partly filled sinkholes in the limestone, which were scattered throughout the reservoir floor, allowing the escape of water through interconnected openings and channels in the limestone. In the latter case, reservoir water leaked through large sinkholes that formed on the floor of the reservoir, which consists of an alluvium layer about 15 to 20 m thick, overlying karstic limestone conglomerate and marl.

(3) Sinkhole Repairs and Prevention: Depending on the characteristics of the sinkholes formed and the site conditions, remedial measures differ from case to case. At Terzaghi Dam (formerly Mission Dam) in British Columbia (Ref. 22), sinkholes formed on the upstream slope of the dam at the right abutment due to the development of leaks in the grout seal, which was used to plug the gap between the bedrock and the lower end of the rock-supported sheetpiles. The sinkholes were plugged with soft clay and sealed with a plastic membrane. Some self-healing was suspected to have already occurred prior to the sealing operations. At Wolf Creek Dam in Kentucky (Ref. 20), sinkholes, which developed in the limestone foundation, led to excessive leakage after about 17 years of normal reservoir use. Repairs included emergency grouting and a concrete diaphragm wall with the maximum depth of about 85 m constructed from the upstream crest of the embankment to the base of all solution features encountered during exploration and construction. At Tarbela Dam in Pakistan (Ref. 15), extensive sinkholes formed in the upstream impervious blanket due to differential foundation settlement (Fig. 3).

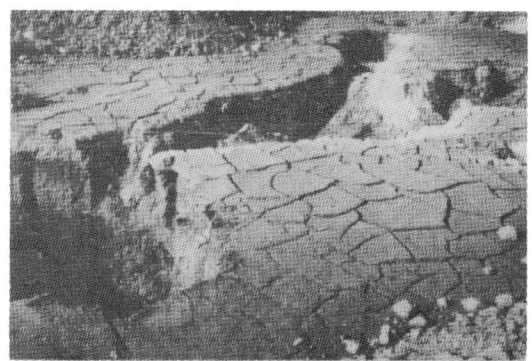
(a) Investigation of sinkholes by trenching after reservoir drawdown

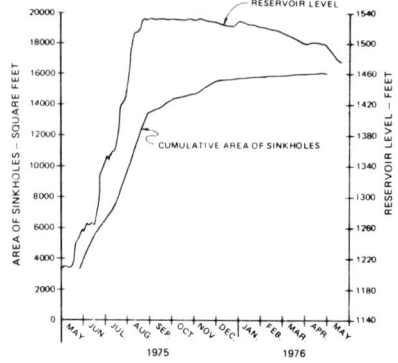
(b) Total area of sinkholes increases with the rise of reservoir level

Fig. 3: Sinkhole formation in the upstream impervious blanket of Tarbela Dam (from Ref. 15)

Backfilling of sinkholes was carried out when the reservoir was emptied as well as when it was filling. Guided by side-scan sonar survey, barge dumping of impervious fill through the water was carried out. Natural sedimentation of silt by the Indus River in critical reservoir areas was also encouraged by constructing training dikes. At Mornos Dam in Greece (Ref. 16), a 2.6 km long, and about 160 m wide asphalt concrete lining, placed on a 2h:1v slope of compacted fill and supplemented by drainage measures, was used to block off the contact of water with the karstic limestone reservoir wall as a preventive measure.

(4) <u>Sinkholes Related to Tailings Dam Construction</u>: Due to economic constraint and shortage of suitable filter materials, the gradation and thickness of filter zones used in rockfill tailings dams are sometimes less than desirable. However, by maintaining a wide tailings beach between freewater in the pond and the tailings dam, the hydraulic gradients across the tailings/filter contacts usually can be reduced to such an extent that no internal erosion will occur. Occasionally, due to unusual runoff events or other circumstances, pond levels may rise rapidly and inundate a substantial portion of the beach, the correspondingly increased hydraulic gradients across the tailings/filter contacts could initiate internal erosion and precipitate piping of tailings and water through the rockfill dam. Because this critical condition occurs near the water surface, and because the rockfill dams are usually quite erosion-resistant, after escape of certain amount of tailings and water, the drawdown of water levels reduces the hydraulic gradients across the tailings/filter contacts and prevents the continual progression of piping. The most economic remedial measure is lengthening the beach width by spigotting additional tailings and watching closely the water balance of the tailings operation. Figure 4 illustrates this type of short-lived sinkhole problem (Ref. 13) related to the construction of rockfill tailings dams. Ultimately, however, a spillway is used to control the pond level and to maintain an adequate tailings beach upon mine abandonment.

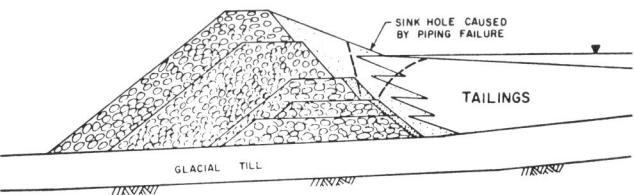

(a) Sinkhole formed during dam construction

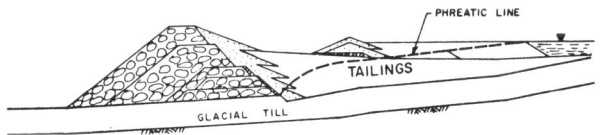

(b) Tailings beach width increased for ultimate dam

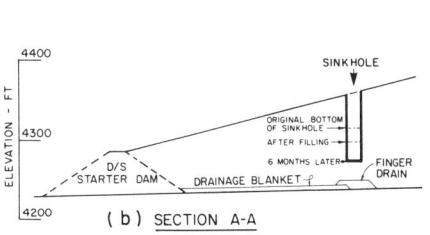

(c) View of sinkhole repair

Fig. 4: Repair of sinkhole for a rockfill tailings dam by reducing hydraulic gradients (from Ref. 13)

A rather unique sinkhole case (Ref. 14) involved a sandfill tailings dam with an underdrainage system consisting of rockfill, finger drains, enveloped by filters, and a gravel drainage blanket. A sinkhole about 4.3 m in diameter and 12 m in depth (Fig. 5) formed suddenly in 1983 above a rockfill drain at a location where the existance of a damaged

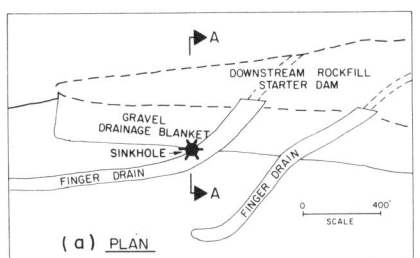

(c) Backfill of sinkhole with filter materials

Fig. 5: Repair of sinkhole for a sandfill tailings dam by improving damaged filter zone around finger drain (from Ref. 14)

section of filter was suspected. The sinkhole is situated far away from the upstream tailings pond. It is formed by the action of the construction water used to sluice in the sandfill rather than by the through-seepage from the pond. Thus, no danger exists for pond-connected piping, and the integrity of the sand dam is not compromised. The hole was backfilled with about 350 m^3 of filter materials, and water was added to encourage the gradual downward movement of the filter materials until they eventually reach and cover the damaged filter section.

Conclusion

Sinkhole development in a dam-reservoir environment is a complicated process which involves many factors, as discussed in this paper. A clear understanding of the interrelation of these factors and thorough site investigations can lead to early recognition of potential sinkhole problems for a given project. This recognition, in turn, can result in the avoidance of unsuitable sites or the implementation of appropriate design measures to prevent and/or mitigate the potential problems.

References

1. ASCE 1972, "Foundation and Abutment Treatment for High Embankment Dams on Rock," Am. Soc. of Civ. Engr., Vol. 98, No. SM 10, Oct.
2. ASCE/USCOLD 1975, "Lessons from Dam Incidents, U.S.A.," Am. Soc. of Civil Engineers.
3. Babb, A.O. and Mermel, T.W. 1968, "Catalog of Dam Disasters, Failures and Accidents," Bureau of Reclamation.
4. Bertram, G.E. 1940, "Experimental Investigation of Protective Filters," Harvard Univ. Soil Mech. Series No. 7.
5. Burwell, E.B. Jr and Moneymaker, B.C. 1950, "Geology in Dam Construction," Application of Geology to Engineering Practice (Berkey Volume), Geo. Soc. of Am. pp. 11-43.
6. Cantwell, B.L. and Whitfield, L.M. 1984, "Underground Mining Near Large Australian Dams, "Water Pwr. and Dam Const. pp. 20-24, Apr.
7. Casagrande, A. 1950, "Notes on the Design of Earth Dams," Journal of Boston Soc. of Civ. Engr., Vol. 37, pp. 405-429, Oct.
8. Casagrande, A. 1969, "Discussion of a paper on Earth and Rockfill Dams by Wilson, S.D. and Squire, R." Proc. of 7th Int'l. Conf. on Soil Mech. and Fdn. Eng., Vol. III, pp. 278-283, Mexico.
9. Hirschfeld, R.C. and Poulos, S.J. (Ed.) 1973, "Embankment-Dam Engineering, (Casagrande Volume)" John Wiley & Sons.
10. ICOLD 1974, "Lessons from Dam Incidents," International Commission on Large Dams.
11. Independent Panel to Review Cause of Teton Dam Failure 1976, "Failure of Teton Dam," U.S. Dept. of Interior and State of Idaho.
12. Int'l Assoc. of Eng. Geol. 1973, "Sinkholes and Subsidence Engineering - Geologic problems Related to Soluble Rock," Proc. of Symposium, Hannover, Germany, Sept.
13. Klohn, E.J. 1972, "Tailings Dams in British Columbia," Geotechnical Practice for Stability in Open Pit Mining, Proc. of 2nd Int'l Conf. on Stability in Open Pit Mining, Vancouver, pp. 151-172.
14. Klohn, E.J. 1984, "The Brenda Mines' Cycloned-Sand Tailings Dam", Proc. of Int'l Conf. on Case Histories in Geotechnical Engineering, St. Louis, May, Vol. 2, pp.953-977.
15. Lowe, J.III 1978, "Foundation Design - Tarbela Dam," 4th Nabor Carrillo Lecture, Mexican Society for Soil Mechanics.
16. Schewe, L.D., 1977, "Mornos Dam (Greece) - Dam Construction in Sliding Endangered Flysch and Karstified Limestone," World Dams Today '77, The Japanese Dam Foundation.
17. Sherard, J.L., Woodward, R.J., Gizienski, S.F. and Clevenger, W.A. 1963, "Earth and Earth-Rock Dams," John Wiley & Sons.
18. Sherard, J.L. and Decker, R.S. (Ed.) 1977, "Dispersive Clays, Related Piping, and Erosion in Geotechnical Projects," Am. Soc. for Test. and Mat'l., STP 623.
19. Sherard, J.L., Dunnigan, L.P., and Talbot, J.R. 1984, "Basic Properties of Sand and Gravel Filters" and "Filters for Silts and Clays," Am. Soc. of Civ. Engr. Vol. 110, No. GT6, pp. 684-718, June.
20. Simmons, M.D. 1982, "Remedial Treatment Exploration, Wolf Creek Dam, Ky.," Am. Soc. of Civ. Engr. Vol. 108, No. GT7, pp. 966-981, July.
21. Terzaghi, K. 1929, "Effects of Minor Geologic Details on the Safety of Dams," Am. Inst. of Min. and Metal. Engr., Tech. Publ. 215, pp. 31-44.
22. Terzaghi, K. and Lacroix, Y. 1964, "Mission Dam - An Earth and Rockfill Dam on a Highly Compressible Foundation," Geotechnique, Vol. 14, No. 1, pp. 13-50, Mar.
23. Wahlstrom, E.E. 1974, "Dams, Dam Foundations, and Reservoir Sites," Elsevier Sci. Pub. Co.
24. Wilson, S.D. and Squier, R. 1969, "Earth and Rockfill Dams," Proc. of 7th Int'l Conf. on Soil Mech. and Fdn. Eng., State-of-the-Art Volume, pp. 137-223, Mexico.

4. Environmental / societal impact of sinkholes

The influence of urbanization on sinkhole development in central Pennsylvania

ELIZABETH L.WHITE, GERT ARON & WILLIAM B.WHITE *The Pennsylvania State University, University Park, USA*

ABSTRACT

The karsted limestone valleys of central Pennsylvania contain two populations of sinkholes. Solution sinkholes occur in the Champlainian limestone units along the margins of the valleys. Solution sinkholes are permanent parts of the landscape and although a nuisance to construction, do not present other problems. The second population is the suffosional or soil-piping sinkholes. These occur on all carbonate rock units including the Beekmantown and Gatesburg dolomites that comprise the two principal carbonate aquifers in the valley. Suffosional sinkholes are the principal land use hazard.

Suffosional sinkholes are transient phenomena. They occur naturally but are exerbated by runoff modifications that accompany urbanization. Suffosional sinkholes are typically 1.5 to 2.5 meters in diameter depending on soil thickness and soil type. The vertical transport of soil to form the void space and soil arch that are the precursors to sinkhole collapse is through solutionally widened fractures and cross-joints and less often through large vertical openings in the bedrock. The limited solution development on the dolomite bedrock combined with soil thickness, seldom greater than two meters, limit the size of the sinkholes. All aspects of suffosional sinkhole development are shallow processes: transport, piping, void and arch formation, and subsequent collapse take place usually less than 10 meters below the land surface.

Factors exerbating sinkhole development include pavement, street, and roof runoff which accelerate soil transport. Such seemingly minor activities as replacing high grass and brush with mowed grass is observed to accelerate sinkhole development. De-watering of the aquifer is not a major factor in this region.

Introduction

The broad limestone valleys of central Pennsylvania are subject to increasingly severe sinkhole problems as the land use pattern of farms and small villages gives way to a pattern of urban development. These problems have been most in evidence near State College and its surrounding townships. Extensive modifications of natural runoff patterns and extensive use of the several carbonate aquifers as municipal water supplies has modified the natural pattern of sinkhole development. This study is concerned with the mechanisms of sinkhole formation, and with the contrasts of natural sinkhole development in the still rural parts of the area with sinkhole development as modified by urbanization.

Geologic Setting

Some 2000 meters of carbonate rocks are exposed in the folded Appalachian Mountains of central Pennsylvania. These extend from the Cambrian Warrior limestone to the upper Ordovician Champlainian series. For hydrogeologic purposes, the carbonate units can be roughly separated into the Gatesburg dolomite aquifer (the lower part of the section) the Beekmantown dolomite aquifer (much of the middle portion of the section), and the karstic Champlainian limestones (the upper part of the section) (Parizek et al., 1971).

The carbonate rocks are folded into broad and complex anticlines which have been eroded to form valleys. The corresponding synclines expose the resistant Tuscarora quartzite and Bald Eagle sandstones which cap the long parallel mountain ridges. As a result, the Champlainian limestones outcrop along the flanks of the ridges where they receive runoff from the sandstone and shale on the ridge flanks. The Beekmantown dolomite and associated units make up much of the valley floors, and the Gatesburg formation, a resistant sandy dolomite, often produces a topographic high along the structure axis of the major anticlinal valleys (Fig. 1).

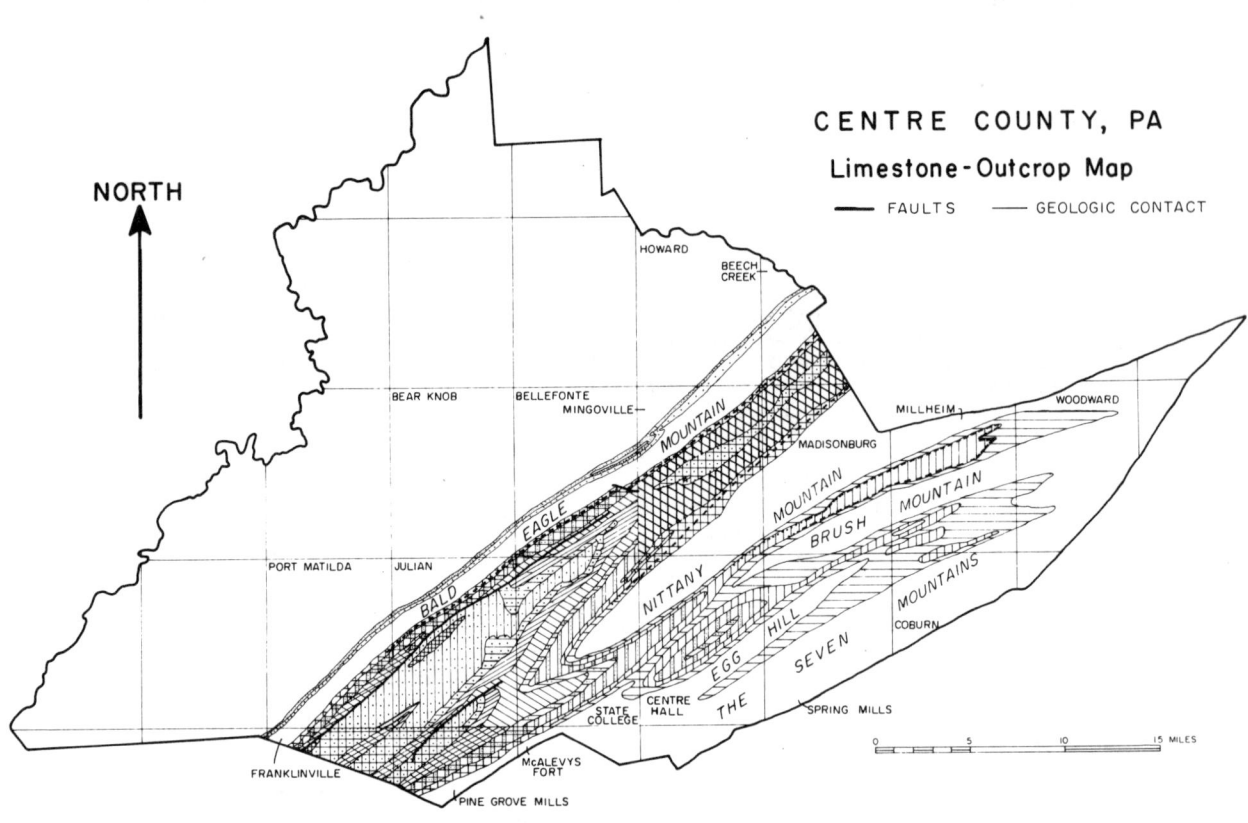

Figure 1(a): Map of Centre County showing carbonate rock outcrops. Grid gives names of USGS 7.5 minute topographic maps. Map drawn by G.O. Dayton (Dayton and White, 1979) based on the limestone resource map of Pennsylvania (O'Neill, 1964).

The Champlainian limestones are highly cavernous, drainage from the mountain flanks sinks at the limestone contact and flows through predominently strike-oriented conduits to gravity springs. The outcrop band of Champlainian limestones contain well-developed karst topography including sinking streams, deep solutional sinks, and occasional collapse sinks. The Beekmantown dolomite is a high grade fracture aquifer. Although solutionally widened joints and fractures provide paths of high permeability for the storage and movement of groundwater, caves are rare. There is little surface expression of karst on the Beekmantown. Much of the water supply for the Borough of State College is derived from the Beekmantown formation. Likewise, the Gatesburg dolomite is a good aquifer with water moving through joints and fractures and also through sandy and silty horizons. The Pennsylvania State University draws its water from wells drilled in the Gatesburg.

Anatomy and Natural History of Suffosional Sinkholes in Central Pennsylvania

Three genetic types of sinkhole are found in central Pennsylvania. Solution sinks are bowl-shaped depressions in the bedrock although they are usually mantled with soil. Collapse sinkholes are formed by collapse or subsidence of bedrock cavities, shallow caves exposed by the general lowering of the valley uplands. Finally, there are soil piping or suffosional sinkholes more related to movement of soil than solution of bedrock. Solution and collapse sinkholes are permanent features of the landscape on human time scales and although they are a nuisance to land development, are not a hazard. Some geomorphic properties of solution and collapse sinks in the Appalachians are discussed by White and White (1979). Suffosional sinkholes are the main concern of this study.

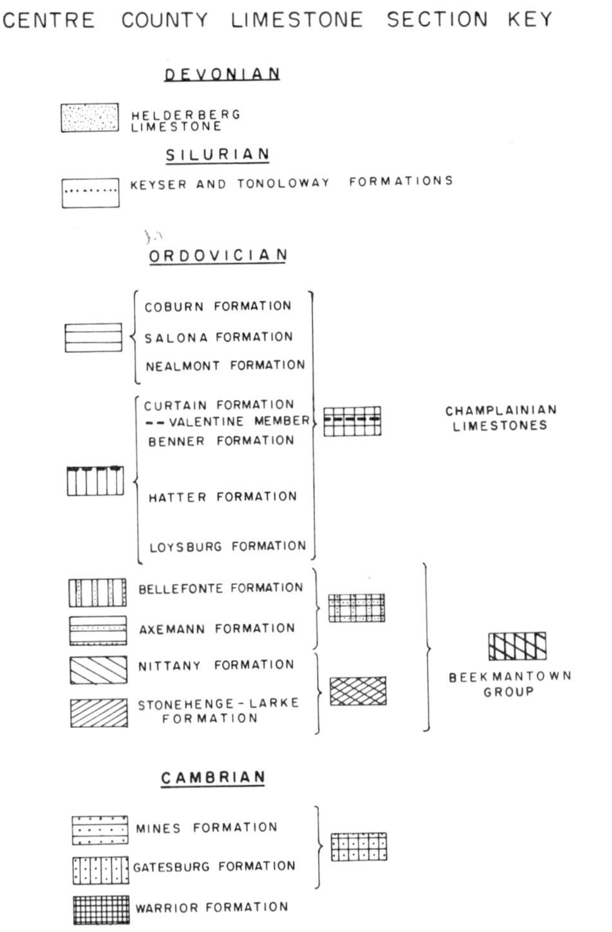

CENTRE COUNTY LIMESTONE SECTION KEY

DEVONIAN
- HELDERBERG LIMESTONE

SILURIAN
- KEYSER AND TONOLOWAY FORMATIONS

ORDOVICIAN
- COBURN FORMATION
- SALONA FORMATION
- NEALMONT FORMATION
- CURTAIN FORMATION -- VALENTINE MEMBER
- BENNER FORMATION   } CHAMPLAINIAN LIMESTONES
- HATTER FORMATION
- LOYSBURG FORMATION
- BELLEFONTE FORMATION
- AXEMANN FORMATION
- NITTANY FORMATION   } BEEKMANTOWN GROUP
- STONEHENGE-LARKE FORMATION

CAMBRIAN
- MINES FORMATION
- GATESBURG FORMATION
- WARRIOR FORMATION

Figure 1(b): Limestone section showing grouping into three carbonate units.

Figure 2 illustrates the essential features of suffosional sinkholes. A bedrock drain is hypothesized although it is rarely visable. The drain must be of sufficient size to carry clastic material. In the few examples that have been exposed in road cuts, the drain is an irregular solution chimney 10 to 40 cm in diameter or a solutionally widened fracture a few centimeters wide but of indefinite extent. The drain serves to concentrate the flow into the otherwise impermeable limestone or dolomite. Velocities are a maximum near the inlet to the drain and soil is first removed from this region, creating a cavity between the bedrock and overlying soil, the roof of which is supported by an arch of cohesive soil (Fig. 3). With time, more and more soil spalls from the roof and is washed down the bedrock drain, enlarging the cavity and the soil arch. Eventually the cavity becomes so large that the stresses in the roof exceed the cohesive strength of the soil and the arch collapses to form the sinkhole. The cover is important. Root mats of plants often serve as the final support for the roof of the cavity as do frozen ground in winter and also various kinds of pavement.

Unlike the solution and collapse sinks which are almost entirely restricted to the Champlainian limestones, suffosional sinks occur in all of the carbonate rock units. Many of those that occur in and around State College are in the Beekmantown dolomite. It is a general observation in Pennsylvania that most of the caves and related solution features are in limestone units, seldom in dolomites. Suffosional sinks are an exception to this rule and occur with the same frequency in all rock units. The dolomites have an abundance of solutionally widened fractures which are the zones of high permeability for the aquifer (Lattman and Parizek, 1964). These fractures are sufficient for soil transport and thus for the initiation of sinkholes. Unlike sinkhole collapses in many other regions, the central Pennsylvania examples are seldom associated with pits and cavities in the bedrock.

Most of the observed suffosional sinkholes have diameters from one to three meters and are from 1 to five meters deep. The final collapse of the soil arch is initiated by a circular shear zone. A plug of soil drops into the void space creating a very sharp lip to the sink immediately after collapse (Fig. 3-B). Bedrock ledges are sometimes but not always exposed at the bottom of the sink indicating that relatively little soil remained in the roof of the arch before collapse.

Some disposal system for the sediment must be present or the fracture system would simply choke up. In better developed karst, the sediment is known to be transported by underground streams and returned to the surface through springs. The transport system is rarely seen in central Pennsylvania. Kings Cave in Mifflin County is an example where the bedrock fracture could be explored 14 meters to an underground stream channel.

Effects of Urbanization and Land Use

Human activities in populated areas affect the natural cycle of soil piping and sinkhole collapse through two major mechanisms: increased hydraulic gradients through the lowering of the natural water table and modifications of stormwater runoff patterns.

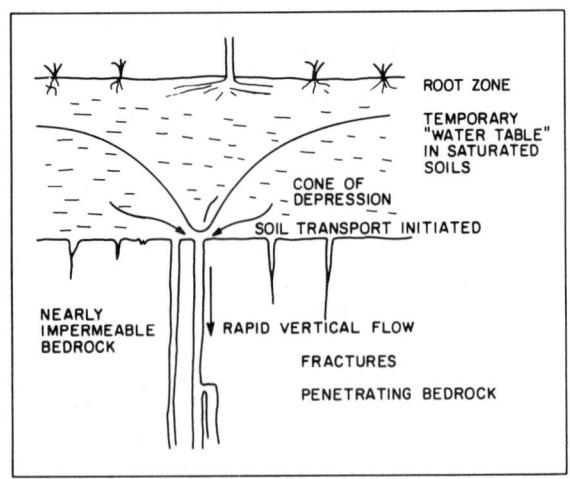

Figure 2: Sketch showing conditions for suffosional sinkhole formation. Saturation of soil during period of storm runoff provides the local 'water table' while the open fractures provide the drain with subsequent drawdown and soil transport.

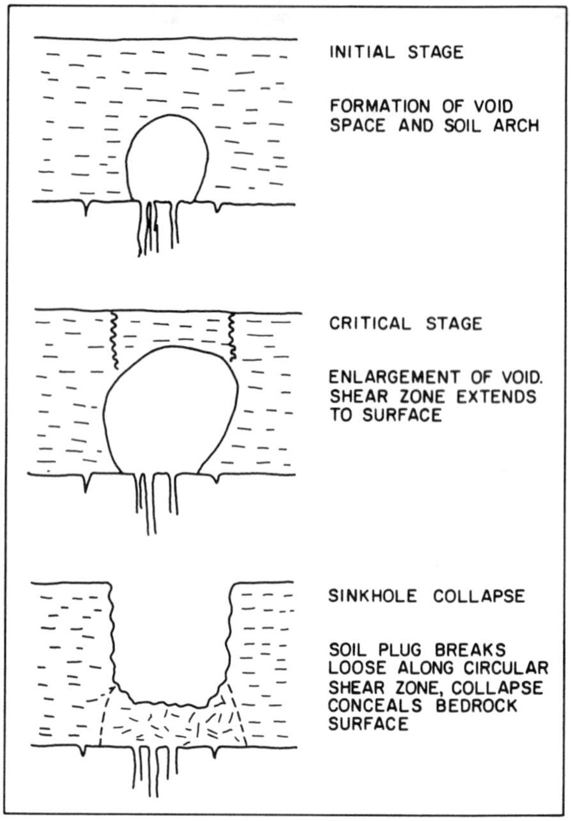

The effects of lowered water tables have received the most attention (e.g. Foose, 1953, 1967, 1979; Jennings 1966; Powell and LaMoreaux, 1969; Newton and Hyde, 1971; Newton et al., 1973). Activities that result in lowered water tables occur in the Nittany Valley. Water is supplied for the Borough of State College by a well field in the Beekmantown dolomite aquifer which, with a yield of 2.5 M gal/day, has a drawdown of ~20 meters. There is deep mining of the chemical grade Valentine limestone unit of the Champlainian series on both east and west sides of the Nittany Valley. Solution cavities occur at depth and mine flooding is a problem during periods of high runoff but those sinkholes that have formed seem to be related to mine subsidence (e.g. the shaft described in the Centre County Cave Survey as 'Peeping Pigeon Pit') rather than soil piping. De-watering does not seem to be a major cause of suffosional sinkhole formation in central Pennsylvania. The reason is the great depth to water table, often 10 to 100 meters. Most of the suffosional sink activity takes place completely in the vadose zone and the additional lowering of the water table has little effect.

Figure 3: Sketch showing the evolution of a suffosional sinkhole. (a) Initial stage of void formation near dominent fracture inlet. (b) Shortly before collapse a shear zone develops between the void and the surface. (c) The collapse sink in which the soil plug falls, masking the bedrock inlets.

The most important factor influencing sinkhole formation in this area is stormwater runoff and its modification by urbanization. In the un-modified rural valleys, infiltration is diffuse and spread over a large area by the plant cover. Specific features of man's activities that modify runoff behavior include:

(i) Roofs and paved areas such as streets, driveways, and parking lots. These areas of zero infiltration and flashy overland flow concentrate runoff into very localized areas. Much of the runoff from the Nittany Mall parking lot is discharged into an existent sinkhole on the corner of the property. The runoff from the University Park airport runway had discharged into a gently rolling swale area. Several years ago, the runoff was concentrated into a man-made gully which continues to the stream (about one km away). Along the entire route of the gully to the stream, large sinkholes have now developed.

(ii) Installation of blind stormwater drains to alleviate ponding on streets. There are typical storm inlet grates extending the width of the street draining into a box culvert from which the exit is a short length of pipe leading into the adjacent soil. Injection of stormwater runoff directly into the soils has caused soil piping, undermined some basements of adjacent dwellings, and caused flooding in other basements.

(iii) General soil subsidence removes support from water mains which sometimes crack causing additional piping and sometimes sinkhole collapse. One such sink formed under Ridge Avenue where a cavity had enlarged to 5 meters in diameter and 7-10 meters deep roofed only by the asphalt pavement before it finally failed. Other sinkholes and undermining of foundations have been blamed on cracked water mains.

In all cases, the modification of normal diffuse infiltration by urbanization results in the build-up of a perched water body in the soil above the bedrock. Enhanced hydraulic gradients increase the rate of soil transport into the subsurface with concurrent arch and cavity formation and, in some cases, sinkhole collapse.

Amerlioration of Suffosional Sinkhole Activity

Given that soil piping sinkholes are a natural process in the karstic carbonate terrain of the central Pennsylvania limestone valleys and that they are exerbated by urbanization and other land development, what can land use planners do about them?

It is amply demonstrated that the usual 'cure' of simply filling in the sinkholes does not work. There is a sinkhole under the main north-south street through State College that slumps and is repaved with frustrating regularity. We have been watching a sinkhole on dolomite at the edge of the University golf course that is filled with soil every summer and just as regularly collapses again when the ground thaws in the spring. It is argued in this paper that the suffosional sinkhole activity in the central Pennsylvania region is aggrevated by increased stormwater runoff, not by modified ground water levels. The only real solution is to prevent the localization of stormwater infiltration and if this solution is to be retro-fit onto existing development it involves difficult and expensive drain systems. It is easier and cheaper to take protective steps as an early part of land development.

Sinkhole prone areas constitute a class of hazard that has aspects in common with floodplains. The land may seem perfectly suited for development and surface evidence of incipient sinkhole formation may be sparse. In urbanized areas, land is expensive and pressure for development is high. Zoning boards and Regional Planning Commissions would be advised to place legal restraints on development of sinkhole prone areas in the same way that they have placed restraints on development on floodplains. The risks to flood plain development are statistical in time; risks to development on sinkhole prone areas are statistical in space but for the individual property owner whose home is condemned, the risks are very real.

Sinkhole prone areas can be identified by examination of geologic, topographic, and soil maps. Existing sinkholes, thick soils, closed depressions that funnel runoff to the subsurface, and underlying limestone units within dolomitic carbonate rock are all suggestions of possible sinkhole activity. Because suffosion sinks occur naturally, the area can be examined carefully for scars and shallow depressions that locate former locations of these short-lived features.

Sinkhole development turns out to be surprisingly sensitive to the cover on the soil. If the most serious sinkhole prone areas can be allowed to grow a thick cover of grass, brush, and trees, the root system will do much to support the soil and the plant cover will disperse stormwater and prevent the localized inputs that lead to soil piping. Such areas can be natural zones within parklands and can be used, where appropriate, as wildlife habitat.

If discharge of runoff from roofs, streets, and pavements into sinkhole prone areas cannot be avoided, adequate storm drain systems can do much to avoid soil piping. Storm water from all sources, including roofs and driveways of private dwellings, should be connected into an integrated stormwater system. If discharge of stormwater into the ground is necessary, it should be drained directly into bedrock fractures which should then be protected with a cap of crushed stone covered with gravel (Fig. 4). Blind culverts and french drains that discharge stormwater directly into the soil should be avoided.

The guiding principle that runoff must be directed to the subsurface without giving it an opportunity to pipe away soil can be applied to the filling of existing sinkholes (Fig. 5). Because soil arches and underlying soil cavities may have stopped some distance upward before collapse, the floor of the typical suffosional sinkhole is a plug of loose soil which is subject to later piping if the sinkhole is merely filled. The soil plug should be excavated to bedrock and the sink then filled with coarse rock properly overlain with a layer of gravel to provide a transition zone between the coarse rock and the fine soils above. This provides a high permeability path that will conduct runoff to the fractures in the bedrock without piping additional soil.

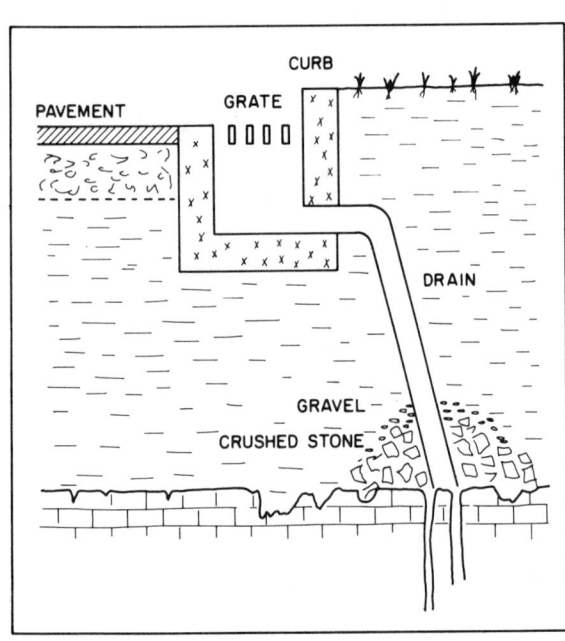

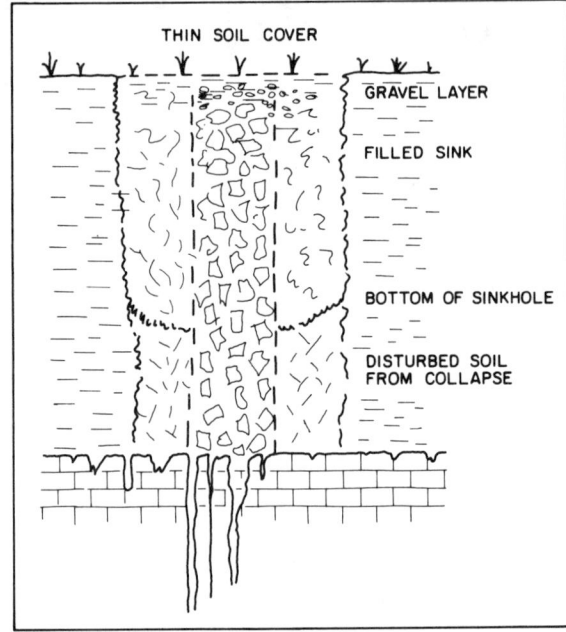

Figure 4: Sketch showing extension of storm inlet drain to bedrock fracture thus reducing soil piping by high velocity flow during storm runoff.

Figure 5: Sketch showing the filling of an existing sinkhole with crushed stone or disgarded broken concerete to produce a high permeability pathway to bedrock. capping of gravel is necessary to prevent soil piping through the stone.

REFERENCES

Braker, W.L. (1981) Soil Survey of Centre County, Pennsylvania. Soil Conservation Service, U.S. Department of Agriculture, 162 pp.

Dayton, G.O. and W.B. White (1979) The Caves of Centre County, PA. Mid Appalachian Region, National Speleological Society Bull. 11, 126 pp.

Foose, R.M. (1953) Ground Water Behavior in the Hershey Valley, Pennsylvania. Geol. Soc. Amer. Bull. 64, 623-645.

Foose, R.M. (1967) Sinkhole Formation by Groundwater Withdrawal: Far West Rand, South Africa. Science 157, 1045-1048.

Foose, R.M. and J.A. Humphreville (1979) Engineering Geological Approaches to Foundations in the Karst Terrain of the Hershey Valley. Bull. Assoc. Eng. Geol. 16, 355-381.

Jennings, J.E. (1966) Building on Dolomites in the Transvaal. Civil Eng. in South Africa 8, 41-62.

Lattman, L.H. and R.R. Parizek (1964) Relationship Between Fracture Traces and the Occurrence of Ground Water in Carbonate Rocks. J. Hydrol. 2, 73-91.

Newton, J.G. and L.W. Hyde (1971) Sinkhole Problem in and Near Roberts Industrial Subdivision, Birmingham, Alabama. Alabama Geol. Surv. Circ. 68, 42 pp.

Newton, J.G., C.W. Copeland, and L.W. Scarbrough (1973) Sinkhole Problem Along Proposed Route of Interstate Highway 459 Near Greenwood, Alabama. Alabama Geol. Surv. Circ. 83, 63 pp.

O'Neill, B.J., Jr. (1964) Atlas of Pennsylvania's Mineral Resources Part 1. Limestones and Dolomites of Pennsylvania. Pennsylvania Geol. Surv. Bull. M50, 40 pp.

Parizek, R.R., W.B. White, and D. Langmuir (1971) Hydrogeology and Geochemistry of Folded and Faulted Rocks of the Central Appalachian Type and Related Land Use Problems. The Pennsylvania State University, Earth and Mineral Sciences Expt. Stat. Circ. 82, 181 pp.

Powell, W.J. and P.E. LaMoreaux (1969) A Problem of Subsidence in a Limestone Terrane at Columbiana, Alabama. Alabama Geol. Surv. Circ. 56, 30 pp.

White, E.L. and W.B. White (1979) Quantitative Morphology of Landforms in Carbonate Rock Basins in the Appalachian Highlands. Geol. Soc. Amer. Bull. 90, 385-396.

Sinkhole flooding associated with urban development upon karst terrain: Bowling Green, Kentucky

NICHOLAS C. CRAWFORD *Western Kentucky University, Bowling Green, USA*

ABSTRACT

Sinkhole flooding is a serious problem for Bowling Green, a city of 50,000 built entirely upon the classic Sinkhole Plain of South Central Kentucky. Since storm drains are prohibitively expensive because of the karst topography, the city relies upon the Lost River and other caves to serve as natural storm sewers. In addition to the numerous places where storm water flows naturally into the shallow carbonate aquifer, over 400 drainage wells have been drilled in an attempt to reduce sinkhole flooding. Approximately 40 sinkhole collapses, most of them small, have occurred adjacent to or near these wells. An investigation of storm water runoff from primarily commercial landuse revealed that, although it is contaminating some of the smaller cave streams with lead, chromium, iron, oil and grease, and fecal coliform, it is sufficiently diluted upon flowing into the large Lost River. A hydrogeologic investigation of sinkhole flooding in the Glendale area revealed that by filling sinkholes and then directing storm water to Nahm Sink, the headwaters of the By-Pass Groundwater Basin has been directed into the Buckberry Groundwater Basin. The Bowling Green-Warren County Storm Water Management Program requires flood easements in sinkholes below the flood level of a three-hour, 100-year storm. Developers are required to build retention basins which will retain on site any increase in runoff during a 100-year storm resulting from a change in landuse. The Management Program has been very successful in reducing sinkhole flooding. However, the numerous retention basins have been expensive for developers, difficult to maintain and the majority have experienced sinkhole collapse. Hopefully the Program can be modified to include a watershed systems approach with an increased emphasis on hydrogeologic research and pollution control.

Introduction

Periodic flooding of karst depressions is a serious problem for urban areas located upon sinkhole plains. The problem is particularly serious in the city of Bowling Green, Kentucky, where homes, streets, apartments, and businesses are affected. The city is located entirely upon a sinkhole plain, with underground streams flowing through solutionally enlarged caves in the shallow carbonate aquifer. All precipitation not lost to evapotranspiration travels by way of these streams to springs on the nearby surface-flowing Barren River. The landscape resembles large funnels (sinkholes) which direct storm water runoff into the underlying caves. Storm water ponds at the bottom of some sinkholes and then sinks slowly through the soil into cave streams below. However, most of the larger sinkholes and many of the smaller ones have experienced sinkhole collapses which have created drains permitting storm water to flow directly into the caves. Periodically these drains become clogged only to be opened again by collapses during later floods. This sequence repeated over thousands of years is the process by which most sinkholes have formed. Even before Bowling Green was built, storm water runoff sank directly through numerous sinkhole drains into the caves below. The caves acted as storm drains for this landscape then, and they continue to serve that function today.

The flooding of sinkholes in karst regions is a part of the natural hydrologic system. Flooding occurs during periods of intense rainfall, usually of short duration: 1) when the quantity of storm water runoff flowing into sinkholes exceeds their outlet capacities, and they cannot drain into underlying caves fast enough to prevent ponding, 2) when the capacity of the cave system to transmit storm water is exceeded, and the water must be stored temporarily in sinkholes since it cannot spread out onto flood plains like surface streams, and 3) when there is a backwater effect upon groundwater flow from sinkholes with bottoms lower than the level of surface streams at flood stage. Unfortunately, in the Bowling Green area houses have been built in these natural storage areas (sinkholes). The problem has been greatly aggravated by increased runoff resulting from urban development and by sinkhole filling by developers and landowners (Crawford 1981).

The worst flooding problems in Bowling Green tend to occur in large, shallow sinkholes with large catchment areas (Figure 1). Often individuals who build or purchase homes in such areas fail to recognize them as sinkholes and never consider the chance of flooding, especially since the nearest surface stream may be miles away. Unfortunately, many people believe that a sinkhole must be a steep-walled depression, a "hole" in the ground. In Figure 1, the steep-walled depression near Batsel Avenue is easily recognizable as a sinkhole, but most of the people who built homes on Covington Street did not realize that they were building in the upper portion of that same sinkhole. People normally do not build in the bottom of deep, easily recognizable sinkholes, and some towns built upon sinkhole plains have relatively minor sinkhole flooding problems for this reason. Unfortunately, Bowling Green has mostly large, shallow karst depressions, and consequently flooding is a major problem.

The Sinkhole Flood Plain

The Department of Housing and Urban Development defines the 100-year flood elevation along streams as the flood plain for flood insurance purposes. For Bowling Green, the Department has accepted the sinkhole flood plain as the three-hour, 100-year flood elevation assuming no drainage from the sinkhole (Booker 1978). This definition of the sinkhole flood plain, first suggested by Daugherty (1976), has been a part of the Bowling Green-Warren County Storm Water Management Program for establishing flood easements since 1976.

Sinkhole flooding may not be a problem when a home or business is built, but continued urbanization of the catchment results in greater areas of impervious surface and consequently an increase in storm water runoff. As landuse in sinkhole catchments changes from agricultural to suburban or from suburban to commercial, the depth, area, and frequency of sinkhole flooding increases. Thus a home built in an agricultural catchment may find itself within the sinkhole plain if the landuse changes to suburban. In order to prevent this from occurring, developers in Warren County are required to retain on site any increased runoff during a 100-year rainfall resulting from landuse changes associated with construction.

Kemmerly (1981) agrees that the definition of the sinkhole flood plain should be the 100-year flood contour assuming no outflow, but he recommends that it reflect the anticipated runoff volumes with maximum urbanization (i.e., impervious surface area $\geq 50\%$). For Springfield-Greene County, Missouri, Aley and Thomson (1981) recommend a 24-hour, 100-year flood elevation assuming no drainage and 100% runoff of the rain falling within the area topographically tributary to the sinkhole. Mills, Starnes and Burden (1982) also suggest a 24-hour rainfall for Cookeville, Tennessee. They maintain that for non-karst areas hourly rainfall intensities are the most important, but for sinkholes the time interval should be somewhat longer because they drain much more slowly than do stream channels. The definition of the sinkhole flood plain may therefore vary from one location to another.

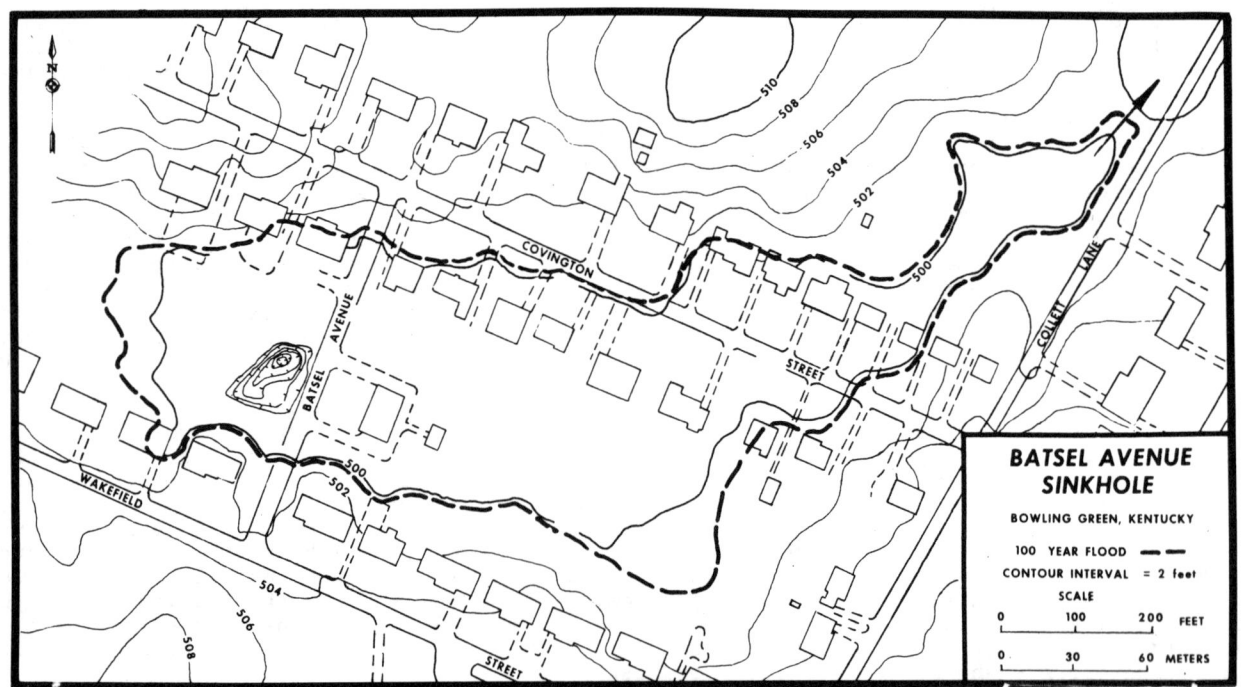

Figure 1. Batsel Avenue Sinkhole three-hour, 100-year flood contour (Sources: Modified from Booker, R. W., and Associates, Inc., 1978).

Towns like Springfield, Missouri and Cookeville, Tennessee also have large areas for growth which do not have karst topography. Considering the problems of sinkhole flooding, groundwater contamination and sinkhole collapse common to sinkhole plains, the cities should not only establish high levels for their sinkhole flood plains but also take other measures, such as, restricting lot sizes to a minimum of three acres (Aley and Thomson 1981) in order to encourage development in the areas of town not having karst topography. Unfortunately, Bowling Green is located entirely upon a sinkhole plain and does not have this option.

Excavation of Sinkhole Drains

The first step in correcting a sinkhole flooding problem is usually to unclog the sinkhole drain by excavating with a backhoe. Although this often reduces future flooding levels, the excavation occasionally blocks the drain further and flood levels increase. If a crevice in the bedrock can be found, a concrete box with a grate is often constructed to prevent the drain from becoming plugged with soil and debris.

Storm Water Drainage Wells

The second corrective step is to drill one or more drainage wells, usually with a cable tool rig. Wells vary in diameter from 15 to 30 centimeters (6 to 12 inches) and in depth from one meter (3 feet) to over 61 meters (200 feet), but most are approximately 30 meters (100 feet) deep. The wells are cased to bedrock which varies from 30 centimeters (one foot) to over 12 meters (40 feet). Although concrete box settling basins with beehive grates prevent most silt and debris from washing down the wells, many still become clogged. Cleaning them out sometimes restores much of the original capacity.

An investigation of drainage wells, funded by the United States Environmental Protection Agency, yielded a total of 400 drainage wells in the Bowling Green area. The most effective wells intersect solutionally enlarged bedding plane partings or joints, and occasionally they hit microcaves or even large cave passages. Other wells, often located only a few meters (feet) away from wells of high capacity, hardly drain at all. Drainage wells help prevent storm water from ponding in streets and yards during normal rains, and some are effective in preventing or greatly reducing flooding of sinkholes with relatively small catchments even during flood-producing rains. In sinkholes with large catchments, wells do not appear to have the capacity to significantly lower the level of flooding.

About forty sinkhole collapses have occurred adjacent to or near Bowling Green's drainage wells, most of them probably collapsing due to improperly installed well casings. The wells are only cased to bedrock and never sealed. One hypothesis is that a large crack often exists where the casing rests on the irregular bedrock surface, and water thereby flows out of the well and saturates the surrounding regolith each time water backs-up. The water is forced into the regolith under perhaps 10 meters (30 feet) of head, creating a lense of perched water around the well. As the water level in the well drops below the crack, piping of the saturated regolith occurs carrying it into the well. This action results in a regolith arch which expands during floods until it collapses all the way to the surface. In some cases the casing may extend only to a large limestone boulder leaving sections of regolith exposed between boulder and bedrock. Perched water along the regolith-bedrock contact flows down the well carrying regolith and leaving behind an arch. In other situations water flowing from a poorly-sealed well may seep into a nearby hole in the limestone. If piping develops it may create a regolith cave with the potential for sinkhole collapse anywhere above it. In many cases water flowing down the outside of the casing also contributes to collapses. Extending the casing into the bedrock about 3 meters (10 feet) and sealing it with concrete all the way to the surface should greatly reduce the number of collapses.

Groundwater Contamination by Urban Storm Water Runoff

The author is investigating potential contamination of the carbonate aquifer under Bowling Green by urban storm water runoff as part of a 208 nonpoint source pollution grant by the Kentucky Division of Water and the Barren River Area Development District. This study includes monitoring the water quality of the Lost River both before it flows under Bowling Green (Blue Hole monitoring station) and after (Lost River Rise monitoring station). A third station has been constructed for studying urban storm water runoff primarily from commercial landuse at By-Pass Cave swallet. Preliminary findings indicate that urban storm water runoff flowing into By-Pass Cave exceeds the surface water criteria for public water supplies in the following areas: a) fecal coliform, b) oil and grease, c) chromium, d) lead and e) iron. Grab samples revealed that ammonia, BOD_5 and total dissolved solids are also high enough to be considered pollutants during the first flush of storm water into the cave. In addition, suspended solids appear to be a significant pollutant, particularly if one considers its potential for clogging cave passages. However, there are no established water quality criteria for suspended solids for comparison.

Preliminary findings indicate that although nonpoint source pollution by urban storm water runoff is a problem for some of the small streams draining commercial and industrial areas of Bowling Green, it is not significantly impairing the water quality of the large

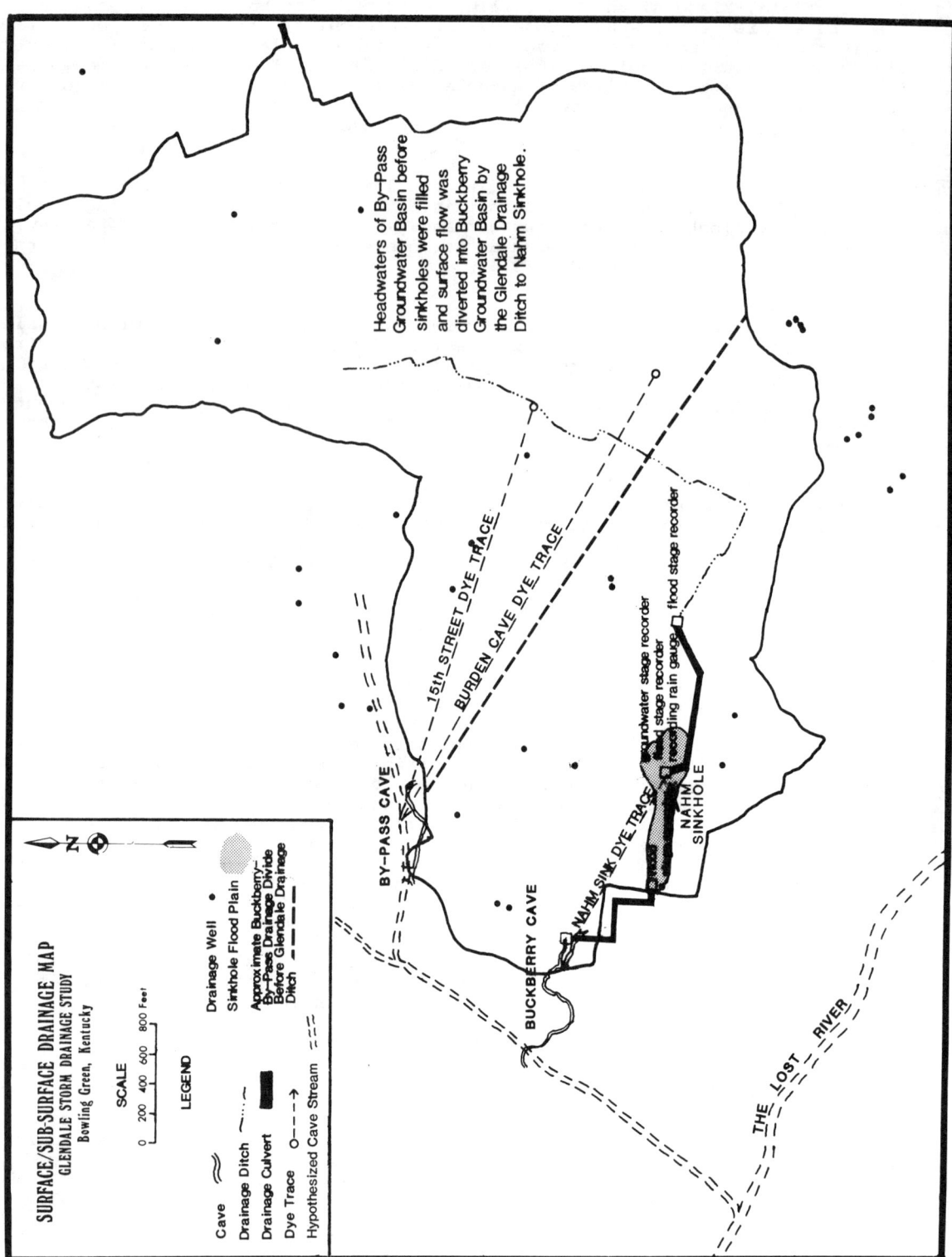

Figure 2. Surface/subsurface drainage map of the Glendale storm drainage study area. Filling of sinkholes and the construction of the Glendale drainage ditch directed storm water which previously flowed to By-Pass Cave into Nahm Sinkhole and Buckberry Cave. The capacity of the Nahm Sinkhole-Buckberry Cave system was greatly exceeded and extensive sinkhole flooding resulted.

Lost River due to dilution by water from the rural area south of town (Crawford 1982).

Attempting to keep urban storm water runoff out of caves under Bowling Green does not appear to be practical, nor possible, even though it carries pollutants into the underground streams. Storm sewers directing storm water runoff to the Barren River have been installed in the downtown area, but their installation throughout the city would be prohibitively expensive because of the karst terrain. Although accepting the use of caves as storm sewers may be difficult, in reality most stream caves receive direct input of runoff from the surface at a rate which is believed to be comparable to surface streams. Even a city-wide storm sewer system could not prevent storm water runoff from getting into underground streams since it sinks naturally throughout Bowling Green. Such a sewer system would be comparable to attempting to prevent storm water runoff from flowing into surface streams draining nonkarstic urban areas. The quality of the subsurface streams in most karst regions will directly reflect the landuse on the surface. Instead of trying to prevent urban storm water runoff from flowing directly underground, perhaps we should attempt to prevent our runoff from absorbing contaminants from our streets and yards. Banning the use of leaded gasoline would be an important first step. Until that can be done, it may be necessary to require runoff from some commercial and industrial areas to be directed into the sanitary sewer or retained and treated on site before being released underground.

Glendale Storm Water Drainage Project:
A Watershed Systems Approach to the Prevention of Sinkhole Flooding

Only in the Glendale area of the city has a watershed systems approach been made in an attempt to solve sinkhole flooding problems. The flooding problem in this area of Bowling Green resulted from increased storm water runoff associated with primarily residential development. Many sinkholes were filled by developers and homeowners, and runoff was directed into adjacent sinks which were usually unable to handle the increased discharge. Overflowing sinkholes during storms produced an unchannelized surface-flowing stream which wound its way through residential property for about 2.5 kilometers (1.6 miles) before completely sinking. In an attempt to alleviate the problem, a concrete-lined storm water channel was constructed to direct storm water to a sinkhole on Nahm Avenue. Since neither the channel nor Nahm Sink had the capacity to convey even a one-year probability storm, they often overflowed.

An intensive investigation of the hydrogeology of the Glendale area was made by the author for G.R.W. Engineers, Inc., and Daugherty and Trautwein Engineers, Inc., (G.R.W. Engineers, Inc. and Daugherty and Trautwein, Inc. 1980). Four stage recorders and a recording rain gauge were installed in the Glendale area and maintained for one year, sinking streams were dye traced, and cave streams were located and surveyed.

The hydrogeologic investigation in the Glendale area revealed three levels of horizontal water movement (Figure 2). 1) The surface runoff, corresponding with surface topography, generally flows south and then west. 2) Shallow caves, believed to be perched upon chert layers only about 5 to 10 meters (16 to 32 feet) underground, deliver sinking surface waters toward the northwest. Most of the storm water runoff from the area sinks and flows by way of shallow caves, such as Buckberry, Burden and By-Pass, to the deeper and larger Lost River. 3) The water table as determined from 29 drainage wells is generally about 15 to 20 meters (48 to 65 feet) below the surface, with a gradient to the southwest towards the Lost River.

The investigation led to the discovery of Buckberry Cave, 427 meters (1400 feet) northwest of Nahm Sink. A dye trace revealed that water flowed from Nahm Sink to a stream discovered in Buckberry Cave. The capacity of Nahm Sink was determined to be approximately 0.2 cms (7 cfs) before overflow. Observations indicated that a constriction between Nahm Sink and Buckberry Cave was responsible for the low capacity of the sink. Another constriction occurred due to the collection of water-borne debris on the trash rack protecting the concrete pipe leading to Nahm Sink.

Most of the construction budget for correcting this problem was spent in the Nahm Sink area. An ingenious, self-cleaning, arrow-shaped trash rack, designed by David Daugherty, was constructed, the storage capacity of the sink was increased by excavation, and a pipe was laid to direct the overflow of Nahm Sink to the stream in Buckberry Cave (Figure 2).

Another solution to the flooding problem was rejected as too experimental, and there were insufficient funds in the budget to attempt something which might not work. Dye traces performed during the investigation at two locations where storm water sinks in the upper half of the surface drainage basin revealed that the subsurface drainage flowed to By-Pass Cave (Figure 2). Previous to urban development, extensive filling of sinkholes (and other holes where storm water flowed underground), and construction of the concrete-lined Glendale drainage ditch, much of the water which has been directed to Nahm Sink in the Buckberry Groundwater Basin must have flowed into By-Pass Cave in the adjacent By-Pass Groundwater Basin. A dye trace of storm water flowing into the entrance to Burden Cave revealed that it flows under the Glendale drainage ditch on its way to By-Pass Cave. If one could determine exactly where it crosses under the channel by cave exploration and survey or geophysical techniques, a shaft could be constructed to direct water from the channel into the cave stream below. It is believed that this would greatly reduce the flooding problems in the Nahm Sink area by diverting this water back into the By-Pass Groundwater Basin where it flowed originally. Perhaps if this relatively inexpensive technique had been tried, the construction in the Nahm Sink-Buckberry Cave area would not have been necessary.

It is unfortunate that a hydrogeologic investigation of the Glendale area could not have been made previous to development. Perhaps the important stream sinks for storm water runoff could have been protected, the headwaters of the By-Pass Groundwater Basin would not have been directed into the Buckberry Groundwater Basin via the Glendale drainage ditch, and the flooding problems which have plagued this neighborhood for over 40 years could have been avoided.

Bowling Green-Warren County Storm Water Management Program

The Bowling Green-Warren County Storm Water Management Program was established in 1976 under the direction of John Matheney, Director of the City-County Planning Commission (Matheney 1984). David Daugherty P.E. assisted the Planning Commission in developing the program (Daugherty 1976) which is primarily designed to prevent future sinkhole flooding problems by preventing development in sinkhole flood plains. Before approving drainage plans for new subdivisions, industrial and commercial sites, and other types of landuse changes, the City-County Planning Commission requires the following.
1) Flood easements in sinkhole bottoms restrict development below a line one foot above the contour correlating with the flood elevation which would result from surface runoff during a three-hour, 100-year probability rainfall event (10 centimeters or 4 inches of rain in three hours in the Bowling Green area). The 100-year flood line is based upon an assumption of zero drainage from the sinkhole bottom and will therefore usually be higher than the actual level of flooding. If the developer can verify the quantity of outflow by field measurements, the 100-year flood line may be lowered accordingly.
2) Downstream areas must not be subject to any flood aggravation as a result of new construction as follows:
   a) if the drainage outlet for the new construction is a surface-gravity system, the increased runoff during a one-hour, 100-year storm must be retained on site;
   b) if the drainage outlet is a sinkhole, the increased runoff during a three-hour, 100-year storm must be retained.

Sinkhole Flood Plain Zoning Restrictions

In general, very few flooding problems exist in the newer areas of town where the plan has been implemented. The sinkhole flood plain zoning restrictions are usually more than adequate. However, some problem sinkholes exist which receive water from beyond their topographic divides. The depth of flooding in these sinks is not directly related to their size. Such sinks need to be identified, the three-hour, 100-year flood contour calculated, and appropriate zoning applied. They include: sinks with ephemeral springs which deliver water from other areas during and after prolonged or high intensity rainfalls, and sinks draining into subsurface streams that back-flood due to constrictions downstream.

During floods water may back up behind a cave constriction (such as a breakdown collapse area) until it has sufficient head to force the floodwater through (Figure 3). Flooding of interconnected sinkholes upstream from the constriction may result while those downstream may drain without ponding even during the largest of floods. The water level in the flooding sinkholes upstream will reach a common level which is not directly correlated with the catchment size of each sink. Eliminating a flooding problem in one sinkhole by cleaning out the sinkhole outlet or by the installation of a drainage well may result in increased flooding of other sinks which are upstream from a cave constriction and lower in elevation (Figure 3). If two sinks are connected by a common conduit, urban development or other landuse which increases runoff into one sink may result in flooding of the other sink some distance away as water backed-up by a cave constriction flows out of the swallet at the distant sink. The sink where the development takes place may or may not flood, depending on its elevation.

Bowling Green is rapidly growing towards the southeast, upstream in terms of the subsurface Lost River. Urban expansion in that direction will increase the flood crest of the Lost River as more storm water runoff is directed underground faster. This may increase the depth of flooding in sinkholes downstream, and sinkholes which have not flooded in the past may flood in the future. An intensive investigation is needed of the effects of increased runoff on areas which are lower in elevation and downstream in terms of the flow paths through the carbonate aquifer upon which the city is built. Flood retention reservoirs to retain increased storm water runoff resulting from changes in present landuse as required by the City-County Planning Commission should help to reduce this potential problem.

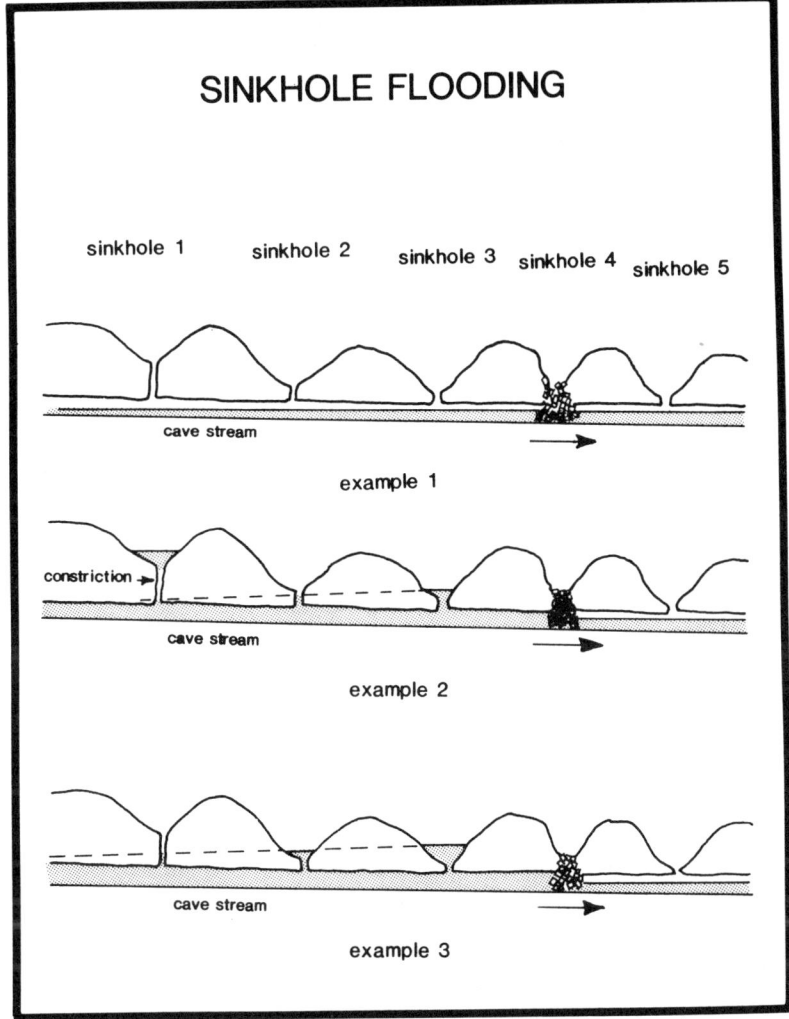

Figure 3. Sinkhole flooding: elimination of constriction in sinkhole No. 1 results in increased flooding in sinkholes Nos. 2 and 3.

Sinks and intersink areas which receive overflow from neighboring sinks should also be identified and zoned appropriately. This type of problem exists in several areas of Bowling Green where flooding affects houses and businesses which are located between sinkholes. In some areas of the sinkhole plain south of Bowling Green long-time residents report that during major floods, such as the flood of 1937, numerous sinkholes overflowed creating wide, slow-moving, surface streams several kilometers in length. These areas need to be identified and zoned accordingly as future urban and suburban development may increase the frequency and severity of this type of flooding.

Storm Water Retention Basins

The restriction of development in flood prone areas by zoning combined with construction of storm water retention basins is a very effective method of dealing with storm water flooding in karst areas. Retention basins: 1) prevent storm water flooding in the local area, 2) retain storm water on the surface thereby relieving pressure on the already overloaded subsurface drainage system, 3) provide a means of filtering storm water through the soil thereby protecting the subsurface drainage system from silt, trash, oil and grease, and some other pollutants, and 4) are far less expensive to construct and maintain than storm sewers which are often prohibitively expensive in karst regions.

Although the many retention basins which have been constructed since 1976 have been very effective in preventing sinkhole flooding, developers complain about the expense of their construction, the nonproductive use of valuable property, and maintenance costs. Also, the high clay content of the soil in the Bowling Green area prevents most of the basins from draining well. The stagnant water is unsightly, unhealthy, and reduces the storage capacity of the basin during subsequent rains. Consequently, drainage wells have been drilled in virtually all basins.

Sinkhole collapses are a problem in most retention basins. The risk of sinkhole collapse is greatly increased when basins are excavated as deep as possible in order to take up less surface area. Drainage wells within the basins also increase the chances for collapse. Although some basins drain more slowly after collapses, most tend to drain faster. Some actually cease to be retention basins in that they no longer retain water. Instead, they become funnels which collect and direct storm water underground.

Of the twenty sinkhole collapses which have occurred within the last five years in the rapidly growing Greenwood area, all but five were in retention basins. During floods if excessive amounts of storm water runoff flow directly into the ground and into the Carver Well Cave system, the elevation of flooding in the cave could be greatly increased. This could result in groundwater rising through solutionally enlarged conduits in the limestone which are normally used by water flowing down from the overlying regolith. If the water level rises above the bedrock-regolith contact, the increased weight and lubrication applied to regolith arches during floods, followed by a rapid decline of support for the arches as the

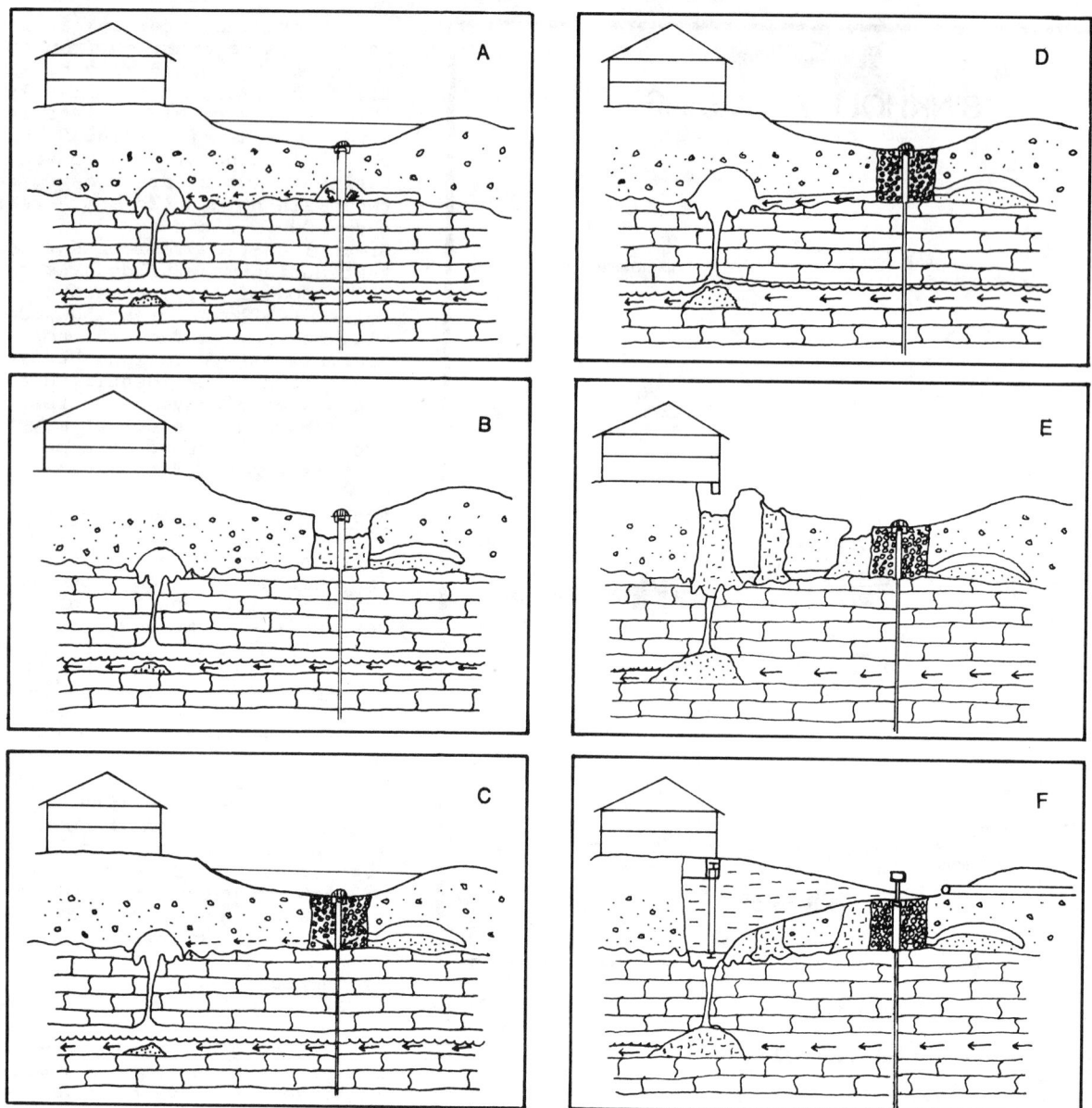

Figure 4. Sketch of hypothesized sequence of events, Greenwood Villa Apartments sinkhole collapse. A) Regolith arches formed at retention basin drainage well and at natural drain under apartment building. B) Initial collapse occurred at retention basin drainage well during August 30, 1984 flood. C) Initial collapse was excavated to bedrock and filled with large rock during summer of 1983. D) During the May 7, 1984 flood, piping of saturated regolith occurred which immediately resulted in collapse. E) Surface collapse occurred in three places due to collapse of the regolith arch over the natural drain and horizontal piping of regolith. F) The collapse was excavated to bedrock under the apartment building. A 1.2 meter (4 feet) thick, concrete slab was poured over a fissure in the bedrock. Two steel I-beams were installed to support the new steel reinforced, 1.2 meter (4 feet) thick building foundation. The excavation was then filled with compacted soil, the retention basin was graded over and a concrete pipe was laid to direct storm water runoff to an existing retention basin 100 meters (327 feet) away. A 10 centimeter (4 inch) diameter plastic monitoring well was placed inside the drainage well before it was plugged with concrete.

water level drops, could result in sinkhole collapses as arches collapse all the way to the surface.

Although a fluctuating water level along the bedrock-regolith contact may be the cause of some collapses, the great majority seem to result from piping associated with water ponded on the surface. In commercial areas, particularly in the Greenwood area, retention basins are being placed dangerously close to buildings. On May 7, 1984 a collapse in a retention basin extended under a nearby apartment building resulting in substantial damage (Figure 4).

Even though retention basins have had several problems in the Greenwood area, the Storm Water Management Program has been significantly successful. Considering the rapid commercial development in an area of shallow sinkholes with large catchments, serious flooding problems undoubtedly have been prevented by the Program.

Conclusions

The cave streams under Bowling Green served as storm drains for storm water runoff before the city was built, and they continue to do so today. Nature has provided a landscape resembling funnels which effectively collects and directs storm water runoff into the underlying carbonate aquifer. Sinkhole flooding results in property damages primarily because: 1) structures have been built within natural sinkhole flood plains, and 2) sinkhole flood plain 100-year flood contours have been raised by: a) unwise landuse practices which have resulted in sinkhole drains becoming clogged with sediment and debris, b) filling of sinkholes which decreases their storm water storage capacities and often clogs their drains, and c) urbanization which increases the impervious area of the catchment and results in an increase in storm water runoff.

The Bowling Green-Warren County Storm Water Management Program is effectively dealing with flooding problems by requiring flood easements in sinkholes below the flood level of a three-hour, 100-year rainfall assuming no drainage from the sinkhole. The Program also requires retention of any increase in runoff during a 100-year rain resulting from a change in landuse. Although the program is successfully reducing flood losses, the numerous small retention basins have taken valuable property out of production, have been expensive for developers to build and maintain, and the majority have experienced sinkhole collapse.

One recommendation is that the Storm Water Management Plan be modified to include a watershed systems approach with an increased emphasis on hydrogeologic research and pollution control (Matheney 1984). The watershed systems approach would identify by hydrogeologic research "true" watersheds, subsurface flow directions, and major caves. Integrated storm water management plans could then be developed for each watershed. Ditches and storm sewers might be used to direct storm water runoff into a minimum number of large sinkholes in each watershed, and drainage wells or vertical shafts might be used to direct non-contaminated storm water runoff directly into cave streams. Retention basins might be planned for contaminated runoff from commercial or industrial areas so that it could be released slowly into the sanitary sewer system or treated before being released into the underground stream. A watershed systems approach based upon sound engineering and hydrogeologic research combined with the present concepts of sinkhole flood plain easements and storm water retention should eliminate most of the problems with the present system. The modified program would: 1) be less expensive, 2) be much easier to construct and maintain, 3) provide a better mechanism for controlling the flow of contaminated runoff into cave streams, 4) allow for a more effective use of the land, and 5) be more equitable in that cost could be spread throughout an entire watershed instead of requiring selected properties to bear a disproportionate cost.

Acknowledgements

Funding for this research was provided by U.S. Environmental Protection Agency, Kentucky Division of Water, Barren River Area Development District, and the Faculty Research Committee and Center for Cave and Karst Studies at Western Kentucky University.

References

Aley, T. and Thomson, K.C., 1981, Hydrogeologic mapping of unincorporated Greene County, Missouri, to identify areas where sinkhole flooding and serious groundwater contamination could result from land development: Report prepared for Greene County Sewer District by Ozark Underground Laboratory, Protem, Missouri.

Booker, R.W., and Associates, Inc., 1978, Study of sinkhole flooding, Bowling Green and Warren County, Kentucky: Report prepared for Federal Insurance Administration.

Crawford, N.C., 1981, Karst flooding in urban areas, Bowling Green, Kentucky: Proceedings of the Eighth International Congress of Speleology, Western Kentucky University, Bowling Green, Kentucky.

Crawford, N.C., 1982, Hydrogeologic problems resulting from development upon karst terrain, Bowling Green, Kentucky: Guidebook prepared for U.S. Environmental Protection Agency Karst Hydrogeology Workshop, Nashville, Tennessee.

Daugherty, D.L., 1976, Storm water management: City-County Planning Commission of Warren County, Kentucky.

G.R.W. Engineers, Inc. and Daugherty and Trautwein, Inc., 1980, Glendale, a storm drainage study, Bowling Green, Kentucky.

Kemmerly, P., 1981, The need for recognition and implementation of a sinkhole-floodplain hazard designation in urban karst terrains: Environ. Geol. 3.

Matheney, J.B., 1984, Bowling Green and Warren County, Kentucky storm water management program review: City-County Planning Commission of Warren County, Kentucky.

Mills, H.H., Starnes, D.D., and Burden, K.D., 1982, Coping with sinkhole flooding in Cookeville: Tennessee Tech. Journal.

Litigious problems associated with sinkholes, emphasizing recent Kentucky cases alleging liability when sinkholes were flooded

JAMES F.QUINLAN *National Park Service, Mammoth Cave, Kentucky, USA*

ABSTRACT

No Kentucky statutes specifically apply to damages to structures which are affected as a consequence of flooding of sinkholes. None of the relevant case law decisions has been published, so the few decisions that have been made are neither binding nor citable in litigation. Nevertheless, it is useful to review the rationale of both the plaintiffs and the defendants in four recent Kentucky cases in which homes were built in or near sinkholes and repeatedly flooded. The plaintiffs have argued that the laws relevant to product liability, rather than real estate, should also be applied to homes that flooded. The tendency to flood constituted "a breach of implied habitability" and the developers had an obligation to provide a suitable dwelling. The value of each house was alleged to have been reduced to zero. The defense, representing the developers, maintained that they had acted in good faith and one argued that the principle of caveat emptor should apply. Where consultants had designed the developments they were not co-defendants. In cases that went to trial, juries have found for the plaintiffs. One case was appealed and settled out of court. Another may be appealed. A third case will probably never go to trial.

Introduction

There are two major types of litigious problems associated with the construction of buildings, roads, and other structures in and near sinkholes: subsidence/collapse and flooding. This paper will discuss the hydrology of these events only as they relate to liability; references are given to the relevant hydrologic literature.

There are no state or federal statutes specifically concerning sinkholes but the case law concerning subsidence and collapse is fairly extensive; it has been repeatedly reviewed. Accordingly, the major emphasis of this paper will be on the rationale of plaintiffs alleging damages caused by flooding.

Guide to Reviews of Case Law

Perhaps the most comprehensive review of divers litigious matters concerning ground water is the synthesis by Coogan (1975). He illustrates and discusses 28 problem cases, only a few of which concern sinkholes. His invocation of rapid solution rates in limestone is unrealistic, but this review is extremely useful to those preparing a groundwater case.

Geologists commonly distinguish between subsidence (relatively slow downward movement or sinking of the land surface induced either anthopogenically or by natural causes) and collapse, which is more rapid and which generally involves more cracking and much greater differential displacement. The law, however, rarely makes this distinction.

The more comprehensive syntheses of case law as applied to subsidence (related to groundwater pumpage in both karst terrains and non-karst terrains) have been made by Morris (1981), Teutsch (1979), and Kenyon (1979). Another useful guide is by Maloney et al. (1980). Morris, Teutsch, and Maloney et al. include review of the two landmark subsidence/collapse cases: Finley v. Teeter Stone Co. (251 Maryland 428, 248 Atlantic Reporter, 2nd ser., 106; reprinted by Tank, 1983, p. 515-526) and Smith-Southwest Industries v. Friendswood Development Co. (Kenyon, 1979). None of these reviews discuss the almost humorous "battle of the bigger pump" in carbonate rocks near Hershey, Pa. (Foose, 1953, 1969).

Three theories have been advocated by litigants in cases contesting subsidence induced by pumping: 1) absolute ownership (the English rule), 2) the doctrine of reasonable use (the American rule), and 3) lack of a right to cause subsidence, as proposed in Restatement of Torts (American Law Institute, 1977, Sect. 818). Recent cases suggest that the courts will provide some protection against damages from subsidence even when the absolute ownership rule is recognized, and that the reasonable use rule is inadequate when applied to subsidence. As pointed out by Maloney et al. (1980, p. 57), it remains to be seen if the strict liability approach of the Restatement of Torts will prevail. Morris (1981), has pointed out that many courts have clouded the subsidence issue by applying water law to determine liability. The problem with any theory or rule of water law is that subsidence generally involves interference not with water rights but with property rights. Accordingly, she opines that the best theory for subsidence damage is, therefore, loss of subjacent support.

I am unaware of any case law associated with liability for leakage of ground water from sinkholes and other impoundments in karst areas (Aley et al., 1972).

I am also unaware of statutes or case laws that are specifically concerned with flooding caused by water contained by sinkholes. There have been several Kentucky lawsuits over flooding of dwellings in and adjacent to sinkholes but, since these cases have not been appealed on grounds relevant to the flooding issues, the cases have not been published and are not citable as precedents in other litigation. Nevertheless, the rationales of the plaintiffs could be used in other cases and they are worth reviewing.

Theories of Liability Advocated by Plaintiffs in Kentucky Cases of Sinkhole Flooding

The geology and hydrology of sinkhole flooding in the Bowling Green area of Kentucky has been incisively analysed by Crawford (1982, 1983, 1984), and will not be summarized herein. Crawford's interpretations and hypotheses could be reasonably applied to many other maturely karsted areas.

A 1983 case in the Pulaski area involved a home which flooded 11 times during a 2-year period. Plaintiffs claimed that there was an "implied warranty of habitability" and cited Crawley v. Terhune (Kentucky, 437 Southwestern Reporter, 2nd ser., p. 743-745), and McDonald v. Mianecki (398 Atlantic Reporter, 2nd ser., p. 1283-1295). Crider (1982), has reviewed the development and scope of this implied warranty in the U.S. Since 1957 it has been adopted by most states, either judicially or by statute. This warranty is briefly reviewed also by Smith and Wenzel (1982). The defendents in the Pulaski case were the developers who had designed the subdivision.

In an earlier case, in the Ft. Knox area, a sinkhole was blocked by sediment deposited as a consequence of erosion caused by loss of vegetation and as a consequence increased runoff from a shopping mall. Numerous homes bordered the drainage axis and were flooded. Successive continuancies have prevented this case from being heard. The plaintiffs have all been transferred, and this case will probably never go to court. The same attorney, however, did win an earlier sinkhole flooding case in which he alleged both negligence and breach of implied contract of habitability. The developer had ignored the specific drainage recommendations of the consultant who designed the subdivision.

In the most recent case, in Bowling Green, the plaintiffs alleged an implied warranty of suitability for intended use, citing an Oregon case involving erosion of an oceanfront lot, Beri, Inc. v. Salishan Properties, Inc. (Oregon, 580 Pacific Reporter 2nd ser., p. 173-179). There were two defendents, the developer (a farmer who had hired an engineering firm to do all design work and then turned the subdivision over to a broker; he never met the buyer of any lots or made any representations about the suitability of the property), and the builder (who had allegedly been warned by a neighbor about flooding of the homesite in question, the last lot on which a home was built; the builder could have easily modified the design of the home so that it would have been a few feet higher and thus flood-proof). The consulting firm that designed the development before there were any city or county drainage regulations was not involved in the trial. The plaintiffs, a civil engineer and his wife, recognized that there was possible flooding of the home and, before buying, had a clause added to the contract specifying that the builder would drill a "dry well" for drainage if the house flooded during the first year of occupancy. It did, and a dry well, albeit a less than properly designed and properly completed dry well, was drilled. The home flooded several times after the dry well was drilled.

The builder, experienced in home construction in a city where sinkhole flooding was already a widely recognized problem (Crawford, 1982, 1983, 1984), was found by the jury to be not liable for the flooding. The developer, claiming honorable ignorance and caveat emptor, was held to be liable. The case will probably be appealed. It has state-wide and national implications for the liability of people who sell land for development.

Testimony for the plaintiffs included a calculation by a consultant of the height and frequency of predictable flooding of the sinkhole in which the home was located. This calculation, based on the widely used and generally valid methodology of the Soil Conservation Service (1975), could not -- except by exceedingly improbable coincidence -- give a valid estimate of flooding in a sinkhole. The complexity of flooding in a karst area, discussed by Crawford, and the fact that the SCS methodology is based on data relationships derived from non-karst areas, makes any calculation of sinkhole flooding unreliable.

Alternative Theories of Liability for Sinkhole Flooding

I am unaware of the application of case law relevant to floodwater (diffuse surface water) to problems of sinkhole flooding. As discussed by Maloney et al. (1980, p. 586-673), and Clark and Martz (1967), civil law rule provides that the upper owner has an easement on the lower owner's land for the water to drain in its natural manner. The common enemy rule gives the lower owner the right to take any measures necessary to keep the water off his land, even to the extent of diverting the water back onto the land of the upper owner. Some states have abandoned both rules in favor of the tort-oriented rule of reasonable use advocated by the Restatement of Torts (American Law Institute, 1977, Sect. 822-31 and 833).

A plaintiff might allege fraud and/or negligence, if either or both can be shown to be probable or certain. A defendent might cautiously invoke an "Act of God" but since hydrology is so much more quantifiable today then it was 30 to 100 years ago, this defense may be refuted on the grounds that certain consequences were forseeable.

An Additional Guide to Literature Relevant to Sinkhole Flooding

A new monthly, six-page newsletter, The Flood Report, an analytical review of issues relating to the National Flood Insurance Program, has regularly published analyses of recent litigation concerning flooding. Volume 1 has included only one article specifically discussing sinkholes. But many other articles, about one per month, have been very relevant to analysis of liability for flooding of sinkholes. The Flood Report is available for $125.00/year from: Flood Law Publishers, 5028 Wisconsin Ave., N.W., Suite 101, Washington, D.C. 20016.

References

Aley, T.J., Williams, J.H., and Massello, J.W., 1972, Groundwater contamination and sinkhole collapse induced by leaky impoundments in soluble rock terrain: Missouri Geol. Survey, Engineering Geology Series, no. 5, 32 p.

American Law Institute, 1977, Restatement (Second) of Torts. St. Paul, Minn., American Law Institute, v. 4.

Clark, R.E. and Martz, C.O., 1967, Classes of water and character of water rights and uses, in Clark, R.E., Ed., Waters and Water Rights. Indianapolis, Ind., Allen Smith. v. 1, Sect. 50-55.

Coogan, A.H., 1975, Problems of groundwater rights in Ohio: Akron Law Review, v. 9, p. 34-115.

Crawford, N., 1982, Hydrogeologic problems resulting from development upon karst terrain, Bowling Green, Kentucky: Karst Hydrogeology Workshop Guidebook, Center for Cave and Karst Studies, Department of Geography and Geology, Western Kentucky University, 34 p.

------------ 1983, Karst flooding in urban areas: Bowling Green, Ky.: International Speleological Congress, 8th (Bowling Green, Kentucky), Proc., v. 2, p. 763-765.

------------ 1984, Sinkhole flooding associated with urban development in karst areas: Bowling Green, Kentucky: Multidisciplinary Conference on Sinkholes, 1st (Orlando, Fla.), Proc., (this volume).

Crider, G.L., 1982, Indiana's implied warranty of fitness for habitation: Limited protection for used home buyers: Indiana Law Journal, v. 57, p. 479-498.

Foose, R.M., 1953, Groundwater behavior in the Hershey Valley, Pennsylvania: Geol. Soc. America, Bull., v. 64, p. 623-645.

------------ 1969, Mine dewatering and recharge in carbonate rocks near Hershey, Pennsylvania: Geol. Soc. America, Engineering Geology Case Histories, no. 7, p. 45-60.

Kenyon, T.F., 1979, Friendswood Development Company v. Smith-Southwest Industries: There may be hope for sinking landowners: Baylor Law Review, v. 31, p. 108-120.

Maloney, F.E., Plager, S.J., Ausness, R.C., and Canter, B.D.E., 1980, Florida Water Law: Water Law Program, University of Florida, Gainesville, Fla., 762 p.

Morris, J., 1981, Subsidence: An emerging area of the law: Arizona Law Review, v. 22, p. 891-?? (reprinted in Public Land and Resources Law Digest, v. 18, p. 309-335).

Smith, N.K., and Wenzel, M.R., 1982, (Implied Warranty of Habitability): Indiana Law Review, v. 15, p. 328-331.

Soil Conservation Service, 1975, Urban hydrology for small watersheds. U.S. Dept. Agriculture, Technical Release no. 55.

Tank, R.W., 1983, Legal Aspects of Geology. New York, Plenum, 583 p., (esp. p. 515-526).

Teutsch, J., 1979, Controls and remedies for groundwater-caused land subsidence: Houston Law Review, v. 16, p. 283-331.

Toxic and explosive fumes rising from carbonate aquifers: A hazard for residents of sinkhole plains

NICHOLAS C. CRAWFORD *Western Kentucky University, Bowling Green, USA*

ABSTRACT

A spill or leak of hazardous chemicals upon a sinkhole plain may quickly sink into underlying cave streams and become a threat to water supplies and aquatic life. Upon vaporizing, it may become concentrated in the cave atmosphere and rise into homes on the surface. Bowling Green, Kentucky, built entirely upon a sinkhole plain, has frequent problems with toxic and explosive fumes. In 1981 five homes in the Riverwood area had to be evacuated because gasoline fumes, rising from caves into their basements had reached explosive levels. In the fall of 1982 the large Lost River Cave beneath Bowling Green became filled with fumes. The toxic chemicals benzene and methylene chloride were found in the Lost River, and it is believed that they were vaporizing into the cave atmosphere. The U.S. Environmental Protection Agency Emergency Response Team treated the source of the chemicals, a spring south of the city, until their concentrations were considered safe. In December 1983 potentially toxic and explosive fumes became a problem in the Forest Park area of the city. The fumes are affecting the residents of about ten homes as they rise from basements, crawl spaces, drainage wells, and sinkholes. The U.S. Environmental Protection Agency Emergency Response Team and the U.S. Center for Disease Control are presently investigating possible health problems associated with the fumes, and the City Fire Department is monitoring their combustion levels. The following investigations are recommended: 1) methods of storing and transporting hazardous chemicals in karst regions, 2) emergency response techniques for spills and leaks of hazardous chemicals in karst regions, 3) identification of major traps for floating chemicals in the caves under Bowling Green.

Introduction

Bowling Green is located upon a sinkhole plain in the karst of South Central Kentucky. Karst is characterized by closed depressions called sinkholes, infrequent surface streams, and an integrated subsurface drainage system through caves. Sinkhole plains are extremely vulnerable to groundwater pollution. Virtually anything capable of percolating through the soil or being transported by storm water runoff into sinking streams may contaminate the underlying aquifer. Carried along by underground streams, contaminates may be transported several miles in only a few hours. Contamination problems are particularly serious when they involve toxic or explosive chemicals. Not only are the chemicals a threat to water supplies and aquatic life, but upon vaporizing they may become concentrated in the cave atmosphere and rise into homes on the surface.

Explosive Fumes: Riverwood

Explosive and toxic fumes rising from the caves into homes are not a new problem for Bowling Green. In 1969, five homes on Riverwood Street had a problem with gasoline fumes in their basements. Again in the spring of 1981 the same homes on Riverwood had to be evacuated as gasoline fumes reached explosive concentrations in their basements. Homes on Chestnut Street, Nashville Road and other areas have also had problems with gasoline fumes rising from the underlying caves (Figure 1).

Leaking underground gasoline tanks belonging to auto service stations are believed to be the major source of fumes in the caves. The Bowling Green area has an estimated 1000 buried tanks, most of them containing gasoline. Sometimes the steel tanks rust through, occasionally they are ruptured by nearby blasting, and quite often leaks develop along the pipes and fittings attached to the tanks.

Gasoline floating on a subsurface stream can travel several miles from the site of a leak or spill in just a few hours, rapidly filling cave passages with explosive fumes. The fumes may then rise into homes built over sinkholes, up water wells and storm water

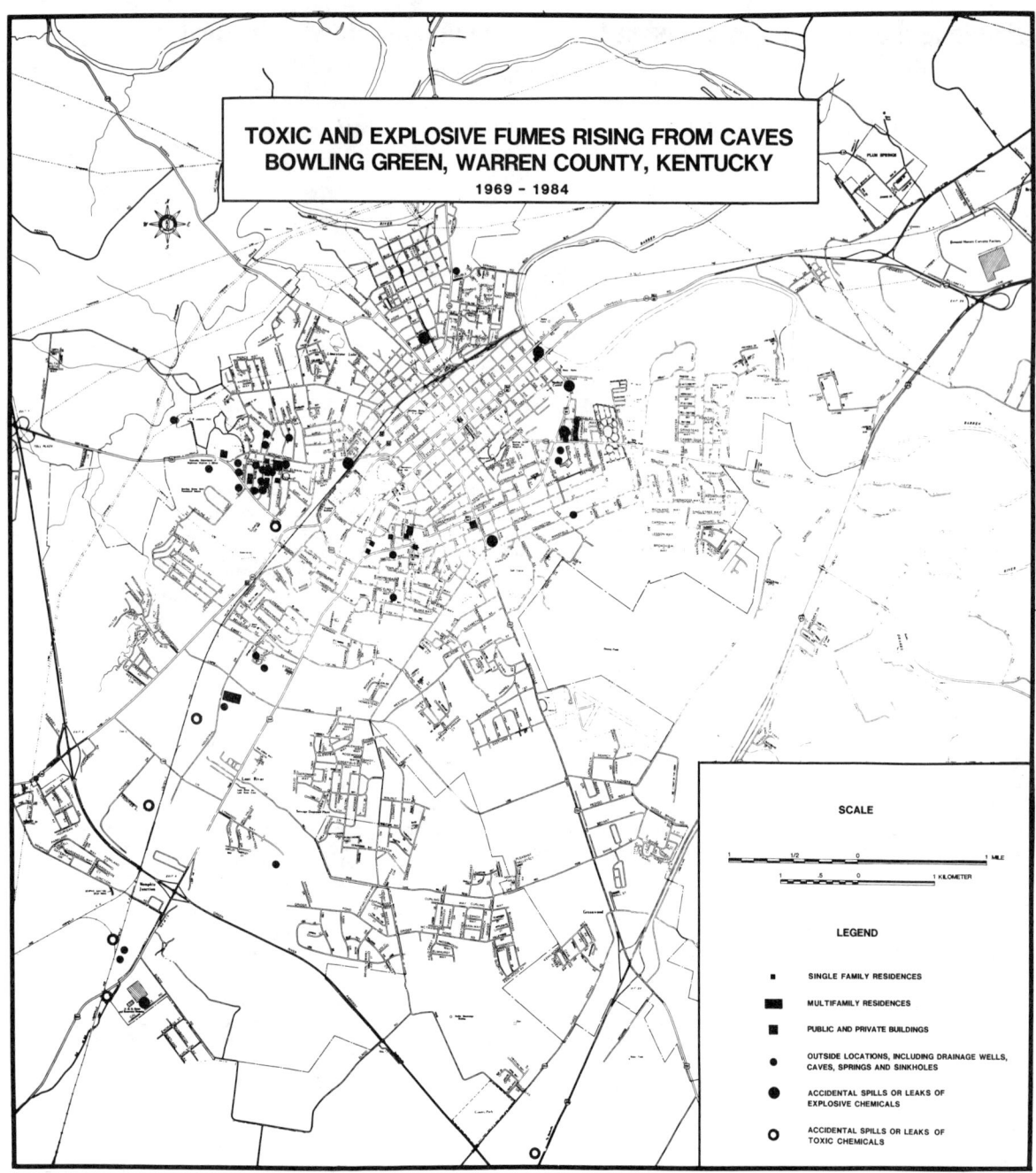

Figure 1: Toxic and explosive fumes rising from caves in Bowling Green, Kentucky

drainage wells, up basement drains and even up abandoned wells and natural openings used for waste disposal.

On one occasion when homes on Riverwood Street were evacuated in 1981, a trench was dug in an attempt to find the cave stream and obtain a sample of the gasoline. Since the backhoe excavated many large blocks of Lost River Chert (a white flint resembling limestone) mixed with the soil, it is fortunate that the cave and gasoline were not located! A check of drainage wells in the vicinity by the author revealed high readings on a combustion meter for two storm water drainage wells in the nearby Fairview Plaza shopping center parking lot. If someone had discarded a cigarette near one of the wells, it is conceivable that the explosion could have travelled through the cave system destroying houses some distance away. An underground explosion in the sewer system of Louisville, Kentucky in 1981 traveled along the sewer system for eleven blocks with estimated damages exceeding forty-three million dollars. The potential for such a large explosion in the caves under Bowling Green in probably remote but certainly possible (Crawford 1982).

Toxic Fumes: Keith Pond

In addition to gasoline fumes, toxic fumes are also a problem. Benzene, a carcinogen, and methylene chloride, a suspected carcinogen, were detected in the Lost River by the Kentucky Division of Water in the fall of 1982, and total organics as high as 15 ppm were measured on May 6, 1983, in the cave air on a portable organic vapor analyzer by the U.S. Environmental Protection Agency. The problem was first reported in September, 1982, by cave explorers who encountered a very strong "paint thinner" or "kerosene" type of odor and observed a red and gray scum on the water at several locations in the cave. As of August 1984, the fumes are still quite strong in the cave at times, and very few people have gone into the worst sections of the cave since the fall of 1982. It appears that benzene, methylene chloride and perhaps other chemicals are vaporizing in the almost constant $14.7^{\circ}C$ cave atmosphere. A potential health threat exists due to chronic exposure to these toxic chemicals. However, only on one occasion is there any evidence of fumes getting into residences on the surface. On April 27, 1983, residents of an apartment building complained to the Fire Department that fumes were getting into their apartments from the nearby Big Bertha entrance of Lost River Cave.

A major source of the toxic chemicals is believed to be a small spring which flows into the Keith Pond south of Bowling Green. The pond overflows into a sinkhole, and the contaminated water then flows through State Trooper Cave, into the Lost River and then under Bowling Green. Dye traces by the author for the U.S. Environmental Protection Agency, Emergency Response Team, have shown that Rhodamine WT dye injected into the ground above the buried tanks of a nearby chemical company flows to the Keith Pond. The chemical company removed its tanks during the spring of 1983 and was ordered by EPA to remove the chemicals from the contaminated soil which surrounded the tanks. However, the soil may still be contaminated, and following rains percolating groundwater may deliver more chemicals from the site down into the cave streams in the underlying limestones.

Toxic and Explosive Fumes: Forest Park

Beginning in December, 1983, the Bowling Green Fire Department, the State Fire Marshall's Office, and Western Kentucky Gas Company began responding to complaints of fumes in, under, and about homes in the Forest Park section of Bowling Green. The fumes were investigated on several occasions and measurements were taken on combustible gas indicators which revealed that the fumes were not present in sufficient concentrations to explode but were potentially explosive. Residents affected were advised to open windows in their basements or to vent their crawl spaces. The author was asked in May, 1984, by Rod Raby of the State Fire Marshall's Office to assist with his investigation. He was also asked to prepare a proposal for dealing with the problem for submission to the City Commissioners by the City Manager (Crawford 1984).

As of August 31, 1984, the investigation has revealed the following in the Forest Park Area:
    a) three homes with fumes in basements,
    b) two homes with fumes in crawl spaces,
    c) five homes with fumes entering through windows from nearby
       sinkholes or drainage wells,
    d) four homes and one business where residents can at times
       smell fumes in their yards,
    e) two apartment buildings where fumes rising from a nearby drainage
       well could at times enter buildings if doors and windows were open,
    f) nine drainage wells with fumes,
    g) two improved sinkhole drains (both near homes) with fumes, and
    h) two sinkholes with fumes

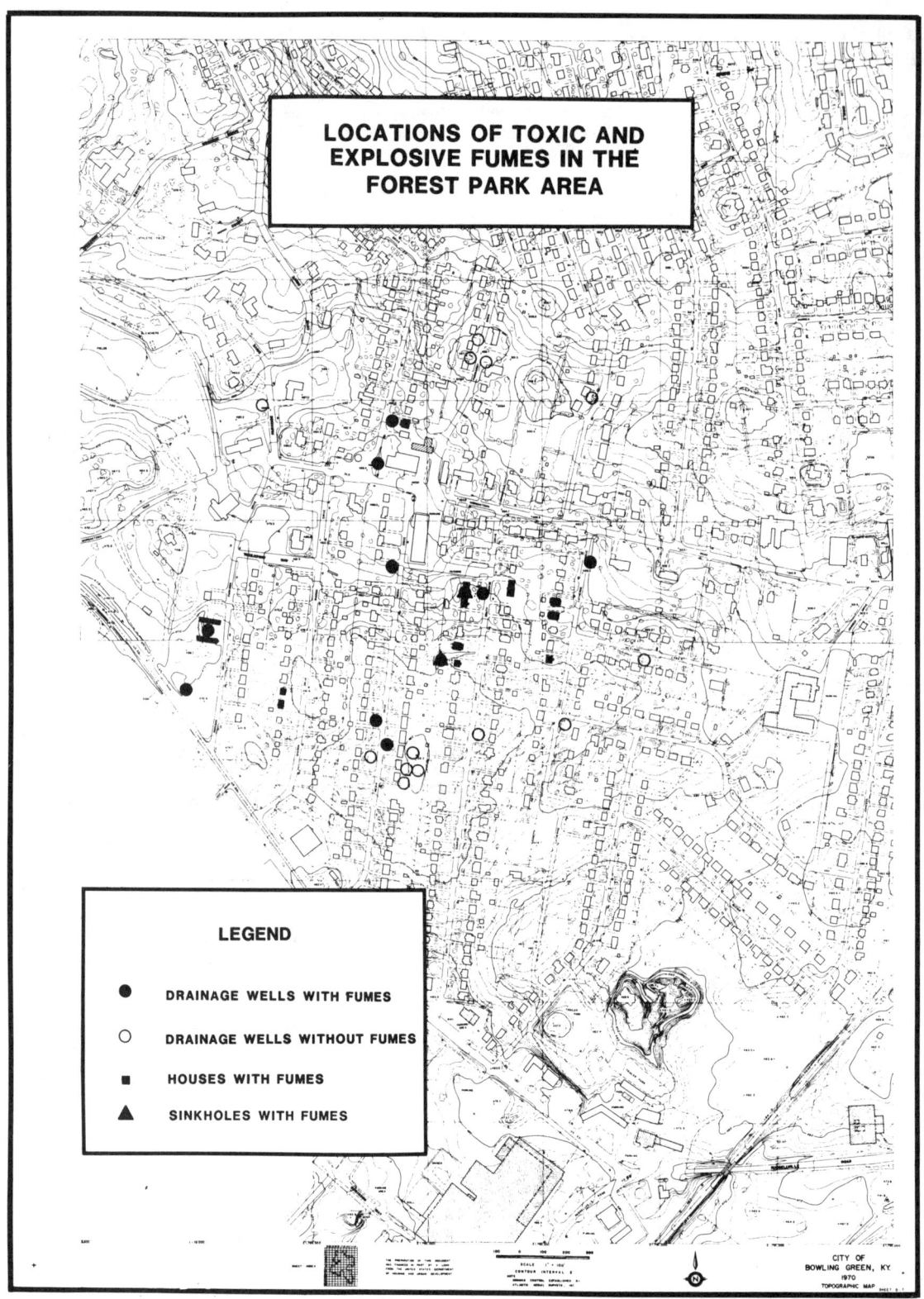

Figure 2: Toxic and explosive fumes in the Forest Park area, Bowling Green, Kentucky.

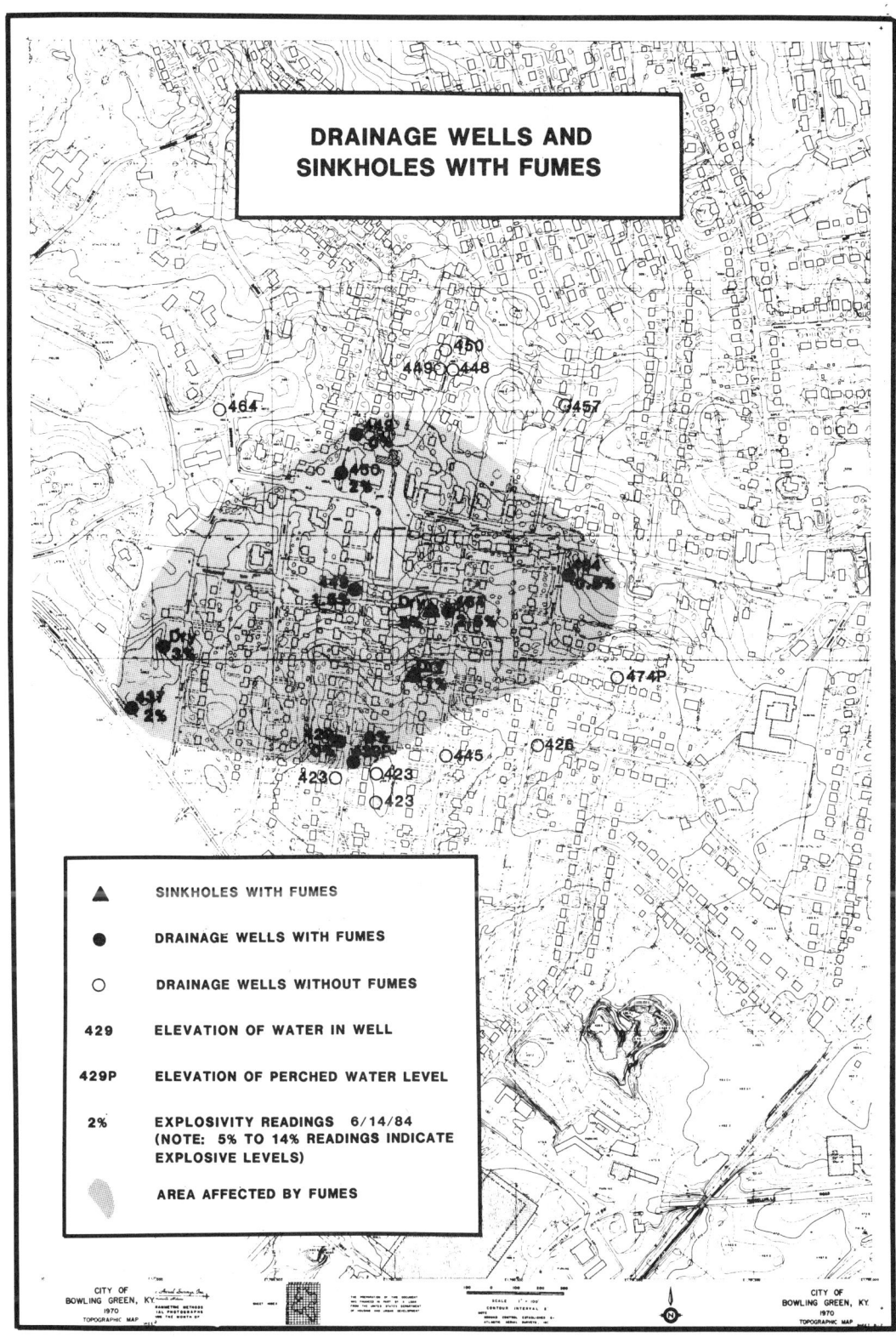

Figure 3: Drainage wells and sinkholes with fumes, Forest Park area, Bowling Green, Kentucky.

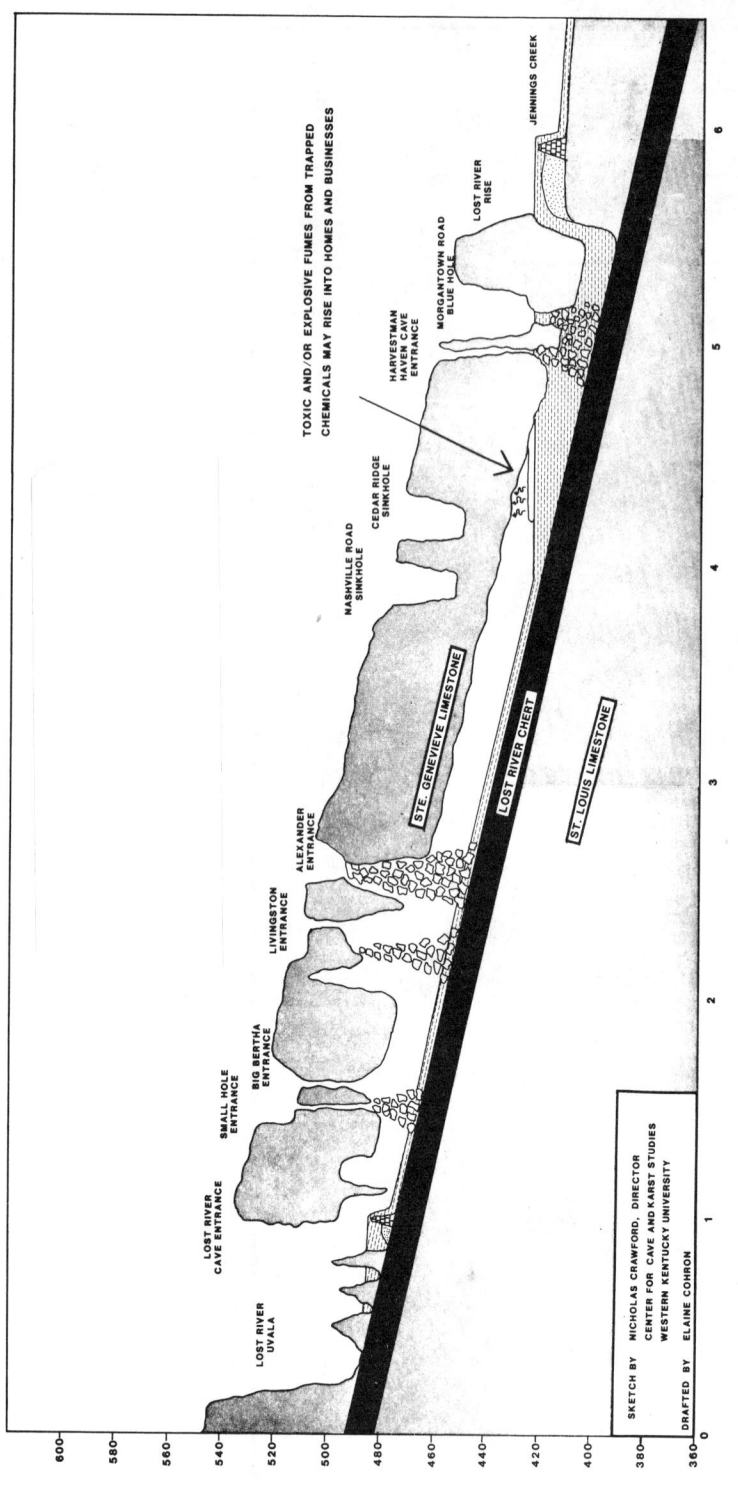

Figure 4: Generalized profile of Lost River Cave under Bowling Green, Kentucky showing potential trap for floating toxic and/or explosive chemicals.

Figure 2 shows the location of ten homes and two apartment buildings where the fumes are considered to be a problem. Drainage wells with fumes are also shown in Figures 2 and 3. Figure 3 shows the water elevation in each drainage well and the explosivity reading taken the morning of June 14, 1984. From this data the general area affected by the fumes has been delimited. Of course the area may change somewhat as more data is collected.

On June 24, 1984, the U.S. Environmental Protection Agency, Emergency Response Team took over the investigation. Since their preliminary investigation indicated that the fumes were potentially toxic containing toluene, xylene, alaphatic hydrocarbons, alkyl benzenes and possibly methylene chloride, a health investigation in cooperation with the U.S. Center for Disease Control was made the first priority. Air samples collected from a bedroom and from under 125 homes were analyzed. Additional air samples and urine samples of residents from nine homes with fumes were taken, and as of August 31, 1984, the laboratory analysis has not yet been completed. The combustion potentials of fumes rising from 9 drainage wells and 2 sinkhole drains are being monitored by the Bowling Green Fire Department. Readings as high as 4 percent were taken in June (5 to 15 percent is the combustion range), but they have lessened to about 1 percent.

Conclusions

Spills and leaks, and sometimes deliberate dumping, of toxic and explosive chemicals occur frequently in a city as large as Bowling Green. With an estimated 1000 buried tanks, with truck and rail tankers traversing the city virtually 24 hours a day, one has to expect accidental spills and leaks. For example, within the past year three significant spills from underground tanks have occurred, each involving thousands of gallons. A 15,141 liter (4,000 gallon) spill of gasoline and a 22,712 liter (6,000 gallon) spill of diesel fuel have both flowed into the Lost River Cave under the city.

In most areas when there are spills much of the liquid can be recovered. It does not sink rapidly into the ground, and if it flows into a stream it often floats and thus can be found, trapped, and pumped back into a tanker truck and recovered. This is usually not the case in karst regions such as Bowling Green, where spills quickly sink into the ground and may be carried several miles in a matter of hours by an underground stream. The stream may flow upon the floor of an air-filled passage for many miles, but it will usually completely fill the cave passage before issuing from a spring. The Lost River, which flows under the southwestern portion of Bowling Green, is a good example (Figure 4). The point were the surface of the stream meets the ceiling in a natural trap for floating liquids such as gasoline. The fumes may then intensify in the cave and rise into homes. During floods the Lost River may rise 20 meters (64 feet), thus displacing the floating chemicals upward and upstream. As the water level goes down a "bathtub ring" is left in the mud on the cave walls and ceiling over extensive areas of the cave, and small stagnant pools with floating chemicals may be left above the stream level. This was observed by the author as he went into the cave to take water samples following the August 30, 1982, flood. The floating chemicals were found to be benzene and methylene chloride mixed with diesel fuel. It is hypothesized that when fumes rise into homes for extensive periods, as they have on Riverwood Street and in the Forest Park area, it is due to floating chemicals trapped against the cave ceiling (Figure 4) at the point where the cave passage becomes completely water-filled.

The author has proposed to do the following investigation if funding can be procured (Crawford 1984).
1) Investigation of the methods of storing and transporting hazardous chemicals. Recommendation of "best management practices" for karst areas. Bowling Green probably cannot require techniques of storage and transportation that are different from the state and federal requirements. However, companies and individuals do not want to lose their expensive chemicals, and they fear potential lawsuits if their losses cause damage to others. If they can be shown that special precautions are needed in karst regions, it is likely that they will take them.

2) Investigation of emergency response techniques for spills and leaks of hazardous chemicals in karst areas. In 1981 when five homes on Riverwood Street were ready to explode, the emergency response was to use a backhoe in an attempt to dig an opening into the cave. One does not use a steel backhoe to dig through flint into a cave full of gasoline fumes! What should be the emergency response when a cave extending for miles under an urban area fills with gasoline fumes-- the fumes possibly rising into someone's basement, perhaps to be ignited by a pilot light? Hopefully, in conjunction with the State Fire Marshall, the local Fire Department and other state and local

agencies, emergency response procedures could be written which
could potentially save lives and property.

3) Prepare maps of groundwater flow under Bowling Green showing a)
subsurface streams, b) caves, and c) potentiometric surface.
The major stream caves under Bowling Green would be located,
explored and mapped. Geophysical techniques, exploratory drilling,
and other techniques would need to be used. The major traps for
floating chemicals could be identified and wells could be drilled
above them so that they could be pumped out in the event of a
spill or leak. Also, knowledge of where the cave streams are
located would be very useful for storm water management, and in
dealing with another problem unique to karst regions, sinkhole
collapse. With virtually no knowledge of where the cave streams
which drain Bowling Green are located or their capacity to handle
storm water runoff, individuals and the city blindly drill
drainage wells and direct storm water runoff underground
anywhere possible. This policy may in some cases be simply
transferring a flooding problem from one location to another, and
it is definitely contributing to sinkhole collapses.

References

Crawford, Nicholas C., 1982, Hydrogeologic Problems Resulting from Development upon Karst
Terrain, Bowling Green, Kentucky: Guidebook for U.S. Environmental Protection Agency
Karst Hydrogeology Workshop, Nashville, Tennessee.

_____, 1984, Potentially Toxic and Explosive Fumes Rising from Caves under
Bowling Green, Kentucky: A Proposal Submitted to Bowling Green City Commissioners.

Characterization of the shallow groundwater system in an area with thin soils and sinkholes (Door Co., WI)

JAMES H.WIERSMA, RONALD D.STIEGLITZ, DEWAYNE L.CECIL & GLENN M.METZLER *University of Wisconsin-Green Bay, USA*

ABSTRACT

Door County, Wisconsin is a region of karst topography underlain by Silurian dolomite bedrock. Numerous sinkholes intercept much of the surface runoff and act as sites for direct groundwater recharge. The clay-rich and impermeable Upper Ordovician Maquoketa Formation separates the dolomite aquifer from the deeper aquifers and appears to be a factor in groundwater circulation and karst formation. Thin glacial drift and Quaternary materials overlie the dolomite and are hydrologically connected with it.

The interactions of surface and groundwater, and the role of solution features in water interchange were studied in a small drainage basin. This basin contains several large sinkholes and a nearby spring complex. Mapping identified many additional sinks and swallets in surface drainage routes. Water flowing into two sinks was traced and found to have a residence time of several hours. Water flowing into sinkholes and from the spring was sampled to identify the quality and seasonal trends in composition of the shallow ground water. Water quality parameters monitored include: magnesium, sodium, potassium, chloride, phosphorous, nitrate and ammonia, nitrogen, alkalinity, pH, turbidity, and specific conductance. Nitrate levels were found to increase 5 to 6 times during periods when there was zero input through sinkhole recharge sites. Nitrate levels approached the 10 mg/ℓ NO_3-N limit set by the U.S. Public Health Service for drinking water.

In this basin sandy soils are most susceptible to sink development, whereas clay-rich soils have a lesser number of sinks. It appears, however, that a network of bedrock solution features exists under all soils. The loss of soil into sinkholes has impacted groundwater quality and reduced agricultural productivity through a reduction in tillable acreage and water retention capacity.

Introduction

The northeastern part of Wisconsin is an area in which the physical environment and landuse pressures combine to result in water quality problems. This is particularly true on the Door Peninsula, (see Figure 1), where scenic features such as high bluffs, sandy beaches, and sheltered harbors attract a growing number of tourists. Cottages and second homes line the shorelines of Lake Michigan, Green Bay, and the inland lakes. Numerous large and elaborate condominium developments have been built which add to water supply and waste disposal problems. In addition, dairy farming operations and fruit growing--primarily apples and cherries--are potential sources of chemicals, nutrients, and organic wastes that can contribute to surface and groundwater problems.

Most of the Door Peninsula is underlain by dolomite bedrock of Silurian age. The western side of the peninsula is characterized by a prominent but discontinuous escarpment formed by the edge of dolomite layers which are inclined gently toward the east and southeast. Solution of the bedrock has been extensive and a great variety of karst features have developed (Stieglitz, 1984). The greatest density of such features is found in areas just east of the dolomite outcrop edge where the bedrock is shallow and the escarpment provides relief for groundwater circulation. Initial mapping of known sinkholes has identified three regions with a high density of sinkholes.

Site Description

In order to begin to understand the interactions of surface and groundwater, and the role of solution features in water interchange within this complex area, a small drainage basin with a few known sinkholes was chosen for detailed study (Stieglitz et al., 1983). The watershed selected was the Sawyer Creek basin located approximately 6.5 km west of the city of Sturgeon Bay in Door County, Wisconsin. The study was centered in sections 3 and 4, T.27N, R.25E, Sturgeon Bay West 7.5' quadrangle.

The headwater area of the watershed is relatively flat and lies about 40 meters above the level of Green Bay. The channel of Sawyer Creek passes across the outcrop edge of the dolomite through a reentrant in the escarpment where it falls over a 3 to 4 m high ledge.

Southeast of the escarpment the stream is intermittent. A series of springs discharge into the channel near the base of the escarpment forming a permanent stream that follows approximately 2.5 km to Sawyer Harbor.

Southeast of the escarpment, the stream passes across Lower Silurian Mayville Dolomite. This formation is composed of dense, finely-crystalline, thin wavey bedded, and well jointed dolomite. This unit and the overlying Middle Silurian dolomites form a prime aquifer system in eastern Wisconsin (Sherrill, 1978). The Upper Ordovician Maquoketa Formation underlies the Mayville Dolomite and outcrops in a narrow lowland between the escarpment and the shore of Green Bay. The Maquoketa Formation is mostly blue-green shale of low permeability which separates the dolomite aquifer above from the deeper aquifers, principally Cambrian sandstones.

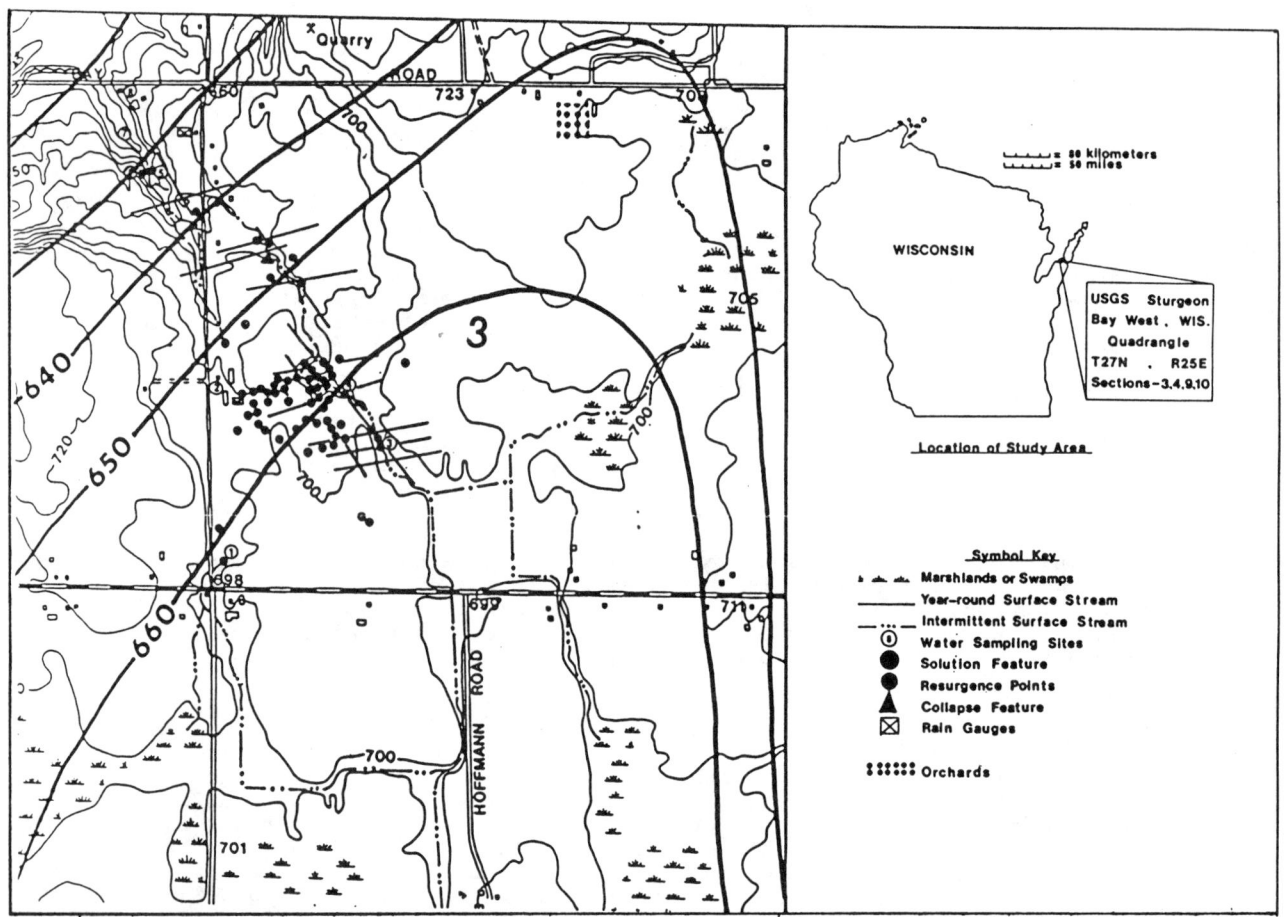

Figure 1: Location of sample sites, mapped sinkholes, and joint trends in the study area, Sawyer Creek Watershed, Door County, WI. Groundwater level contours from Sherrill, 1978

Surface drainage in Sawyer Creek is directed northwestward across the escarpment toward Green Bay, however, the channel only flows during spring snow melt. Shallow groundwater also flows northwestward. Numerous swallets occur in the stream channel. During spring recharge these sinks intercept water until the shallow aquifer is full and the stream flows throughout its entire course. A high percentage of the streams total annual discharge occurs in a very short period in spring. As snow melt proceeds a complex flow pattern develops in which some of the lower swallets become resurgent points where shallow ground water returns to the stream channel. Later, when snow melt is nearly complete the rate of the groundwater reservoir discharge exceeds the recharge rate and the system of swallets sequentially dries up beginning with the downstream swallets. Eventually, the surface flow from the spring complex at the escarpment front contributes the total flow in Sawyer Creek.

Objectives

The objectives of this study were to map solution features, especially sinkhole formation density and to determine their impact and potential impact on land use and water quality. The Sawyer Creek area contains several known sinkholes and a spring complex purported by local residents to be connected to at least one of the sinkholes that receives surface water. Hence, one objective was to confirm the suspected hydraulic connection between sinkholes that accept surface water and the spring discharge point. A further objective was to determine the influence that sinkhole recharge has on the water quality of the upper portion of the groundwater reservoir. (This aquifer is frequently tapped by private wells for domestic supply.) Furthermore, the identification of seasonal differences in water quality was necessary to provide information about the processes whereby recharge occurs and their relative seasonal importance. The final objective was to determine if sinkholes are an important mechanism for soil loss. This is a particular concern because much of the region is devoted to agriculture and has very thin soils (0 to 7 m, with 0.3 to 2.0 m a more typical value).

Study Methodology

Field mapping was conducted to determine the type and density of solution features as well as the pattern of jointing in the bedrock.

Two sinkholes that receive surface drainage that were suspected to be important recharge of the spring complex were studied by chemical tracer techniques. Lithium salt solutions were introduced at the sinkhole recharge points on separate dates using the sudden injection method. Lithium bromide and lithium acetate solutions of approximately 7g Li/ℓ were used in the two studies to confirm hydraulic connection. Lithium salts were chosen as tracers because of the absence of lithium in the environment, their low toxicity, high water solubility, and ease of detection. The intensity of the 670.8 nm emission line of lithium was measured to determine lithium concentration using Perkin-Elmer Model 460 Atomic Absorption instrument in the emission mode.

The location of the sampling points used to study the seasonal variation in water quality are indicated on Figure 1. One prominent sink near the intersection of County Highways C and M (number 1), and another large sink acting as a swallet, in the channel of Sawyer Creek (number 4) served as sample collection points. The two sinkhole recharge points are located only about 1 km from each other, but receive water from two distinctively different regions. The southeast sinkhole is located in a creek bed that drains land with heavy agricultural use. The south sinkhole receives drainage from a swampy region. Because the spring complex (number 5) consists of a wide fan shaped discharge with several bedding planes, three different positions in the complex were sampled at 5-minute intervals over a period of four hours after the tracer was introduced. The sampling positions were located approximately 10 meters apart with a 0.3 meter range in elevation.

To ascertain seasonal changes in water quality the inputs to the sinkholes and the spring complex were sampled 4 to 20 times over the course of about a year. These water samples were subjected to the following physical and chemical analysis: soluble reactive phosphorus, total phosphorus, nitrate plus nitrate-nitrogen (reported as nitrogen), ammonia-nitrogen, magnesium, sodium, chloride, alkalinity, turbidity, and pH. Methods used for the analyses are found in Standard Methods for the Examination of Water and Wastes, fifteenth edition and EPA publication (EPA-60014-79-020), Methods for Chemical Analysis of Water and Wastes.

Results and Discussion

Mapping Studies

Field mapping has revealed a variety of solution features ranging from small scale surface karren to large sinkholes or dolines (Figure 1). Joint trends were measured in areas of exposed bedrock. The dominant set strikes at N73°E and a more variable set strikes from N40°W to N50°W. Some of the joints have been greatly widened by solution action. Sinkholes are common in the area and new ones open each year, usually in spring. Sinks are found throughout the area, however, several concentrations occur where the land has become unfit to farm. Sinks located in areas of deeper soils are aligned along the approximate trend of mapped joint sets shown in Figure 1. Although sinkholes appear to be concentrated in sandy soil areas, it appears certain that bedrock solution features exist under all soil types, but that the clay rich soils more effectively plug the subsurface openings.

Tracer Studies

The lithium concentration-time curves resulting from the tracer studies for the in-stream swallet (number 4) are shown in Figure 2. The shape of the curves indicates that the water experiences little longitudinal mixing which would be indicative of relatively open conduit flow. The fact that the entire stream of water flowing from the fan-shaped spring complex contained lithium indicates that the swallet in the creek bed and ground water moving below it is a major route for the ground water. If this were not the case, that is, if there were other underground streams reaching a confluence at the spring complex, one would expect major differences in the lithium distribution across the width of the spring complex. Samples collected from the West sample site in the spring complex, however, were diluted compared to the samples collected at the East and Center positions in the spring complex. A second tracer study at the C and M site (number 1) indicated that it was also hydraulically connected to the spring complex. Flow from this sink is likely responsible for the dilution observed at the West sampling point.

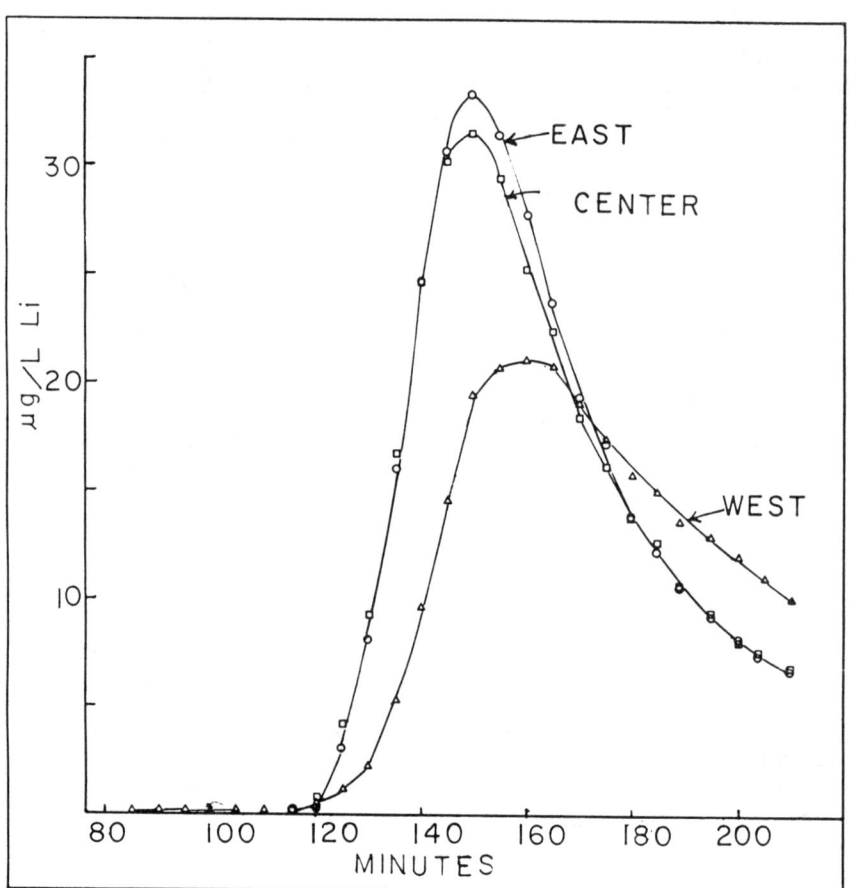

Figure 2: Lithium levels in three different sampling points in the spring complex.

Water Quality Studies

Comparison of physical and chemical water quality parameters in the two sinkhole recharge points and the spring complex give clues to process controlling water quality in karstic regions. A summary of the comparison of major chemical parameters is given in Table 1. A detailed description of more noteworthy seasonal differences observed between the inputs and the spring discharge follows.

Nitrate nitrogen levels varied considerably between the recharge points and the spring discharge throughout the course of the study. Figure 3 shows nitrate nitrogen levels plotted against Julian days; sampling began at the end of March 1982 and continued through January 7, 1983. The data clearly show that the sinkhole inputs do not account for the levels of nitrate-nitrogen measured in the spring discharge. There are two periods when the nitrate-nitrogen concentration increases dramatically in the spring discharge--first in early summer, and second in early-to-mid winter.

Since sink input water spends such a short time underground--3 to 5 hour retention times were observed in the two tracer studies--these sharp increases in nitrate concentration are somewhat surprising. Apparently during periods of high flow, spring runoff or storm events, the nitrate-nitrogen concentrations in the recharge and discharge water more closely coincide because of the dilution of a relatively small volume of high-nitrate, subterranean water with relatively large volumes of low nitrate surface water recharge. But, during periods of zero surface water recharge through the sinks, nitrate-nitrogen levels climbed substantially. Most likely infiltration water continues to slowly recharge the aquifer and carries with it a high level of nitrate to the surface of the groundwater reservoir. Since the spring-complex drains the surface of the aquifer, these high levels are observed in the spring discharge. Hence, the quality of shallow groundwater, especially with regard to nitrate, is highly dependent upon precipitation events. Nitrate levels during dry periods could climb to levels that are of concern to human health (10 mg/ℓ NO_3-N) and the well-being of animals. Fertilizer application rates and timing no doubt have an influence on the level of nitrate found in the infiltration water. Furthermore, if such areas are more intensively farmed or fertilized, one would predict an increase in the nitrate levels in ground water.

Ammonia-nitrogen levels in this study were substantially lower than nitrate-nitrogen levels. The maximum concentration of 0.9 mg/ℓ versus 7.0 mg/ℓ nitrate-nitrogen was recorded in surface water sampled in early spring. Agricultural runoff is the most likely source of ammonia in this basin. Levels in the spring discharge water never exceeded 0.4 mg/ℓ and in almost every case levels were lower than the levels found in the recharge water on the same date. Differences in ammonia concentration on some occasions were much greater than could be explained by simple dilution. This seems to indicate that some ammonia-nitrogen is retained by the aquifer either through sorption or it may be associated with suspended solids which also were observed to be lost in the aquifer matrix on several occasions. It is possible that some ammonia that is retained in the aquifer matrix and may be converted to nitrate through nitrification and subsequently discharged as nitrate.

Table 1. Summary of Water Quality Parameters - Sawyer Creek Sinkhole Study - 1982

	C and M Sink (#1) $\overline{X}$	(N)	Sawyer Creek Swallet (#4) $\overline{X}$	(N)	Spring Complex (#5) $\overline{X}$	(N)
Nitrate-Nitrogen (mg/ℓ)	0.89	14	0.81	4	3.16	19
Ammonia-Nitrogen (mg/ℓ)	0.19	14	0.05	4	0.14	19
Total Kjeldahl-Nitrogen (mg/ℓ)	1.24	8	----	--	1.23	3
Total Phosphorus (mg/ℓ)	0.064	13	0.056	4	0.105	11
Soluble Reactive Phosphorus	0.027	13	0.034	4	0.045	19
Magnesium (mg/ℓ)	32.7	14	35.0	4	38.5	19
Sodium (mg/ℓ)	5.00	14	5.22	4	6.45	19
Potassium (mg/ℓ)	2.14	9	----	--	3.10	5
Alkalinity (mg/ℓ as $CaCO_3$)	262.6	10	193.5	4	286.2	18
Chloride (mg/ℓ)	8.65	14	12.6	4	13.1	19
Specific Conductance (μmhos/cm)	494	13	395	4	583	15
Turbidity (NTUs)	1.4	5	5.3	4	5.4	7

Total phosphorus concentrations at the recharge sites experienced a large pulse in the spring that is attributable to agricultural runoff. But as with ammonia there was no corresponding pulse observed in the spring discharge indicating retention of phosphorus in the aquifer matrix. This phosphorus is probably associated with the suspended fraction. Soluble reactive phosphorus levels were on the order of 25 percent of the value of total phosphorus but may range between 50 to nearly 100 percent of the total phosphorus level when total phosphorus is particularly high.

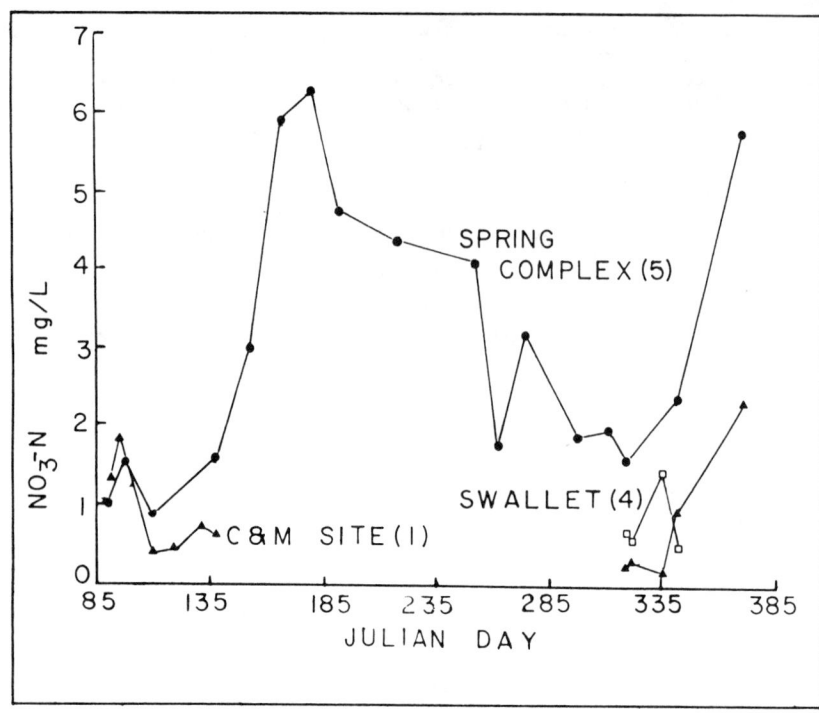

The levels of magnesium, sodium, potassium, alkalinity, chloride and specific conductance experienced small increases as the water moved through the aquifer to the spring discharge. Limited turbidity measurements show both deposition in the aquifer and on one occasion the turbidity levels in the discharge were more than twice the level found in recharge water indicating resuspension of earlier deposited materials. The sequence of events that leads to resuspension was not determined.

Figure 3: Comparison of nitrate-nitrogen levels in the two surface recharge sites, (1) and (4), and the spring discharge (5).

References Cited

Environmental Protection Agency, 1976. Methods for Chemical Analysis of Water and Wastes (EPA-60014-79-020).

Sherrill, M. G., 1978. Geology and Ground Water in Door County Wisconsin with emphasis on contamination potential in the Silurian dolomite. United States Geological Survey, Water Supply Paper 2047, 38 p.

Standard Methods for the Examination of Water and Waste Water, 15th edition, 1980. American Public Health Association.

Stieglitz, R. D., 1984. Karst landforms of eastern Wisconsin. Geol. Soc. Amer., North-Central Section, Abstracts with Programs, V. 16, No. 3, p. 200.

Stieglitz, R. D., J. H. Wiersma, L. D. Cecil, and G. M. Metzler, 1983. Bedrock control of groundwater and surface water interactions in southern Door County, Wisconsin. Geol. Soc. Amer., North-Central Section, Abstracts with Programs, V. 15, No. 4, p. 213.

Altura Minnesota lagoon collapses*

E.CALVIN ALEXANDER, Jr. & PAUL R.BOOK** *University of Minnesota, Minneapolis, USA*

ABSTRACT

In April 1976, a series of karst sinkholes opened in the holding lagoon of the Altura, Minnesota Waste Treatment Facility. This major failure was preceded by minor sinkhole formation during the construction of the facility in 1974. Subsequent detailed field mapping of the region around the community revealed at least 23 sinkholes not shown on existing maps. The distribution of the sinkholes as well as post-failure investigations of the lagoon indicate that catastrophic collapse is related to the presence of a thin, poorly indurated, jointed sandstone overlying a thick carbonate unit. The sandstone served to collect solutionally aggressive vadose water and to concentrate that water onto specific areas of the underlying carbonate. The resulting differential solution produced voids into which the overlying materials collapsed.

Introduction

Southeastern Minnesota is an area characterized by productive farms, small community centers, and a few moderate-sized cities. It is also an active karst area with characteristic geomorphic and ground water quality problems. Sinkholes develop over widespread areas. In most of the region drinking water has been shown to nearly exceed maximum acceptable standards for several parameters, and many individual wells exceed the drinking water standards. Growing concern by local and state leaders over increasing ground water problems prompted support for research in hydrogeology throughout the region.

The history of public works development on karst is all too often one of structural failure and/or ground water contamination caused by underestimating the design limitations inherent to karst regions. Yevjevich (1981) reviews many such problems on a national and international scale. Design experience from non-karst areas has proven to be an inadequate preparator for construction in karst regions. Evaluation of the past failures and their causes from a hydrogeologic standpoint helps to unravel the complexities inherent in karst regions and can help to avoid future problems. This paper documents the hydrogeologic environment and the consequences of one karst-related failure.

Background

Altura is a small town (population 354 in the 1980 census) in southeastern Minnesota. The major local industry, a turkey processing plant, increases the load on Altura's waste treatment facility by a factor of 10 during the 6 months of the year when the plant operates. The town's 1954-vintage filtration plant was seriously overloaded by the processing plant's wastes. During the late 1960s the Minnesota Department of Health received numerous complaints about the malodorous red effluent discharged from the old plant. That effluent sank into carbonate bedrock a short distance from its discharge point.

During the period 1971 through 1974, consulting engineers for the city of Altura designed and built an aerated pond system to treat the community's waste water. The system consists of two 1.1 ha primary aeration ponds, a 5.0 ha effluent storage pond, a chlorination and dechlorination system, and associated pumping system (Ellison, 1977). In the spring of 1974 precipitation waters, accumulated in the nearly completed effluent storage pond, drained suddenly into two sinkholes which developed in the pond bottom (Beaton, 1974). The estimated locations of these two collapse pits (Liesch, personal communication 1984) is shown on Figure 1 (Alexander, 1980). The sinkholes were excavated, backfilled with clay, and the clay liner in the pond upgraded. A hydrogeologic investigation was initiated and the ponds went into operation in the fall of 1974 (Beaton and Meyer, 1980).

The consulting hydrologist, investigating the potential for additional sinkhole development, conducted hammer seismic and electrical resistivity studies of the site. This study

---

* This is publication number 1074 of the School of Earth Sciences, Department of Geology and Geophysics, University of Minnesota, Minneapolis, MN 55455.

** Present address: Minn. Poll. Cont. Agency, 1935 W. Cnty Rd B2, Roseville, MN 55113 U.S.A.

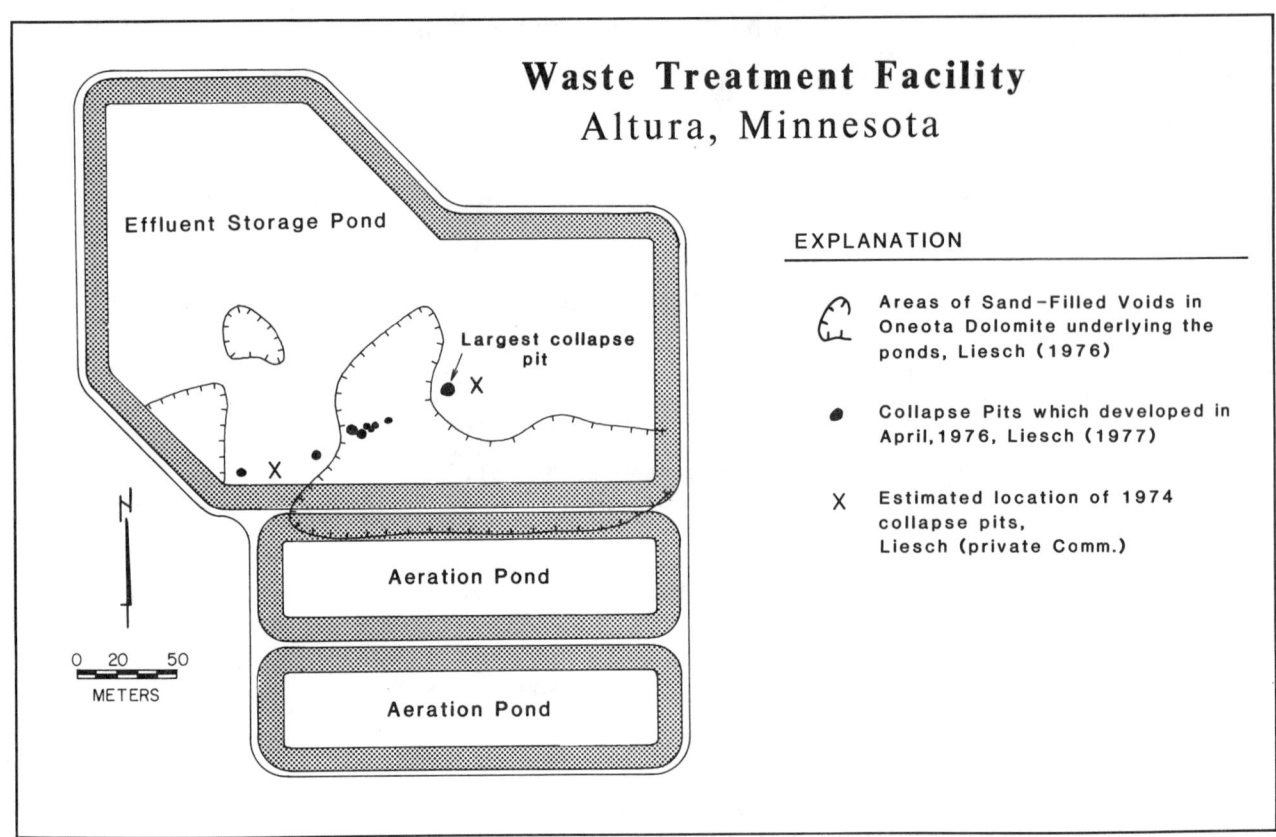

Figure 1: Plan view of the disabled Altura, Minnesota, Waste Treatment Facility showing the locations of the collapse sinkholes in the effluent storage pond.

revealed extensive areas of sand-filled voids in the underlying Oneota Dolomite (Liesch, 1974, 1976). Liesch recommended that three deep monitor wells be installed around the site and that an ongoing monitoring program be initiated. He concluded that any undetected voids would not be of sufficient size to cause a substantial collapse and a hazardous loss of water from the storage ponds or lagoon (Liesch, 1976). The monitor wells were never installed.

The treatment facility processed the waste stream from Altura for a year and a half before the effluent storage pond filled to its design depth of 3.0 m. The subsequent events are described in Ellison (1976). On April 27, 1976, the planned discharge of the effluent pond's contents, 1.41×10^8 l, into a nearby dry ravine was initiated. As soon as the discharge was underway it appeared to the operator that the pond level was dropping faster than could be accounted for by the discharge. The discharge gate was closed at 4:00 PM on April 28, but the level in the pond continued to drop. At 8:00 AM on April 29 discharge was resumed in an effort to minimize the loss through the unknown leaks.

By the morning of May 7, the effluent storage pond was completely drained. The cause of the leaks was clearly visible. A line of nine sinkholes had opened across the bottom of the pond. At 2:37 PM on May 7, 1976 the tenth and largest sinkhole was observed to develop by sudden collapse. The location in the storage pond of these sinkholes is shown in Figure 1. The largest sinkhole was approximately 3.6 m in diameter and approximately 7.0 m deep as measured to the top of the bedrock rubble.

Ellison (1976) calculated that just under 7.6×10^7 l were lost through the sinkholes. The remaining effluent, discharged into the dry ravine, reportedly disappeared underground within a few hundred feet. Whichever route was taken, the effect was the same -- the entire content of the effluent storage pond had entered into the area's ground water system in about 9 days. No adverse health effects were reported. A makeshift monitoring of Altura's municipal wells and nearby private wells, initiated after the pond was drained, detected no evidence of the effluent.

With the effluent storage pond out of operation the immediate problem became what to do with the effluent from the two remaining aeration ponds. It was decided to bypass the storage pond and to discharge, on a semi-continuous basis, the partially treated effluent from the primary aeration pond into the dry run. The effluent averaged about 2.0×10^5 l/day and rose to about 1.26×10^6 l/day when the turkey plant was in operation (Ellison, 1976). This effluent sinks underground before reaching the South Fork of the Whitewater River. The original NPDES permit for the site specified BOD_5 and TSS limits of 5 mg/l each for the effluent discharged into the dry run. The partially treated effluent from the two primary aeration ponds was several times the original limits. The Minnesota Pollution Control Agency (MPCA) then raised the BOD_5 and TSS effluent limits to 25 and 30 mg/l respectively (Breimhurst, 1977). This decision was based on the observation that the effluent sank underground and therefore would not have a significant effect on the Whitewater River (Anderl, 1977).

During December 1976 and January 1977 a series of ten exploration holes were bored in the bottom of the effluent storage pond (Liesch, 1977). These borings ranged from 18 to 30 m and extended through the New Richmond Sandstone into the top of the Oneota Dolomite. All 10 holes penetrated 'voids', which ranged from 5 cm to 1.1 m thick, in the Oneota Dolomite. Liesch (1977) concluded that an apparently integrated system of voids, penetrated at various depths by most of the bore holes, caused a continuous loss of circulation (drilling fluid) during the coring operations in the lower levels of the dolomite. The drilling confirmed the location of the sand-filled voids identified in the 1974 geophysical survey.

Liesch (1977) and Ellison (1977) outlined a number of alternative solutions to the problem caused by the failed storage pond. Beaton and Meyer (1980) and Beaton (1980) outlined the subsequent considerations which resulted, in June 1980, in an agreement to rehabilitate the storage pond by 1) sealing the existing sinkholes, 2) by building a new dike to isolate the northwestern portion of the pond, 3) to line the northwestern portion of the pond with a 20 mil polyvinyl chloride membrane, and 4) to use the resulting smaller, sealed pond for effluent storage as per the original design. To date no action has been taken.

It was in the preceding context that we undertook, in the fall of 1980, a hydrogeologic study of the Altura area. Our study had two primary goals: 1) to document the karst hydrogeologic conditions which led to the double failure of the effluent storage pond, and 2) to determine the subsurface flow path of the partially treated effluent flowing from the facility. The first goal involved detailed mapping of the area's geology, hydrology and karst features and a dye trace from the sinkhole in the pond. The second goal was accomplished by the dye trace of the sinking effluent. The results of the dye traces will be reported elsewhere.

Hydrogeology
The most detailed published bedrock geologic mapping of the area around Altura is Sloan and Austin's (1966) 1:250,000 scale St. Paul sheet. The region's hydrogeology and Paleozoic lithostratigraphy have been mapped at 1:500,000 scale by Broussard and others (1975) and by Mossler (1983). We have mapped the bedrock geology of the Altura 7.5-minute USGS topographic quadrangle at 1:24,000 scale (Book, 1983). Figure 2 is adapted from a portion of Book's (1983) map.

Altura is underlain by a series of lower Ordovician and Cambrian sandstone and carbonate strata which regionally dip very gently to the southwest. The town is situated on an elevated rolling ridge which is more than 100 m above deeply incised stream valleys to the northwest, east and west. The lowest stratum of concern to this study is the Cambrian Franconia Formation. Although the Franconia sandstone is not exposed in the region shown in Figure 2, it is not far below the surface of the deeper valleys and is an important aquifer in the area.

The Franconia is overlain by the St. Lawrence Formation, which is about 18 m thick in the study area and is exposed in the lowest portions of the valleys. The St. Lawrence is a silty to sandy carbonate unit with sporadic thin layers of interbedded shale. Outcrops are highly jointed. Outside the study area, the joints can be seen extending into the underlying Franconia Formation. In the study area the joints visibly extend upwards into the overlying Jordan sandstone. Although the St. Lawrence is mapped as a regional confining bed in Minnesota (Kanivetsky and Walton, 1979), it is leaky at best in Winona County. Book and others (1983) have recently shown, in a dye trace a few miles from Altura, that dye injected into the Oneota Formation emerges from Franconia springs. Several of the springs

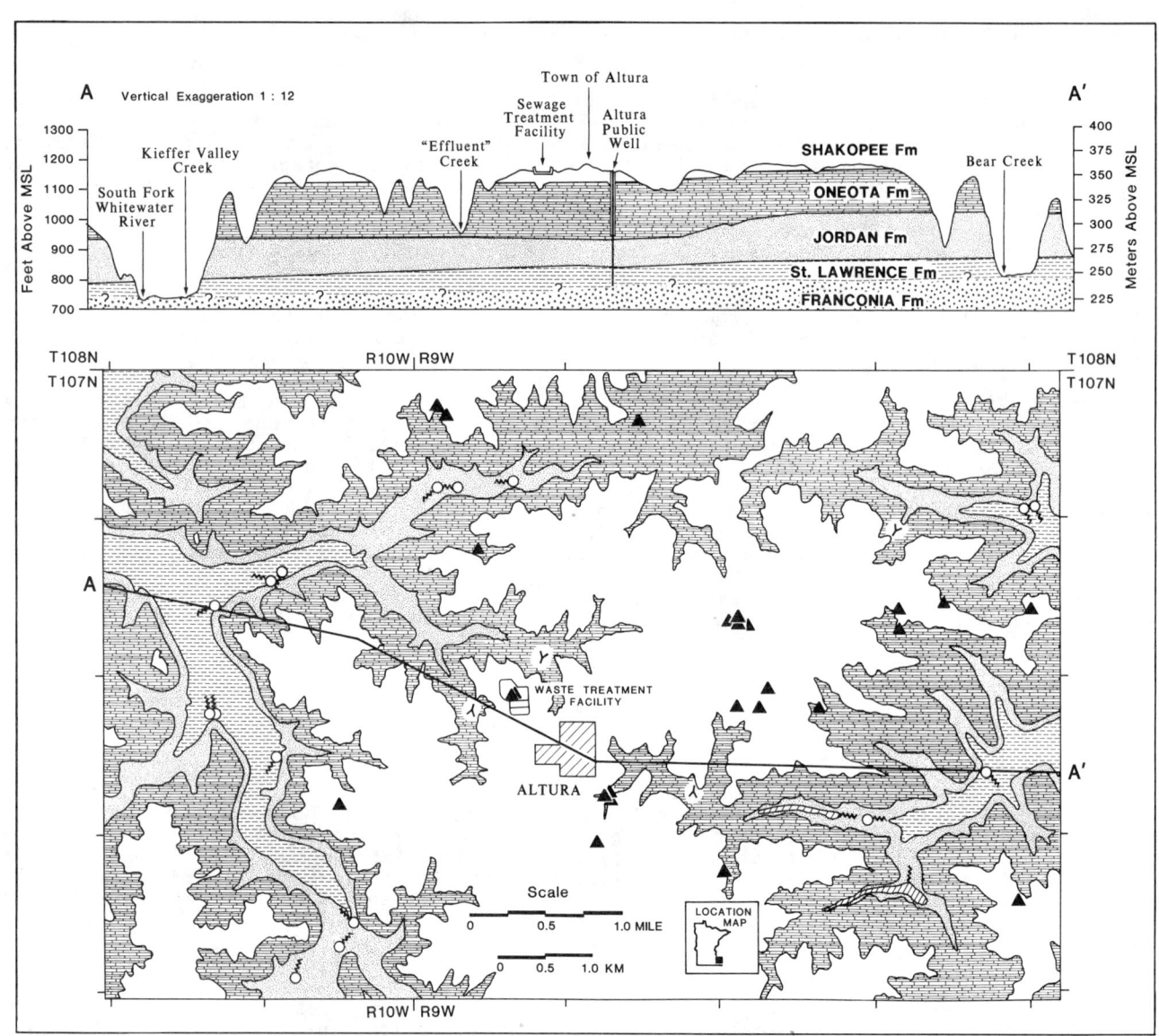

Figure 2: Bedrock geology of the area around Altura, Minnesota. Solid triangles are sinkholes. Open circles and crosshatched areas with squiggly tails are springs and seeps respectively. 'Ys' are caves.

shown in Figure 2 emerge from the St. Lawrence. In the study area, the Crystal Springs State Fish Hatchery receives in excess of 6000 l/min from two St. Lawrence springs.

The uppermost Cambrian unit is the Jordan Sandstone. The Jordan averages 30 m in thickness and crops out extensively in the study area. Directly above the St. Lawrence contact, the Jordan is a massive, upward-grading, fine- to coarse-grained friable sandstone. Upward in the unit, it becomes progressively more indurated with carbonate and siliceous cements, first forming lenses and concretions and then well-bedded, highly lithified strata. Joints are common throughout the Jordan and springs in the well-lithified portions tend to discharge directly from joints. In the more friable lower part, springs are often a combination of discrete flow from joints and diffuse flow from numerous seeps. The Jordan is the major source of water for upland wells in the study area.

The Ordovician Prairie du Chien Group conformably overlies the Jordan. The Prairie du Chien is composed of the lower Oneota Dolomite and upper Shakopee Formation. The Oneota is nearly 61 m thick in the study area and is a fine- to medium-grained, thick- to thin-bedded dolomite. The Oneota is a prominent cliff-former and the break in slope from the rolling uplands to the steep valley walls is usually at or just below the top of the Oneota. Both drill cores and outcrops reveal that the dolomite is highly jointed and has undergone extensive solution. The dolomite is vuggy to cavernous particularly in the upper portion. Only a few springs, confined to discharge from well-developed joints, have been mapped in the Oneota. No wells in the study area have been found which are finished in or rely solely on the Oneota as a water supply. Many older wells are open holes through the Oneota, however.

The Shakopee Formation is subdivided into the lower New Richmond Sandstone member and the upper Willow River Dolomite member. The latter is present only at the highest elevations of the study area and does not enter into the following discussions. The New Richmond Sandstone of the Shakopee Formation is a fine- to medium-grained quartzose sandstone with infrequent interbedded medium-grained arenaceous carbonate beds. This sandstone unit is friable, easily eroded, and does not form many outcrops. It does, however, form the first bedrock unit on much of the uplands in Figure 2. The New Richmond rarely exceeds a thickness of 6.1 m. It is extensively jointed and a few high springs emerge from the New Richmond/Oneota contact. The New Richmond is not a significant aquifer in the study area but is used to a minor extent south of the study area.

The entire region was glaciated at least once by a pre-Wisconsinan advance and scattered patches of drift can be found throughout the region. A blanket of Wisconsinan loess was deposited on the area, and varying thicknesses of recent alluvium and colluvium have collected on the valley floors and in karst solution cavities.

Karst Features
We have located and mapped four caves in the region shown on Figure 2. All four caves are small, phreatic maze caves in the Oneota (Alexander, 1981, 1983a, 1983b). The locations of these caves are shown on Figure 2. Sediment-filled solution cavities are common features in outcrops and quarry walls which expose the Oneota. The karstification of the Oneota probably began during the Ordovician and has continued intermittently until the present. U/Th disequilibrium dating of a speleothem in Skunk Hollow Cave (Alexander, 1983a), one of the caves shown on Figure 2, indicates that the cave is older than $(116\pm4)\times10^3$ yr (R. Lively, pers. comm., 1983).

We have located 23 sinkholes (in addition to the ten in the bottom of the effluent storage pond) in the area covered by Figure 2. Eight of these sinkholes are currently open, while the others have been filled to return the land to agricultural use. The filled holes were identified through interviews with the local landowners. At least five of the sinkholes developed catastropically in the spring of the year in response to unusually wet conditions. It appears that the sinkholes tend to develop through the New Richmond Sandstone into the underlying Oneota. The New Richmond appears to be an integral part of the sinkhole development phenomena in this area.

Water-Quality Considerations
The deeply incised stream valleys northwest, west and east of Altura serve to isolate the hydrologic system beneath the town. The Kieffer Valley, northwest of Altura, is an intermittent stream system into which the effluent discharge flows. It is used primarily for livestock grazing, but two residences near the mouth of the valley draw domestic water from shallow sandpoint wells in the valley alluvium. Kieffer Valley is a tributary to the South Fork of the Whitewater River whose valley forms the west side of the ridge on which Altura is built. The Whitewater River is a major trout fishery and recreational stream. To the east is Bear Creek Valley, a residential/agricultural area. Several local homesteads draw water from shallow wells in the valley.

The waste treatment facility is on the crest of a knoll immediately northwest of Altura (Fig. 2). The town has three municipal wells, each of which produces water primarily from the Jordan Sandstone. Wells No. 1 and No. 3 are on the west side of town 0.61 km from the treatment facility. Well No. 2 is on the east side of Altura about 1.07 km from the lagoons. Well No. 1 is cased to only 14 m. It is evidently in direct hydraulic connection with the surface, shows evidence of surface contamination, and is used only in emergencies. Public health investigations (see for example Mierau, 1975) indicate that although Well No. 2 is cased and grouted through the Oneota into the Jordan, the well exhausts and draws large volumes of air as the static water level fluctuates. Well No. 2 exhibits seasonal fluctuations in nitrates and coliform bacteria, indicating surface connections and potential contamination. Well No. 2 is shown diagramatically in the cross section of Figure 2. Well No. 3, though cased and grouted into the Jordan, hydraulically intersects with Well No. 1

and therefore with the surface. This connection was determined when an attempt to seal Well No. 1 was undertaken. Four yards of clay were dumped into Well No. 1 and within an hour the previously sediment-free Well No. 3 turned muddy (Mierau, 1977).

Liesch (1976) reported that the municipal wells of Altura produce a cone of depression in the Jordan large enough to intercept any downward leakage from the treatment facility. The partially treated effluent from the treatment facility and any leakage from the site, enters the local ground water system. That effluent must be reaching: 1) the municipal wells, 2) the local private wells, 3) local springs and seeps discharging to the Whitewater River or Bear Creek, or 4) some combination of 1), 2), and/or 3).

Discussion

The goal of this study was to evaluate the hydrogeologic environment connected with the collapses of the storage lagoon at Altura. The goal was accomplished by detailed geologic mapping and by an inventory of karst features. Interviews with local landowners revealed information on the 23 sinkholes shown in Figure 2, and that catastrophic sinkhole formation is a common occurrence around Altura. The sinkholes range in size from meters to tens of meters across and deep. Most are small and are immediately filled to return the land to its original use. They are also too small and/or too transient to be shown on regional scale geologic or topographic maps. Anecdotal information indicates that the sinkholes most often form in the spring, particularly after heavy rains. As can be seen in Figure 2, most of the sinkholes develop through the New Richmond Sandstone into the underlying Oneota Dolomite.

Our detailed mapping therefore substantiates Liesch's (1974) interpretation that development of sinkholes in the Altura lagoon was dependent on the presence of the New Richmond Sandstone. Expanding on Liesch's (1974) explanation of this phenomenon the following model emerges: 1) Surface water percolating through the soil layer becomes charged with CO_2 and thereby can dissolve carbonates. 2) The sandstone unit serves to collect the carbonate-undersaturated soil water and <u>concentrates</u> its downward migration along joints and fissures. 3) The solutionally agressive waters are concentrated into specific areas of the carbonate beneath the joints in the sandstone. 4) Selective dissolution and widening of interconnecting Oneota joints creates voids in the dolomite. 5) The growing Oneota voids are bridged by increasingly unsupported and incompetent sandstone beds. 6) Loading of the sandstone, naturally by precipitation or artificially by construction, ultimately leads to collapse. 7) Surface runoff or impounded water was then concentrated into the collapse depression and mechanical erosion moves sediments through the underlying karst openings further enlarging the surface collapse.

Note: The installation of bentonite clay liners may actually aggravate this phenomenon, despite the low permeability of the clay. Ion exchange reactions in the bentonite will tend to reduce the calcium and magnesium ion activities as water passes through the liner. Thus waters seeping from a bentonite-lined structure will probably be considerably more solutionally aggressive to carbonates than was the soil percolation.

Williams and Vineyard (1976) compiled data on 97 recorded catastrophic collapse features in Missouri karst since 1930. Twenty-four of the collapses were attributed to artificially altered surface drainage and ten were caused by impoundment structures. The Altura collapse fits Williams and Vineyard's pattern. The hydrogeology of the Altura site is particularly susceptible to such collapses. There is good reason to believe that the collapses could eventually expand to the two remaining primary treatment ponds as the conditions which led to the original collapses continue to exist.

Camin (1978) discusses the history of the West Plains, Missouri, sewage treatment lagoons. Sinkhole collapse began during the construction of the facility in 1964. Additional collapses occurred in 1966 and 1974 at various locations in the lagoons. A major collapse in May of 1978 contaminated the local aquifer and resulted in 800 cases of flu-like illnesses among people who drank contaminated ground water. The most damaging collapse did not occur, therefore, until the facility had been in operation for 14 years. The sequence of events at West Plains and Altura has been strikingly similar so far.

Summary

The waste treatment lagoon failures at Altura, Minnesota, are not unique. The failures are only one of many cases in which water retention structures have collapsed in karst hydrogeologic regimes. Although sinkholes were not visible at the site prior to construction and are not shown on available topographic and geologic maps of the surrounding area, sinkholes were present a short distance in any direction from the site. These sinkholes, had they been identified, could have provided an indication that the region was an active karst. A clear indication of the site's instability was given by the initial collapses in 1974 during the construction phase of the lagoons and Liesch's (1974, 1976) subsequent geophysical investigations. These warnings were not heeded and the disabling 1976 collapses occurred. The presence of a thin, jointed, poorly indurated sandstone overlying a thick carbonate unit is particularly prone to catastrophic collapse. There is a substantial risk of future collapses under the existing aeration ponds.

Final Note

The 1980 plan to rehabilitate the effluent storage pond has been changed. The current plan (S. Higuchi, pers. comm., 1984) is to: 1) abandon use of the effluent storage pond, 2) fill the sinkholes in the pond bottom, and 3) construct a building to house a waste water clarifying device within the abandoned pond. The turkey processing plant is now pretreating its effluent to decrease the seasonal loading on the entire facility. Altura's waste treatment facility will ultimately cycle waste water through the two existing aeration ponds, the new clarifying device, a chlorination/dechlorination system and then continuously discharge the final effluent into the dry run west of the facility.

Acknowledgments

This study was supported through a grant from the Legislative Commission on Minnesota Resources of the State of Minnesota.

References

Alexander, E.C. Jr., 1980, Geology Field Trip. in Alexander, E.C., Jr. (ed.), An Introduction to Caves of Minnesota, Iowa and Wisconsin: National Speleological Society Convention Guidebook No. 21, 180 p.

Alexander, C., 1981, Minor caves near the Altura Waste Treatment Lagoon: Minnesota Speleology Monthly, v. 13, no. 1, p. 3-4.

Alexander, C., 1983a, Skunk Hollow Cave, Winona County: Minnesota Speleology Monthly, v. 15, no. 2, p. 30-31.

Alexander, C., 1983b, Hog Hollow Cave, Winona County: Minneosta Speleology Monthly, v. 15, no. 7, p. 81-82.

Anderl, W.H., 1977, Field investigation on Altura Minnesota, July 12, 1977: Minnesota Pollution Control Agency Office Memorandum to J.F. McGuire, dated August 16, 1977, 1 p.

Beaton, P.T., 1974, Altura, Minnesota pond collapse: Minnesota Pollution Control Agency Office Memorandum to L.C. Barbie, dated June 18, 1974, 3 p.

Beaton, P.T., 1980, Letter to E.C. Alexander, Jr., dated August 25, 1980, 5 p.

Beaton, P.T. and Meyer, G.W., 1980, Summary of the events leading to the current rehabilitation proposal for the Altura pond system: Minnesota Pollution Control Agency Memorandum to T. Hoffman, dated July 18, 1980, 3 p.

Book, P.R., 1983, Bedrock geology of the Altura 7.5' quadrangle: Unpublished manuscript map.

Book, P.R., Dalgleish, J.B., and Alexander, E.C., Jr., 1983, Semiquantitative tracing of groundwater flow through fractured karst and jointed sandstone aquifers (abs.): EOS, Trans. Amer. Geophys. Union, v. 64, no. 18, p. 229.

Breimhurst, L.J., 1977, Letter to R.J. Ellison, dated August 26, 1977, 2 p.

Broussard, W.L., Farrell, D.F., Anderson, H.W., Jr., and Felsheim, P.E., 1975., Water resources of the Root River watershed, southeastern Minnesota: U.S. Geol. Survey Hydrologic Investigations Atlas HA-548, 3 plates.

Camin, K.Q., 1978, Report of West Plains, Mo. Lagoon incident, May 5 - June 22, 1979: U.S. Environmental Protection Agency, Region 7, Kansas City, Mo. Report dated June 30, 1978, 45 p.

Ellison, R.J., 1976, Report on sinkholes and leakage at Altura, Minnesota waste treatment facility: Letter to P.T. Beaton, Minnesota Pollution Control Agency, dated July 19, 1976, 80 p.

Ellison, R.J., 1977, Report on proposed rehabilitation of the Altura waste water treatment facilities: Ellison-Pihlstrom Report dated Oct. 18, 1977, 29 p.

Kanivetsky, R. and Walton, M., 1979, Hydrogeologic map of Minnesota bedrock hydrogeology: A discussion to accompany State Map Series S-2: Minnesota Geol. Survey, 11 p.

Liesch, B.A., 1974, Hydrogeologic investigation of the Altura area for Altura, Minnesota: Report dated November 1974, 18 p.

Liesch, B.A., 1976, Evaluating pollution-prone strata beneath sewage lagoons: Public Works, v. 107, no. 8, p. 70-71.

Liesch, B.A., 1977, Test drilling investigations at Altura, Minnesota waste treatment facility: Report dated April 7, 1977, 10 p. plus maps.

Mierau, L.W., 1975, Report on investigation of municipal water supply, Altura, Minnesota: Minnesota Department of Health Report dated April 8, 1975, 7 p.

Mierau, L.W., 1977, Notes from a 21 June, 1977 telephone conference with J. Winkelman: Minnesota Department of Health Files, Rochester, Minnesota, 2 p.

Mossler, J.H., 1983, Paleozoic lithostratigraphy of southeastern Minnesota: Minnesota Geol. Survey Miscellaneous Map Series M-51, 8 plates.

Sloan, R.E. and Austin, G.S., 1966, Geologic map of Minnesota: St. Paul Sheet: Minnesota Geol. Survey.

Williams, J.H. and Vineyard, J.D., 1976, Geologic indicators of catastrophic collapse in karst terrain in Missouri: in Zwanzig, F.R. (ed.), Transportation Research Record 612: Subsidence over Mines and Caverns, Moisture and Frost Action, and Classification: National Academy of Sciences, p. 31-37.

Yevjevich, V. (ed.), 1981, Karst Water Research Needs: Water Resources Publications, Littleton, Colorado, 260 p.

Part 1: The applicability of the Florida Mandatory Endorsement for Sinkhole Collapse Coverage - Legal aspects

WILLIAM G.SALOMONE *Consulting Civil Engineer, Lakeland, Florida, USA*

ABSTRACT

Florida Statute §627.706 was enacted to provide sinkhole collapse coverage. The legislative intent of the Statute was to protect the property owner from the unexpected damage to structures and personal property arising from the occurrence of sinkholes. However, since the Statute was enacted, litigation involving considerable expense to the insurance companies and property owners has occurred because of various interpretations of the Statute. The technical and legal interpretations do not appear always to be the same. This paper discusses some of the interpretations accepted by the court. This paper addresses the role of the engineer, geologist and hydrologist in the technical interpretations of the Statute. In an attempt to reduce litigation and associated court costs, recommendations are made to resolve some of the vagueness and ambiguities in the Statute and still maintain the Statute's flexibility and intent to protect public welfare.

Introduction

In November 1969, Rules of the Insurance Commissioner were adopted to provide for a pool concept or participation plan whereby insurers were required to provide coverage for insurable sinkhole losses if requested by the landowner. The Florida legislature first mandated coverage for insurable sinkhole losses through Section 627.351(2) Florida Statutes (1979). Rule No. 4-23.07, Florida Administrative Code, was intended to carry out the legislative purpose of the statute which was to cover insurable sinkhole losses.

The Commissioner defined "sinkhole collapse" as follows:
> Direct loss by sinkhole collapse means actual physical damage to the property covered arising out of or caused by sudden settlement or collapse of the earth supporting such property only when such settlement or collapse results from subterranean voids created by the action of water on limestone or similar rock formations.

(Zimmer v. Aetna Ins. Co. 1980).

Florida Mandatory Endorsement for Sinkhole Collapse Coverage is now provided pursuant to the latest Florida Statute §627.706. With the passage of Florida Statute §627.706 in 1981 in the First Regular Session of the 7th Legislature on April 7th in the Florida House of Representatives (HB147, Chapter 81-280), the Florida legislature required every insurer who was authorized to transact property insurance in Florida to make available coverage for insurable sinkhole losses on any structure, including contents of personal property contained in the policy. Section 627.706, Florida Statutes, 1980 Supplement herein known as Statute is presented in Table I. (Emphasis has been added to certain words for discussion purposes.)

Since the enactment of the sinkhole insurance statute, litigation involving considerable expense to insurance companies and property owners has occurred because of various interpretations of the Statute. Although this paper does not discuss specific trial court decisions (the lowest Florida court), this paper reflects the author's experience as an expert witness and consultant regarding the applicability of the Florida Mandatory Endorsement for Sinkhole Collapse Coverage to both litigated and non-litigated cases. Many of the interpretative questions of the Statute are reflected in the Florida District Court of Appeal's (court above trial court) case of Zimmer v. Aetna Ins. Co. (1980). In Zimmer, the homeowner contended that the lower court erred in construing the insurer's policy as excluding sinkhole loss when the earth supporting their house did not "suddenly" settle or "suddenly" collapse. The District Court of Appeal, Fifth District, defined "sudden" to mean "coming or occurring unexpectedly, unforeseen or unprepared for," and the court defined settlement to mean "a gradual sinking of a structure, either by the yielding of the ground under the foundation or by the compression of the joints or material." The court ruled in favor of the homeowner. The court held that the legislative intent was to 1) cover insurable sinkhole losses, 2) include settlement and collapse and 3) not limit losses to those of a "sudden" overnight nature, the type that amounts to an immediate catastrophe, but also to include those losses where the loss occurs over a more extended period. This court decision illustrates some of the vagueness and ambiguity in the Statute. This paper discusses refinement of the Statute to resolve some of the vagueness and ambiguity emphasized in Table I.

The author has presented his recommendations on refinement of the Statute in light of mostly Florida Supreme Court opinions on how to resolve interpretative problems in contracts for insurance. Items discussed in the paper herein include how to provide for the reasonable expectations of the parties, the importance of plain, clear, unequivocal and unambiguous language, interpretative guidelines, appropriate definitions to be included in the Statute and recommendations that should be included in the Florida Mandatory Endorsement for Sinkhole Collapse Coverage.

Legislative Intent

The Legislature in its statutes sometimes uses terms and phrases that may leave the individual involved with interpreting the statutes perplexed as to the meaning of such terms. The absence of any legislative intent should not leave the individual contracting parties in limbo with regard to the meaning of statutory terms (Calio v. Equitable Life Assurance Society of United States, 1964). In Calio, the meaning of the term, "termination of employment" was ambiguous in the insurance contract. The Florida District Court of Appeal, Third District recommended guidelines when legislative intent was absent in some of the terms. The author has labelled the threefold test that follows a "Rationality Test":

1. Where there is a complete absence of legislative expression as to the meaning of statutory terms, individuals involved should be permitted to reasonably provide definitions.
2. The unclear terms of a statute may be agreed upon between contracting parties, but only to the extent necessary to avoid ambiguity and
3. The definitions provided by the involved parties may not be such as to violate, obstruct or restrict the spirit, meaning and clear intention of the legislature.

In Table I, certain words have been emphasized because of possible interpretation problems. The words, "sudden" and "settlement" have been interpreted by the court in Zimmer. The word, "structure," in subsection (1) conflicts with the word, "building," in subsection (2); the words, "structural damage," defining "loss" in subsection 2 conflicts with the words, "actual physical damage, defining "sinkhole loss" in subsection (3); criteria for "arising out of" or "caused by" are omitted and the legislative intent for the previously mentioned words is omitted. In applying the Rationality Test, the author would change the words, "building" and "structural," in subsection (2) to "structure" and "actual physical," respectively. Therefore, subsection (2) would read, "Loss shall mean actual physical damage to the structure. Contents coverage shall apply only if there is actual physical damage to the structure." By making this modification, the potential for litigation on the issue of what constitutes structural damage would be reduced. If there was a conflict between the parties on what constitutes actual physical damage, the "finder of fact" (judge in a non-jury trial or the jury in a jury trial) in a court of law would decide the actual physical damage based upon the evidence. The finder of fact then would award the appropriate remedy. Before examining the criteria for "arising out of" or "caused by" and providing reasonable definitions for words such as "structure," "settlement" and "collapse" that will not interfere with the legislative intent to provide coverage for insurable sinkhole losses, the author must examine the importance of determining the intention of the parties in a contract for insurance.

After any ambiguity in the statute has been resolved by applying the "Rationality Test," ambiguity in the contract for insurance must be resolved in light of the intentions of the parties. The Florida District Court of Appeal, Fourth District in Commerce Nat'l Bank in Lake Worth v. Safeco Ins. Co. of America (1971) has confirmed the opinion of the California Supreme Court in Gray v. Zurich Ins. Co. (1966). In Gray, the insured was accused allegedly of maliciously, brutally and intentionally assaulting a person. The insurer refused to defend the insured under the insurance policy because the insurer claimed that defending a person for intentional torts was excluded from the policy. The court in Gray held that in interpreting an insurance contract, the intent and reasonable expectations of the parties in entering into the agreement must be considered. Consequently, the insurer's contract form, the insured's knowledge and understanding as a layperson and the insured's normal expectations of the extent of coverage of the policy must be considered in resolving ambiguities in the policy. The Supreme Court of California in Gray held in favor of the insured after considering the reasonable expectations of the insured.

The Florida Legislature requires that the insurer assume the risk of insurable sinkhole losses. The insurer expects to know the extent of that risk and to know in clear, plain, unambiguous and unequivocal language the scope of the insurer's responsibility to the insured under the Statute. Having a sinkhole insurance statute that clearly expresses legislative intent and defines clearly the responsibility of the insurer is most desirable. Using the "Rationality Test" to determine legislative intent is one alternative to coping with vagueness and ambiguity in the statute, but it is more desirable to have a clear and unambiguous statute. To clarify the Statute (Section 627.706) from the Zimmer case, "sinkhole loss" needs to be redefined. Moreover, "settlement," "collapse" and "structure" need to be defined. Criteria for what causes the "sinkhole loss" need to be established. In considering the modification proposed by Beck (1984) for the definition of the words "sinkhole" and "sinkhole loss," the author proposes the following:

(1) "Sinkhole Loss" shall mean actual physical damage to the covered structure including the land below the covered structure and the contents of personal property contained in the covered structure arising from, or caused by, settlement or collapse of the earth supporting such structure only when such settlement or collapse results from subterranean voids created by the action of water on limestone or other soluble rocks, or from the movement of overlying sediments into such voids.

(2) "Sinkhole" shall mean a localized sinking of the land surface, which may be shallow or deep and may occur either rapidly or gradually, but which, by definition, is ultimately caused by the dissolution of underlying rock strata, the collapse of the roof of dissolved voids in the rock strata, or the gradual or sporadic downwashing of overlying sediments into dissolved voids in the rock strata.

In definition (1) the author has not used the word, "property" by itself. This word, "property," appears to be too broad a definition which will subject the insurer to liability that was not intended by the Legislature. The legislative intent appears to require insurance of the "structure" and "personal effects" contained within the structure. Therefore, personal property shall be defined based on the Black Law Dictionary (5th Ed. 1979) definition as:

(3) "Personal property" shall mean personal effects other than land.

In addition, the author has included in "sinkhole loss" actual physical damage to the land below the covered structure. Since the land below the covered structure can be damaged from a sinkhole, insurance protection for the land below the covered structure should be provided. If insurance protection for the land below the covered structure was not provided, an insured would receive compensation for a damaged structure but may not have land on which to build the structure. Sinkhole insurance for agricultural loss (for example, citrus groves) should be provided in a separate section of the Florida Statutes.

In definition (2), the author has not used such words as "subsidence" because the addition of new words leads to ambiguities and potential conflicts between the parties. Moreover, the Supreme Court of Florida in Pafford v. Standard Life Ins. Co. of Indiana (1951) has emphasized the importance of using the "natural" meaning of words. The word "sinking" is clear and unambiguous to the layperson. In Pafford, the insured was killed in a plane crash in a commercial air carrier. The insurer claimed under an exclusion in the policy that the insured was participating in aviation or aeronautics. The Supreme Court of Florida held for the insurer. When the meaning of a contract is plain, clear and unambiguous, the court is not at liberty to modify the contract by interpretation though the court may modify judge-made rules for better conformity with justice. In definition (2), the words "rock strata" have been used consistently throughout the definition. Interchanging the words, "limestone" with "rock strata" leads to confusion of the layperson. By reading definition (2) with definition (1), it is implied that the "rock strata" are limestone or other soluble rocks. In Swigert v. American Bankers Ins. Co. of Florida (1971), the Florida District Court of Appeal, Third District used implication to decide whether "tools of the trade" were on or off the premises. The court held that when an insurer makes reference to an item in an exclusion (no liability for business tools while off the premises) by implication there would be liability when the business "tools of the trade" were on the premises. Consequently, the "business tools" stolen from the employer's truck which was parked on the insured's premises were covered under the policy.

Further examination of definition (1) and (2) reveals that in defining "sinkhole" four types of earth movement have been identified. The four types are 1) movement (in general) of overlying sediments, 2) sinking of the land surface, 3) settlement of the earth and 4) collapse of the earth. Sinking is downward movement. Settlement and collapse are special types of movement. The author proposes the following definitions based on Zimmer and the opinions of the Supreme Courts of Kansas and Ohio:

(4) "Movement" shall mean any movement whether it be up, down or sideways.
(5) "Settlement" shall mean a gradual sinking as evidenced either by the yielding of the ground under the foundation of the covered structure or by the compression of the joints or material of the covered structure.
(6) "Collapse" shall mean a falling down, falling together or caving into an unorganized mass.

In Steward v. Preferred Fire Ins. Co. (1970), the Kansas Supreme Court used definition (4) to determine if a house that sank into an old mine shaft was covered under the policy. The court held that the house loss was excluded from coverage because the policy excluded loss from earth movement. In Zimmer v. Aetna Ins. Co. (1980), the Florida District Court of Appeal, Fifth District, used definition (5). The author has defined "settlement" in general as a gradual sinking. As previously discussed in the introduction to this paper, the court in Zimmer defined settlement as a "gradual sinking of a structure." By using the general definition for "settlement," gradual sinking, settlement can pertain to either the covered structure or earth under the covered structure. Settlement can be proven by providing evidence that shows (1) the ground under the foundation of the covered structure has yielded or (2) distress in the covered structure by the compression of joints or material of the covered structure has occurred. Cracking of walls or uplifting of floors would be one form of distress in a structure. Large voids under the foundation of the covered structure would be evidence of ground yielding. As defined in Zimmer, the word "sudden" has not been used in definition (1) for sinkhole loss. The word "sudden" as defined as "unexpected" by the Zimmer court expands the liability of the insurer which the author feels was not the legislative intent. Settlement from clays, waste (man-made) clays, shrink-swell clays or organics can be unexpected or unforeseen by the layperson (insured), but the legislative intent of the Statute was to cover only insurable sinkhole losses. Therefore, the author omits the word "sudden" from modifying "settlement." In Olmstead v. The Lumbermens Mut. Ins. Co. (1970), the Ohio Supreme Court used the plain, common and ordinary sense definition of the word "collapse." The Ohio Supreme Court used definition (6) to determine the liability of the insurer when damage of a structure occurred when adjacent property was excavated. The court held that when words used in a policy of insurance have a plain and ordinary meaning, it is neither necessary nor permissible to use construction unless the plain meaning would lead to an absurd result. The court held for the insurer based upon the specific facts of the case. The author has not used any modifying adjectives such as "rapid" in defining collapse. Although collapse has a connotation of "rapid" which was probably the legislative intent, to add a modifier to collapse like "rapid" would subject definition (6) to conflict between parties. How rapid does earth movement have to be to be classified as "collapse." The "Winter Park type sinkhole" would be classified as a collapse according to definition (6) whether it occurred in one day or several weeks.

The author proposes the Black Law Dictionary (5th Ed. 1979) definition for the word "structure" as follows:

(7a) "Structure" shall mean any construction, or any production or piece of work artificially built up or composed of parts joined together in some definite manner.
(7b) "Covered structure" shall mean a structure identified specifically in the sinkhole insurance policy.

By making the definition of "structure" general, the insurer and insured have the flexibility to agree on what type of structures are to be insured under sinkhole insurance. However, the liability of the insurer is limited by requiring both parties to agree in writing specifically what structures are to be "covered structures" under the policy. In establishing the criteria for defining "arising from or caused by," the author proposes the following "Foreseeability Test" to define "caused by":

(8a) The settlement or collapse resulting from subterranean voids created by the action of water on limestone or other soluble rocks, or from the movement of overlying sediments into such voids was the cause-in-fact, the indirect, proximate or direct cause of the actual physical damage to the covered structure including the land below the covered structure and the contents of personal property contained in the covered structure OR
(8b) "BUT FOR" the settlement or collapse resulting from subterranean voids created by the action of water on limestone or other soluble rocks, or from the movement of overlying sediments into such voids, it is reasonably probable that the actual

physical damage to the covered structure including the land below the covered structure and the contents of personal property contained in the covered structure would not have occurred and

to define "arising from":

(8c) A sinkhole is the contributory cause for the actual physical damage to the covered structure including the land below the covered structure and the contents of personal property contained in the covered structure. The contributory cause is foreseeable by the reasonably prudent soil mechanics and foundation engineer, geologist and hydrologist under circumstances which put the covered structure including the land below the covered structure and the contents of personal property contained in the covered structure in the orbit of influence of a sinkhole.

If a conflict develops between parties with regard to whether actual physical damage to a covered structure "arises from" or is "caused by" a sinkhole, the foreseeability test (8a, 8b and 8c) should be applied by the professionals engaged by the insurer and insured. If either 8a, 8b or 8c is valid, insurable sinkhole loss has occurred. It is imperative that the professional(s) be proficient in all three of the following specialties: soil mechanics and foundation engineering, geology _and_ hydrology. The manifestation and effect from a sinkhole only can be evaluated when soil mechanics and foundation engineering, geologic and hydrologic principles are considered. If the professional(s) is not adept at all three specialties in applying the Foreseeability Test proposed by the author, a false conclusion can result in whether sinkhole insurance is applicable. This false conclusion would lead to litigation which would result in needless expense to both parties. As an aid to the insurer and insured, the author has provided a general guideline that the reasonably prudent soil mechanics and foundation engineer, geologist and hydrologist can use in applying the Foreseeability Test in evaluating whether sinkhole insurance is applicable. This general guideline is presented in Table 2. The author hopes that by providing the above-mentioned definitions contained within this paper, the legislative intent is clear, unambiguous and in plain language and that the insurer understands his responsibility to the insured who expects to be compensated for insurable sinkhole losses. The Supreme Court of Florida emphasizes the importance of implementing a statute into a contract for insurance in clear, plain, unequivocal and unambiguous language.

Contract For Sinkhole Insurance

The Florida Mandatory Endorsement for Sinkhole Collapse Coverage is the contract for insurance provided to the property owner. This Endorsement should reflect the legislative intent of the Statute. The Supreme Court of Florida is emphatic about the importance of clear, plain, unequivocal and unambiguous language in contracts for insurance. In Hartnett v. Southern Ins. Co. (1965), the Florida Supreme Court held that there is no reason why insurance policies cannot be phrased so that the average person can clearly understand what he is buying. If these contracts are drawn in such a manner that it requires the proverbial Philadelphia lawyer to comprehend the terms embodied in the policy, the court should and will construe the policy liberally in favor of the insured and strictly against the insurer to protect the buying public who rely upon the insurer in such transactions. In Hartnett, a definition for the word "theft," was omitted from the policy. Since no definition was provided, the court liberally interpreted the policy to protect the insured. In The Practorians v. Fisher (1965), the Florida Supreme Court explained its rationale for this policy of liberally construing the policy in favor of the insured when ambiguities arise in the interpretation of a contract for insurance. In The Praetorians, the method of forfeiture on a life insurance policy was ambiguous. The court liberally construed the policy in favor of the insured. The court held that life insurance policies are prepared by experts in this complex field and the interplay of various policy provisions is intricate and difficult for the layperson to understand. For this reason, the public interest requires that a policy be interpreted by the courts in the manner most favorable to the insured. The insurer should state the terms of the policy in clear, unambiguous language. The insurer should rely on the natural meaning of words. As previously mentioned, the Florida Supreme Court in Pafford expressed its preference to the natural meaning of words. In Rigel v. Nat'l Casualty Co. (1954), the Florida Supreme Court held that the court does not interpret a policy that is clear and unambiguous. The court should not add meaning to language that is clear. In Rigel, the court held that carcinoma of the intestine was covered in the policy even though carcinoma of the breast clearly was excluded in the policy. The court should not add meaning to the language of the policy so that carcinoma anywhere in the body would be excluded by the policy. In summary, if the insurer states the terms of the Florida Mandatory Endorsement for Sinkhole Collapse Coverage in clear, plain, unequivocal and unambiguous language in light of the legislative intent of Florida Statute § 627.706, the court probably will not interfere with the terms agreed upon between the parties and the intention of the parties. However, if ambiguities develop in interpreting the Endorsement, the court will consider the sinkhole insurance policy in its entirety to determine the intention of the parties. In N.Y. Life Ins. Co. v. Kincaid (1939), the Florida Supreme Court had to consider a disability insurance policy in its entirety before finding that the insured was entitled to disability benefits to the time of his death. In Queens Ins. Co. v. Patterson Drug Co. (1917), the Florida Supreme Court held that in interpreting the different provisions of a contract for insurance, the court must interpret the contract so that effect to each provision can be made if possible. In Queens Ins. Co., the insurer was liable under a fire insurance policy even though the insured sold gasoline which he kept outside of the building. The Florida Supreme Court in L'Engle v. Scottish Union & Nat'l Fire Ins. Co. (1904) held that absurd interpretations should be avoided. Reason and probability should serve as a guide when two interpretations are possible in determining the intention of the parties. In L'Engle, the insurer refused a claim on a fire policy because the insurer claimed that the policy was void when the insured procured another contract for insurance. The Florida Supreme Court held that "total concurrent insurance" was ambiguous but did imply that additional concurrent insurance was permitted. If the phrase "concurrent insurance" only pertained to the original policy, this conclusion would have been absurd. The Florida Supreme Court reversed the lower court decision and granted a motion for a new trial. The Florida Supreme Court has demonstrated the importance of preparing and phrasing a contract for insurance in clear, plain, unequivocal and unambiguous language to avoid interpretation by the court. Consequently, the author proposes the recommendations shown in Table 3 to be included in the Florida Mandatory Endorsement for Sinkhole Collapse Coverage in an attempt to reduce the liability of the insurer and clarify the expectations of the insured. Hopefully, these recommendations combined with the modification of Florida Statute §627.706 presented in Table 4 will aid in reducing litigation.

Conclusions

The author has provided a modification to Florida Statute §627.706 to clarify the legislative intent to require coverage for insurable sinkhole losses. The author hopes that the modification is clear and unambiguous and that the language is plain and ordinary so that the insurer understands clearly the insurer's responsibility to the insured. The Florida Supreme Court has stressed the importance of contracts of insurance that are clear, unequivocal and reflect in plain and ordinary language the intention and expectations of the parties. The recommendations pertaining to the Florida Mandatory Endorsement for Sinkhole Collapse Coverage hopefully clarify 1) the liability that is being undertaken by the insurer with regard to sinkholes, 2) the insured's understanding of the nature of what a sinkhole is and the type of damage associated with sinkhole loss and 3) the earth movement which is excluded by the policy. Hopefully, the recommendations presented in this paper will aid in reducing litigation by making each party understand clearly their responsibility to the other party and by providing a means by which each party can achieve their reasonable expectations from the contract for insurance. We should not forget that LITIGATION IS FAILURE that burdens all parties with needless costs. There really is NO WINNER. Moreover, the author hopes that in the future the technical community can provide to the insurer a better understanding of the risk that the insurer undertakes when a particular structure is covered by sinkhole insurance.

Table of Authorities

1. Journal of Florida House of Representatives, First Regular Session of the 7th Legislature, House Bill 147, April 7, 1981, page 23.

2. Laws of Florida, Chapter 81-280, House Bill No. 147, page 1302.

3. Zimmer v. Aetna Ins. Co., 383 So. 2d 992 (Fla. 5th D.C.A. 1980).

4. Calio v. Equitable Life Assurance Society of United States, 169 So. 2d 502 (Fla. 3d D.C.A. 1964).

5. Commerce Nat'l Bank in Lake Worth v. Safeco Ins. Co. of America, 252 So. 2d 248 (Fla. 4th D.C.A 1971).

6. Gray v. Zurich Ins. Co., 65 Cal. 2d 263, 54 Cal. Rptr. 104, 419 P.2d 168 (Cal. 1966).

7. Beck, B. (1984), letter correspondence dated April 30, 1984 to Dr. William G. Salomone.

8. Black Law Dictionary, 1979, 5th ed., pp. 1096, 1276.

9. Pafford v. Standard Life Ins. Co. of Indiana, 52 So. 2d 910 (Fla. 1951).

10. Swigert v. American Bankers Ins. Co. of Florida, 247 So. 2d 737 (Fla. 3d D.C.A. 1971).

11. Steward v. Preferred Fire Ins. Co., 206 Kan. 247, 477 P.2d 966 (Kan. 1970).

12. Olmstead v. The Lumbermens Mut. Ins. Co., 23 Ohio App. 2d 185, 261 N.E. 2d 671, affd. 22 Ohio St. 2d 212, 259 N.E. 2d 123 (Ohio 1970).

13. Hartnett v. Southern Ins. Co., 181 So. 2d 524 (Fla. 1965).

14. The Praetorians v. Fisher, 89 So. 2d 329 (Fla. 1956).

15. Rigel v. Nat'l Casualty Co., 76 So. 2d 285 (Fla. 1954).

16. N.Y. Life Ins. Co. v. Kincaid, 186 So. 675 (Fla. 1939).

17. Queens Ins. Co. v. Patterson Drug Co., 74 So. 807 (Fla. 1917).

18. L'Engle v. Scottish Union & Nat'l Fire Ins. Co., 37 So. 462 (Fla. 1904).

Statutes

Florida Statutes (1979) Section 627.351(2).

Florida Statutes (1980 Supplement) Section 627.706.

Florida Statutes (1982 Supplement) Section 627.706.

Codes

Florida Administrative Code, Annotated, 1982, Rule 4-23.07, Vol. 2, pp. 69, 70.

Florida Administrative Code, Annotated, 1983, Annual Supplement, Rule 4-23.07, pg. 9.

TABLE 1

REPRODUCED FROM CHAPTER 81-280, LAWS OF FLORIDA
House Bill No. 147

An act relating to insurance; amending s. 627.702(7),
Florida Statutes, 1980 Supplement, amending provisions authorizing property insurers to repair or replace damaged property at their own expense; creating a new section §. 627.706, Florida Statutes, 1980 Supplement, requiring property insurers to make available coverage for insurable sinkhole losses; providing an effective date.

Be It Enacted by the Legislature of the State of Florida:
Section 1. Not reproduced.

Section 2. Section 627.706, Florida Statutes, 1980 Supplement is created to read:

627.706 Sinkhole Insurance --

(1) Every insurer authorized to transact property insurance in Florida shall make available coverage for insurable <u>sinkhole losses</u> on any <u>structure</u>, including contents of <u>personal property</u> contained therein, to the extent provided in the form to which the sinkhole coverage attaches.
(2) "Loss" shall mean <u>structural damage</u> to the <u>building</u>. Contents coverage shall apply only if there is <u>structural damage</u> to the <u>building</u>.
(3) "Sinkhole Loss" shall mean <u>actual physical damage</u> to the property covered <u>arising out of</u> or <u>caused by sudden settlement</u> or <u>collapse</u> of the earth supporting such property only when such <u>settlement</u> or <u>collapse</u> results from subterranean voids created by the action of water on limestone or similar rock formation.
(4) Every insurer authorized to transact property insurance in Florida shall make a proper filing with the department for the purpose of extending the appropriate forms of property insurance to include coverage for insurable sinkhole losses.
(5) This section shall apply to new and renewal policies or contracts delivered or issued for delivery in this state on or after October 1, 1981.

Approved by the Governor July 2, 1981.
Filed in Office Secretary of State July 2, 1981.

TABLE 2
PROPOSED GUIDELINE FOR REASONABLY PRUDENT SOIL MECHANICS AND FOUNDATION ENGINEER, GEOLOGIST AND HYDROLOGIST TO EVALUATE THE APPLICABILITY OF THE FLORIDA MANDATORY ENDORSEMENT FOR SINKHOLE COLLAPSE COVERAGE

The following guideline should serve only as a framework for the type of site investigation a reasonably prudent soil mechanics and foundation engineer, geologist and hydrologist should perform to evaluate the applicability of the Florida Mandatory Endorsement for Sinkhole Collapse Coverage. The reasonably prudent professional still is responsible for providing a comprehensive site investigation for each particular site.

It is imperative that the professional(s) engaged by the insurer or insured be proficient and carefully consider soil mechanics and foundation engineering, geologic and hydrologic principles in making a proper evaluation of the applicability of the Endorsement.

The author feels that the following guideline can be implemented at reasonable expense to the insurer or insured when considering the type of expense encountered through litigation if an improper evaluation is made.

The following guideline is recommended:
1. Obtain a legal description of the site.
2. Inspect the covered structure.
3. Review pertinent background information (if available) provided by the attorney.
4. Review any available technical information on the site (e.g. boring logs, ground-water measurements, laboratory tests).
5. Review old and recent airphotos and United States Geological Survey maps of the site to compare topographic features before and after development of the local area.
6. Survey neighborhood to evaluate history of the area.
7. Observe surface runoff patterns during heavy rainfall.
8. Obtain structural drawings (if possible) of the covered structure.
9. Review published information (flood plain maps, geologic reports).

Make a preliminary recommendation to client to verify if client wants to proceed.
10. Drill additional borings as required.
11. Install observational wells at site to monitor ground-water fluctuation.
12. Evaluate the seasonal fluctuation of ground water in the site area.
13. Review rainfall data over 10's of years to evaluate patterns of heavy rainfall.
14. Perform additional laboratory tests on site soils.
15. Obtain sinkhole data of site area from government agencies.
16. Review well log data of the local area.
17. Plot subsurface cross sections.
18. Calculate settlement of covered structure based upon subsurface cross sections.
19. Monitor movement of covered structure.
20. Compare calculated and actual settlement of covered structure.
21. Prepare geologic contour maps of rock surface over local area.

Make final recommendations to client.

TABLE 3
RECOMMENDATIONS FOR INCLUSION INTO THE FLORIDA MANDATORY ENDORSEMENT FOR SINKHOLE COLLAPSE COVERAGE

The following recommendations are proposed to be added to the Florida Mandatory Endorsement for Sinkhole Collapse Coverage:
1. A layman's pamphlet on sinkholes prepared by the Florida Sinkhole Research Institute shall be provided to the insured with the sinkhole insurance policy.
2. The title of the sinkhole insurance policy shall be changed to read, "Florida Mandatory Endorsement for Sinkhole Loss Coverage." The word "collapse" shall be removed.
3. As shown in Table 4, Proposed Modification, Section 627.706, Florida Statutes, 1980 Supplement, definitions for "loss", "sinkhole loss," "sinkhole," "structure," "covered structure," "personal property," "movement," "settlement" and "collapse" shall be added.
4. If a conflict arises and can not be resolved between the insurer and insured on what constitutes "actual physical damage," the "finder of fact" (judge in a non-jury trial or the jury in a jury trial) in a court of law shall decide the

"actual physical damage" based upon the evidence. The "finder of fact" then would award the appropriate remedy.

5. If a conflict arises and can not be resolved between the insurer and insured on whether the actual physical damage to the covered structure including the land below the covered structure and the contents of personal property contained in the covered structure arose from or was caused by a sinkhole, the insurer and insured shall engage a reasonably prudent professional(s) who is proficient in all three of the following specialties: soil mechanics and foundation engineering, geology and hydrology. The professional(s) shall apply the Foreseeability Test presented in Subsection 14, Proposed Modification, Section 627.706, Florida Statutes, 1980 Supplement (Table 4 presented within the text of this paper) to evaluate whether insurable sinkhole loss has occurred. If either part of the Foreseeability Test (14a, 14b or 14c presented in Subsection 14) is valid, insurable sinkhole loss has occurred.
6. Settlement or collapse of the covered structure including the land below the covered structure and the contents of personal property contained in the covered structure arising from or caused by the compression of clays, waste (man-made) clays, shrink-swell clays and organics shall be excluded from this sinkhole insurance policy.
7. "Time of Occurrence" of sinkhole loss shall mean the time when the complaining party (insured) has notified the insurer in writing that "sinkhole loss" has occurred.
8. For the purpose of defining liability, the insurer shall not be liable to the insured if the "time of occurrence" of "sinkhole loss" occurs outside the "time period of coverage" of the sinkhole insurance policy.
9. The "time period of coverage" shall be stated clearly within the sinkhole insurance policy. Any "grace period" for deficient premium payments shall be stated clearly and shall be considered as part of "time period of coverage."
10. Notice of termination of the sinkhole insurance policy shall be made in writing.
11. "Sinkhole loss" shall not be covered by the aforesaid sinkhole insurance policy when the "time of occurrence" of "sinkhole loss" occurs after the date of notice of termination.
12. The aforesaid sinkhole insurance policy shall not be transferrable to a new owner of the covered structure.
13. The insurer of the original sinkhole insurance policy shall not be liable for "sinkhole loss" when the "time of occurrence" of "sinkhole loss" occurs within the "time period of coverage" of the new sinkhole insurance policy.

TABLE 4
PROPOSED MODIFICATION
SECTION 627.706, FLORIDA STATUTES, 1980 SUPPLEMENT

627.706 Sinkhole Insurance
(1) Every insurer authorized to transact property insurance in Florida shall make available coverage for insurable sinkhole losses on any structure (herein known as covered structure) including the land below the covered structure and the contents of personal property contained in the covered structure to the extent provided in the form to which the sinkhole coverage attaches.
(2) The insurer shall have the responsibility to prepare and phrase the contract for sinkhole insurance in clear, plain, ordinary, unequivocal and unambiguous language that is understood easily by the layperson.
(3) "Loss" shall mean actual physical damage to the covered structure including the land below the covered structure and the contents of personal property contained in the covered structure. Coverage of the land below the covered structure shall apply even if actual physical damage to the covered structure has not occurred. Contents' coverage shall apply only if there is actual physical damage to the covered structure.
(4) "Sinkhole Loss" shall mean actual physical damage to the covered structure including the land below the covered structure and the contents of personal property contained in the covered structure arising from, or caused by, settlement or collapse of the earth supporting such structure only when such settlement or collapse results from subterranean voids created by the action of water on limestone or other soluble rocks, or from the movement of overlying sediments into such voids.
(5) "Sinkhole" shall mean a localized sinking of the land surface, which may be shallow or deep and may occur either rapidly or gradually, but which, by definition, is ultimately caused by the dissolution of underlying rock strata, the collapse of the roof of dissolved voids in the rock strata, or the gradual or

sporadic downwashing of overlying sediments into dissolved voids in the rock strata.
(6) "Structure" shall mean any construction, or any production or piece of work artificially built up or composed of parts joined together in some definite manner.
(7) "Covered Structure" shall mean a structure identified specifically in the sinkhole insurance policy.
(8) "Personal Property" shall mean personal effects other than land.
(10) "Settlement" shall mean a gradual sinking as evidenced either by the yielding of the ground under the foundation of the covered structure or by the compression of the joints or material of the covered structure.
(11) "Collapse" shall mean a falling down, falling together or caving into an unorganized mass.
(12) Movement shall mean any movement whether it be up, down or sideways.
(13) If a conflict arises and can not be resolved between the insurer and insured on whether the actual physical damage to the covered structure including the land below the covered structure and the contents of personal property contained in the covered structure arose from or was caused by a sinkhole, the insurer and insured shall engage a reasonably prudent professional(s) who is proficient in all three of the following specialties: soil mechanics and foundation engineering, geology and hydrology. The professional(s) shall apply the Foreseeability Test presented in Subsection 14 of Section 627.706, Florida Statutes to evaluate whether insurable sinkhole loss has occurred.
(14) To evaluate whether insurable sinkhole loss has occurred, the reasonably prudent professional proficient in the specialties presented in Subsection 13 shall use the following Forseeability Test to define "caused by":
  (a) The settlement or collapse resulting from subterranean voids created by the action of water on limestone or other soluble rocks, or from the movement of overlying sediments into such voids was the cause-in-fact, the indirect, proximate or direct cause of the actual physical damage to the covered structure including the land below the covered structure and the contents of personal property contained in the covered structure OR

  (b) "BUT FOR" the settlement or collapse resulting from subterranean voids created by the action of water on limestone or other soluble rocks, or from the movement of overlying sediments into such voids, it is reasonably probable that the actual physical damage to the covered structure including the land below the covered structure and the contents of personal property contained in the covered structure would not have occurred and
  to define "arising from":
  (c) A sinkhole is the contributory cause for the actual physical damage to the covered structure including the land below the covered structure and the contents of personal property contained in the covered structure. The contributory cause is foreseeable by the reasonably prudent soil mechanics and foundation engineer, geologist and hydrologist under circumstances which put the covered structure including the land below the covered structure and the contents of personal property contained in the covered structure in the orbit of influence of a sinkhole.

If either part of the Foreseeability Test (14a, 14b or 14c) is valid, insurable sinkhole loss has occurred.
(15) Settlement or collapse of the covered structure including the land below the covered structure and the contents of personal property contained in the covered structure arising from or caused by the compression of clays, waste (man-made) clays, shrink-swell clays and organics shall be excluded from Section 627.706.
(16) Every insurer authorized to transact property insurance in Florida shall make a proper filing with the department for the purpose of extending the appropriate forms of property insurance to include coverage for insurable sinkhole losses.
(17) This section shall apply to new and renewable policies or contracts delivered or issued for delivery in Florida on or after October 1, 1981.

Acknowledgements

The author expresses his appreciation to Representative Bobby Brantley, Longwood, Florida for his review of this paper. Representative Brantley introduced sinkhole insurance bill HB147 in the Florida House of Representatives in 1981. The author also expresses his appreciation to Dr. Thomas H. Patton, Geologist and Attorney at Law, for his review of this paper.

The manuscript was typed by Donna Ornowski.

Part 2: The applicability of the Florida Mandatory Endorsement for Sinkhole Collapse Coverage – Case history, foundation settlement of a residential structure – Was it a sinkhole?

WILLIAM G. SALOMONE *Consulting Civil Engineer, Lakeland, Florida, USA*

ABSTRACT

The Winter Park Sinkhole created publicity for sinkhole activity in the State of Florida. In fact, the State of Florida requires insurance companies to provide sinkhole coverage to residential homeowners. However, to receive payment on a sinkhole claim, the homeowner must have sustained "sinkhole loss" according to the definition presented in Florida Statute §627.706.

In this paper, the author presents an analysis used to evaluate the applicability of the Florida Mandatory Endorsement for Sinkhole Collapse Coverage clause taken from Florida Statute §627.706 for a residence located in Lakeland, Florida. Pertinent geologic and hydrologic information was reviewed, borings were drilled, laboratory tests were performed and ground water was monitored during the author's investigation. From the investigation, the author concluded that although the residence was generally within an ancient sinkhole, the foundation settlement appeared to be influenced by compression of a peat layer found under part of the residence. It appears the compression of the peat occurred because of the induced loading on the peat from: (1) seasonal ground-water fluctuation, (2) the structural loading of the residence on the foundation soils and (3) the placement of fill before construction of the house. Consequently, the Florida Mandatory Endorsement for Sinkhole Collapse Coverage clause was not applicable.

Introduction

The Winter Park Sinkhole created publicity concerning the potential for sinkhole activity in the State of Florida. In fact, the State of Florida has now required insurance companies to provide sinkhole coverage for residential homeowners. The Florida Mandatory Endorsement Document (FMED) for Sinkhole Collapse Coverage includes the following definition for sinkhole collapse:

> Sinkhole collapse, meaning only actual physical damage to such property arising out of, or caused by sudden settlement or collapse of the earth supporting such property and only when such settlement or collapse results from subterranean voids created by the action of water on limestone or similar rock formations.

This paper presents the analyses used to evaluate the applicability of the above-mentioned definition for sinkhole collapse found in FMED for sinkhole collapse coverage with regard to a residence in Lakeland, Florida which was subjected to significant foundation settlement. This analysis was made on behalf of an insurance company in cooperation with the developer who owned the residence.

Purpose

The author was asked to evaluate the applicability of the Florida Mandatory Endorsement Document for Sinkhole Collapse Coverage with regard to the significant settlement that occurred on the residence identified as Lot 42 in Lakeland, Florida.

Scope of Work

To evaluate the feasibility of a sinkhole collapse causing the significant settlement on Lot 42, the author performed the following investigative program:
1. Reviewed pertinent information regarding the geology, the potential for sinkholes, and ground-water level fluctuations in the vicinity of Lot 42;
2. Evaluated the subsurface conditions by drilling borings;
3. Installed piezometers to evaluate the ground-water levels; and
4. Evaluated the engineering characteristics of the peat by laboratory testing.

The literature review and information retrieval included obtaining and reviewing USGS maps regarding the geology of the area in question, soil conservation maps showing surficial soils in the vicinity of Lot 42, pertinent well log data, air photos, and pertinent sinkhole maps of the Lakeland/Bartow area. Moreover, geological reports were reviewed to evaluate potential for ground-water fluctuation in the area around Lot 42.

An excerpt from the U.S. Geological Survey Map, Lakeland Quadrangle, dated 1975, was reviewed. Drilling logs of wells provided by the Southwest Florida Water Management District were reviewed for general subsurface stratigraphy.

Upon review of the above-mentioned information, a subsurface investigation was performed on Lot 42. The drilling program included the drilling of four borings to a depth of approximately 610 meters (200 feet). The borings were drilled near the four corners of the residence. Four piezometers were installed near the midpoint of each of the four sides of the residence to depths ranging from 76 to 152 meters (25 to 50 feet). Rock was not encountered; therefore, no rock coring was performed. Upon review of the subsurface stratigraphy, evaluation of the engineering characteristics of the soils found within the borings was made.

House Description

The house located on Lot 42, south of Lake Bonny in Lakeland, Florida, has subsequently been demolished and removed, and a new wood frame house had been placed before the performance of the author's investigation. However, the original house was a concrete block framed home with a net bearing pressure on the wall footings on the order of 50 kPa (1,000 psf). The developer used an allowable soil bearing pressure not to exceed 90 kPa (1,800 psf) under conventional shallow wall footings.

Subsurface Conditions

The elevation of the existing ground surface has been assumed to be approximately Elevation +131 (City Datum). The subsurface conditions generally consist of approximately 12.2 meters (4 feet) of fill consisting of brown to black, loose, fine to medium sand with a trace of silt and vegetation. The fill is underlain by approximately 4.6 to 39.6 meters (1 1/2 to 13 feet) of gray to brownish gray, loose, fine to medium sand with a trace of silt. A 1.5 to 199.6 meter (1/2- to 65 1/2-foot) layer of dark brown, very soft peat underlies the gray to brownish gray sand. The peat is thickest in the northwest corner of the house, where the significant settlement occurred. It thins out to the east and to the south. In Boring 1, drilled in the northwest corner, the peat becomes medium stiff at a depth of 190.5 to 251.5 meters (62 1/2 to 82 1/2 feet). The peat is approximately 1.5 meters (6 inches) thick in Boring 4, 6.1 meters (2 feet) thick in Boring 3, 199.6 meters (65 1/2 feet) thick in Boring 1, and approximately 15.2 meters (5 feet) thick in Boring 2. Borings 2, 3 and 4 are in the southwest, northeast and southeast corners of the residence, respectively. A 7.6 meter (2 1/2-foot) layer of dark brown, medium dense fine sand was found interbedded within the peat layer in Boring 1. Alternating layers of gray, loose, fine to medium sand and gray, fine to medium silty sand, ranging in thickness from 117.3 to 312.4 meters (38 1/2 to 102 1/2 feet) underlies the peat layer. With depth, these sand layers grade into a white, loose silty fine sand above the silt/clay layer. Some yellow sand was encountered within these interbedded layers of sand and silty sand. Within this sand layer, lost circulation of drilling mud and artesian pressure were encountered in Boring 1. The sand layer is underlain by interbedded layers of silt and clay, ranging in thickness from 146.3 to 341.4 meters (48 to 112 feet). The yellow-brown sandy, clayey silt is loose to medium dense. The yellow-brown silty clay is soft to medium stiff. At various times, the silt and clay layers grade into orange and white silt and clay, and with depth the color grades into green. Lost circulation was also encountered within this layer of silt and clay, and some artesian pressure was noted in Boring 2 in the upper part of the silt/clay strata. The silt/clay strata is underlain by a gray, loose to medium dense, fine to medium silty sand and gray, loose to medium dense, fine to coarse gravel. The gravel was penetrated in the front of the house in Borings 3 and 4. The ground water was generally located at a depth of approximately 32 meters (10 1/2 feet, Elevation +121.5) from the existing ground surface.

Laboratory Testing

To evaluate the compressibility of the soft peat encountered under Lot 42, a one-dimensional oedometer test was performed on a sample taken from Boring 1 at a depth of approximately 67 meters (22 feet). From the consolidation test, it appears that the soft peat is highly compressible, with a compression index on the order of 0.63. The water content of the peat is on the order of 740 percent, and the dry density is on the order of 112 kilogram per cubic meter (7 pcf). The pre-consolidation pressure is on the order of approximately 43 kPa (860 psf).

Engineering Analysis

To evaluate the possibility of whether a sinkhole collapse caused the significant settlement on Lot 42, the author's engineering analysis and investigative program included a review of pertinent geological and hydrological information, drilling borings, evaluating ground-water levels, and performing laboratory testing.

Potential for Sinkholes. The lakes in central Florida and Polk County are widely held to be of sinkhole origin. A review of geology and hydrology of Polk County will provide an understanding of the sinkhole origin of many of Polk County lakes (Stewart, 1966).

For the purpose of this paper, it is important to realize that Lake Bonny, with its irregular or complex arcuate shorelines, is a coalescent group of small sinks called valley sinks. Although the site was located generally within these ancient valley sinks, a review of the maps provided by the Department of Public Safety in Bartow, Florida and discussions with representatives of the Public Safety Council who investigate complaints about sinkholes in the Bartow/Lakeland area indicate that no recent sinkhole has been encountered specifically on the residential site. From the author's investigation, it appears that the closest known sinkholes to the residential site developed in 1976 just south of Lake Parker. In approximately 1969, there was a sinkhole four blocks north of Crystal Lake on Combee Road. In approximately 1966, there was a sinkhole on the east side of Lake Hollingsworth.

Peat Strata. The deposition of peat in the southern part of Little Lake Bonny appears to be compatible with the geologic origin of Lake Bonny, which was formed by a group of valley sinks with aquatic grass and other vegetation growing around the periphery of the sinks. The lowland depression indicated by the topographic contours would support this hypothesis. Moreover, the 1941 air photos of Little Lake Bonny indicate a high moisture area south of the lake. This could serve as an indicator of deposition of organic soils. The U.S. Geological Survey map, Lakeland Quadrangle, dated 1975 also shows a lowland depression south of Little Lake Bonny. There appears to be clear evidence that peat has been deposited (the boundary of which is unknown) south of Little Lake Bonny.

Ground-water Fluctuation. From information supplied by the developer, the ground-water table was measured at approximately Elevation +127 in 1981. The ground surface elevation for Lot 42 and nearby vicinity was assumed to be between +131 and +132, City Datum (for U.S. Geological Survey datum, add 0.86 to City Datum). Generally, this elevation (Elevation +127) for the ground water appears compatible with the water level in Little Lake Bonny (Elevation 126.86) recorded in August 1978, and presented on the paving and drainage drawings provided by the developer. Records containing lake levels for Little Lake Bonny for 1979 and 1981 were not available.

The ground-water table in the four piezometers on Lot 42 was measured on May 25, 1982. The ground-water table was approximately Elevation +122. The lake level of Little Lake Bonny was measured during the author's investigation as Elevation +123 on May 25, 1982. The water levels of the piezometers and the lake level appear to be in agreement. Rainfall data for the period between 1979 and 1982 was plotted. The data indicate seasonal fluctuation in the amount of rainfall.

Comparison of the 1981 and 1982 ground-water levels appears to indicate:
1. The water level under the residential site generally follows closely the water level of Little Lake Bonny;
2. Ground-water observations will depend upon the season of the year in which the observations are made;
3. Seasonal fluctuations of the ground water exist in the vicinity of the residential site.

Settlement Analysis

From the information furnished to the author by the insurance company and the developer, the author understands that settlements on the order of 38.1 centimeters (15 inches) occurred in forty days. However, to the author's knowledge, over the life of the structure (1 1/2 years), settlement measurements were not plotted and recorded periodically from the time of house construction to the time of demolishment of the house. Consequently, no total amount of settlement was recorded.

To evaluate the feasibility of compression of the peat from induced loading on the peat, a settlement analysis was performed. Based upon a review of the compaction test data provided by the developer, the total weight of the 12.2 meters (4 feet) of fill placed before the construction of the home was assumed to be 1856 kilogram per cubic meter (116 pounds per cubic feet). This is approximated by a loading on the foundation soils of 23.2 kPa (464 psf). If we consider a net bearing pressure of 50 kPa (1000 psf) under the wall footings, a fill surcharge loading of 23.2 kPa (464 psf), and a compressive index of the peat on the order of 0.63, settlements on the order of 63.5 centimeters (25 inches) are possible. The settlement (on the order of approximately 48.3 centimeters, 19 inches) caused by the placement of the fill probably occurred before the house was constructed. However, after the house was constructed, more significant settlement probably occurred from seasonal fluctuation of the groundwater. A drop in water table of 15.2 meters (5 feet, Elevation 127 to Elevation 122) is approximated by an induced concentrated loading on the peat layer on the order of 15.6 kPa (312 psf). This would result in additional settlements on the order of 40.6 centimeters (16 inches) occurring over a relatively short period of time. This is compatible with the observed events (38.1 centimeters, 15 inches of settlement in forty days). From rainfall data for the period between 1979 and 1982, it is important to realize that the soil under the house went through wet-dry-wet cycles before being demolished. The house was constructed during a dry season. During the life of the structure (approximately 1 1/2 years), the house went through a wet cycle and dry cycle before being demolished during a wet cycle. The cracks in the house were noticed after the dry cycle when a drop in water table would have induced more loading on the peat and would have caused additional compression of the peat.

In summary, the settlement analysis indicates that the settlement under the house was caused by compression of the peat.

Conclusions

Based upon the author's investigation, it appears that the Florida Mandatory Endorsement Document for Sinkhole Collapse Coverage is not applicable with regard to the significant settlement of the house on Lot 42. Rather, the significant settlement appears to have been caused by compression of a peat layer. It appears the compression of the peat occurred because of the induced loading on the peat from:
1. Seasonal groundwater fluctuations.
2. The structural loading of the house on the foundation soils and
3. The placement of fill on Lot 42 before the construction of the house.

Reference
1. Stewart, Herbert G., Jr. (1966), "Ground-Water Resources of Polk County," Report of Investigations No. 44, Florida Geological Survey.

Acknowledgements

The author acknowledges Auto Owners Insurance Company and Bromwell Engineering, Inc. for permission to publish the results of the author's investigation.

5. Case histories: Remedial engineering of sinkholes

Remedial measures associated with sinkhole-related foundation distress

JOHN E.GARLANGER *Ardaman and Associates, Inc., Orlando, Florida, USA*

ABSTRACT

Sinkholes in Central Florida result from the phenomenon known in geotechnical engineering as "piping". Piping occurs when a subterranean conduit is eroded backward from a location where seepage is discharging from an unconsolidated soil deposit. When the subsurface erosion finds its way through an opening in the bedrock surface into the overlying soils, a void is created within the overburden resulting in surface subsidence or collapse. To arrest the subsidence and restore foundation support, the geotechnical engineer must locate the opening in the rock surface and plug it with grout. Case histories are presented which describe the grouting and other remedial measures employed to arrest the subsidence and restore foundation support at three different sites. The size of the sinkholes at the three sites ranged from 1.5 meters in diameter and less than 0.3 meters deep to over 20 meters in diameter and greater than 15 meters deep. The cost of remedial action ranged from $20,000 to over $200,000.

Introduction

Although the infamous Winter Park sinkhole, because of its immense size (100 meters in diameter, 30 meters deep), focused world-wide attention on Central Florida for a few days in May, 1981, and was another reminder of the awesome power of nature, it was certainly not without precedent. Dozens of smaller sinkholes occur annually in the Central Florida area. Some of them, like the one shown in Figure 1, are relatively small and cause little, if any, property damage. Others, like the one shown in Figure 2, are considerably larger and, as shown, can result in complete loss of property.

The type of sinkhole which occurs in Central Florida results from the phenomenon known in geotechnical engineering as "piping" (Terzaghi, K., 1931, Terzaghi, K. & R.B. Peck, 1948). Piping occurs when a subterranean conduit or tunnel is eroded backward from a location where groundwater is discharging from an unconsolidated soil deposit, e.g., at a spring. Once erosion begins, it proceeds backward along the line of maximum hydraulic gradient toward the source of the seepage.

For a sinkhole to occur as a result of piping, sufficient hydraulic gradient must be present to begin the erosion process (usually somewhere within an underground limestone cavity), the flow within the cavity system must be sufficient to carry away the eroded material, the subterranean tunnel must be eroded back up through an opening in the rock surface into the overlying soils, and sufficient soil must be carried down into the cavity system to create the subsidence which appears at the ground surface.

At the present time it is not possible to predict where or when piping will begin in a subterranean cavity or whether the subsurface erosion will extend upward into the overlying soil deposits. However, when subsidence does occur and threatens adjacent or overlying property, geotechnical engineers are often called upon to design remedial measures for arresting the subsidence and restoring foundation support.

The purpose of this paper is to describe the types of remedial measures which have been used in the Central Florida area when sinkhole subsidence has occurred. Three case histories will be presented. Each case study will describe the physical characteristics of the subsidence, the investigation undertaken to define subsurface conditions within the area of the subsidence, the types of remedial measures that were used to prevent further subsidence and restore foundation support, and the post-repair monitoring used to evaluate the success of the remedial measures.

Case History No. 1 - Residence

On May 12, 1977, a sinkhole occurred beneath the front wall of a house in Belle Isle, Florida, a small suburb south of Orlando. As shown in Figure 3, the foundation and first few rows of masonry block had fallen into the sinkhole along with the front porch slab. The sinkhole was approximately 11 meters in diameter and the subsidence was measured at just over 10 meters, a relatively small sinkhole when compared with the approximately 30-meter depth to bedrock generally encountered near this location.

Figure 1. Small erosion sinkhole near Brooksville, Florida.

Figure 2. Large erosion sinkhole near Casselberry, Florida.

Figure 3. Sinkhole described in Case History No. 1.

Figure 8. Grade beam designed to span Case 1 sinkhole.

Figure 9. Case 2 sinkhole during grouting operations.

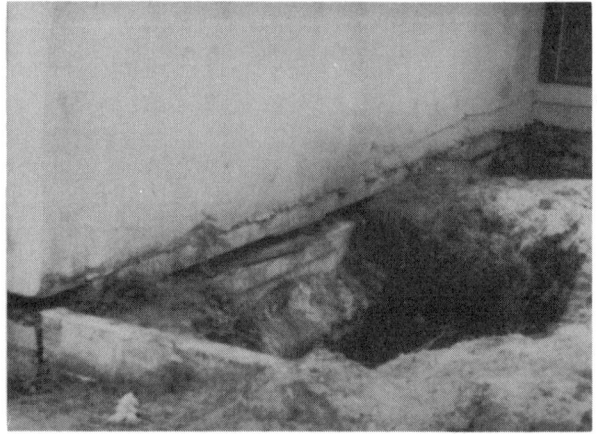

Figure 15. Subsidence described in Case History No. 3.

The homeowner's insurance company retained the writer's firm within several hours of the occurrence to make recommendations concerning possible remedial action. The house was insured for approximately $90,000. To prevent the front wall of the house from collapsing as the sinkhole continued to enlarge, it was necessary to fill the crater with clean sand fill and provide temporary shoring for the wall. Fortunately, fill was available on an adjacent lot and it was possible to fill the crater within a matter of hours. The shoring consisted of wooden bearing plates and mechanical jacks installed directly beneath the wall. Because the fill continued to subside, the jacks were adjusted on a regular basis as the subsurface investigation and grouting program was completed. The measured settlement of the fill with time is shown in Figure 4.

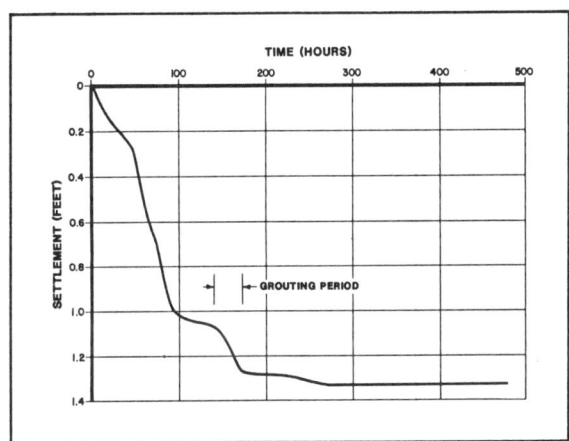

Figure 4. Settlement at center of Case 1 sinkhole vs. time after backfilling.

The boring layout for the subsurface investigation is presented in Figure 5 and the results presented in Figure 6. As shown, all three of the test borings, including the one drilled at the approximate center of the sinkhole, encountered a 3-meter thick layer of clay at a depth of about 15 meters. The clay was medium stiff and highly plastic. The boring near the center of the sink encountered what appeared to be loose, ravelled conditions just below the bottom of the clay and encountered a cavity just below what was expected to be the bedrock surface.

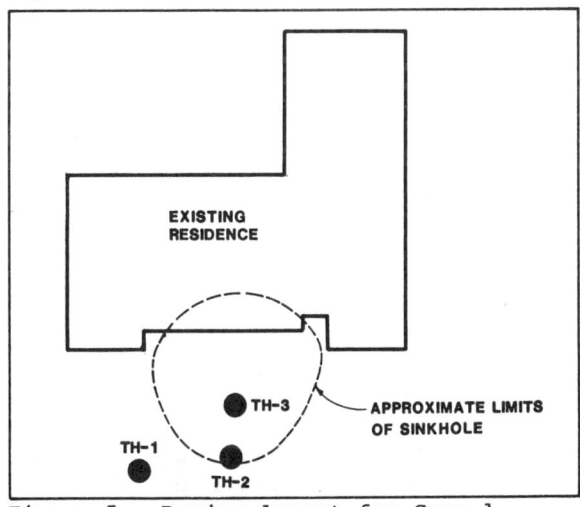

Figure 5. Boring layout for Case 1 subsurface investigation.

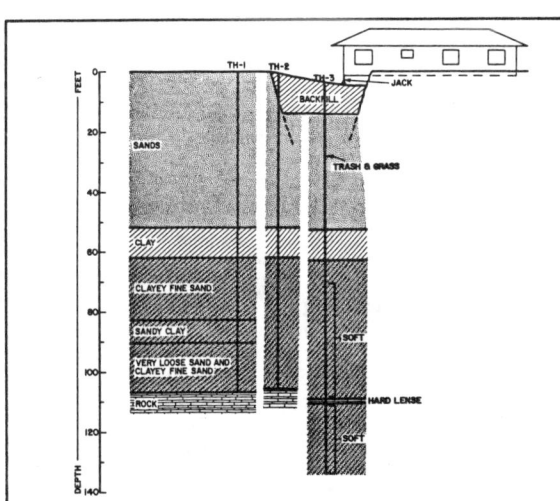

Figure 6. Test boring results for Case 1.

Figure 7 illustrates the writer's interpretation of subsurface conditions just before and after the sinkhole collapsed. Subsurface erosion, beginning within the cavity system, (where the piezometric level was 9 to 12 meters below the groundwater table) worked its way up through a breach in the bedrock surface and eventually up through the clay layer. Enough sand was washed down into the cavity system to create a large void in the overburden. When the slightly cemented sands above the void were no longer capable of supporting the roof of the void, the collapse occurred.

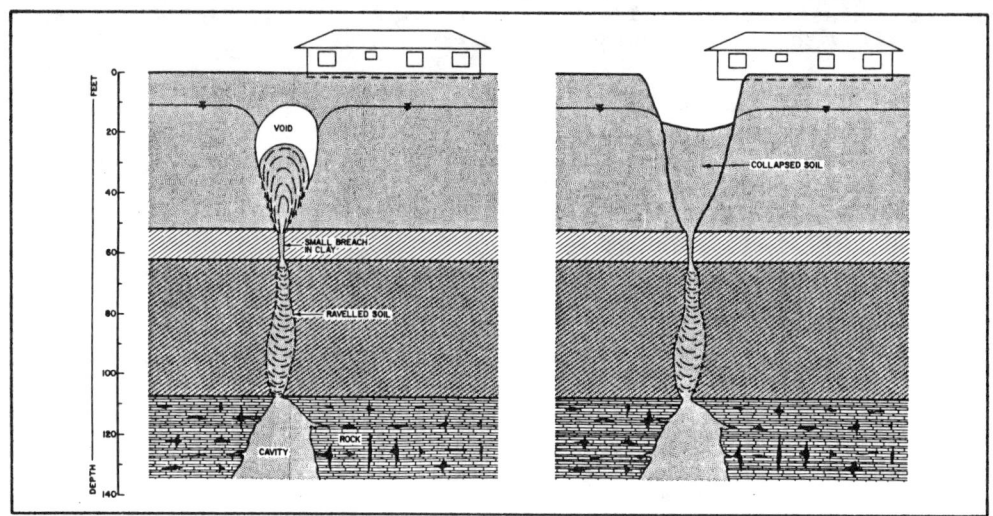

Figure 7. Interpretation of subsurface conditions before and after Case 1 sinkhole collapse.

To arrest the subsidence, cement grout was pumped into the bedrock cavity just below the top of the rock. The objective of the grouting was to plug off the breach in the bedrock surface. Approximately 45 m^3 of grout was pumped in stages into the top of the cavity before any significant grout pressure was developed. After completion of the grouting program, the settlement, as shown in Figure 4, essentially stopped.

Foundation support for the front wall of the house was provided by a grade beam designed to span the 11-meter sinkhole should future subsidence occur. The undisturbed sand beneath both ends of the grade beam was grouted with a chemical grout prior to pouring the beam (Figure 8, photograph page). The front porch slab was cantilevered from the grade beam.

The total cost of the remedial measures including engineering services and cosmetic repairs to the interior and exterior of the house, was about $50,000. There has been no further subsidence or any noticeable movement of the house during the past seven years.

Case History No. 2: Industrial Plant Site

Figure 9 is a photograph taken during grouting operations of a much larger sinkhole which occurred on March 18, 1981 within a large industrial complex located near Lakeland, Florida. This sink was approximately 20 meters in diameter and was threatening an adjacent structure. Figure 10 shows the layout of the test borings used to investigate subsurface conditions prior to planning the grouting program. The sinkhole was partially filled with soil prior to drilling the test borings. A cross section through the sinkhole is presented in Figure 11. The overburden soils consist of slightly silty to silty sands overlying clayey, phosphatic sands. The bedrock consists of clayey limestone. A rather large, vertical cavity approximately 6 meters wide at its widest point extended for over 18 meters down into the bedrock surface. The difference in hydraulic head between the groundwater table and the piezometric level in the underlying limestone aquifer within the general vicinity of the subject site was approximately 21 meters. Water level readings obtained in cased boreholes during the field investigation are shown in Figure 12. As shown, the head difference within the collapsed area was less than 6 meters (20 feet). As in the first case history, subsurface erosion had worked its way back through a breach in the bedrock surface and a large quantity of overburden soils was washed into the underlying cavity system. During the initial field investigation, a portion of the backfilled area collapsed a second time indicating that the area was still active.

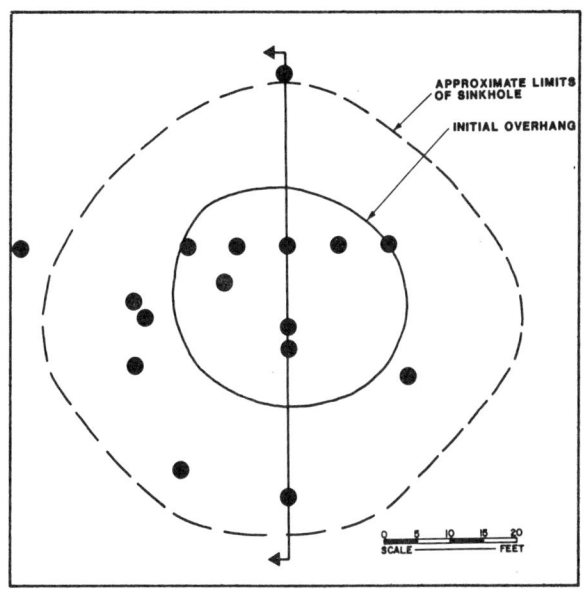

Figure 10. Boring layout for Case 2 subsurface investigation.

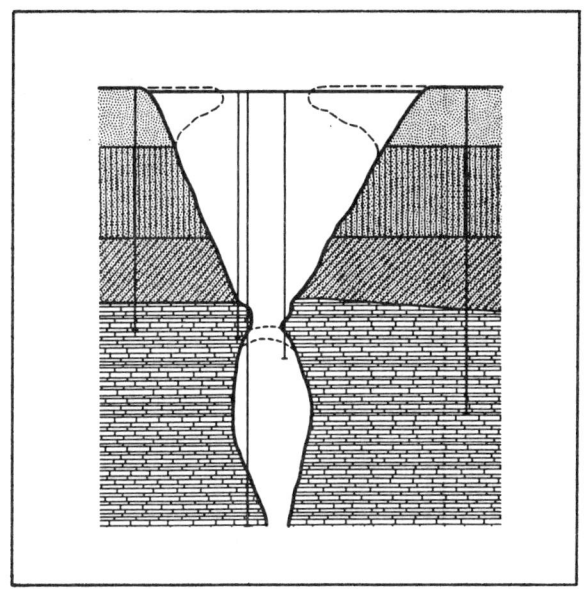

Figure 11. Cross section through Case 2 sinkhole.

Grouting with low slump concrete was initiated on April 25, 1981 at the locations shown on Figure 13. Over 280 m^3 of concrete grout was pumped into the breach. Borings completed after the grouting program revealed the conditions shown in Figure 14. A concrete plug was successfully installed within the throat of the breach. Water levels within the subsidence area increased to approximately the level of the adjacent groundwater. As shown in Figure 12, piezometers installed below the plug indicated that the water level had decreased to the piezometric level expected for the underlying aquifer. No further subsidence has been observed and water levels have remained relatively constant. The cost of the remedial measures associated with this sinkhole, including engineering services, was in excess of $200,000.

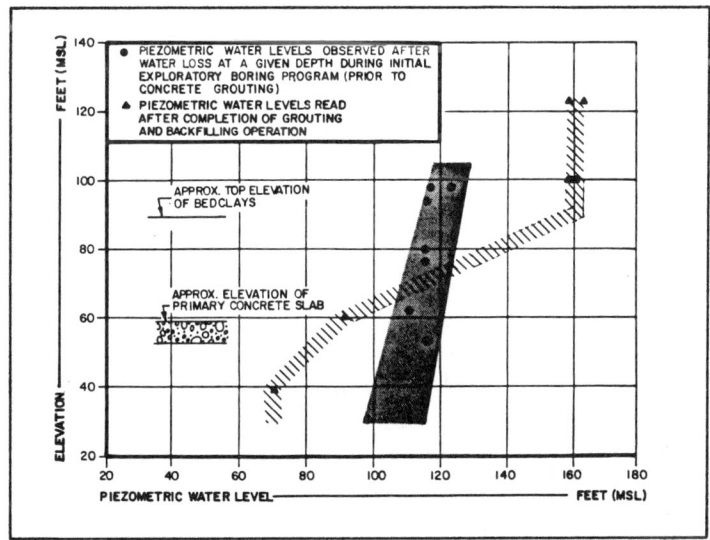

Figure 12. Piezometric water level vs. elevation prior to and subsequent to grouting.

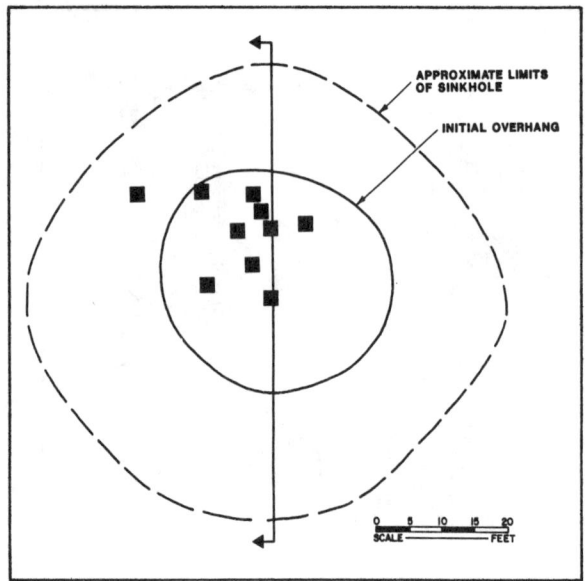

Figure 13. Plan of grouting locations for Case 2 sinkhole.

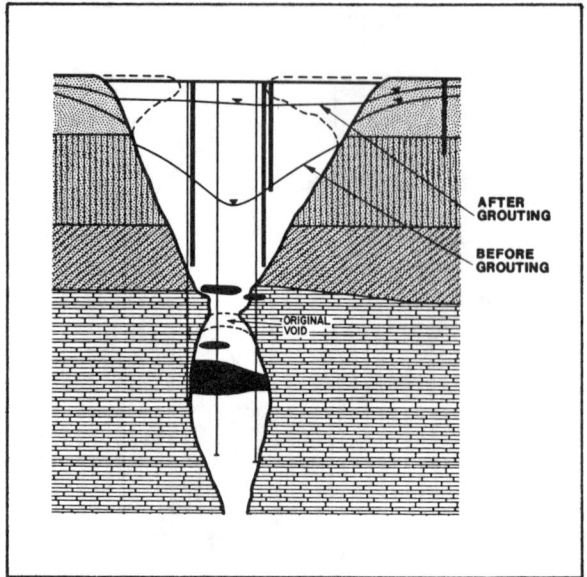

Figure 14. Cross section of Case 2 sinkhole showing grouting results and water levels before and after grouting.

Case History No. 3: Residence

This case history is somewhat unique because there was no apparent difference between the water table in the surficial sediments and the piezometric level in the underlying limestone, i.e., the limestone aquifer was under water table conditions. It was also different from the previous cases because the center of one of the two sinkholes that occurred at this site was located beneath the structure and was difficult to access. The sinkholes themselves were quite small: 1 to 2 meters in diameter and 0.2 to 0.3 meters deep. Figure 15 shows the magnitude of the subsidence. A test pit was excavated adjacent to the subsidence area to expose the foundation.

Visual observations in the vicinity of the subject site disclosed near surface limestone overlain by a thin veneer of sand. Several limestone springs were located near the subject structure. Visual observations also revealed that the downspout from the roof gutter discharged onto the ground surface at the location of one of the depressions and during heavy rains the roof gutter overflowed into a raised flower bed overlying the other depression. Apparently, ponded water within the flower bed and surrounding the downspout created enough hydraulic gradient within the sand to initiate piping within the underlying limestone cavities.

The subsurface investigation and grouting program at this site were performed simultaneously. The boring/grout hole location plan is shown on Figure 16. A total of 26 m^3 of concrete grout was pumped beneath the two depressions. Prior to performing the grouting program, it was necessary to move the east wing of the house away from the distressed area to provide equipment access.

After the grouting had been completed, the homeowner decided to expand the size of the structure by substituting a two-story addition for the original breezeway. A layout of the house pre- and post-subsidence is shown on Figure 16. The owner also insisted that the rear wall of the addition had to be located, as shown, across one of the depressions. Consequently, the wall footing for this wall was designed to bridge a 2-meter depression should future subsidence occur. The roof drainage was also redesigned and water from the downspout was carried away from the house. The only monitoring performed at this site was a visual inspection made two years after the remedial work was completed. There was no evidence of any subsidence or distress around the entire structure. The cost of the remedial measures associated with this case, not including the new addition, was just over $20,000, most of which was for moving the east wing.

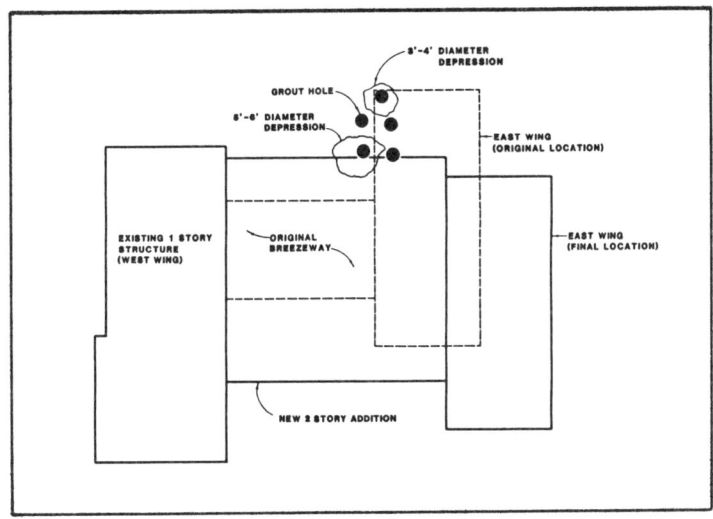

Figure 16. Plan of depressions and grout holes for Case No. 3.

Conclusions

All of the case histories involved the subsurface erosion or piping of unconsolidated sediment from the overburden into solution cavities in the underlying limestone. The movement of soil into the cavity system in all three cases was arrested by sealing off the breach in the bedrock surface through grouting. None of the three sites has experienced a reoccurrence.

Acknowledgements

The writer gratefully acknowledges the assistance of his staff and colleagues in the preparation of this paper. The work described in Case History No. 2 was carried out under the direction of the writer's associates: Dr. A.E.Z. Wissa and Dr. N.F. Fuleihan. The manuscript was typed by Mrs. Evelyn Dodge and the graphic presentations were prepared by Mr. Robert Felter.

References

Terzaghi, K. (1931). "Underground Erosion and the Corpus Christi Dam Failure", Engineering News-Record, 107 pp. 90 92.

Terzaghi, K. & R.B. Peck (1948). Soil Mechanics in Engineering Practice. John Wiley & Sons, Inc., New York.

A geological survey's cooperative approach to analyzing and remedying a sinkhole related disaster in an urban environment

ROBERT CANACE & RICHARD DALTON *New Jersey Geological Survey, Trenton, USA*

ABSTRACT

A sinkhole resulting from the break of an underground water main severely undermined the foundation of a private residence in densely populated Phillipsburg, New Jersey. After the condemnation and removal of the affected house, the town had to face the issue of the security of the remaining homes in the neighborhood. Through disaster relief funds the underground utilities and the street were completely replaced; this investment, too, had to be considered in the event of a reoccurrence of soil failure.

The township contracted the services of a geotechnical consultant. The consultant provided geophysical services, performing seismic refraction and terrain conductivity surveys. Backhoe excavations sited on the basis of these surveys revealed little about additional trouble spots.

The New Jersey Geological Survey offered its assistance and made an initial recommendation to undertake a detailed subsurface investigation. Through the generosity of a local drill manufacturer, the engineer for the town was able to arrange subsurface borings recommended by the Survey. The borings revealed the presence of a deeply dissected dolomitic bedrock surface and, more importantly, significant voids in the stiff, silty- to gravelly-clay overburden; some voids were in excess of 10 feet in height.

A remedial plan was designed in concert with the engineer and other geotechnical firms experienced in the grouting of voids. The remedial approach attempted to strike a balance between the magnitude of the void problem, the town's responsibility to public safety, and cost.

Figure 1: Damage caused by the "Thomas Street Sinkhole," Phillipsburg, New Jersey, June 16, 1983 (photo courtesy of Rick Rader)

Sinkhole Swallows House

The "Thomas Street Sinkhole," as it has come to be known, in Phillipsburg, New Jersey, occurred immediately after and adjacent to a major leak in a 36 centimeter (14 in.) water main. Interestingly, the incident was first reported to be a burglary (Phillipsburg Police Report). The residents notified police at 2:20 AM on the morning of June 16, 1983, that intruders were making noise in their basement. Upon investigating, the officers detected an odor of gas in the basement and witnessed, "a large crater which covered the entire basement floor as far as we could see towards the street." During the next hour the officers reported that they could hear wood creaking and could see the aluminum siding crinkling and bending. By 5:52 AM it had been reported that "the entire building sunk into a large crater"; sidewalk was lost as well (fig. 1).

The residence was condemned by the town and demolished two days after the incident by a private contractor retained by the residents. Two days after demolition it was reported that the sidewalk across the street from the site was sinking; small sinkholes appeared in other places along the street, as well.

Immediate Responses

The first step in dealing with the sinkhole involved remediating the apparent source and the immediate, visible threat and damage it caused. The water to the broken main was shut-off within 3 hours of the break, whereupon ground subsidence ceased. Individuals were evacuated from the affected home and neighboring homes. The town next decided that the sunken house should be removed. The house site was eventually backfilled with a mixture of crushed 3/4 inch stone and fines. During demolition of the house other areas of subsidence were similarly filled.

Damage to utilities on Thomas Street were so severe that the town immediately addressed the issue of the need to repair or replace water, sewer and gas lines. A grant of $202,500 was obtained from the Federal Small Cities Block Grant Program administered by the New Jersey Department of Community Affairs. The money was used for the complete reconstruction of street, curbing, sidewalk and utilities along a 650 foot section of Thomas Street.

To address the issue of possible additional threats to life and property, beyond those immediately visible at the surface, the township contracted the services of a geotechnical consultant.

Geophysical Investigation by Consultant

The consultant opted to investigate subsurface conditions using geophysical methods and backhoe excavations. An electromagnetic terrain conductivity (EM) survey and seismic refraction traverses were performed in the vicinity of Thomas Street. A Geonics EM-34 terrain conductivity meter was used by the consultant to define 21 "anomalous areas warranting further investigation." Seismic refraction traverses were performed at the anomalous areas to determine subsurface velocities and "locate discontinuities in the subsurface materials."

Six backhoe pits were dug by the town at locations selected from the above geophysical surveys. The pits ranged in depth from 1.8 to 4.3 meters (6 to 14 ft.) and revealed only a single small void with a dimension of .3 meters by .6 meters (1 ft. by 2 ft.). The geophysical investigation failed to detect large voids beneath the street which were later identified in the subsurface boring program, detailed below.

Role of Geology in the Decision-Making Process

After examining the consultant's findings and reviewing available information on the geology of the affected area (Drake, 1967), the New Jersey Geological Survey made a firm recommendation that subsurface borings be performed. The recommendation to undertake a subsurface investigation was based on the geologic framework of the impacted area. The likelihood was great that additional voids would have been caused by the pipe break, due to the specific geologic unit underlying Thomas Street. A geologic reconnaissance was the first step undertaken by the Survey. Outcrops in nearby railroad cuts and quarries permitted identification of the geologic unit as the Limeport Member of the Allentown Formation.

The Allentown Formation is a thick sequence of oölitic dolomite of Cambrian age (Wherry, 1909). The formation is one of five carbonate formations which constitute the Kittatinny Supergroup (Drake and Lyttle, 1980), which appears on the New Jersey State Geologic Map as the Kittatinny Limestone (New Jersey Geological Survey Atlas Sheet 40). The Allentown Formation has been further subdivided into two members — the Limeport Member and the Upper Allentown Member (Howell, 1950; Dalton and Markewicz, 1972).

The Limeport is the lower member of the Allentown Formation and consists of 122 to 214 meters (400 to 700 ft.) of fine to coarse crystalline, cyclically bedded, alternating light and dark dolomite beds from 0.3 to 1.8 meters (1 to 6 ft.) in thickness (Markewicz and Dalton, 1976). Distinguishable units can be recognized within the member, consisting of: (1) a lower, fine-grained cyclically-bedded oölitic and cryptozoan-bearing dolomite; (2) a medium to coarse-grained cryptozoan and oölitic dolomite containing silt and sand, and; (3) an upper fine-grained unit.

The Limeport Member is characterized by a well developed solution channel network, notable caverns, major springs, sinkholes and a pinnacle-and-trough weathering profile on the bedrock surface. The member's large number of thin beds provide many potential solutioning sites. The alternation between fine and coarse crystalline beds results in an intensive degree of solutioning, as coarse-grained dolomites tend to weather in preference to fine-grained dolomites (Siddiqui and Parizek, 1971; Dalton and Markewicz, 1972). A study of caves in New Jersey revealed that the greatest total passage length and greatest average passage length of known caves occurs in the Limeport (Table 1). Indeed, the largest cave system in New Jersey is located in the Limeport Member (Dalton, 1976). The state's largest springs occur in the member, with the largest of these typically occurring near its contact with the Upper Allentown Member.

Drill hole data generated during the study of the Spruce Run Reservoir, located 14 miles east of Phillipsburg, further demonstrates that the solubility of the Limeport is high in comparison to the other Cambro-Ordovician carbonate formations (fig. 2). Logs of drill holes along the dam axis reveal that, of the Kittatinny units encountered, the Limeport exhibits the highest percentage of void footage per foot of hole drilled (Table 1).

Table 1: Results of subsurface borings at the Spruce Run Reservoir Dam site, Clinton, New Jersey and relationship of known caves to formation (Modified from Dalton and Markewicz, 1972, Table 4.)

Formation	No. of drill holes	Void per total depth	Average passage length of caves (ft.)
Epler	66	4.1	40
Rickenbach	31	5.8	89
Allentown (Upper Member)	33	2.4	14
Allentown (Limeport Member)	32	7.0	197
Leithsville	9	6.5	167

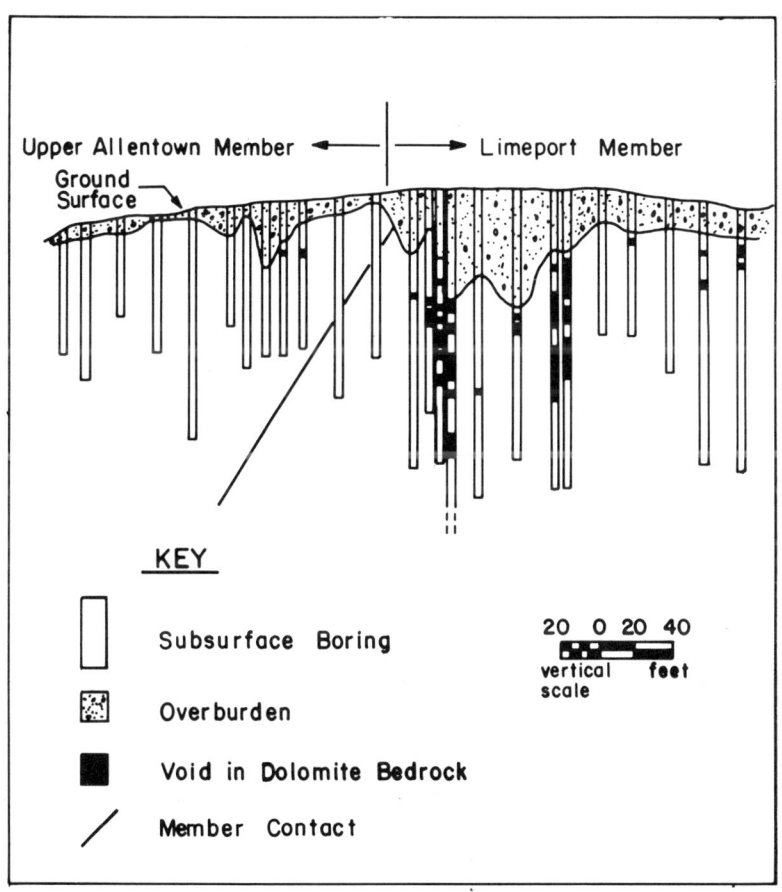

Figure 2: Subsurface profile of a portion of the Spruce Run Reservoir dam site, showing contrast in intensity of weathering between the Limeport and Upper Allentown Members of the Allentown Formation. (Source: Markewicz, 1958-1961)

Subsurface Borings

The Geological Survey offered its services to the town to review the site conditions and assess potential future problems. We felt that the consultant's choice of using a geophysical technique to locate voids was unsuitable in this case, for two principal reasons: (1) the large number of potential sources of interference in the developed setting and their possible effect on the EM readings, and; (2) the critical need for a precise determination of subsurface conditions in view of the insecurity felt among residents of the neighborhood, versus the inexact nature of geophysical techniques in locating voids. The Survey, therefore, thought it was imperative to undertake a detailed subsurface exploratory program consisting of a grid of subsurface borings. Subsurface borings constitute the most objective and thorough method of evaluating the stability of the ground.

The Geological Survey recommended that a small "air-track" drill machine, such as a quarry drill, should be used for the borings. The quarry type drill machine is ideal for this type of investigation, because: (1) it is small and can be used in the close quarters caused by overhead power lines, narrow streets and construction equipment; (2) the machine works quickly and can install many borings in a short time, which is important when working on a low budget, and; (3) borings can be drilled without the use of water, so as to avoid causing additional sinkholes. Drilling services were provided on an entirely voluntary basis by Ingersoll Rand Corporation of Phillipsburg. The company supplied a drill technician and an experimental hydraulic drill from its Sandseddy Rock Drill Development Center in Pennsylvania to perform the test borings. Initially, the Geological Survey recommended that borings be performed on 15 meter (50 ft.) centers and that holes be drilled until 1 meter (3 ft.) of competent bedrock was encountered. The capabilities of the machine and the willingness of the operator permitted borings to be drilled to depths greater than planned. This additional drilling permitted a more complete evaluation of bedrock conditions than originally anticipated.

Subsurface conditions were interpreted from the behavior of the drill machine and the presence or absence of cuttings. Heavy reliance was placed on the experience of the drill operator. No cores or split spoon samples were taken, as this would have involved greater time and imposition on the generosity of the company. The presence of voids was interpreted on the basis of: (1) free drops of the drill string; (2) the loss of air pressure from the compressor; (3) lack of return to surface of cuttings; and/or; (4) the vibration and sound of the drill machine, especially when drilling through bedrock. The bottom of a void was defined by the return to "normal" of these factors.

One hundred borings were performed in two phases as part of the program (fig. 3). All borings were located on public right-of-way, as this is where the town felt its responsibility began and ended. Borings were drilled initially on 15 meter (50 ft.) centers. Additional holes were drilled for comparison where major voids were found and in areas removed from the voids. Drilling revealed a highly irregular dolomite surface and several large voids in the silty- to gravelly-clay overburden (fig. 4). The largest and most numerous voids were encountered in the vicinity of the water main break (fig. 3). Several voids were in excess of 3 meters (10 ft.) in height. A few cavities were encountered in the bedrock, but these were partially or wholly mudfilled and did not appear to pose a major threat to the surface. All drill holes were backfilled with a mixture of dry sand and Portland cement in order to avoid creating conduits to the subsurface for stormwaters.

The origin of the large voids is believed to have been related to the initial sinkhole occurrence. Borings not in the immediate area of the sinkhole revealed small and/or isolated voids, generally not more than 0.6 meters (2 ft.) in height. Two larger voids were encountered upgradient from the sinkhole; the origin of these was not readily apparent. We felt, based on the more than 100 borings in the immediate area, that the size and concentration of voids in the overburden near the sinkhole was atypical. In contrast, voids found at greater distances from the water main break were scattered and usually small. These findings tended to confirm the hypothesis that the large voids were part of the initial wash-out and collapse.

The "Dental Plan"

After the completion of the subsurface boring program, the town decided that it had to face the question of its responsibility for remedial action beyond road and utility repair. The feeling was expressed by the town that it was in the compromising position that no matter what course of action it took, it would be held responsible for a reoccurrence of sinkholes in the immediate area. The Geological Survey recommended that a grouting program be undertaken to stabilize the identified soil voids. In this way, the town would be in the position of having been thorough in investigating potential soil problems related to the initial sinkhole and attempting to avert their later repercussions.

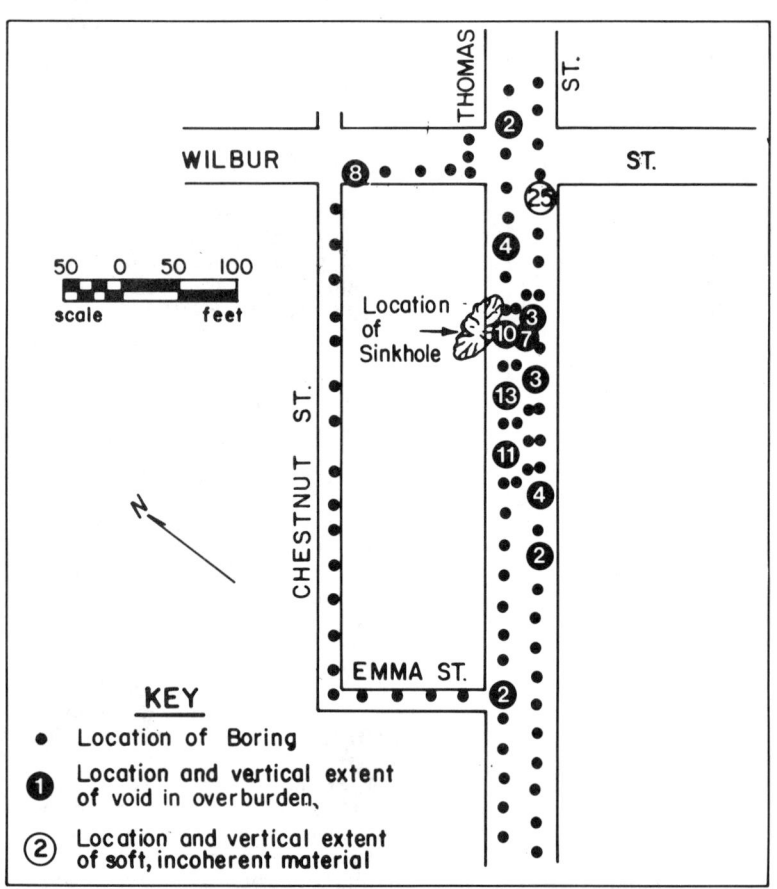

Figure 3: Location of subsurface borings and location of and vertical extent of voids in overburden, Thomas and adjacent streets, Phillipsburg, NJ.

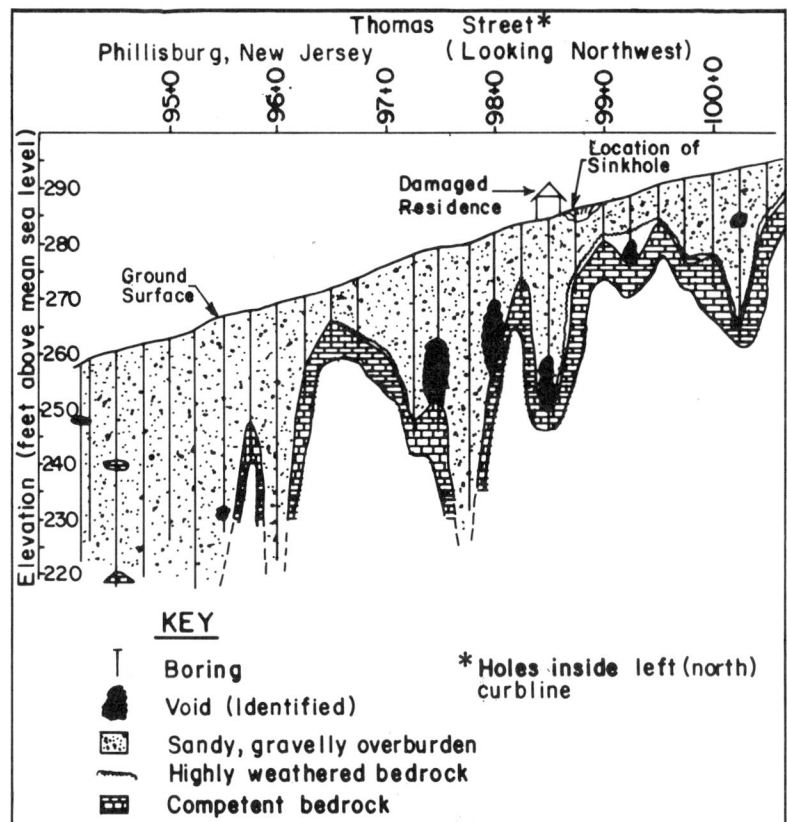

Figure 4: Geologic profile from subsurface borings along Thomas Street, Phillipsburg, N.J.

The question arose naturally as to the threat posed by the voids identified through drilling. In recommending that the township pursue remedial action, the Survey pointed to the following considerations: (1) several of the voids were in excess of 3 meters (10 ft.) in height, collapse of these could result in a significant displacement in the land surface; (2) the interrelationship and therefore the overall size of the voids was not fully known, even though additional borings were installed between initial 50 foot centers; (3) the recently installed street and utilities represented a significant investment in taxpayer dollars and remedial action for underlying voids was a means of helping to protect that investment, and; (4) surveying of reference points indicated that subsidence had taken place in houses adjacent to the sinkhole in the year following the incident. The greatest subsidence was recorded on the corner of a neighboring house away from the sink, indicating possible subsurface movement related to existing voids as opposed to settlement of backfill at the sinkhole (fig. 5).

Working in conjunction with the engineer for the township, several experts in grouting were contacted and various approaches were discussed. From these discussions a series of options was outlined and presented to the town, ranging from doing nothing to an elaborate barrier-and-grout scheme often employed in grouting abandoned mines. The township, of course, was seeking a combination of maximum effectiveness and minimum cost.

The subsurface boring program provided an unexpected additional benefit in that it permitted a reasonable estimate of anticipated grout take, providing the engineer and grouting firms with a working base from which to calculate bids. The bids included the use of chemical additives to limit the grout take to a reasonable volume and were based on the town supplying Portland cement, fly ash and a limited amount of labor. The town was then able to approach the Department of Community Affairs with a good estimate of its funding needs for additional remedial work. The town was granted a supplemental grant of $75,000 to grout the voids. A remedial approach was agreed upon by the engineer, the town and the Survey.

The planned remedial approach, which has come to be called the "Dental Plan", consists of targeting those voids identified in the boring program as first priority for grouting. Ingersoll Rand has once again agreed to provide drilling services. The strategy to be followed involves attempting to intercept the voids encountered in the exploratory program. Careful surveying of bore holes performed during the subsurface investigation will permit resetting the drill rig over or near known voids during the remedial work. Since the overburden is stiff and unsaturated, grouting will take place through an open hole. Grout take will be limited through the use of sodium silicate. Once the known voids are filled, additional holes will be drilled, splitting centers between voids. Additional voids will be grouted within the allowance of the project's budget.

All remedial work will be confined to the right-of-way owned by the town. The township holds that its responsibility ends at that point.

Summary and Conclusions

The Phillipsburg sinkhole was a classic case of the disturbance by human infrastructure of a susceptible carbonate environment. The impact of a broken water main was severe in this case, because the precise geologic setting in which it occurred contained the elements of ground instability. Knowledge of the geology of the disaster site permitted an assessment of the likelihood of damage beyond what was immediately visible. Because it was thought damage might be pervasive, an in-depth subsurface investigation through the use of borings was recommended. The subsurface borings uncovered potential hazards which would otherwise have gone undetected.

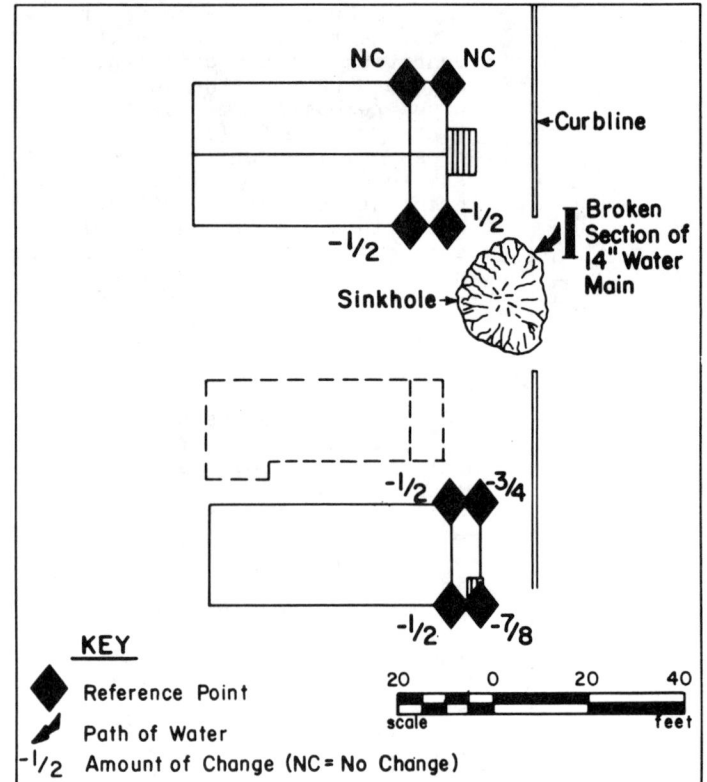

The tread-mounted, air-driven, quarry-type drill machine proved versatile in conducting the subsurface investigation. The rig was maneuverable in a difficult work setting, provided rapid information on subsurface conditions and, because it uses air instead of water to return cuttings, did not worsen sinkhole related problems. In addition to pinpointing the location of voids in the unconsolidated overburden, the borings permitted estimates to be made of the cost of the remedial program. This enabled the town to apply for relief funds with reasonable estimate of the total cost in hand.

Although geophysical techniques have been used extensively to attempt to delineate subsurface voids, their use should be weighed against the urgency of a given problem, their interpretability, and the presence of outside sources of interference.

Figure 5: Amount of settlement, in inches, of surveyed reference points, from June 21, 1983 to May 5, 1984. (Source: Tom Kasopius, Town of Phillipsburg)

References

Dalton, Richard, and Markewicz, Frank J., 1972, Stratigraphy of and characteristics of cavern development in the carbonate rocks of New Jersey: Bulletin of the National Speleological Society, 1972, v. 34, pp. 115-128.

Dalton, Richard, 1976, Caves of New Jersey: Bulletin 70, New Jersey Geological Survey, 51 p.

Drake, Avery A., 1967, Geologic map of the Easton quadrangle New Jersey-Pennsylvania, United States Geological Survey, Map GQ-594, scale 1:24,000

Drake, Avery A., and Lytlle, Peter T., 1980, Alleghanian thrust faults in the Kittatinny valley, New Jersey, in Manspeizer, Warren, ed., in field studies of New Jersey geology and guide to field trips, 52nd annual meeting of the New York State Geological Association, Rutgers University, Newark, New Jersey, pp. 92-114.

Lewis, J. Volney, and Kümmel, Henry B., 1901, Geologic map of New Jersey, Atlas sheet no. 40, New Jersey Geological Survey, Trenton, N.J., scale 1:250,000.

Markewicz, Frank, J., 1958-1961, Field notes, maps, and construction profiles of the Spruce Run Reservoir site, New Jersey Geological Survey, unpub.

Markewicz, Frank J., and Dalton, Richard, 1976, Stratigraphy and applied geology of the lower Paleozoic carbonates in northwestern New Jersey, in guidebook for the 42nd annual field conference of Pennsylvania geologists, October 6-8, 1977, Harrisburg, Pennsylvania.

Siddiqui, S.H. and Parizek, Richard R., 1971, Evaluation of Measured Errors by Repeated Pumping Tests, Journal of Hydrology, v. 13, pp. 182-191

Town of Phillipsburg, New Jersey, Bureau of Police, docket no. 0683-4687, report of June 16, 1983, ff.

Wherry, E.T., 1909, The early Paleozoics of the Lehigh Valley district, Pennsylvania, Science, n.s., v. 30, p. 416

Sinkhole activity in the vicinity of the Sunny Point, N.C., Military Ocean Terminal

EARL F.TITCOMB, Jr. & JACK M.KEETON *US Army Corps of Engineers, Savannah, Georgia, USA*

ABSTRACT

On 1 December 1976, two springs and a sinkhole appeared along the Army-owned railway three miles north of the Sunny Point Military Ocean Terminal. The sinkhole occurred in the section of the railway that parallels the downstream toe of an earth dam which impounds the Boiling Spring Lake. Following the appearance of the first sinkhole, two additional sinkholes developed in the rail bed about one-half mile south of the location of the first sinkhole. These same problems with sinkholes had occurred previously in this general area in 1962. Exploratory borings revealed an underground cavity of large extent beneath the railway in the vicinity of the 1962 sinkhole. The geology, causes, exploratory programs, remedial actions, and present status of this problem are addressed.

Location and Topography

The Sunny Point Military Ocean Terminal (MOTSU) is located 17 miles south of Wilmington, N.C., along the Cape Fear River. (See Figure 1) The area lies within the Atlantic Coastal Plain, is generally characterized by low relief. Higher sandy interfluvial areas are cut by easterly flowing creeks, frequently with wide swampy flood plains. In the vicinity of the town of Boiling Springs, north of Orton Pond and on the MOTSU, numerous small, circular, closed depressions, often water filled, are found. Three large northeast flowing creeks are found north of MOTSU, Orton Creek, Allen Creek and Town Creek. A large spring (Boiling Springs) flows into Allen Creek about two miles north of MOTSU. Orton Pond is a large shallow lake impounded during revolutionary time by the construction of a low earth dike. The terminal is served by a 21 mile long, Army - owned railroad which leads north from the terminal, across Allen and Town Creeks, to Leyland, N. C.

Geology

The area within and around MOTSU is underlain by gently dipping beds of sand, clay, and limestone. The formations, in descending order, are: undifferentiated sands and clays of Pleistocene age, quartz sands, clays and shell beds of the Waccamaw Formation of Pliocene age and limestone of the Castle Hayne Formation of Eocene Age. The Pleistocene sediments and the Waccamaw are similar in lithology and have not been separated during our investigations at and near MOTSU. The Castle Hayne upper surface is unconformable and is generally underlain by the hard limestone. This harder rock is generally 5 to 15 feet thick, and often contains many small finger size interconnected vugs. The limestone is light gray to buff in color, occasionally with a greenish "salt and pepper" appearance due to grains of glauconite. Fossils are usually abundant. This hard limestone changes with depth to a soft calcareous silt and with further depth this sequence is repeated. The top of the Castle Hayne near Boiling Springs Lake is at 0 msl while in the Cape Fear River near the MOTSU wharf area it is found at about -51. In the vicinity of Boiling Springs Lake and along Cape Fear the Castle Hayne is immediately overlain by a clayey sand or clayey silt of moderately low permeability.

Sinkhole Occurrence (1976)

In December 1976 while on a routine track inspection, the Army railroad maintenance personnel heard water flowing downstream of the railway embankment. This was in an area immediately downstream of the Boiling Springs Dam where the rail line paralleled the dam axis. The source of the water was determined to be a spring approximately 200 feet downstream of the rail line or 425 feet downstream of the dam. Further investigations revealed another spring between the railway embankment and the dam. Shortly thereafter a small depression formed on the upstream (dam) side of the railway embankment. This depression became larger and expanded into a 10 foot diameter sinkhole 8 feet deep. The owner of the dam opened one of the spillway gates and began lowering the lake. This reduced the flow from the downstream spring and the flow from the spring between the dam and railroad stopped altogether. However, sinkholes began to develop on the upstream face of the dam and as the water lowered, other sinkholes in the dam face were exposed. Dye introduced into the dam sinkhole was observed about 1/2 hour later exiting from the spring. As the lake receded, the flow from the spring decreased and finally ceased. These sinkholes caused much concern to the MOTSU personnel since the rail line carries over 90% of the cargo shipped from the terminal. Our conclusion was that the springs and sinkholes were caused by water flowing out of the lake, under the dam and railway, and exiting from the spring. The extent of the damage to the railway embankment was immediately investigated by exploratory borings. See Figure 2 for locations.

Conditions Found

There were no large solution cavities encountered in the exploratory borings; however, there were numerous small cavities measuring from 0.3 to 0.5 of a foot in thickness found in the borings in the vicinity of the sinkholes. The largest cavities encountered were 1.0 foot in thickness. These small 0.5' to 1.0' solution cavities were restricted to an area about 100 feet long straddling the railway sinkhole. This upper

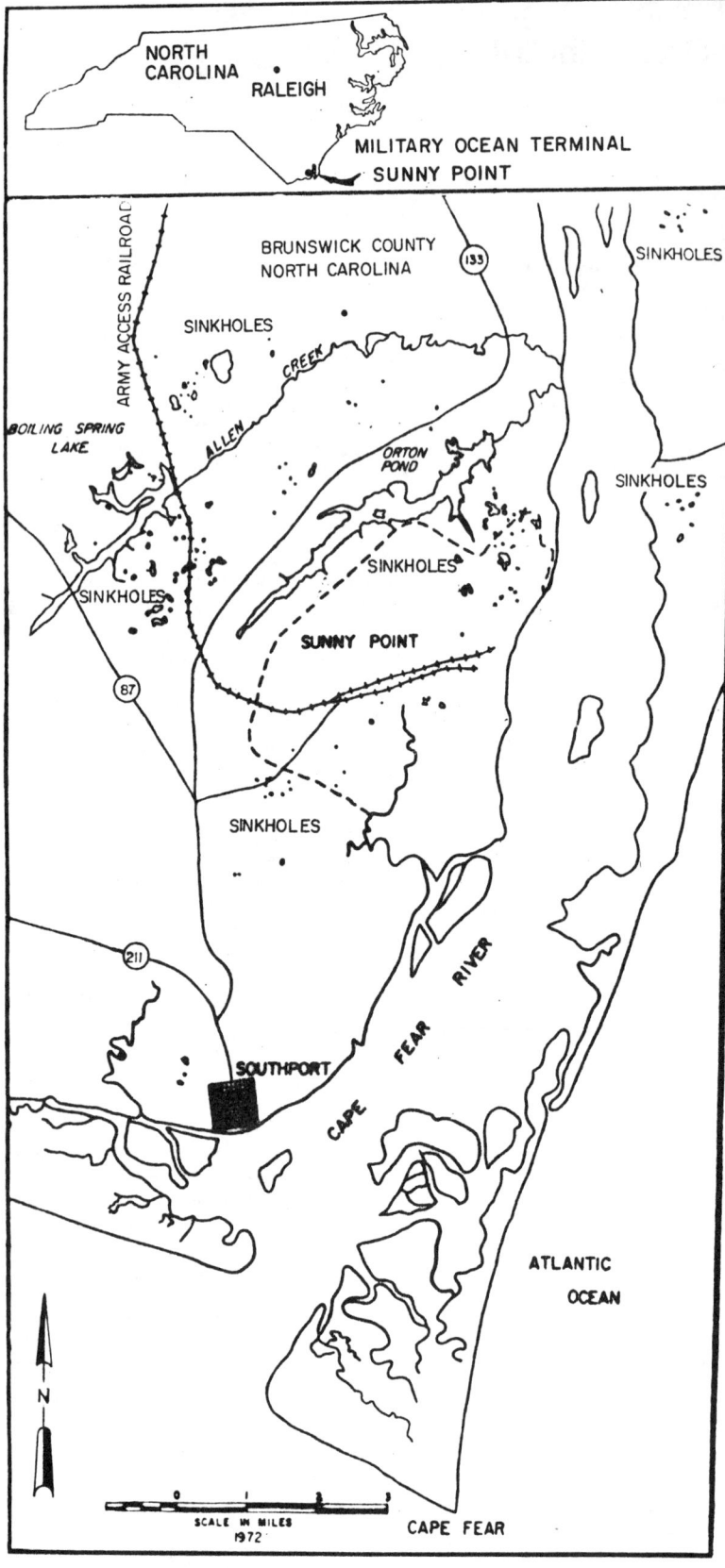

Figure 1: Vicinity Map

5 to 7 feet of the limestone in the 100 feet reach, however, was characterized by a zone of numerous small "finger-like" solution cavities. It is in this "solution zone" that much brown staining was encountered. Also, in this zone, some of the rock was found to be broken and crumbly in many of the borings. Underlying this 5.0 to 7.0 foot "solution zone" is a soft white limestone. This "second" limestone extends to the bottom of the borings. There were no solution cavities or brown stains encountered in this "second" zone. The top of the Castle Hayne limestone occurs fairly uniformly at about elevation 0.0 feet msl in this area. Four holes were drilled across the sinkhole that occurred upstream of the access railroad. One boring was drilled in the sinkhole, and a very loose sand was encountered in the overburden. During the splitspooning of the hole, the weight of the hammer forced the splitspoon a total of 15.0 feet. When the splitspoon was recovered, no sample remained in the spoon, indicating either a void or very loose overburden. The top of rock in this boring was at about the usual elevation 0.0 feet msl. No cavities were encountered although the 7.0 foot "solution zone" was present with the brown staining. A second boring was drilled 5.0 feet south of the first. In this boring the weight of the drill rods themselves pushed the splitspoon sampler about 8 feet to the top of rock at elevation -2.4 feet msl. Again no sample was contained in the splitspoon when it was recovered. Between elevation -2.4 feet and -7.2 feet msl the limestone was stained, crumbly, and solutioned, although no large cavities were encountered. A third boring was drilled 5.0 feet south of the second. Top of the rock was encountered at about elevation 0.0 msl in this hole, and the solution zone was again present in the upper 6.0 feet of the limestone, with a small cavity about 1.0 foot in thickness being found at 0.0 msl. The fourth boring was drilled over the sink area 5.0 feet north of the first; no cavities or soft zones were encountered. The information from these borings indicated a depression in the rock surface near the sinkhole location. A total of 13 holes were drilled on 100 foot centers along the railway immediately downstream of the dam, and with the exception of one boring, there was no loss of water during drilling. There was a loss of water during the drilling of one boring; however, no large cavities encountered.

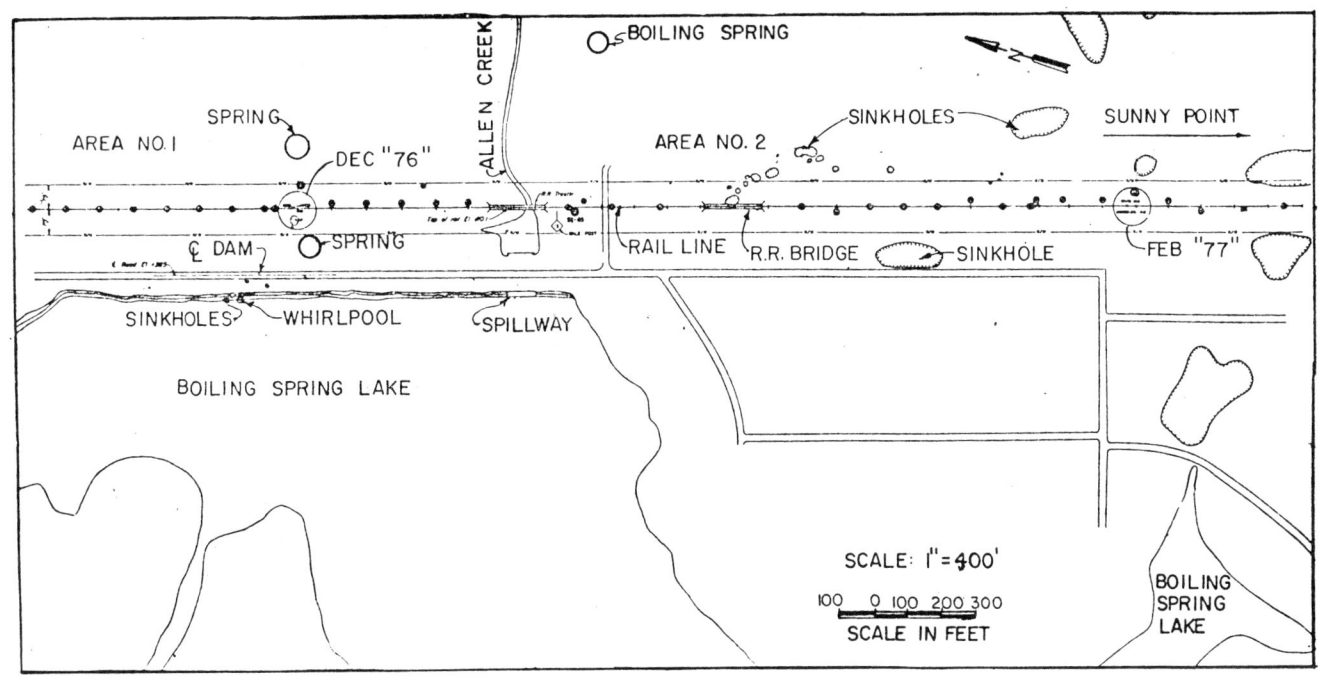

Figure 2: Site Plan

Sinkhole Occurrence (1977)

About the time of the completion of the above investigations (Feb 1977), another sinkhole appeared 2,600 feet south of the December 76 occurrence, 32 feet from the railway. Again an exploratory program was begun; twenty-one borings were made along a 100 foot section near the sinkhole.

Conditions Found

Unlike the limestone in the first area, the cavities and "solution zones" are not restricted to the upper zone of hard limestone, but are also found in the upper 10 feet or so of the underlying "soft" zone. There were 27 small cavities encountered ranging from about 0.5 to 2 feet in both limestone zones. The largest cavities found in this area were 5 feet in thickness; these cavities were encountered 17 feet and 10 feet respectively below the top of rock.

The immediate response to the two sinkhole occurrences described above, in order to prevent any further collapse, was to grout the exploratory borings. Over 3,000 cubic feet of grout were pumped into the southern area shown on Figure 2, while the area immediately below the dam only took 450 CF. In addition to this program the exploratory boring program was extended for the entire rail line east of the dam and about 2,000 feet south. In addition to the boring program, a micro-gravity survey was made, color and enhanced imagery air photographs were run, and a thorough research of the past sinkhole activity in the area was made.

Past Sinkhole Activity

While there was substantial evidence of past sinkhole activity in the area, the only recent occurrences in Brunswick County had been in the immediate area of Boiling Springs Dam and Lake. The dam was constructed in 1960, four years after the completion of the rail line. In 1962 a sinkhole developed overnight 10 feet from the rail line, 600 feet south of Allen Creek. Investigations revealed a large, open, water-filled cavity with a maximum thickness of 14.0 feet at a depth of about 40 feet. The solution at that time was to construct a caisson supported land bridge over the cavity area. In 1963 a spring appeared downstream of the rail line near the site of the December 1976 occurrence. At that time the dam owner lowered the lake and found sinkholes in the upstream face of the dam. These were backfilled with a sandy clay. There were no documented occurrences of other activity in the dam area between the 1964 backfilling and the December 1976 occurrence. There was, however, during this period a growth of sinkholes near the land bridge. The area under the bridge collapsed and a series of sinkholes "grew" southeast of the bridge. No measurements of the size or rate of growth of these sinkholes were made; however, after the draining of the lake in 1976-1977, there was very active sinkhole development observed during February and March 1977.

Mechanism of Sinkhole Formation

Our studies revealed no active sinkhole development anywhere in the area except in the vicinity of Boiling Springs Dam and Lake. We concluded that the construction of the Boiling Springs Dam, and subsequent impoundment of the Lake, was the factor responsible for the problem being experienced along the access rail line. The cause and effect was easily established in the area immediately downstream of the dam since the dye injection and effects of plugging of the lake sinkholes were readily apparent. The effect on the area south of Allen

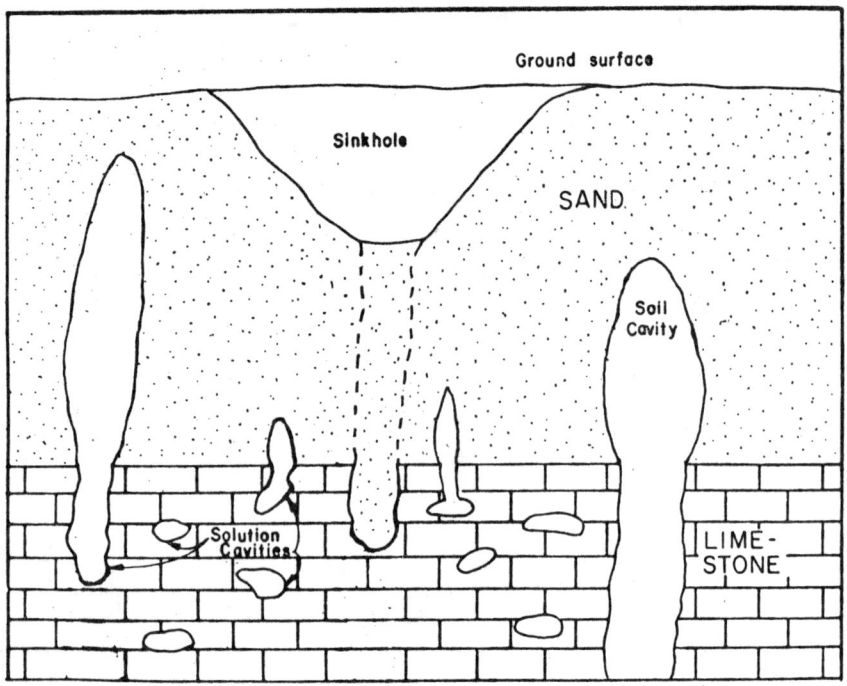

Figure 3: Idealized drawing showing soil cavities and sinkhole formation. This is described as "hourglass erosion," by George F. Sowers in Mechanisms of Subsidence Due to Underground Openings, in Alabama Transportation Research Record 612.

Creek however was not as obviously related to the Lake. Our studies indicated that the mechanism of collapse was probably as shown on Figure 3. The small, clay-filled cavities were subjected to increased head due to the impoundment of Boiling Springs Lake. The increased velocity eroded and removed the clay filling and eventually caused solution of the limestone. The coalescing of the small voids and cavities creates larger cavities. The cavities near the limestone - soil interface eventually intersect with the loose sands being carried down into the open cavity. With time a cavity is formed in the sands which eventually enlarges to a point that the overlying sand bridge collapses, creating the surficial expression. The collapse can be triggered by rapid lowering of the water table as occurred in Dec 76 - Jan 77, causing the obvious sinkhole growth observed near the land bridge.

Long Term Solutions

The Army's primary concern is the creation of these sudden collapses, especially in that area downstream of the dam. Here the rail line embankment is approximately thirty feet high so that a derailment could have catastrophic consequences. Numerous solutions were considered, ranging from moving the rail line, bridging the entire area partial bridging, grouting, to a posture of what might be called benign neglect. The costs of all of the positive actions were in the order of millions, so that the final solution was a compromise which provides a large measure of protection but does not eliminate entirely, the potential for damage should a sinkhole appear suddenly under the track. It was decided to widen the rail line embankment for the reach immediately below the dam approximately 1200 feet and to install a 100 foot long multiline grout curtain across the area of the December 1976 sinkhole. The widening would prevent a derailed train from rolling down the embankment. The grout curtain was to fill any voids in the limestone and also to cause collapse of any sand cavities. The owner of the dam also installed a short grout curtain in the dam, giving added protection against seepage from the lake. It was decided that Area 2 south of Allen Creek would receive no treatment since there was no way to identify those areas particularly susceptible to sinkhole development. However, materials were stockpiled to provide backfill in case of collapse, as well as track building materials to provide a temporary bypass while remedial repairs were being made. This program was carried out in 1978-1979.

Since 1979 there has been only one new sinkhole develop along the rail line. This sinkhole occurred on 11 July 1983 at the extreme south end of our previous investigations. Drilling began immediately. Conditions found were similar to those in the vicinity of the February 1977 sinkhole which was located about 250 North. Cavities were found both at the top of the limestone and lower down in the rock. In this area ten holes were drilled and grouted along a 20 foot section. Grout takes were high; over 2,000 C.F. of cement grout were pumped into this limited area.

Present Status

Since the July 83 occurrence the Sunny Point Rail Line - Boiling Springs area has been quiescent. The potential for future problems will probably exist for as long as the dam exists. No economically feasible solution was available to provide the 100% degree of protection for the rail line that was desired by the Army. We feel that while there is an element of risk in continuing to operate the rail line, it is a risk which, by constant observation of the rail embankment and cut, can be minimized to an acceptable level. The Boiling Springs area is characteristic of many other Karst regions of the world in being environmentally sensitive. Cause and effect relations tend to be indicated readily, and man-made actions at one place are commonly reflected by adverse reactions elsewhere in the system. Past actions by man in the Boiling Springs area have resulted in undesirable consequences (sinkholes) which will undoubtedly be again active in the future. Unfortunately our ability to predict the timing and magnitude of these sinkholes is not yet developed to the degree needed to prevent or ameliorate the consequences.

Evaluation, repair and stabilization of the Boling Sinkhole FM 442, Wharton County, Texas

BOYD V.DREYER *Cook-Joyce, Inc., Austin, Texas, USA*
CLYDE E.SCHULZ *Department of Highways and Public Transportation, Wharton, Texas, USA*

ABSTRACT

Abrupt ground subsidence occurred along FM 442, approximately 3 miles east of Boling, in Wharton County, Texas, in August, 1983. The subsidence created a depression of approximately 250 feet in diameter with a maximum depth of about 25 feet. Roadway collapse, pavement breakage, and inundation of the sinkhole by subsurface waters forced the closure of FM 442 to thru-traffic.

The vicinity of the sinkhole is underlain by alluvial deposits of the Colorado River drainage basin and unconsolidated Miocene, Pliocene and Pleistocene sediments to a depth of about 500 feet. The sediments consist largely of sands, silts, clays, and gravels. The Boling Salt Dome and its associated caprock occur directly beneath the section of unconsolidated sediments. Major and minor axes of the dome are about 5 miles and 3.5 miles, respectively, and the sinkhole occurred approximately 1 mile east-northeast of the dome's center.

An investigation was conducted in the immediate area of the sinkhole to determine the cause of the subsidence and also to determine the feasiblity of dewatering the sinkhole. Four monitoring wells were installed around the perimeter to evaluate the potential recharge from shallow sand units. Two 300-foot holes were drilled and geophysically logged to determine the stratigraphy adjacent to the sinkhole, and two slope indicators were installed in order to evaluate further earth movements. A total of six million gallons of saline water were removed and disposed of from the sinkhole.

Based on information provided by the geologic and dewatering investigation, the Texas State Department of Highways and Public Transportation elected to fill the existing sinkhole and reconstruct the roadway. A total of 3,500 cubic yards of rock borrow and 26,000 cubic yards of soil borrow were placed in the depression. Construction began in February, 1984, and the roadway was completed in May. A monitoring program to evaluate subsequent earth movements will continue into 1985.

INTRODUCTION

During a rain shower on 15 July 1983, water was observed to be flowing across FM 442. This was reported to the Texas State Department of Highways and Public Transportation (SDHPT) office in Wharton, Texas, which responded by placing "high water" warning signs at the site of the observed flow. On 8 August a citizen advised SDHPT that he had perceived the roadway to be sinking. A reconnaissance inspection indicated a 2-ft to 3-ft depression along the roadway, and on 9 August a SDHPT survey crew measured the roadway to be 3.26 ft below original grade at the lowest point within the depression.

On 11 August 1983, additional subsidence was accompanied by the encroachment of collecting water onto the roadway, formation of cracks in the roadway pavement approaching the depression, and pavement buckling near the center of the depression. A SDHPT survey crew measured up to 0.8 ft of additional subsidence. Sometime between 11:00 P.M. and midnight that evening, ground subsidence increased dramatically, resulting in the rapid collapse of a circular area about 250 ft in diameter with a depth of approximately 25 feet. The depression was immediately filled with saline water forced upward from the subsurface. The center of the collapse was located approximately 75 ft on the north side of the centerline of FM 442.

It was recognized that the roadway could be completed by means of constructing an embankment, bridge or selecting an alternative alignment. The appropriate design would depend to a large degree on the stability of the sinkhole and underlying materials. Knowledge of the location of faults and voids along with the rate of lateral earth movements associated with the sinkhole was considered to be critical in determining the appropriate design recommendation along with an assessment of the ground-water conditions of the sinkhole. The following sections describe the general geologic conditions of the area and detail the geologic/hydrogeologic evaluation and the remedial design measures for repair of the roadway.

GEOLOGIC/HYDROGEOLOGIC SETTING

General Geology

The project area is located within the Gulf Coastal Plain province of Texas, approximately 40 miles inland of the Gulf of Mexico. The land surface in the FM 442 vicinity is generally flat with ground elevations between 65 ft and 85 ft msl. The geology of the area is characterized by 500 ft of unconsolidated to moderately compacted clastic sediments overlying the caprock and salt of the Boling salt dome.

The Boling Dome is a large salt dome that is roughly elliptical in shape and has major- and minor-axis dimensions of about 5 mi and 3.5 mi, respectively. The Boling Sinkhole is situated on the northeastern shoulder of the dome, where the dome's caprock is encountered at an elevation of about -500 ft msl (Figure 1).

Fresh to very saline ground water occurs in unconsolidated strata overlying the cap rock, within the calcite cap rock layer, and possibly within portions of the anhydrite cap rock layer. The availability of fresh ground water directly above the central portions of the dome is limited, as the base of fresh water is rather shallow. On the flanks and away from the dome, freshwater is available at greater depths.

Unconsolidated strata above the cap rock are differentiated into the lower Evangeline Aquifer and the upper Chicot Aquifer. The distinction between the Evangeline and Chicot aquifers is somewhat arbitrary, as both units display similar lithologic characteristics

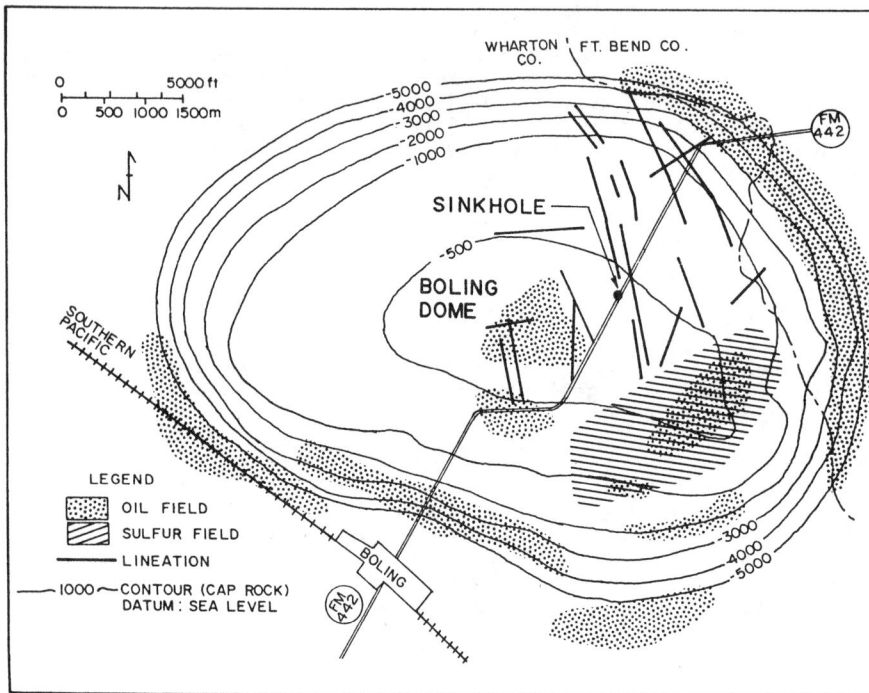

FIGURE 1 GEOLOGIC FEATURES-BOLING DOME, WHARTON AND FT. BEND COUNTIES

regionally, and can have direct hydraulic interconnection. The units are commonly separated on the basis of differing hydraulic properties. Value ranges for several aquifer properties obtained from aquifer pumping tests in Wharton County are listed below.

	Chicot Aquifer	Evangeline Aquifer
Transmissivity (ft^2/d)	3,000 - 46,000	1,130 - 2,680
Hydraulic Conductivity (ft/day)	25 - 212	10 - 11
Specific Capacity (gal/min/ft)	15.5 - 149	7.9*

Source: (Loskot et al., 1982)   *Single datum

The contact between the Chicot and Evangeline aquifers lies at a depth of about 160 ft in the vicinity of the sinkhole. The base of fresh water (<1,000 ppm total dissolved solids) within the area is reported to be at a depth of about 110 ft.

Mineral Resource Operations
Numerous oil and gas wells have been drilled in the area of the Boling Dome. Gulf Production Company drilled the B. Monroe Well No. 1 in 1927 near the area where the sinkhole occured. The driller's log for the well reported penetrating a void of 106 feet at a depth of 679 to 785 feet before the drill stem began binding. The void reportedly occured in the gypsum section of the dome.

Oil and gas production is primarily confined to the southwest, north, northeastern and southern flanks of the dome (Figure 1). Oil is produced from super caprock sands, caprock, basal Miocene sands, Heterostegina lime, Marginulina and Frio sands. The accumulation of oil is controlled by faulting, lenticular sand bodies, structural re-entrants and differential uplift of various stratigraphic units adjacent to the salt or dome material. Both radial and peripheral faults are encountered at the Boling Dome.

In addition to the oil and gas production, the Boling Dome is the largest salt dome sulfur mining area of the Texas Coastal Plain. Sulfur production is mostly confined to the southeast portion of the dome area (Figure 1). Sulfur is encountered at depths of 900 to 1200 feet with the sulfur bearing zone reaching thicknesses of up to 200 feet with production values up to 50 percent sulfur (Ellison, 1971).

Sulfur is currently being produced by Texas Gulf Sulphur Company southeast of the sinkhole. Previous production has existed east and northeast of the sinkhole. Sulfur has been commercially extracted from the cap rock using the Frasch solution process since the late 1920's.

SINKHOLE EVALUATION

Aerial Photographic Interpretation
High and low altitude black and white photography of the area in which the sinkhole occurred was obtained from SDHPT. Low altitude photographs along FM 442 were taken by SDHPT after the sinkhole occurred. In addition to the black and white photography, color infrared photographs were obtained from Texas Natural Resources Information System (TNRIS).

The detailed examination of the aerial photographs revealed numerous lineations throughout the area. These lineations are shown on Figure 1. Lineations noted may represent fractures or fault planes that exist in the sedimentary strata above the cap rock and may penetrate the cap rock, anhydrite, and possibly the salt. The fault or fracture planes could provide paths for fluid migration and possible solutioning of the more soluble materials such as the anhydrite (gypsum) and salt.

Dewatering and Geologic Assessment
A preliminary assessment of the ground-water conditions of the sinkhole recognized two sources of potential recharge to the sinkhole. These sources were recharge from surrounding surficial sands and upward recharge through the collapsed zone. In order to assess dewatering requirements, a pumping test was recommended and subsequently conducted.

A 3000 foot 8 inch pipeline was constructed to carry the water from the sinkhole to an existing saltwater disposal system operated by Texas Gulf Sulphur Company. Four shallow monitor wells were installed to depths ranging from 30 to 40 feet around the periphery of the sinkhole to monitor water level decline in laterally adjacent deposits during the pump test.

A pumping test was conducted utilizing a high volume pump with a total capacity of 1000 gpm. Pumping began on 1 December 1983 with the pumping rates, water level in the sinkhole, and water levels in the surrounding wells being monitored with time. A total of six million gallons of water was removed from the sinkhole during the period of 1 December thru 8 December with no significant recharge occurring from surrounding sand units or upward through the collapsed zone.

In addition to the pump test, two holes were drilled and geophysically logged to a depth of approximately 300 feet. Based on the information provided by these borings, two additional holes were drilled, geophysically logged and instrumented to determine the rate of any residual lateral earth movements associated with the sinkhole. The two 300 ft. holes were drilled and geophysically logged at Stations 163 +27.5 29.3 L and 168 +99 27.4 L. Based on information gained from these borings and the surface expression of the sinkhole, two slope indicators were installed at Stations 163 +43.3 30.8 L (SI-1) and 167 +44.5 32.2 L (SI-2). SI-1 was installed to a depth of 52 m (170.6 ft) and SI-2 was completed at a depth of 56 m (183.7 ft.). The center of the sinkhole was located at Station 166 +00 75 L.

Design of Sinkhole Fill

In December, 1983, a contract was let to fill the sinkhole with a combination of stone and earth borrow. Included in the contract were provisions for cutting an outfall channel to carry rainfall runoff away from the sinkhole area into a natural stream. The contract amount was approximately $245,000 which was considerably less than the engineer's estimate. Actual work began in mid-February 1984, and the project was completed at the end of May 1984.

The theory of the sinkhole fill was to build a working platform in the bottom and then to complete the fill with normal roadway embankment procedures. Since the stability of the sinkhole bottom surface was questionable, it was decided to cover the portion of the bottom which would fall under the roadway embankment with a 6 to 8 foot layer of stone borrow. Stone borrow was specified to be a durable, graded stone sized to pass an 18 inch square mesh sieve, 30 to 50 percent of which would pass a 12 inch square mesh sieve, 0 to 15 percent to pass an 8 inch square mesh sieve, and 0 to 5 percent to pass a 6 inch square mesh sieve. A further requirement of the stone borrow was that neither the breadth nor thickness of any piece should be less than one-third its length.

The contractor furnished limestone which was processed to meet the grading requirements. This material was trucked to the sinkhole site from a nearby rail delivery point. It was then dumped at the edge of the sinkhole and pushed with a dozer over the edge into the sinkhole. As the stone borrow layer thickened, the dozer was able to start operating on top of it, enlarging the working platform by continuing to push the stone forward over the edge as more stone was delivered. As the stone layer was placed across the bottom of the sinkhole, the approximate 2 foot layer of muck was displaced to one side.

With the 3,500 cubic yards of stone borrow in place under what would be the roadway embankment, the earth haul from a nearby borrow pit was begun using 20-yard scrapers. Earth borrow was used to finish displacing the muck to a small area at the edge of the sinkhole where it was pushed up into the bank to dry and be absorbed into the earth fill. The remainder of the earth fill was completed employing standard density control measures and specifications for earth embankments. A nuclear density-measuring device was used to monitor the density requirements. SDHPT standard specifications require a density of 98 to 102 percent of a laboratory density (Test Method Tex-114-E) for swelling soils (those with a plasticity index of 20 or more). For non-swelling soils (soils with a plasticity index of less than 20) the requirement is a density of not less than 100 percent of the laboratory density.

The total amount of earth borrow including the channel excavation which was placed in the sinkhole was 26,000 cubic yards. This quantity filled the sinkhole to the rim of the crater leaving the surface sloped slightly toward the center of the sinkhole where rainfall runoff will pass under the roadway through two 30 inch concrete pipes and via an outfall channel approximately 1100 feet long to a natural creek. The presence of the nearby creek with a flow line elevation low enough, allowed provision for drainage away from the sinkhole in the above manner. Otherwise, considerably more fill would have had to be added to bring the area to its original elevation and restore the original drainage pattern. This would have appreciably increased the project cost.

The roadway structure itself consists of a one course surface treatment over 8 inches of compacted limestone base material on approximately 6 inches of lime stabilized subgrade.

Monitoring

Slope indicator measurements were made weekly during a three month period following installation and have continued on a biweekly basis since that time. In addition, elevations of the roadway and land surface have been monitored since the occurrence of the sinkhole.

Soundings of the sinkhole floor taken between 17 August and 5 October 1983 showed an additional subsidence after the catastrophic collapse of up to 1.7 feet, tapering at the edges of the sinkhole to approximately 0.6 feet. Between the time of installation of the slope indicators SI-1 and SI-2 in early December 1983, and the virtual completion of the rock and earth fill in mid April 1984, the elevations of the tops of the well casings dropped 0.02 and 0.15 feet respectively. By the first of June 1984, the tops had dropped another 0.05 and 0.19 feet. The rate of settlement slowed after that for the next two months so that at the end of July 1984, the SI-1 casing top remained stable with no additional settlement and the SI-2 casing top dropped an additional 0.06 feet. Total settlement through July 1984 was 0.07 feet for SI-1 and 0.40 feet for SI-2.

Corresponding maximum lateral movements encountered in the slope indicators are summarized below:

	Depth	1-19 thru 2-16-84	2-16 thru 6-27-84
SI-1	17 m (55.8 ft)	4.5 mm (0.80 in)	8.3 mm (0.33 in)
	43 m (141.1 ft)	4.7 mm (0.19 in)	8.6 mm (0.34 in)
	48 m (157.5 ft)	4.6 mm (0.18 in)	7.7 mm (0.30 in)
SI-2	35 m (114.8 ft)	15.7 mm (0.62 in)	82.0 mm (3.23 in)
	41 m (134.5 ft)	6.5 mm (0.26 in)	25.3 mm (1.00 in)
	53 m (173.9 ft)	4.7 mm (0.19 in)	12.1 mm (0.50 in)

Profiles of the completed roadway taken since 4 June 1984 indicate that a point near the center of the former sinkhole experienced a maximum roadway settlement through 15 August 1984 of 0.25 feet.

CONCLUSIONS

The sinkhole which developed in the area of FM 442 is believed to have resulted from natural geologic processes that have been acting for long periods of time. In particular, it is believed that the lineations which cross FM 442 near the sinkhole represent possible avenues for dissolution of the soluble materials such as the cap rock, anhydrite (gypsum), and salt. In support of this activity is a reported void in Gulf Production Company's B. Monroe Well #1 drilled near the sinkhole in 1927. The reported cavity is believed to have developed along a fault plane and may represent the cavity which caused the collapse of FM 442.

Although the occurence of the sinkhole is viewed as a unique feature, it probably represents only one of many that have occurred within the area throughout time. The frequency of such an occurrence is unpredictable and an event such as the one which occurred may not be witnessed again for hundreds of years. Filling of the sinkhole allowed for a relatively rapid recompletion of the roadway. Monitoring data obtained since the filling of the sinkhole indicates that settlement of the roadway is relatively minor.

ACKNOWLEDGEMENTS

This paper is based, partly, on data derived from a continuing evaluation of the geologic conditions of the Boling Sinkhole being conducted for District 13 of the Texas State Department of Highways and Public Transportation by Espey, Huston and Associates, Inc., Austin, Texas.

REFERENCES CITED

Ellison, S.P. 1971. Sulfur in Texas: Univ. of Texas at Austin, Bureau of Economic Geology, Handbook No. 2, 48 p., 14 figs., 2 pls.

Loskot, C.L., W.M. Sandeen, and C.R. Follet. 1982. Ground-water resources of Colorado, Lavaca, and Wharton counties, Texas: Texas Department of Water Resources, Rpt. 270, Austin, 240 p., 32 figs.

Engineering problems associated with sinkholes in the Valley of Virginia

H.G.LAREW & E.O.GOOCH *E.O.Gooch and Associates, Charlottesville, Virginia, USA*

ABSTRACT

The Valley of Virginia lies between the Blue Ridge Mountains on the east and the Appalachian Province on the west and extends more than 300 miles from Winchester on the north to Bristol on the south. It is underlain by highly deformed sedimentary rocks of the Cambrian, Ordovician, Silurian, Devonian and Mississippian periods. Sinkholes and solution phenomena are present and pose problems in the location and construction of buildings, highways, dams and waste storage facilities in the study area. The authors, who have been confronted with these problems for over 25 years, describe how sinkholes and solution features are identified and have been dealt with on a number of engineering projects.

The limestone, dolomite, shale and sandstone bedrocks which underlie the Valley of Virginia have been greatly deformed in past geological times so that most of the strata dip rather steeply to the southeast and strike northeast-southwest, i.e., roughly parallel to the Blue Ridge Mountains. There are many exceptions to this general rule, however. As one begins to study and focus in on a small area to be occupied by an industrial complex, a waste lagoon or a pipeline, for example, he may find that the extreme folding and faulting that has occurred in times past have greatly distorted and displaced these once horizontally bedded rocks in the study area.

During and following periods of great deformation, the agents of weathering and erosion were also active. A detailed discussion of these mountain building, weathering and erosional processes is beyond the scope of this paper. However, the surface and near surface conditions that have resulted from this interplay of natural forces is of great interest to the geologist and geotechnical engineer.

Today we find the Valley composed of a series of parallel valleys with intervening ridges. The valleys are usually underlain by limestones and shales while the ridges are normally composed of sandstones. The valleys are gently rolling to hilly while the ridges are often steep and rough.

Most, but not all, engineering structures occupy the valleys while water tanks and transmitting towers, for example, are usually located on the hills or ridges.

The presence of sinkholes, solution topography, crevices, caves, mudseams, irregular rock profiles, sinkholes filled with fill, colluvial and soft sedimentary soils is a normal occurrence in the limestone formations found in the Valley. Some limestone formations are more prone to the presence of these features than others.

The presence of these features, i.e., sinkholes, crevices, mudseams, etc., on a given site is sometimes most difficult to determine.

Existing sinkholes and depressions are located by the use of air photos, topographic maps and an on-site inspection of the property, for example. In the field they appear as steep-sided depressions varying in size from a few square feet to a large football stadium. They are often overgrown with brush and trees and property owners often use them to dispose of rocks, boulders and assorted fill material.

There are a few instances known to our firm where these depressions have developed suddenly as a result of the collapse of the roof of an underground cave or cavern. However, we believe that this phenomena is more prevalent where the limestone formations are more nearly horizontally bedded. Our observations lead us to believe that many of the depressions and sinkholes in the Valley have resulted from a slow erosional migration of the typically red-brown-yellow, residual, silty clays and clayey silts into and along solution channels that have developed in the rather thin-bedded steep-dipping limestone formations. This conclusion results from our having core drilled into and adjacent to numerous depressions and sinkholes.

In many locations there may be no surface evidence of underlying caves, caverns, etc. at a site, yet that possibility exists based upon the type of geological formation present and/or the presence of surface evidence nearby. In these cases some of the borings should be extended into the underlying bedrock. Even then some of the smaller cavities in the rock may go undetected. For example, we have recently been called in to consult on two projects which had been carefully drilled by other firms yet numerous small channels or openings were present in the inclined limestone bedrock. These went undetected, even when much of the overlying soil had been removed, during the site grading operations, and footings were in place. However, subsequent heavy rains deposited surface water on the sites and this water quickly sought out these now shallow channels or openings in the rock and soon carried much of the thin soil cover into these openings with the resulting undermining of many of the footings.

In another instance we were asked to investigate a lagoon in which an opening appeared when the lagoon was being filled to check for water tightness after construction and before placing it in operation. No test borings were taken prior to construction although the site was underlain by limestone and there was evidence of solution activity in the area. The site was adjacent to a large stream and the underlying residual soils and bedrock were covered with alluvial deposits and apparently there was no surface evidence of any solution activity beneath the lagoon. However, test borings later revealed the presence of a rather large solution opening in the underlying limestones in the problem area beneath the lagoon. This opening had to be grouted before the lagoon could be placed in operation.

It should be noted, however, that had test borings been taken prior to construction, the presence of the solution opening could have gone undetected, depending upon the spacing and location of the test borings.

When surface depressions or sinkholes are observed during the early planning phases of a project one can often move and shift buildings, tanks, lagoons, etc. off of the questionable areas. For example, ponds and lagoons can often be shifted to areas underlain by shales that are free of the caverns and solution channels.

However, for roads and parking lots, for example, and for buildings that cannot be shifted or moved, a thorough subsurface investigation of the sinkhole and cavern is made and special measures are taken to prepare the sinkhole for the compacted fill and structure that will subsequently be placed upon the area.

In a few instances we have encountered old sinkholes that have been filled with sedimentary deposits from nearby or former streams. Quite frequently the surface evidence of this filled-in depression is absent. However, the soils in these depressions are usually quite soft and compressible. Buildings placed on conventional footings on these deposits will settle excessively and evidence of this can sometimes be found in nearby existing buildings. Special studies and foundations are required if these areas cannot be avoided.

Another feature of the limestone bedrock profile in the Valley is its uneven surface. It is more the rule than the exception to find a very irregular rock surface. Quite frequently, for example, the limestone may be seen outcropping in various parallel bands at the surface, yet a few feet away one may drill to depths of 4.57 to 18.28 meters (15 to 60 feet), for example, without encountering bedrock. This irregular rock surface creates problems for the foundation engineer who prefers to place all of his footings on similar materials in order to reduce differential settlements. Quite frequently one "winds up" undercutting the rock in footing excavations, excavating all of the soil to the uneven rock surface or preparing special foundation designs such as a cantilever section which was used to support the overhanging portion of a large water tank.

In many cases, one will be involved in costly rock excavation or in deeper excavations that may also require special footing designs with dowels placed into the sloping rock surfaces. As an alternate solution, the deeper excavations to the uneven rock surface are sometimes backfilled with an open graded crushed stone and the footings are then placed upon the surface of the densified stone.

In some places we find that the deeper pockets of soil that overlie the uneven rock surface become wetter and softer with depth. The split spoon sampler may push into these soils during the standard penetration test. This raises the questions: (1) Are these soft soils sedimentary? (2) Are they normally consolidated? (3) How much will they consolidate under the loads and pressures that will be brought to them by the structure that is to be placed above them?

The geologist and geotechnical engineer must be prepared to obtain the necessary soil samples and to perform the necessary laboratory soil tests and analyses to determine whether these soft soils will settle or not and how much they will settle.

Finally, where it becomes necessary to employ driven piles as a foundation type, we find that the required length of pile can vary greatly where the rock surface profile is uneven. For example, the length of piles to support a press in one bay of an industrial building was estimated to be 7.62 meters (25 feet)--based upon soil test borings carried to the rock surface. When the decision was made to shift the press into another bay 12.19 meters (40 feet) away the architect did not ask for additional borings. We learned about the change in location when he called and reported that piles had been driven to a depth of 18.28 meters (60 feet) under the press and no bottom had been found as yet. Piles some 22.86 meters (75 feet) in length were finally required to reach practical refusal on the underlying rock surface.

Sometimes the pile tips will encounter an inclined mudseam between rock layers and will follow the seam to great depths. Whether it is true or not we do not know but the story is told of a highway bridge contractor who while driving a pile into one of these mudseams discovered that the pile was coming out of the ground a few meters (a few yards) from the point of entry. With each hammer blow the pile would rise higher in the air. Being an alert and conservative Virginian he proceeded to cut sections from the emerging end of the pile and to weld them to the end being driven.

When it becomes apparent that a pile may have encountered and be following a mudseam, additional borings may be necessary. Moreover, additional offset piles with spanning beams may be required to support the loads that were to have been carried by the unsatisfactory pile.

In conclusion, we have presented a few of the numerous ways in which sinkholes and solution activity manifests itself in engineering construction work in the Valley of Virginia. In most cases the geologist and geotechnical engineer can identify the problems in advance during the planning stages. However, there are some situations that will be missed no matter how carefully the geotechnical studies are made. In these cases the geologist and geotechnical engineer can render valuable service by investigating the nature and scope of the problem and by proposing and evaluating schemes for solving the problem.

Maturation of the Winter Park sinkhole

S.E.JAMMAL *Jammal and Associates, Inc., Winter Park, Florida, USA*

ABSTRACT

The Winter Park sinkhole developed between May 8-13, 1981 as a conical hole in overburden, with central aven leading to a deep subsurface cavity. Initially it was dry and about 106 meters (348 feet) in diameter and 30 meters (98 feet) deep. It destroyed various facilities and utilities and caused major problems for the City of Winter Park. Since 1981 the hole has stabilized and its surroundings have been restored to a usuable state. It contains a small lake.

Environmental Conditions in Central Florida

Central Florida receives about 130 centimeters (51 inches) of rain each year with about 70% (36 inches) lost to evapotranspiration. Thus there is a substantial flow which passes into or over the ground. Since upper levels are sands, much infiltration occurs, with recharge of the limestone aquifer through fractures, sinkholes, solution pipes, or porous formations. Western Orange County, in which Winter Park lies, is a fine recharge area. The Floridan Aquifer in the limestone has a piezometric level of about +14 meters (46 feet) (MSL). It is overlain by the Hawthorn aquiclude. A non-artesian aquifer lies above the Hawthorn.

Over geologic ages solution cavities have occurred. That subsurface cavern development is extensive is shown by the many large springs and lakes. The countryside, along with Winter Park, is dotted with lakes due to various types of subsidence in the past.

Most water use in Central Florida is derived from subsurface wells drawn from the Floridan or overlying aquifer. Thus there are many local disturbances of the subsurface hydrology, which become of particular importance in the normally dry months of April and May. Pumpage is a factor of importance in sinkhole formation. In addition, numerous drainage wells, installed over many decades for stormwater disposal, affect the local flow nets as well as the chemical condition of the water inserted into the aquifer.

Geology of Sinkhole Site

The Florida Platform consists primarily of sedimentary, marine deposits formed on continental shelves. These deposits consist essentially of fragmental and pasty marine limestones, sandstones, and shales. Since formation there has been repetitive uplift, resulting in extensive fracturing, thus accentuating cavern formation.

At the sinkhole site, below ground surface at about 27 meters (89 feet) above Mean Sea Level (MSL), lie about 18 meters (59 feet) of loose to dense deposits of sands, clayey sands, and slightly cemented sands (Pleistocene). Below these lie about 27 meters (89 feet) of loose to very dense clayey sands, with silt, shell, phosphate and dolomite fragments (Hawthorn formation), bedded on limestone of the Ocala Group, this latter layer overlaying the Avon Park limestone formation. Solution cavities would occur in the two limestone beds.

Winter Park and adjacent areas lie at the terminus of at least three faults. The City is in a topographically closed area which means that stormwater flows into the area. Various water supply wells exist, pumping large volumes, as well as drainage wells delivering corrosive water to the limestone. Recharge does occur, particularly in faulted conditions. With cavities in the underlying limestone, developed over geologic ages, interconnected with piping into the Hawthorn layer, overlain by sand, gradual solution, and erosion of sands and clays, can lead to surface layer penetration and a sinkhole. Several sinkholes have occurred in the vicinity since 1961 and sinkhole activity is expected to continue.

The Winter Park Sinkhole

Penetration of the surface sands at the sinkhole site in Winter Park occurred about 8:00 p.m., May 8, 1981. A resident of a nearby house, subsequently lost in the hole, noted a swishing noise and a large sycamore tree disappearing into a hole. Police secured the area overnight. Over the next few days the hole grew to about 106 meters (348 feet) in diameter and 30 meters (98 feet) deep. Initially it was a dry hole, as can be seen in early photography. At that point the central aven through the Hawthorn was well defined. Later the hole filled to the piezometric level of the Floridan Aquifer. During formation, the sinkhole detroyed streets, utilities, recreational facilities, one house, several businesses, and several cars. It seriously interrupted local traffic. There were no personal injuries. Substantial community impact occurred, with demands on police, city and county governments, and news media. The site required surveillance and restricted access.

Investigation - Winter Park Sinkhole
The sinkhole had an immediate detrimental effect on the City of Winter Park, its traffic, businesses, utilities, and a resident, whose house fell into the hole and was destroyed. As the hole grew, it threatened a major traffic artery and physically interrupted two well-traveled City streets, Denning and Comstock. Because of uncertainty, the City was unsure just how much additional damage would occur, immediately or over a long period, since the future stability and growth of the hole were uncertain. Hence it retained Jammal and Associates for initial advice as the hole was growing, then instructed Jammal to investigate and report on the collapse, the possibility of disastrous long-term effects, and proposals for corrective measures. Jammal investigated the sinkhole site for ten months. A detailed, written report was made to the City (Jammal, 1982).

The sinkhole stabilized with vertical banks in various rim areas and steep slopes in others. These were in the surficial sands, hence the sinkhole, after stabilization, still was expected to grow substantially due to rain and weathering as time passed, unless corrective measures were taken. It was found from borings that there was a non-artesian surficial aquifer, conically depressed around the sinkhole. Below this, above the Hawthorn, was a secondary piezometric aquifer, not interconnected with the surficial aquifer, again circumferentially depressed around the hole. Two other areas of depression were found, separated by over 100 meters (a few hundred feet) indicating possible future sinkhole sites.

Technical Conclusions - Winter Park Sinkhole
The Winter Park sinkhole was formed as a result of a long-term erosion and ravelling of overburden material into cavernous limestone, not a roof collapse in limestone. Even though the sinkhole developed rather rapidly, it was a progressive erosion of overburden starting with a small ground depression. The aven was 13 to 17 meters (43 to 56 feet) in diameter and by observation penetrated 9 to 12 meters (30 to 39 feet) of the Hawthorn formation. It was Jammal's opinion that the sinkhole had been forming over a long-time period, but that formation was accentuated over a period of about 50 years, resulting from a progressive decline of the piezometric level of the Floridan aquifer. About 50 years ago this level was about +20 meters (66 feet) (MSL) and now is about +14 meters (46 feet) (MSL). Decline can be attributed both to extended below average rainfall and areal pumping of wells. No evidence was found to contributing factors suggested, such as local plumbing leaks.

Restoration of Sinkhole Site
The sinkhole was in part on City property and in part on private lots, including the commercial businesses along Fairbanks Avenue. Thus the restoration to optimum condition involved the public and private sectors operating independently. The Jammal report had listed seventeen restoration alternatives for consideration.

Public Sector
The City, with Jammal as consultant, developed plans to restore Denning Drive during the Fall, 1982. This work involved filling and restoring the banks in accordance with Jammal recommendations, then repaving the broken street by standard practices. A highway guardrail was added on the sinkhole side. During this restoration the remnants of the swimming pool and masonry bathhouse on the north side were demolished and pushed into the hole ($6,500.00 demolition cost). The sinkhole lake was pumped down to about +19 meters (63 feet) (MSL) by discharging water to Comstock Avenue storm drains to the west. It then was possible to push the rubble and fill down into the center of the hole. Filling on the Denning Drive side involved a mixture of rubble and sand lifts. The final grading was 3 to 1 on the north side and 2½ to 1 on the Denning Drive side, without benches. Slopes were grassed, and intermittent trees were planted. Restoration was completed in Dec., 1982. In this restoration Comstock Avenue to the West was left barricaded.

The City graded the north side up into the ballfield area. It also reworked utilities under City control. For example, an interrupted sanitary sewer was abandoned and the discharge diverted to the Fairbanks Avenue sewer. The gap in the 0.30 meter (12 inch) water main on Denning was restored. All storm drainage was taken away from the sinkhole, as before, so that the only water entering is subsurface, and that which falls on the sinkhole area of surface influence. Restoration cost was about $85,000.00

FIGURE 1: Aerial photography of the sinkhole. Top-May 1981. Bottom-August 1984.

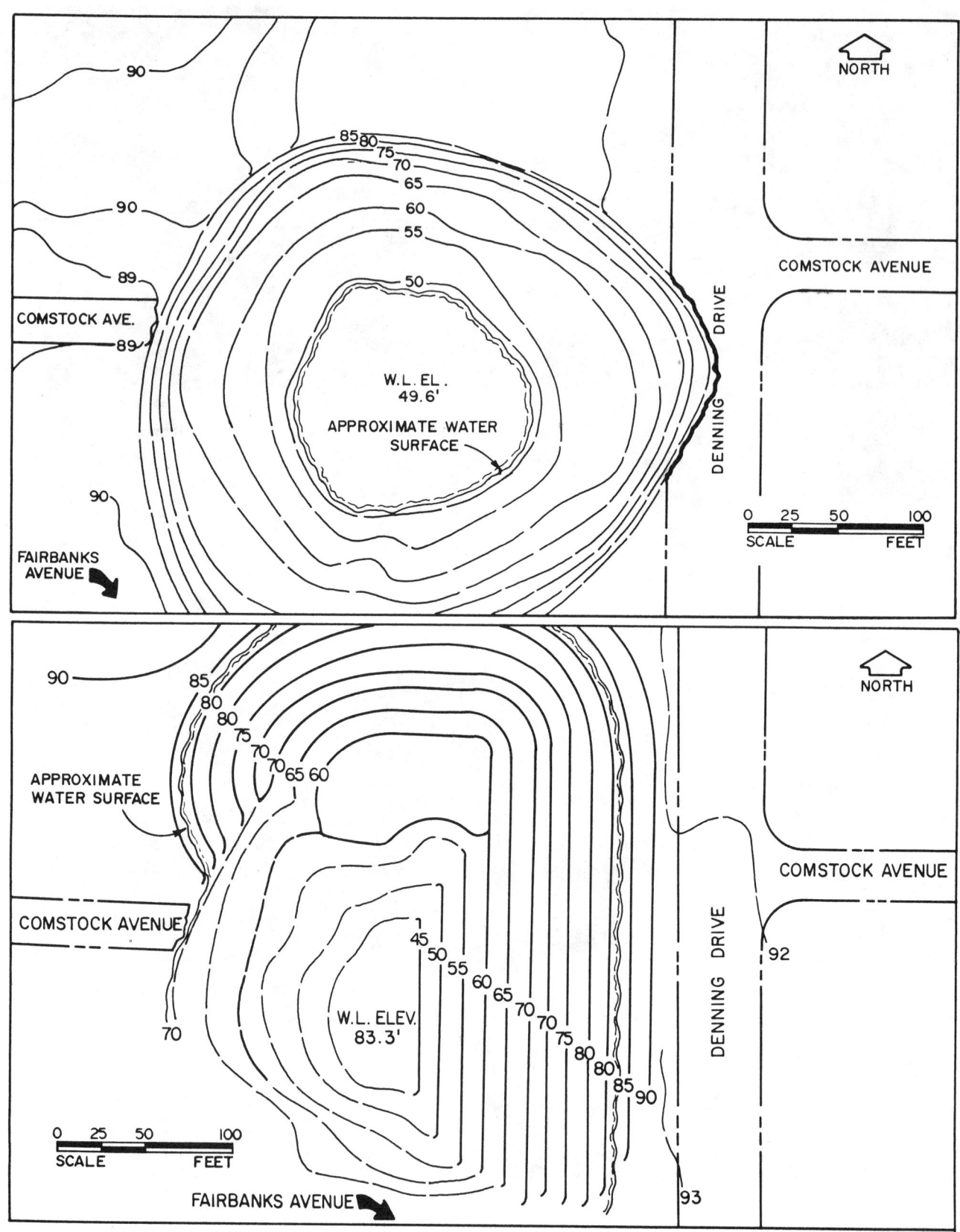

FIGURE 2: Topography of the sinkhole area. Top-May 1981. Bottom-August 1984

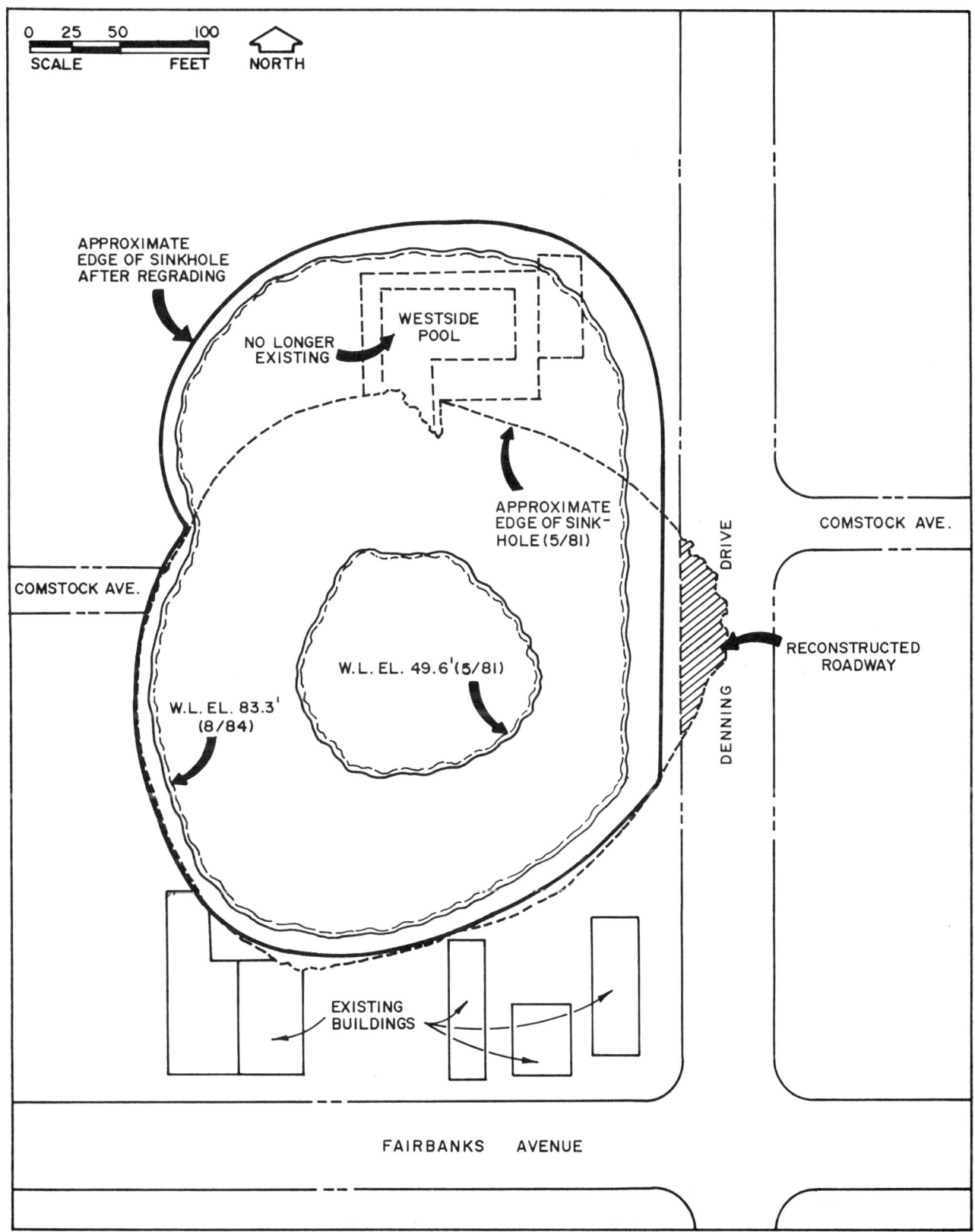

FIGURE 3: Overlay topography.

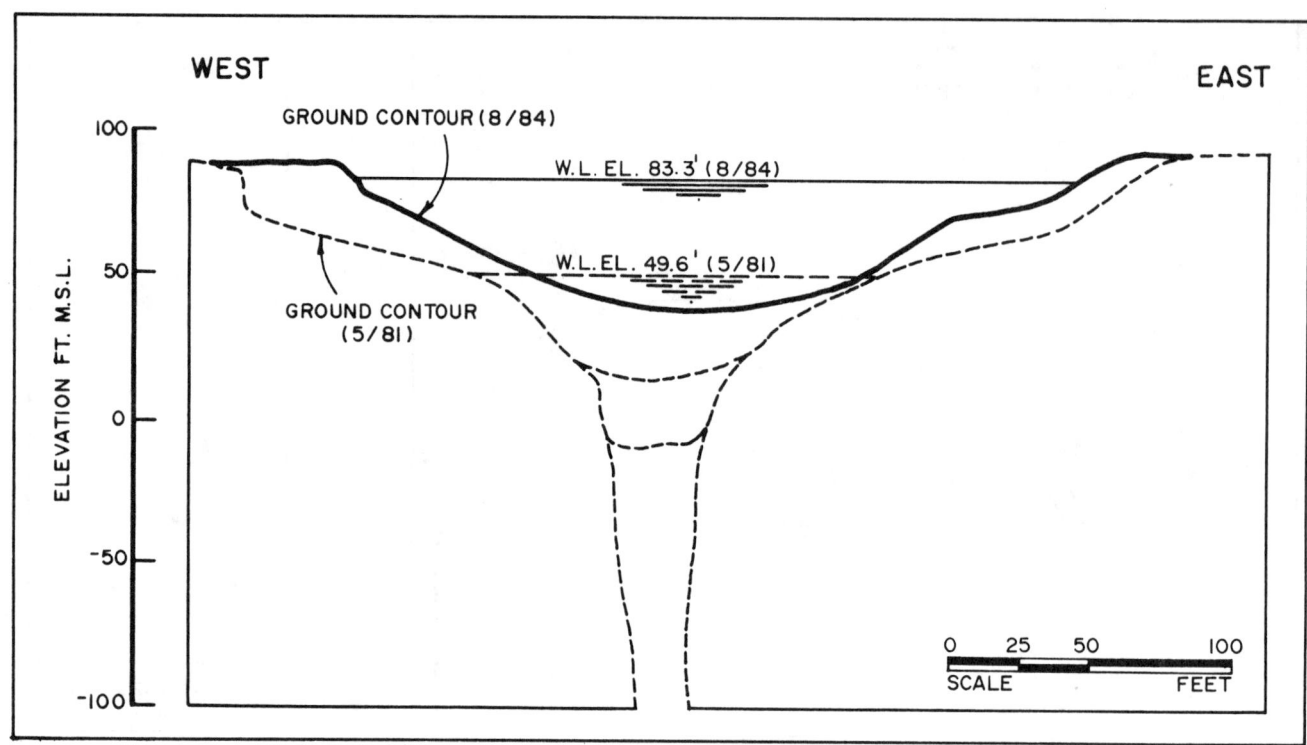

FIGURE 4: West-East vertical profile of the sinkhole.

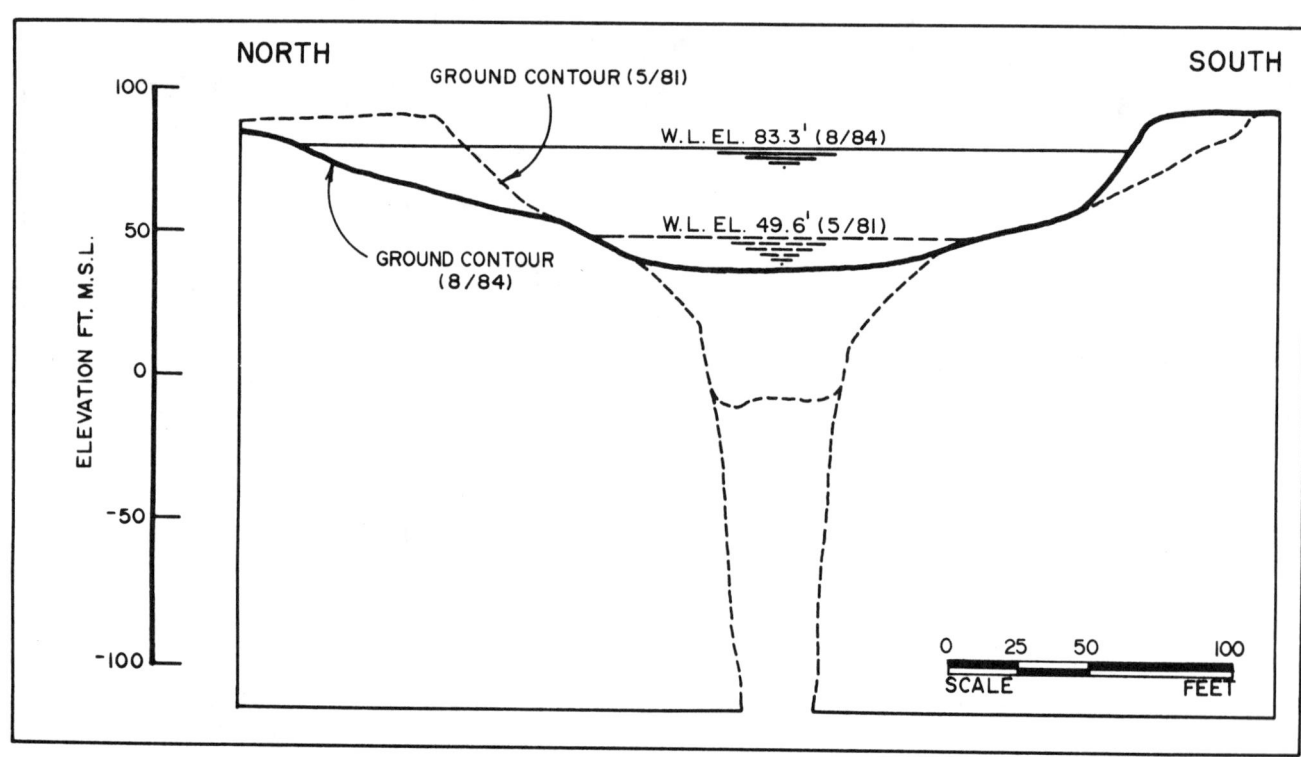

FIGURE 5: North-South vertical profile of the sinkhole.

FIGURE 6: Surface photography looking west. May 1981.

FIGURE 7: Surface photography looking south. August 1984.

6. Engineering in sinkhole-prone areas

Correction and protection in limestone terrane

GEORGE F. SOWERS *Law Engineering Testing Company & Georgia Institute of Technology, Atlanta, USA*

ABSTRACT

Structural features that are hazards to building foundations include pinnacles, slots, and chimney like openings in rock, alternate strata of hard rock and soft clay, dome-like cavities in soil, caves in rock, weak and compressible soils in cone shaped depressions over collapsed domes and cavities, and open sinkholes from soil or rock collapse and soil erosion. Design to overcome them includes avoiding areas of concentrated hazards, correcting the hazards by filling them or collapsing them, bridging over small hazards, reinforcing the rock, bypassing shallow hazards with deeper foundations, and minimizing activation of the hazard-forming processes.

1. Introduction

The solution of carbonate rocks produces a variety of geologic features that are hazards to structures. Much attention necessarily has been paid to the mechanism of development and behavior of these features. Ultimately they must be controlled to prevent injury and structural damage. This paper describes the function and limitation of some of the control and modification measures that have been successful in overcoming the hazards. For simplicity all the carbonate rocks will be termed limestones, although they include limestones, dolomites or dolostones and marbles.

2. Nature of the Defects

The major structural features that are hazards to engineering structures are listed below and depicted in the illustrations.

1. Pinnacles, slots and chimney-like openings in rock, - Fig. 1
2. Alternate strata of hard rock and soft soil, - Fig. 1
3. Dome-like cavities in soil, - Fig. 2
4. Cavities or caves in rock, - Fig. 2
5. Weak and compressible soils in cone-shaped, depressions over collapsed domes or cavities, - Fig. 3
6. Open sinkholes from soil or rock collapse, and from soil erosion. - Fig. 4

The list is incomplete; there are many variations and combinations. However, these illustrate the range in problems encountered.

Engineering design to control or overcome the hazards requires both extensive and intensive goetechnical data:

1. Depth, thickness, and engineering properties of the soil and rock strata.
2. Groundwater levels, their changes and groundwater movement.
3. Location of slots or cavities in the rock and the reflection of these defects in the soil above.
4. The nature and the extent of the defects and the possible changes in the defects produced by nature and any proposed construction.

Investigation of a site to determine these data begins beyond the site boundaries with studies of regional geology, hydrology, and geomorphology. It narrows to the site as the regional picture clarifies. However, even with intensive investigation using closely spaced borings and sophisticated techniques such as ground penetrating radar, there are uncertainties. Thus, while design must be based on good data, it must allow for the unknowns which cannot be detected.

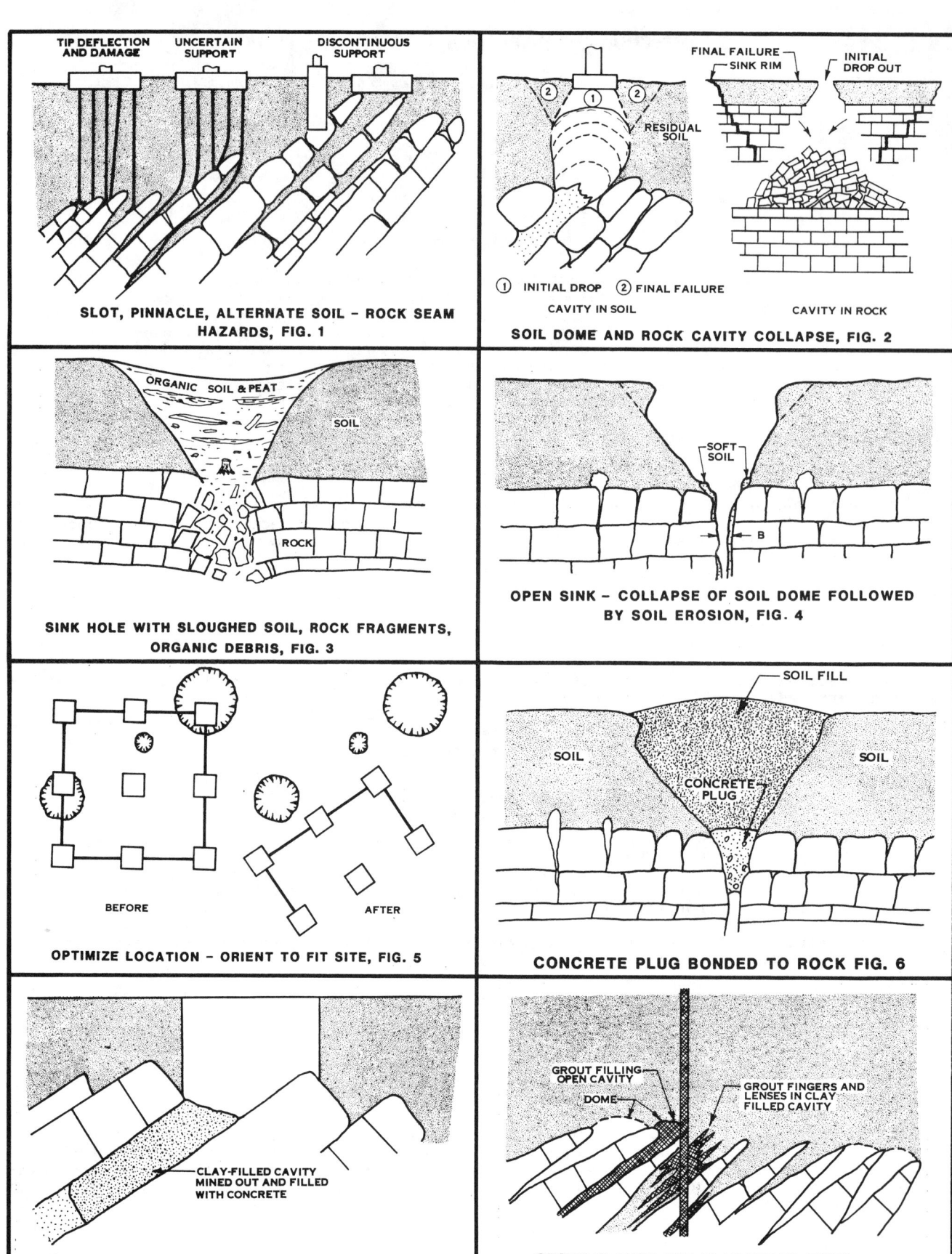

3. Design Approach

There are a number of approaches to design. These can be categorized as follows:

1. Optimize the location on the site.
2. Correct or mitigate defects that are present.
3. Use shallow foundations, modified to overcome the defects.
4. Use deep foundations modified to overcome the defects.
5. Minimize future activation of defects.

Because of the unknowns, many designs include a combination of several. All should include measures to correct or mitigate the existing defects and to minimize activating old defects and developing new defects.

3.1 Optimization - Optimization involves establishing the patterns of defect occurrence by remote sensing, direct observation and physical investigation. Many occur in patterns, reflecting the strike of the formations, joint alignments, or faults and shear zones. The structures are then positioned in areas of fewest defects and are oriented so that there is minimum exposure. This is illustrated in Fig. 5. Unfortunately, not all defects can be identified, and new defects can develop from changes in the environment, especially changes caused by new construction. However, the greatest incidence of new activity is usually in areas of previous defects that reflect the underlying rock structure.

3.2 Correct or Mitigate Defects - The defects can be partially corrected or their effect reduced by eliminating them physically, changing their character, or preventing their enlargement. Filling the cavities with concrete is an obvious measure. Any loose material is excavated and replaced with concrete. It is most effective when the concrete bonds with the rock, Fig. 6 and 7. Grout can be pumped into cavities, either in the rock or in soil, Fig. 8 and 9. When voids in rock are partially filled with soil, the grout extends out in sheets or fingers. In this way it sometimes blocks further erosion solution and cavity enlargement. However, if erosion continues, the grout becomes loose fragments in the enlarged cavity, and any benefit is lost.

Cavities in soil (and sometimes in rock) can be collapsed by high impact compaction, vibration, or explosive detonation. The cavity is partially eliminated, accompanied by a surface depression that must be filled, Fig. 10. Shallow foundations are used. However, new cavities can develop as groundwater continues to circulate. Local void filling in rock with concrete, after mining out the clay filling to a depth of 1.5 to 2 times the void width (dental filling), maintains rock continuity and bearing capacity below footings, Fig. 11.

A graded filter in an open sink provides a pervious flexible fill that can absorb some seepage without changing subsurface drainage, Fig. 12. A geotextile filter in 1 or 2 layers can be effective in the very small throats that develop in porous limestones such as underlie parts of Florida, Fig. 12A. It has been very successful under highway embankments. Preloading of loose soil and debris in the throat of a sinkhole has been successful in reducing settlement when the sink is not active, Fig. 13. However, if a changing environment reactivates rock solution or soil erosion, the consolidated fill can subside or even disappear.

3.3 Foundations on Soil or Rock Surface - Foundations that bear directly on the soil or rock surface have four alternative treatments. The first is to correct any defects below by dental filling described previously, Fig. 10. A second alternative is to bridge over the opening. This can utilize a bridging beam on separated foundations, Fig. 14. Reinforced earth and geogrids have been used to bridge over cavities; however, their long term performance has not been tested. A third alternative is reinforcing the rock by rock bolts or grouted dowels, Fig. 15. Detached blocks are tied together to produce a rigid mass.

The fourth alternative is to build a structure so strong and rigid that it will not fail or deflect catastropically if a dome or cavity develops beneath a critical point, Fig. 16. In this case a maximum unsupported width, B, is selected on the basis of observed nearby sinkhole diameters or the potential dome width (comparable to the soil depth). Such a structure must be monitored regularly to detect the deflection accompanying a developing cavity before it enlarges excessively.

3.4 Deep Foundations on Rock - Deep foundations to rock such as piles or piers (caissons) are sometimes believed to be cure-alls for defects in limestone terranes. However, As Fig. 1 illustrates, there are hazards. Piles can skid on sloping surfaces and the tips can be crippled by uneven bearing. Piles driven into slots and chimneys become excessively long. Even if a long pile eventually reaches good bearing its capacity is reduced by excessive elastic deflection. Some piles may wander off in deep cavities and chimney-like openings. One such pile in Florida is nearly 800 ft. long! Finally, in some cases the pile

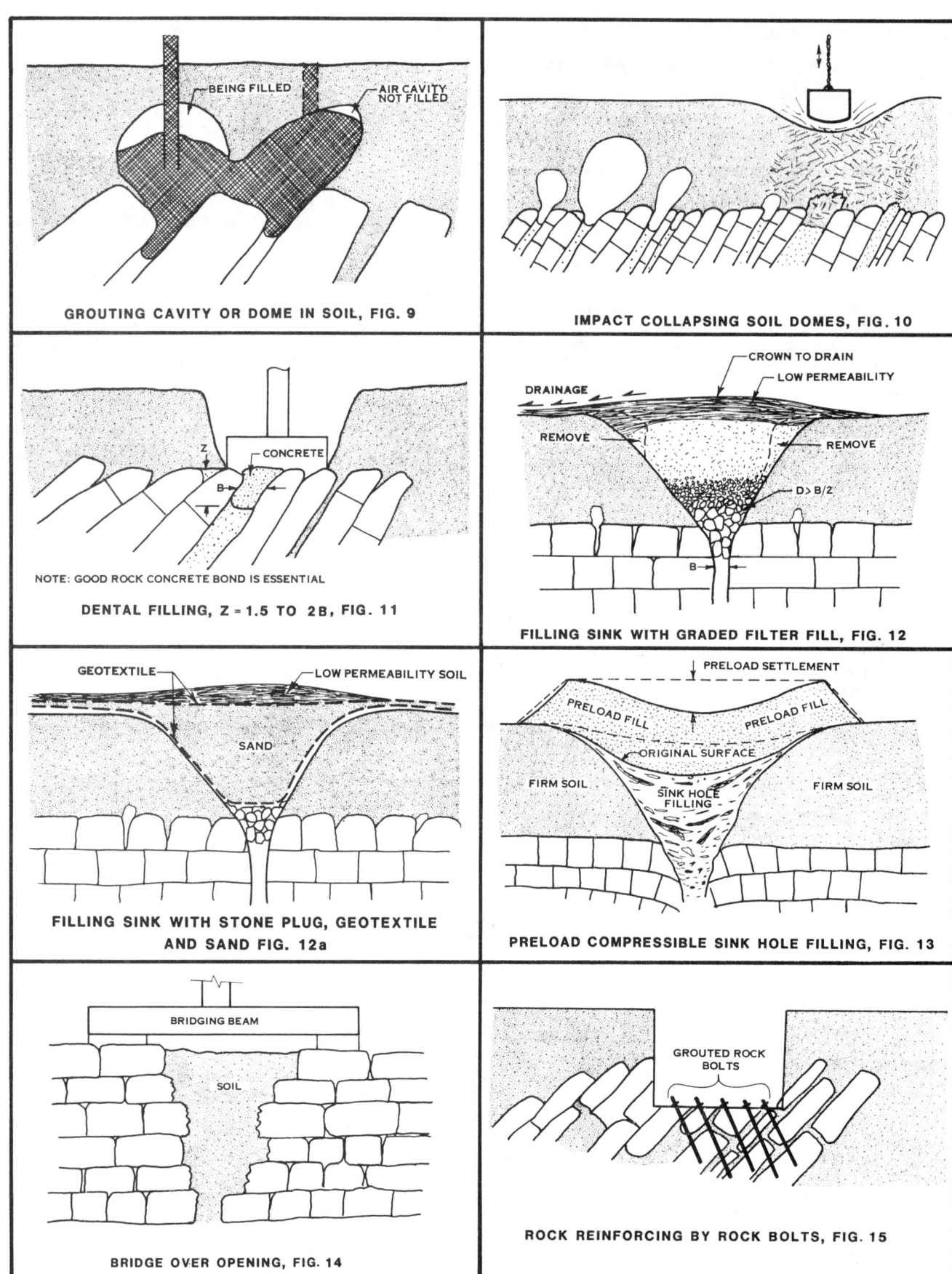

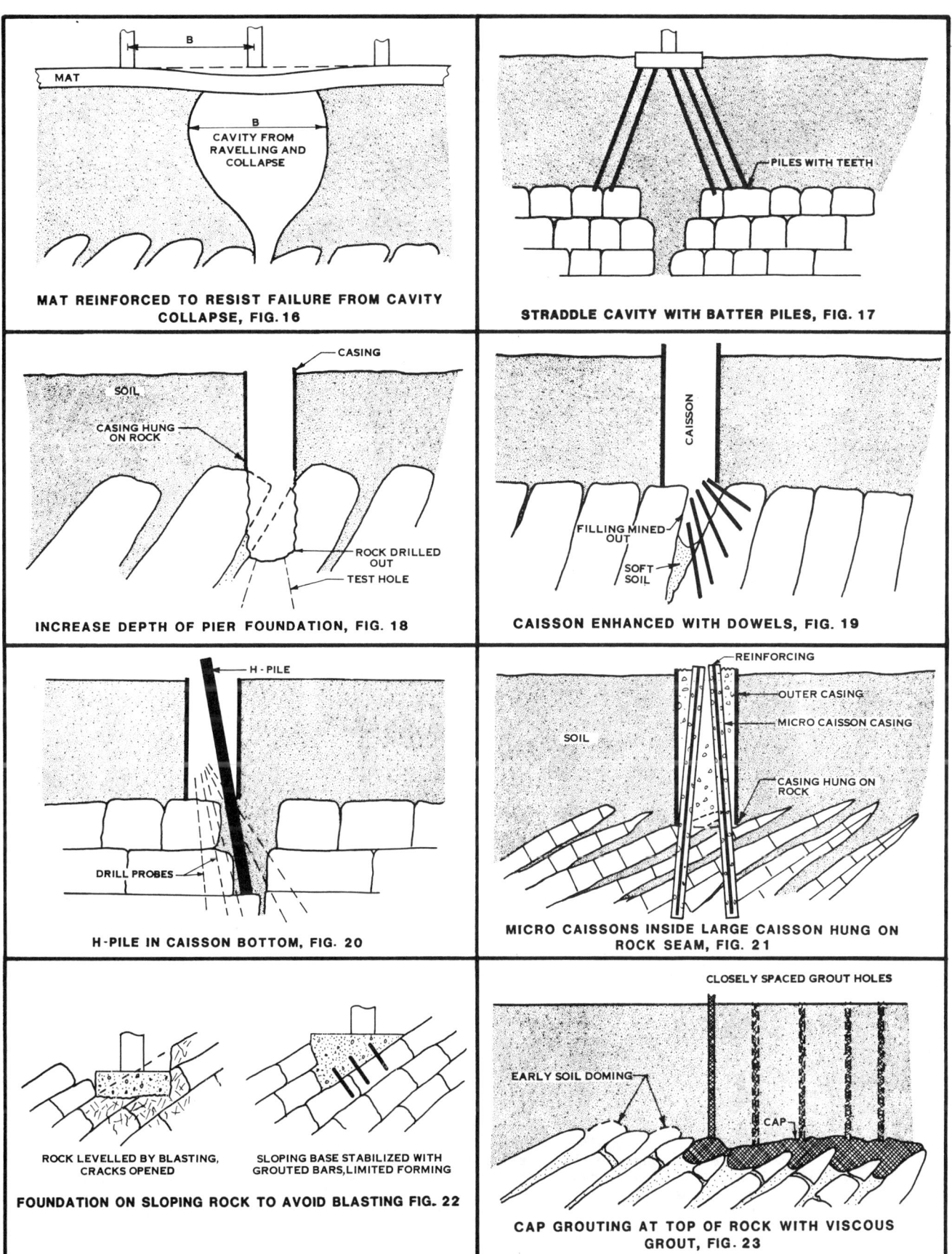

can place a concentrated load on a weak pinnacle that might withstand driving impact but gradually dislodge under continuing load.

Straddling an identified cavity by batter piles is a simple alternative, Fig. 17. The piles must have tips that can bite into the rock so they don't skid across the surface.

Piers (drilled piers, sometimes termed caissons) make it possible to excavate through weak seams and cavities, Fig. 18. Probe holes drilled in the caisson bottom determine that there is sufficient sound rock below.

When the pier casing hangs up on hard rock with vertical cavities, several alternatives are possible. Dowels or rock bolts can be installed in the rock sides of large cavities after partially mining out soil, Fig. 19. H-piles can be driven into narrow cavities, Fig. 20. Drilled probes determine if the rock below is continuous. Micro-cassions drilled from within the large pier, Fig. 21, can extend down through clay filled cavities into continuous rock. They are drilled through lightweight casings that extend up into the larger shaft. They are reinforced to increase their capacity and then concreted with high strength mortar under pressure.

3.5 _Precautions_ - Blasting should be controlled carefully because it opens cracks, increasing both solution and soil erosion. It is better to place foundations on a sloping but clean rock surface instead of a level surface made by blasting. If the slope exceeds about 20 degrees, sliding can be prevented by short dowels set in holes drilled in the rock, Fig. 22. In order to minimize forming the foundation can be non symmetrical and the column not in the centroid. These irregular shapes will not influence the performance of a foundation on rock.

3.6 _Minimizing Activation_ - Regardless of the foundation, it is usually prudent to take steps to minimize activating the solution and erosion that produce defects. Drainage should be diverted away from solution depressions. Industrial wastes must be diverted, especially if they are acid. Groundwater changes and particularly groundwater lowering should be controlled where possible. Blocking the open channels in the rock surface is termed cap grouting, Fig. 23. When cap grouting has been done extensively, defect forming activity has been reduced significantly. However, because much cap grouting has been done where the defect forming processes have been inactive, the effectiveness is difficult to evaluate.

4. Conclusion

Many different measures for correcting defects, overcoming their effect, and minimizing their activity are used. No one measure is entirely successful but most have proved to be effective in many situations. Foundation design in carbonate rock terranes requires minimizing risks through understanding and controlling the mechanisms that produce the defects.

There is some risk inherent in building in such areas. The risk can be minimized but not eliminated. The engineer must warn owners of the uncertainties involved and the need for continuing observation for distress, prompt corrective action and continuing maintenance.

References

M. Herak and V. T. Stringfield, KARST, American Elsevier Publishing Company, Inc., New York, 1976.

G. F. Sowers "Failures of Limestones in Humid Subtopics" _Journal, Geotechnical Engineering Division, Proceedings ASCE_, V101, No. GTB Aug., 1975, p. 771-787.

G. F. Sowers "Settlement in Terranes of Well-Indurated Limestone", _Analysis and Design of Building Foundations_, Lehigh Univ., Bethelem, Pa., Envo Publishing Co., 1975, p. 701.

G. F. Sowers "Foundations Bearing in Weathered Rock", _Proceedings Speciality Conference on Rock Engineering for Foundations and Slopes_, Vol. 2, Univ. of Colorado, August 15-18, 1976 (Copyright 1977, by ASCE New York).

Acknowledgements

Engineers of the Law Engineering Testing Company provided case histories and advice which were distilled into this contribution. My thanks to particularily to D. Thomasson, P. Gooding, D. Bourne, D. Berry, D. Alcott, R. Knott and D. Wheeless. Additional ideas were gained from J. Kellberg, geologist, Knoxville and Dr. S. F. Chan, Kuala Lampur, Malaysia.

Sinkhole and subsidence damage and protective measures

NATH S. PARATE *Tennessee State University, Nashville, USA*

ABSTRACT

Ground Subsidence has long been recognized as a major problem in both Middle and East Tennessee. This subsidence consists of downward and lateral movement of the ground surface, and is a result of loss of support within the low level subsurface materials. This loss of support can result from extraction of solids or liquids from beneath the earth's surface (particularily in the coal mining regions of East Tennessee), or may result from a subsurface erosion of overlying soil into either caverns or slots within the underlying limestone bedrock. Of prime importance to the engineer or geologist, is reducing subsidence damage to man-made structures. The methods for reducing damage will vary according to the subsidence causes and mechanisms.

Introduction

Ground subsidence in Tennessee results from man-made processes (mining activities) in East Tennessee and from natural processes (sinkhole development) in both Middle and East Tennessee. The discussion of sinkhole development will focus on Middle Tennessee since this region of the state contains two distinct geologic regions which allows us to observe a wider range of sinkhole related problems.

Mining Subsidence Causes and Effects

The effect of removing coal, minerals, oil or water from under ground is to create a basin-like depression at the ground surface. The subsurface environment, both soil and bedrock, is subjected to displacement from its natural stable position due to the above activities. This displacement consists of a combination of downward, horizontal & tilting movement of the surface of the ground, and is a result of the loss of support within the various subsurface materials.

The subsidence due to coal mining operations (the primary mining activity in East Tennessee) and its damaging effects have been detailed elsewhere as well as the environmentally disturbing activities. The protective measures required to reduce subsidence effects, both short term and long term, require understanding from a multi-disciplinary point of view.

The cause of subsidence due to coal mining is very complex and is affected differentially by a varity of factors (Fig. 1). These factors include dimensional factors (average depth, width, and length of panel of extraction, extent of waste support, and the surface topography), geologic factors (type or variety of rock underlying and overlying the coal seam, the structural geology of the area being mined, the thickness, strike and dip of the mined seam, and the frequency of fault planes, natural cleavage planes, etc), and rate factors (average rate of face advance and deviation from this average rate of advance). Other factors such as the presence of old abandoned workings in the vicinity and the natural cavitation of certain rock formations will affect the surface movement consequent on mining operations.

The ground movement pattern is a function of the underground mining methods and includes vertical subsidence, lateral movement, surface tilting, and surface curvature.

The surface effect of mining subsidence is to develop a surface through. As a subsidence trough is initially developed at the surface, the central part of this trough or zone moves vertically downward, with the sides exhibiting both horizontal and vertical movement (Fig. 2).

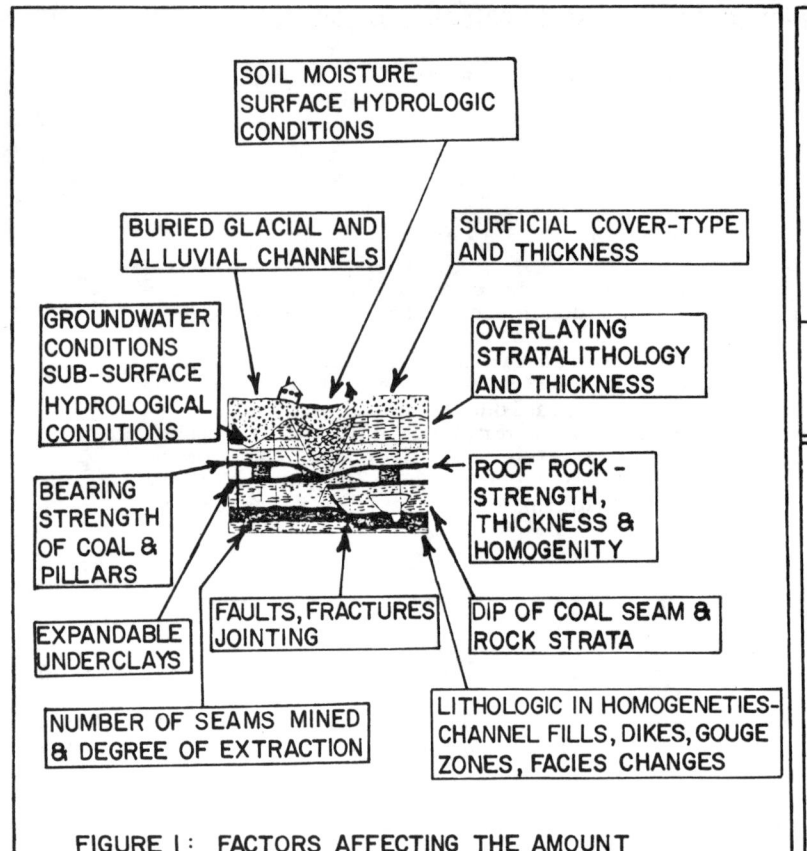

FIGURE 1: FACTORS AFFECTING THE AMOUNT AND RATE OF SURFACE SUBSIDENCE

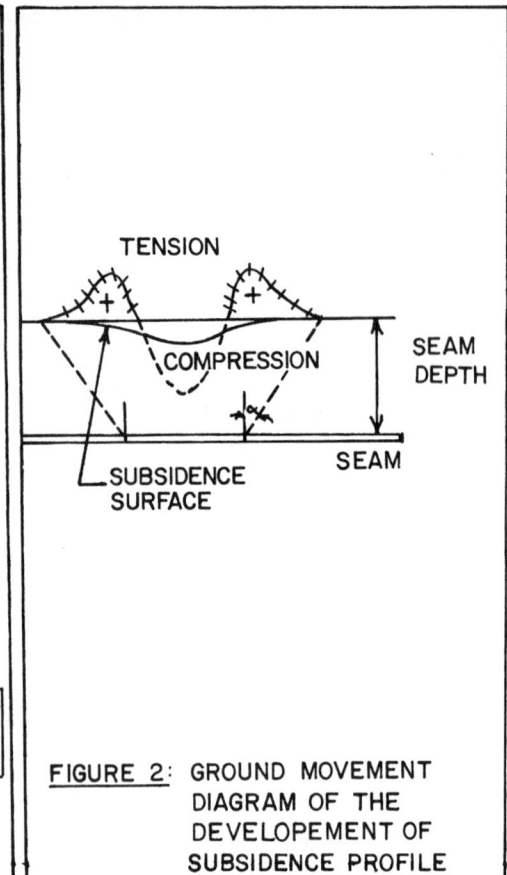

FIGURE 2: GROUND MOVEMENT DIAGRAM OF THE DEVELOPEMENT OF SUBSIDENCE PROFILE

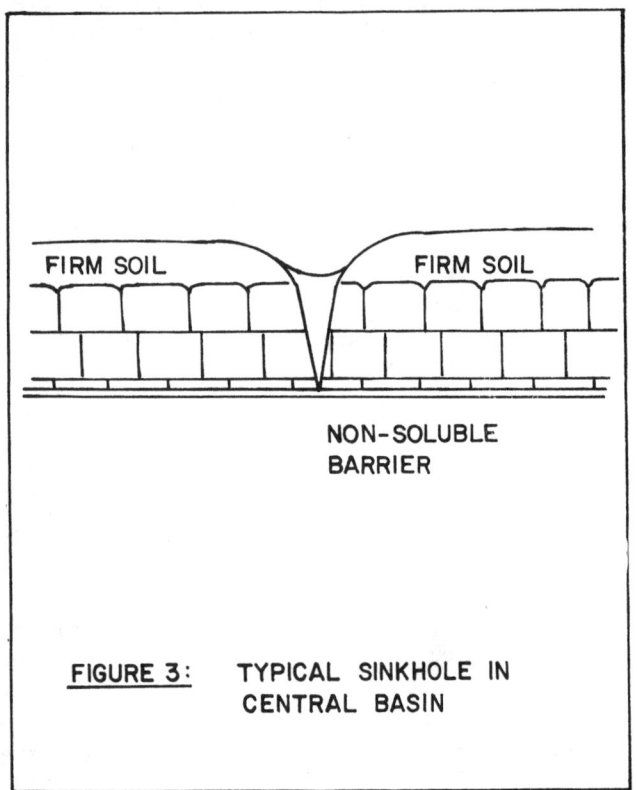

FIGURE 3: TYPICAL SINKHOLE IN CENTRAL BASIN

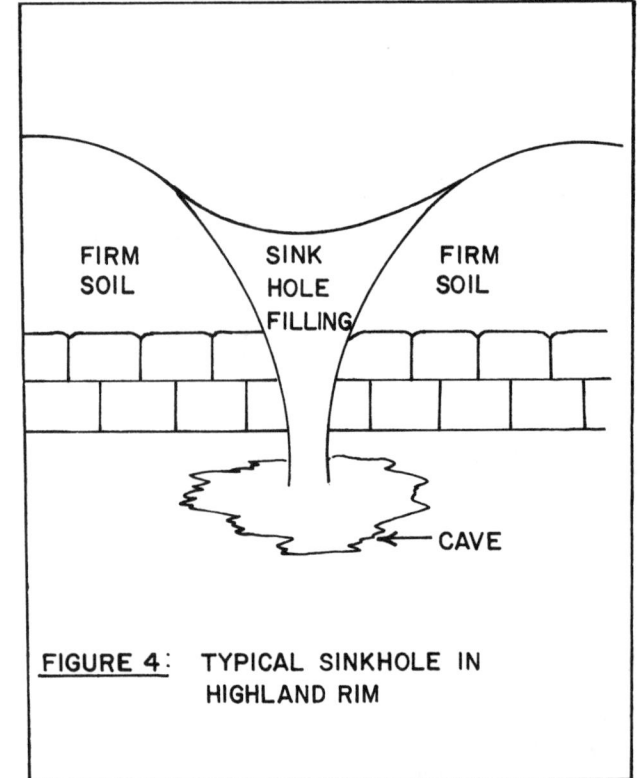

FIGURE 4: TYPICAL SINKHOLE IN HIGHLAND RIM

The amount of the subsidence will depend upon the amount, type and distribution of any backfilling which replaces the extracted coal. Experience has shown that clayey washery waste is the least effective backfill, with sand stowing being the most commonly used backfill. Efforts are in progress to utilize a flyash-cement grouting technique as a filling method.

Uniform vertical subsidence of the ground surface in relatively small increments does not itself generally cause structural damage; but rather differential settlement causes the majority of damage. In the case of most buildings, small changes in the foundation level are ordinarily without importance.

Protective Measures Against Mining Subsidence

Various techniques have been utilized by the mining industry for a number of years to reduce subsidence problems. These techniques include complete or partial filling of the mined area with material brought in, leaving a solid pillar around the area to be protected, systems of partial extraction involving the permanent leaving of 50 or 60 percent of the coal as supporting pillars, and longwall extraction methods designed to reduce strains on ground level structures.

In addition to techniques involving mining operations, subsidence problems can be reduced by design changes in structures built above underground mined zones. These design modifications can include foundations bearing at levels below relatively shallow mining operations, rigidly designing the structure by use of foundation techniques such as raft or mat foundations, or relocating structures to avoid areas of potential subsidence.

Sinkhole Activity and Mechanisms in Middle Tennessee

Sinkhole activity is the most common form of ground subsidence encountered in the Middle Tennesseee area. Middle Tennessee is composed of two distinctly differing geologic regions, and the mechanism, results, and remedial actions required for this subsidence are also distinctly different for each region.

Middle Tennessee is composed of two geologic or physiographic regions: The Central Basin, and the Highland Rim, which encircles the Central Basin. The Central Basin is an elliptical shaped area underlain by rocks of the Ordovician geologic age. These rocks are typically thinly bedded limestones, interbedded with shale. Residual soil derived from the weathering of these rocks are relatively thin silty clays (usually less than 10 feet thick).

The Highland Rim encircles the Central Basin, and is underlain by rocks of the Mississippian geologic age. These rocks are typically fossilliferous thick bedded limestones with a relatively high chert content. Residual soils on the Highland Rim are typically silty clay with chert and average about 10 meters thick.

The mechanism for sinkhole development has similar beginnings for both regions. The process begins with the chemical solution and removal of calcium carbonate along vertical joints and fissures within the limestone bedrock. The chemical solution normally begins at the soil-rock interface and extends downward and can extend laterally along the weaker bedding planes. At this point in the sinkhole development process, the two Middle Tennessee regions begin to show differences. On the Highland Rim, this weathering "slot" normally extends much deeper and has a greater lateral extent than in the Central Basin (Fig. 3 & 4). The development of underground caverns is much more pronounced in the Highland Rim, as exemplified by nearby "Mammoth Cave" in neighboring Kentucky.

Once these open passageways are developed in the underlying limestones, a means for subsurface erosion of the overburden soils exists. Since the "cavern" system is generally much better developed within the Highland Rim bedrock, it should follow that subsurface erosion will also be much more extensive. The opened system in the rock unit will be able to transport a larger volume of eroded soil without "choking" or "silting up".

The difference in soil overburden thickness will also result in a different form of surface expression of sinkholes between the two regions. Soil erosion proceeds upward from the soil-rock interface within an "erosion cone". Since soils from both regions have roughly the same strength parameters, the angle of inclination of the erosion cone should remain constant. The aerial extent of a surface dropout or sinkhole should be directly related to soil overburden thickness, i.e. surface dropouts in the Highland Rim are likely to average three to four times as large (both in depth and diameter) as surface dropouts in the Central Basin.

However, the initial small depressed area in the Central Basin often enlarges significantly laterally with time. This is usually the result of progressive washing or erosion of surface materials into the sinkhole throat, and not necessarily the result of continued subsurface erosion.

Remedial Action for Sinkhole Subsidence

Remedial action to any sinkhole prone site must take into account three major factors: Aerial extent of surface depression, depth to bedrock, and size of opening or cavern in the bedrock unit. In the Central Basin, there are fewer, smaller bedrock openings for subsurface erosion and the surface extent of sinkhole dropouts is much smaller. Therefore, surface dropouts are less frequent and less threatening to buildings and other structures. Many sinkholes can be left untreated when they are not directly adjacent to or beneath structures. When remedial action is required, it typically consists of a shallow excavation to bedrock (within backhoe reach), locating and cleaning the opening into the bedrock, and then plugging the opening with concrete or other cementation material.

In the Highland Rim, remedial action is usually not so simple. The depth to bedrock is usually about 10 meters, well beyond the excavation limits of most construction equipment. Remedial action often includes siting planned structures to avoid sinkhole prone areas. Where relocation of structures is not practical, the most commonly used approaches are deep foundations supported on bedrock or the construction of an inverted filter in a attempt to stop subsurface erosion.

An inverted filter consists of an excavation in the sinkhole area, preferably at the throat if identifiable. The resulting excavation is successively backfilled with crushed stone, each layer becoming progressively finer grained, and capped with a layer of impervious soil. The inverted filter functions by filtering soil solids from moving subsurface water, by controlling drainage into the sinkhole and by slowing the rate of water flow, thereby slowing subsurface soil erosion.

Conclusion

Subsidence can result from either man-made or natural causes. While the surface appearance may be similar for either case, the remedial measures required for engineering purposes requires that the earth scientist understand more fully the mechanisms of each type of subsidence.

Bibliography

Futrell, A. L., Jr. "Sinkhole problems relating to construction in and around Knox County, Tennessee" Masters Project, University of Tennessee, 1981.

Parate, N. S., "A study of ground movement in relation to buildings and surface features," M. Eng. Thesis, Sheff., 1965.

Parate, N. S., "Reducing the effects of mining subsidence on surface features." Coll. Eng., May 1967.

Parate, N. S., Critère de Rupture des Rôches Fragîle Annales de l' I.T.B.T.P., January 1969, No. 253, pp. 148-160, supplement No. 71.

Parate, N. S., "Energy resources and technologies" paper presented at 6th Miami International Conference on Alternative Energy Sources, December 12-14, 1983, Miami Beach, Florida.

Parate, N. S., "Ground movement due to coal mining and it's damaging effects" paper presented at SME-AIME annual meeting, Dallas, Texas, February 14-18, 1982.

Sowers, George S., "Settlement in Limestone Terrain" unpublished paper, 1968 (Courtesy of Law Engineering Testing Company).

Geotechnical considerations in the location, design, and construction of highways in karst terrain – 'The Pellissippi Parkway extension', Knox-Blount Counties, Tennessee

HARRY L.MOORE *Tennessee Department of Transportation, Knoxville, USA*

ABSTRACT

The extension of a limited access, four-lane highway facility, "The Pellissippi Parkway," involves the possible location of corridors in active karst areas of Knox and Blount Counties in East Tennessee. Geotechnical involvement in the location and design phases of this highway development process has resulted in the recognition of several unstable areas of karst and has led to the adoption of remedial and preventive design measures which are to be incorporated in the construction plans. Karst problems identified along the proposed "Pellissippi Corridors" include induced subsidence and collapse, bridging of existing caves and flooding and siltation of existing swallets and ponors. Innovative design concepts and construction methods used in the Pellissippi Parkway Project include rock pads and rock fills, rock backfill, paved ditches, curbs and flumes, overflow channels, and swallet improvement/protection.

Introduction

The construction of highway facilities in karst areas is becoming more common due to urban development. As the spread of construction projects encroaches on karst areas the resulting geotechnical problems will require varied and innovative remedial concepts.

The continued development and expansion along East Tennessee's high technology corridor - - - The Pellissippi Parkway - - - has required the need for it's extension. Located in the western part of Knox County, the Parkway extension is designed to connect the high-tech areas of Oak Ridge and west Knox County to the Knoxville Airport in neighboring Blount County. (Figure 1).

The pecularity of karst areas with characteristic sinkholes, depressions, swallets, and cave systems has intrigued the earth scientist for years. The technical understanding of the development of karst is still a fresh science with new information and understanding continually being developed. However, the treatment of karst areas with respect to man's impact (i. e. highways, commercial and residential development) has been approached with uncertainty in years past. Very little information seems to have been published with regard to the treatment of karst in relation to construction projects. Newton, 1976, Moore, 1981, and Royster, 1984, are technical papers related to highway engineering in karst terrain. Additionally, Sowers, 1976, Newton, 1981, and Foose & Humphreville, 1979, detail specific engineering problems related to karst and karst terrain.

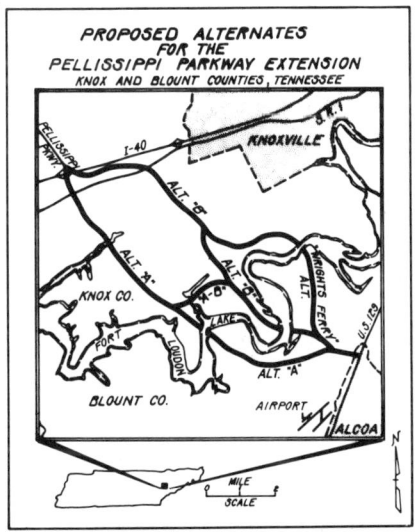

Figure 1: Location map for study area.

Experience has shown that the treatment of karst problems can be tenuous at best. Often the treatment of one kind of karst problem can lead directly to the cause of another kind of karst problem (i. e. treating sinkhole flooding problems can lead to induced collapse problems).

Karst problems are usually dealt with in an after-the-fact maintenance approach. Involvement of geotechnical expertise in the location, design, and construction phases of highway development in karst areas can and often does reduce the incidence of karst related problems. The recognition of potential karst

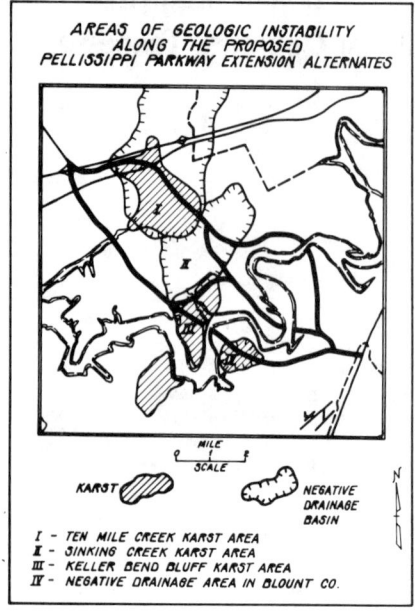

Figure 2: Location map of karst areas within project study limits.

problems before the fact can and often does lead to better engineered projects. Combined with refined karst related remedial concepts the effect of geotechnical expertise can result in a more environmentally compatible facility.

This paper describes the geotechnical aspects of several karst areas that may be impacted by the extension of the Pellissippi Parkway. In addition, proposed remedial design concepts are discussed.

General Geology

Geotechnical involvement during the location phase of the highway project development resulted in the compilation of geologic data detailing karst areas. The locations of four alternates were studied with respect to geologically unstable conditions.

All of the proposed alternates are located in a topographic area composed of alternating ridges and valleys underlain by tilted sedimentary rock strata. Most of the area is underlain by carbonate rock strata containing limestone and dolostone lithologies of Cambrian and Ordovician age. Thick residual clay soils are found along some sections of the alternates while numerous outcrops of bedrock can be found along other sections.

There are three geologic formations composed of carbonate strata that are crossed by the proposed alternates and exhibit characteristic karst conditions. These rock units strike in a NE-SW direction with average dips of $30°$ to the southeast. The three formations are the Knox Group (Cambrian/Ordovician) and the Holston and the Lenoir formations (Ordovician).

Karst

The characteristic karst conditions found along the proposed alternates include depressions, sinkholes, ponors, caves, uvalas, springs and underground drainage systems. The Holston Formation and the Knox Group are very prone to solution activity and contain characteristic karst along the alternates. Existing karst features, induced subsidence and collapse, and flooding were determined to be the main karst problems found along the project.

Four main areas of karst were identified within the project limits (Figure 2). These are as follows:

    I.   Ten Mile Creek negative drainage basin and associated cave system (Knox Co.)
    II.   Sinking Creek drainage basin (Knox Co.)
    III. Keller Bend cave system (Knox Co.)
    IV.   Negative drainage area on Alternate "A" in Blount County.

In addition to existing karst features, there are two main problems associated with karst. These are stability problems (subsidence/collapse features) and flooding.

Typically, active karst areas experience subsidence and collapse features which are usually associated with cave systems. Very often, these days, collapse and subsidence features are man-induced, triggered by 1) altering the natural flow of surface and subsurface drainage, 2) increasing soil moisture seepage pressures, and 3) increasing subsurface erosion. However, natural fluctuation in groundwater tables and subsurface erosion can and often do lead to the occurrence of continued subsidence, new collapse features, and flooding.

In some areas along the proposed extension alternates negative drainage basins have been created due to the formation of depressions and sinkholes which have coalesced into very large areas of negative flow. Three such large scale negative drainage basins are located within the project limits. The Ten Mile Creek and Sinking Creek drainage basins are two such areas of negative flow (see Figure 2). The third smaller negative drainage basin is located along Rankin Ferry Road in Blount County (Alternate "A").

Alternate "B" crosses the Ten Mile Creek and Sinking Creek drainage basins while Alternate "A-B" crosses the Sinking Creek drainage basin. The largest negative drainage basin within the study area is the Ten Mile Creek drainage basin composed of an area approximately 42 sq. kilometers (15.8 square miles) in size and characterized by sinkholes, depressions and cave systems. An extensive cave system (composed of over 4.8 kilometers [3 miles] of mapped passageways) carries Ten Mile Creek underground in a SW direction from the south end of the basin to nearby Fort Loudon Lake where Ten Mile Creek reemerges as a large spring.

Figure 3: Collapse structures such as this along S. R. 72 in Loudon Co., TN, usually occur along unpaved ditches in karst areas.

Figure 4: The prevention of karst related subsidence and collapse of highways is centered around controlling the drainage; I-40, Loudon Co., TN.

The Keller Bend Bluff area along Alternate "A" (at the north side of Fort Loudon Lake) is underlain by the Holston Formation and contains numerous interconnected caves. The caves are developed along joints that are both parallel and normal to the bedding strike. The Keller Bend Bluff area is subject to continued cavern enlargement and subsequent sinkhole enlargement as evidenced by at least three different levels of past solution activity.

Geotechnically related stability problems and related karst features (negative drainage basins, cave systems, and sinkhole development) which will impact the Parkway project might include one or all of the following:

*The increased likelyhood of flooding due to the unpredictable nature of negative drainage systems.

*Stability problems concerning induced collapses and sinkhole enlargement.

*Stability problems concerning toe saturation of embankments due to flooding within sinkholes and depressions.

*Alteration of groundwater conditions by induced siltation or runoff contaminants entering the subsurface via sinkholes, caves, etc.

Karst and Highway Construction - Conceptual Design

The construction of highway facilities across karst areas usually results in the development of collapse features and flooding. The collapse features along with subsidence usually occur along unpaved ditches where a freshly excavated cut interval has approached the soil/rock interface (Figure 3). Occassionally, a collapse will occur beneath the roadway driving surface where subsurface erosion has enlarged a cavity in the soil (Figure 4).

Flooding in karst areas usually occurs when runoff accumulates in depressions, sinkholes, or large negative drainage basins faster than the subsurface can absorb the water. In addition when runoff is channeled and directed from a highway into a sinkhole or depression, then an increase in runoff volume results in flooding of the sinkholes and often floods the adjacent highway. History has shown that development along highways often increases drainage volume and, if located in karst, then flooding usually results.

The purpose of the geotechnical investigation for the Pellissippi Parkway extension was not only to locate and identify specific karst problems but to develop remedial concepts for these problems. Experience with these kinds of karst problems in East

Tennessee has led to the development of geotechnical concepts with which to treat karst areas effectively.

Controlling the drainage is of primary importance in coping with karst problems. Providing the appropriate drainage treatment for a karst condition during design and construction can result in reducing the impact a new highway will have on a karst regime (i.e. channeling runoff into depressions and sinkholes without appropriate design measures can and usually does result in the formation of collapse features and serious flooding).

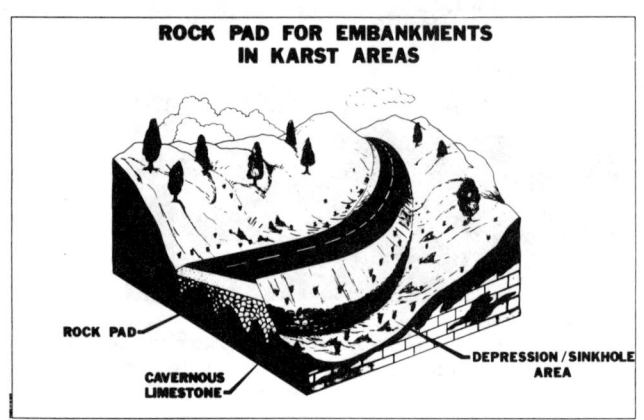
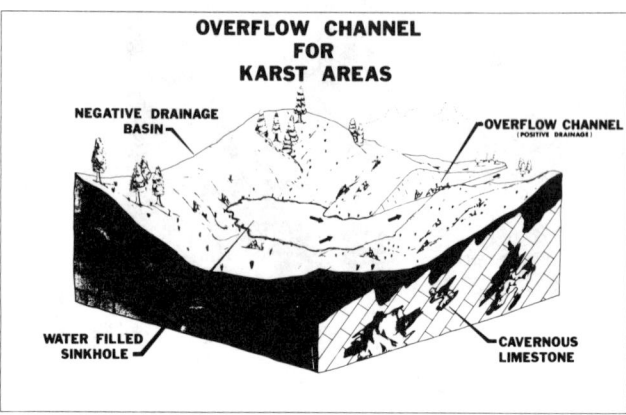

Figure 5: Schematic diagram illustrating the use of rock pads for embankments in karst areas.

Figure 6: Schematic diagram illustrating the use of an overflow channel to relieve negative drainage basins.

A number of innovative and "tried and proven" concepts were applied to the preliminary design of the Pellissippi Parkway Extension alternates. Some of the more effective concepts utilized include:

*Paved ditches - The use of paved ditching for all ditches and channels within a karst area is recommended to prevent or greatly reduce the build-up of groundwater seepage pressures and subsurface erosion which can lead to collapse problems.

*Rock pads - The use of rock pads for bridging depressions and sinkholes and preventing fill-toe saturation during flooding involves constructing the lower lifts of an embankment out of rock fill material. This concept works best when the soil cover is removed down to bedrock before placing the rock fill material (Figure 5).

*Rock backfill - The use of the rock backfill concept is designed to bridge collapse features with a "chunk" rock backfill plug.

*Curbs for fill sections - The use of asphalt curbs will provide for better control of runoff through karst areas assuring that water will flow along established courses reducing seepage pressures and erosion.

*Overflow Channels - Overflow channels are a method which provides a positive flow for runoff in a negative drainage basin. This concept involves the construction of a lined channel or pipe from a negative drainage area to a positive draining system (Figure 6).

*Swallet improvement/protection - This concept involves improving the runoff flow into subsurface cavities by removing debris and trees from around the throat of a swallet and protection of the cavity opening from siltation or clogging by debris laden runoff. Protection methods may involve the use of siltation barriers, debris catchment fences, rip-rap, gabion barriers and concrete structures (Figure 7).

In addition, relocation and the alteration of grades were considered. However, the avoidance of the karst and negative drainage areas located within the proposed corridor limits was not possible without altering the integrity of the Parkway project.

Closing Remarks

Upon the selection of the final alignment for the Pellissippi Parkway Extension, refinement of the proposed geotechnical recommendations outlined in this paper along with additional remedial designs will be required to facilitate a better engineered project. Innovative and cost effective remedial concepts for solving karst related geotechnical problems will require modifications and refinement to insure proper results. Stringent land use restrictions and building codes for the karst areas outlined in this study will be required to insure the success of the karst related remedial design concepts proposed for the extension of the Pellissippi Parkway in Knox and Blount Counties, Tennessee.

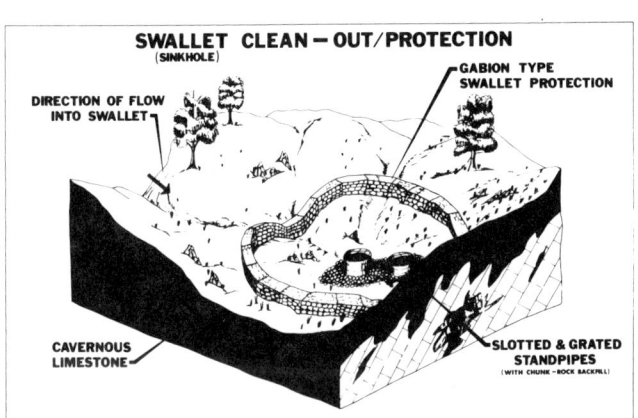

Figure 7: Schematic diagram illustrating the concept of swallet clean-out and protection.

Acknowledgements

The writer wishes to acknowledge the Tennessee Department of Transportation and specifically the Geotechnical Section and the Region I Design Office for their assistance and support in the development of this paper.

References

Foose, Richard M., and Humphreville, James A., 1979, Engineering Geological Approaches to Foundations in the Karst Terrain of the Hershey Valley: Bull. of the Association of Engineering Geologists, Volume XVI, Number 3, Summer, 1979, pp. 355-381.

Moore, Harry L., 1981, Karst problems along Tennessee Highways: An Overview. Proceedings of the 31st Annual Highway Geology Symposium, Austin, Texas, August, pp. 1-28.

Newton, J. G., 1976, Induced and Natural Sinkholes in Alabama - - A Continuing Problem along Highway Corridors. Transportation Research Record 612, pp. 9-16.

Newton, J. G., 1981, Induced Sinkholes: An Engineering Problem: ASCE Journal of the Irrigation and Drainage Division, Volume IR2, June, pp. 175-185.

Royster, D. L., 1984, The Use of Sinkholes for Drainage: Proceedings of the 63rd Annual Transportation Research Board Meeting, Washington, D. C., January, pp. 1-26.

Sowers, George F., 1976, Mechanisms of Subsidence Due to Underground Openings: Transportation Research Record 612, pp. 2-8.

A model study of a proposed concrete road pavement over a potential sinkhole area

J.MARIUS LOUW *University of Stellenbosch, South Africa*
PAUL H.GOEDHART *South African Transport Service*
FREDERICK J.VAN ZYL *Steinhobel and Partners, South Africa*

ABSTRACT

One-tenth scale models simulating the behavior of strips of a proposed specially treated concrete road pavement over a sinkhole reasonably substantiated the proposed design. The prototype comprised a rebar mesh fixed on top of used mine hoist ropes laid at one metre centers in the slab in the longitudinal direction. These continuous ropes were anchored at 10 m centers to one metre deep reinforced concrete beams sunk into the subgrade at right angles to the road center line and casted monolithicly with the slab. This special pavement would extend 150 m past the boundaries of the suspect sinkhole areas. The model test was performed in a 6.0 x 1.5 x 0.7 deep brickwalled tank partially filled with a compacted clayey soil. Two strips of 17 x 200 mm micro concrete were constructed about 0.5 m apart forming in fact two two-dimensional models. A sinkhole was formed by removing a metre length of fill through a side wall of the tank. The live load consisted of two static axle loads applied over the sinkhole center. Failure occured when passive slip surfaces developed through the fill at the anchorage beams.

Introduction

Two sections of road on the goldfields of the Western Transvaal were closed 14 years ago as sinkholes appeared in the dolomitic formations due to the lowering of the watertable by deep gold mining operations. Escalating transportation costs eventually called for a feasibility study to provide new safe roads in these areas.

A geological investigation pinpointed the danger regions. These investigators predicted that the most probable sinkhole diameter at surface level could be between 10 to 15 m with a 2 percent probability of a 50 m diameter hole.

Prototype

A scheme was proposed for reconstructing the roads with a specially reinforced 170 mm continuous concrete pavement to form a "stress-ribbon bridge" over a sinkhole (Ref. 1, 2). It was not intended necessarily to serve as a permanent structure after a sinkhole developed but at least to safeguard road users under such circumstances. Reinforced sections would vary in length from 400 to 2,040 m.

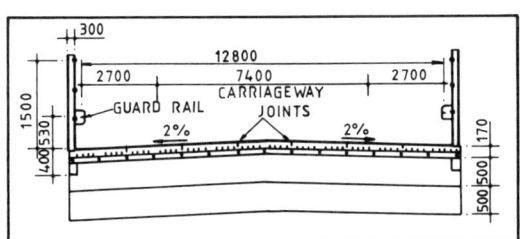 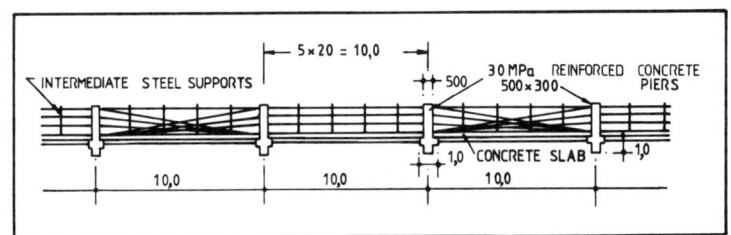

Figure 1.a: Typical cross section of proposed prototype design.

Figure 1.b: Elevation of prototype balustrade.

Fig. 1 shows a typical cross section and elevation of the proposed design. The spacing of the cross-beam anchorages is a compromise between localizing possible passive failures to each anchorage beam and providing an adequate number of anchorages over a relative short length of sound substrata. The sawn joints in the slab are intended to create an articulated suspended structure and hence minimize possible concrete crushing in bending and twisting.

The "bridge" reinforcement comprise 46 mm diameter used mine hoist ropes (Fig. 2) with a rebar mesh fixed on top of it (Fig. 1.a). Reinforced concrete piers protruding upwards from the cross-anchorage beams support longitudinal balustrade ropes which are designed to prevent vehicles from sliding off the bridge in the event of an excessive tilt due to an unsymmetrical cave-in (Fig. 1.b).

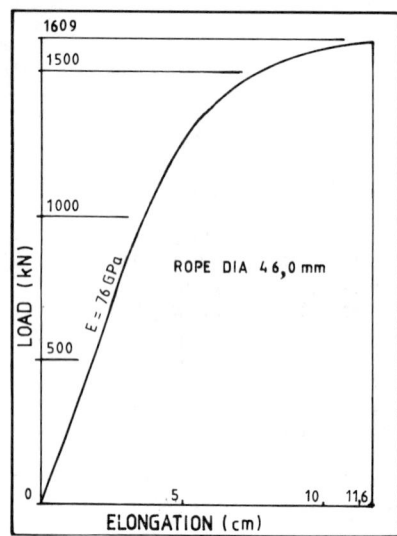

Figure 2: Typical test curve of used mine hoist rope (courtesy National Mechanical Research Institute).

Fig. 3.a and 3.b depict some of the original analytical model analyses considering the nonlinear rope behavior of Fig. 2 and assuming a linear subsoil reaction with a horizontal volumetric modulus of 40 MN/m³.

To incorporate the nonlinear behavior of the anchorages a representative stress-strain relationship of the relevant subsoil would be required. In addition, the bending stiffness of the slab should be incorporated. Alternatively a study of a reasonably scaled model could be undertaken.

Despite the severe limitations of the analytical model studies, the following limits can be derived:
(a) With an infinitely stiff horizontal anchorage resistance ($L_a = 0$, Fig. 3) and conservatively taking the rope alone to provide full cable resistance the minimum factors of safety with respect to rope strength (Fig. 2) would be respectively 3.4 and 1.4 for a 12 m and 50 m diameter hole. A lesser anchorage resistance would cause larger deflections over the sinkhole (e.g. $L_a = 240$ m, Fig. 3) and hence smaller cable forces to support the same vertical loading. These factors of safety are, therefore, lower bound values for the cable action.
(b) From these rather conservative cable forces the lower bound values of the safety factors of the anchorages can also be determined. It is very improbable that fewer than four anchorages ($L_a = 30$ m) would be available under most circumstances. This would require a maximum single anchorage resistance of 115 kN per m of beam for a 12 m hole and 263 kN for a 50 m hole taking w as 15 kN/m ($L_a = 0$, Fig. 3). If the substrata has a friction angle of 12 degrees and a cohesion of 120 kPa as was used in the laboratory model, then the Rankine theory predicts an ultimate resistance of 304 kN per metre of beam. The lower bound values of the factor of safety with respect to an anchorage slip failure are then 2.6 and 1.16 respectively.

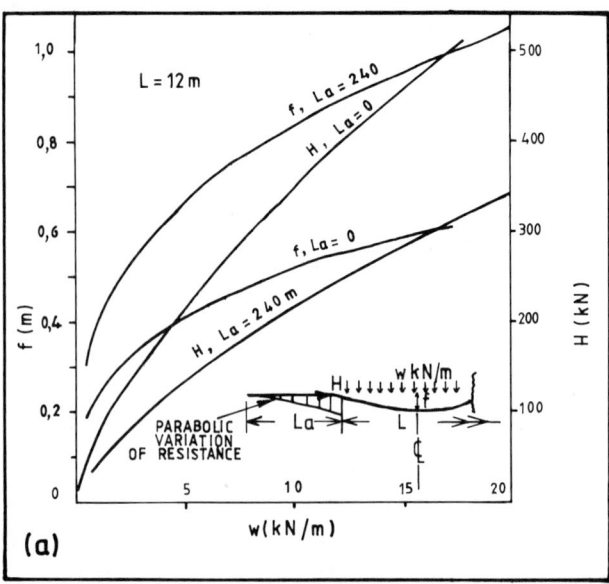

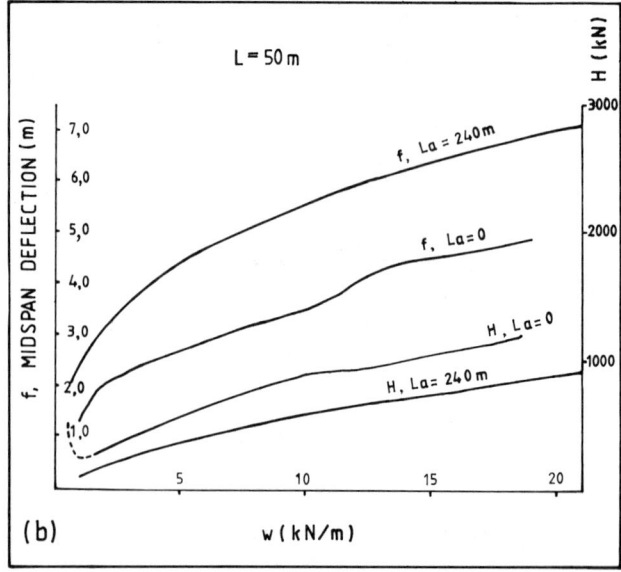

Figure 3: Simple two-dimensional analytical model results.

It is thus obvious that failure would most probably occur due to passive slip failures at the anchorages.

Model

A scale factor of 10 was selected as a compromise between physical size, that is, labor involved and experimental reliability.

The model was constructed in a 6.0 x 1.5 x 0.7 m deep brick-walled tank (Fig. 4). Two layers of 150 mm size concrete cubes were packed on the bottom of the tank to save labor and facilitate creating the model sinkhole. A silty clay available had the following properties: plasticity index = 9.0, liquid limit = 39.0, shrinkage limit 5.0, 18% coarse sand, 27% fine sand, 28% silt, 27% clay and an optimum modified AASHTO dry density of 1 670 kg/m^3 at 14% moisture content. It was compacted in layers into the tank at about 17% moisture content at a final dry density of 88.4% of optimum. Two layers of loose bricks separated the top layers of soil from the longitudinal walls of the tank. Afterwards these bricks would be removed to observe the failure surfaces through the compacted fill at each anchorage. Samples were extracted for triaxial testing yielding an average cohesion of 140 kPa and a 12 degree friction angle.

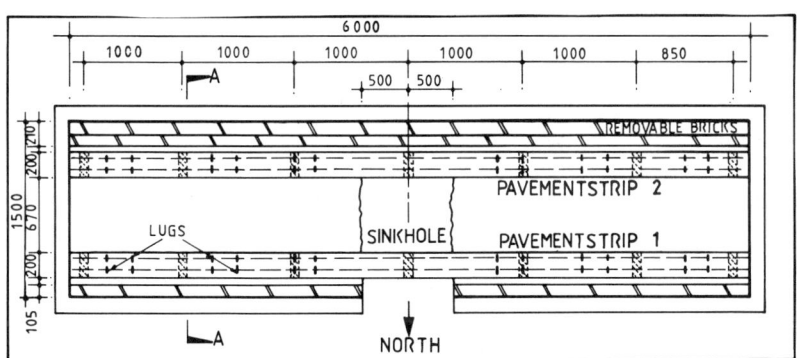

Figure 4.a: Plan view of laboratory model set-up.

Figure 4.c: Section through anchorbeam of laboratory model.

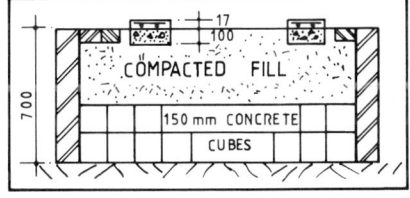

Figure 4.b: Section through laboratory model set-up.

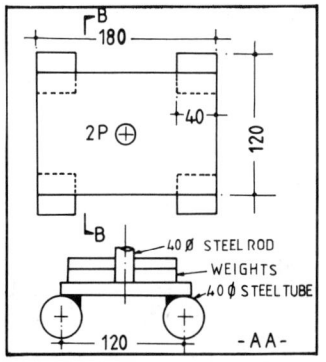

Figure 5: Model of double axle loading.

Two separate model concrete strips, each 17 x 200 mm were cast on top of the levelled compacted fill (Fig. 4.a). The actual thickness varied from 18 to 25 mm. The micro-concrete had a 4.8 mm maximum size aggregate and a strength of 40 MPa (100 mm cube tests at loading of the model). The aggregate had a fineness modulus of 2.94 and the mix comprised 219 ℓ of water, 492.7 kg cement and 1 630 kg aggregate. The mesh reinforcement was 2.0 mm diameter high-tensile wire spaced at 50 mm centers with an ultimate strength of 1.70 kN which reduced to 1.44 kN after indentation to simulate the surface deformations of the prototype rebars. The failure loads for 20 mm diameter and 16 mm prototype bars should be 163 kN and 104 kN respectively. The model ropes were 5 mm diameter 6 x 19 (12/6/1) F steel wire having a 17.0 kN breaking load and modulus of elasticity of 100 GPa as compared with about 1 600 kN and 76 GPa of the prototype (Fig. 2). The mesh was placed on top of the ropes which were clamped to the beam reinforcement (Fig. 4.c). Only three cross-beam anchorages were provided on either side of the intended sinkhole. Movement of small lugs made of 3 mm steel plates epoxied onto the ropes were recorded by dial gages. From this the rope strains were also calculated. Additional weights were placed on top of the concrete strips to obtain the correct mass relationship. The presence of the reference lugs and dial gages unfortunately made a uniform application of these weights along the pavement length impossible.

The sinkhole was formed by removing the concrete cubes and soil through the opening provided in the northern wall of the tank (Fig. 4.a). The applied live load simulated two adjacent axles with four wheelsets of a heavy truck-trailer combination (Fig.5).

Vertical deflections of the pavement strips were recorded at the center of the sinkhole.

Results of Pavement 1

Cracks appeared in the slab at the rim of the sinkhole at an axle load of 45 kg (Fig. 6). At the service axle load of 90 kg a sag of 45 mm occurred and the concrete cracking at the rim enlarged. The relative horizontal movements along the pavement seem to support the assumption used in the analytical model of an anchorage resistance diminishing parabolically along the pavement away from the hole (Fig. 7).

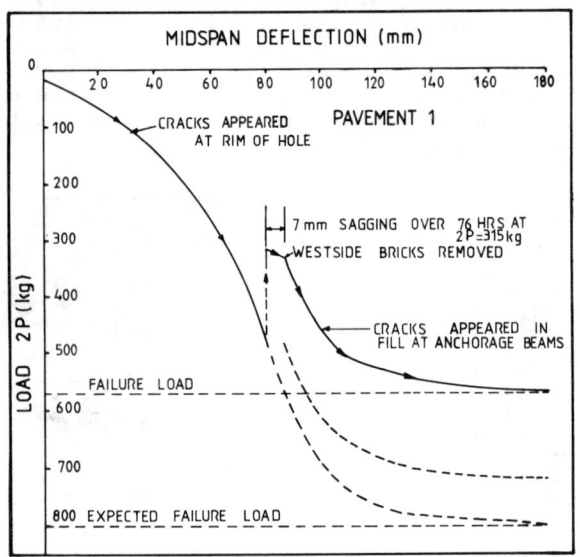
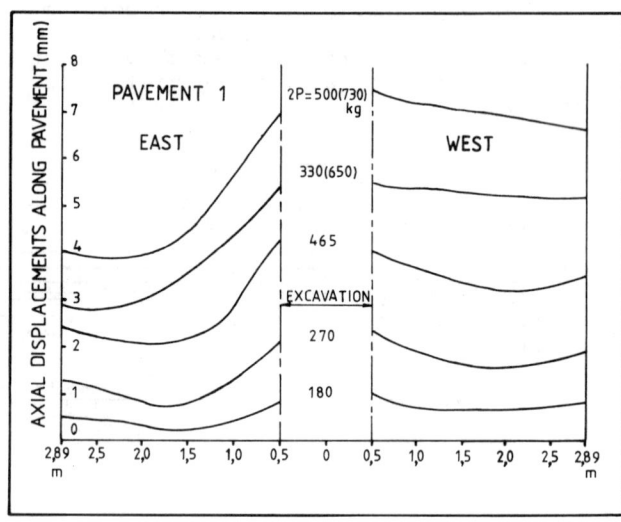

Figure 6: Pavement 1 - Vertical deflection at center of sinkhole vs load.

Figure 7: Pavement 1 - Horizontal translations along pavement vs load.

After obtaining an axle load of 243 kg it was reduced to about 156 kg for 65 hours (Fig. 6). During this period it appeared that consolidation at the anchorages caused a further sag of 7 mm. The bricks along the North-West side were removed before reloading commenced. Cracks soon appeared in the fill around the anchorages at an axle load of 225 kg. The West side subsequently failed at an axle load of 285 kg by sliding towards the hole and sideways into the observation gap left by the removed bricks. If the loading process had not been interrupted, failure might have occurred somewhere between 360 and 400 kg axle load (Fig. 6). The removal of the bricks and thus side support might be the major cause for a lower failure load compared with pavement 2 (Fig. 9).

Near failure a second lineal hinge developed across the slab along the center line of the hole as the concrete crushed in bending. No wires were ruptured.

No rope strains could be detected from the dial gage readings. This was apparently due to the fact that the upward curvature which developed behind each anchorage (Fig. 8) caused compression in the ropes (Fig. 11). As pavement 1 failed at a relative low load, the axial strains seemed to be locally neutralized by these compressive strains in the regions of the dial gages.

Results of Pavement 2

Cracks appeared in the slab at the rim of the sinkhole at a 50 kg axle load. At the service axle load of 90 kg the pavement deflected 37 mm at the centre of the hole (Fig. 9). The deflection increased almost linearly with the axle load except for two "jumps", one at 75 kg axle load and again at 275 kg. The first one coincided with a "hinge" which suddenly developed directly below the load (Inset, Fig. 9). The latter one presumably was caused by the beginning of passive slip surfaces at the anchorages. Again the West side translated more than the East side and eventually began to slide towards the sinkhole at a steady rate under an axle load of about 550 kg while large cracks appeared around the anchorages (Fig. 8). Those slip surfaces starting at the anchorage nearest to the sinkhole eventually reached the hole and the sides of the hole began to cave in at 565 kg axle load. At 475 kg the upward curvature behind each anchorage was very prominent (Fig. 8) and substantial local compression developed in the ropes (Fig. 11). The rope forces increased rapidly as the bending resistance of the slab spanning across the sinkhole peaked and diminished as lineal concrete hinges developed at the rim and at the center of the hole (Fig. 11).

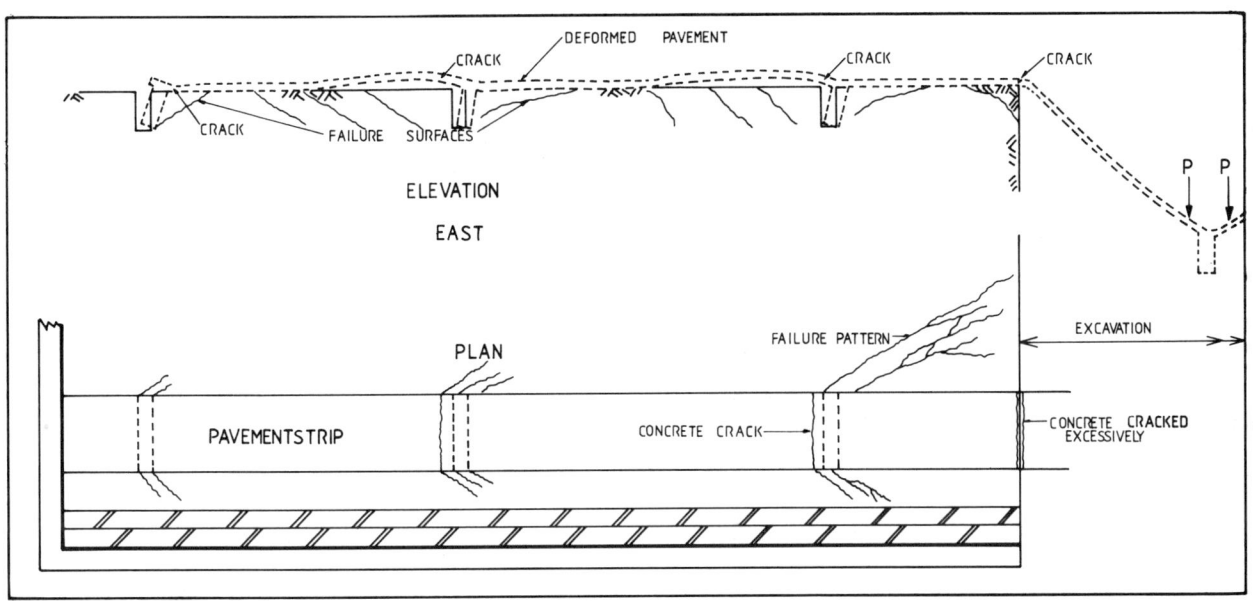

Figure 8: General failure pattern.

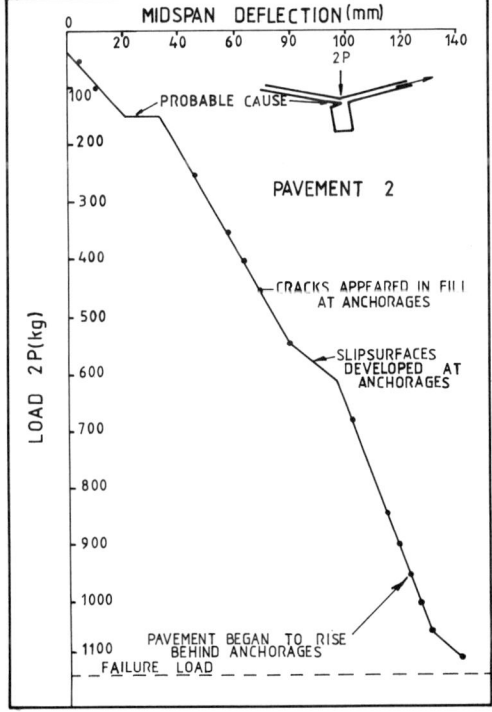

Figure 9: Pavement 2 - Vertical deflection at center of sinkhole vs load.

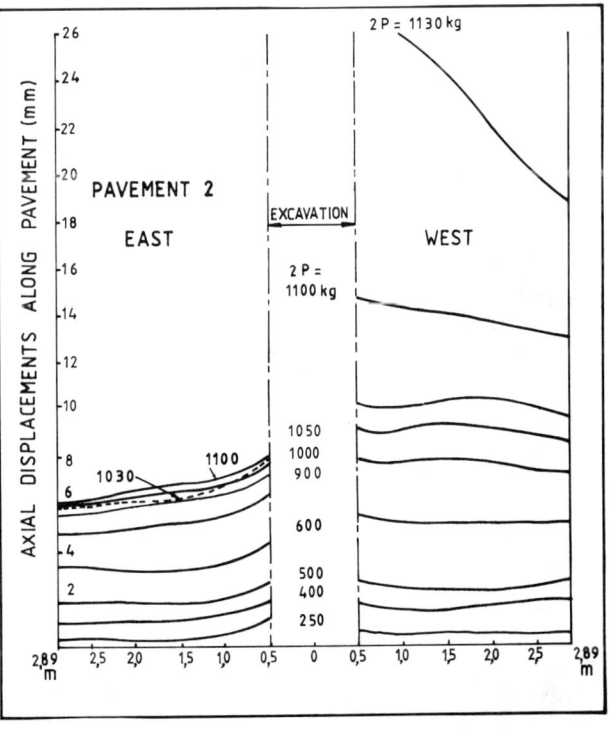

Figure 10: Pavement 2 - Horizontal translations along pavement vs load.

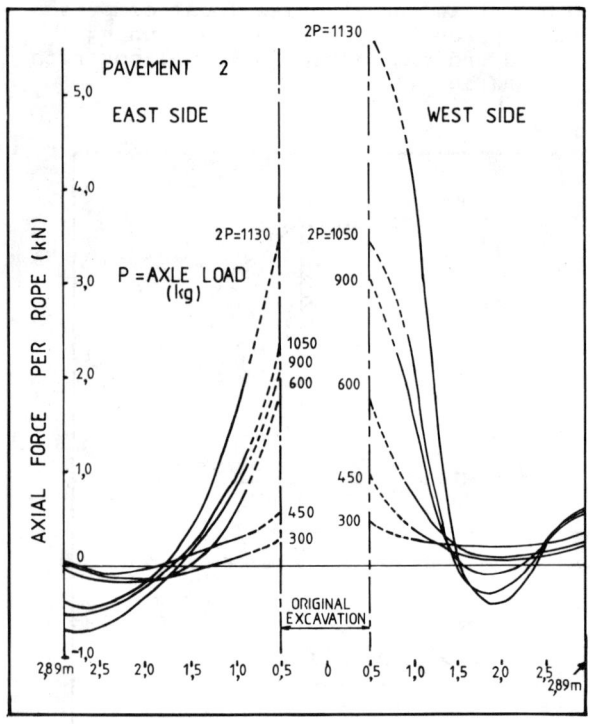

Figure 11: Pavement 2 - Rope forces along pavement vs load.

Conclusions

The live load on the physical model was not a true simulation of that on the analytical model. However, the moment over the center of the holes in the two models were of the same order of magnitude. The vertical deflection at the center of the hole as well as the cable force are dependent on this moment. Furthermore, a slight increase in the hole size due to local deformations of the rim of the physical model justified the comparison with the 12 m analytical one. Therefore, several valid conclusions pertaining to the proposed design but based on the limitations of these models can still be drawn:
1. Both models predicted a passive anchorage failure mode.
2. The factor of safety against such failure was at least 3.2 for the 10 to 12 m hole and 1.2 for a 50 m diameter hole.
3. In neither case would the respective factors of safety relating to failure of a wire rope be less than the above-mentioned figures.
4. The vertical deflections of the two test pavements at service load over a 10 to 12 m hole agreed well, predicting 410 mm for the prototype. This would still form a reasonable slope for any vehicle to negotiate.
5. This measured vertical deflection was much less than that calculated with the analytical model which neglected any bending resistance of the slab. Over the past two years further research on quarter scale models of such cable-supported slabs over a 14.7 m prototype hole proved the importance of this bending moment and the complicated interaction between rope force and reinforcement in such relative short spans. Moreover, even with a nominal mesh reinforcement of 0.14% very little concrete cracking occurred at service load. A minimum factor of safety against wheel punching of 4.7 was obtained and about 17.6 for the ropes. At punching failure there was none of the severe concrete crushing phenomenon as had been observed in the 1/10-th scale models. Even then the predicted slab cracking was still within acceptable practical limits. Jointing the slab was also found to be unnecessary. Increasing the rope spacing for this size of hole to 2.44 m reduced these safety factors respectively to 2.7 and 6.3. The results of these latest investigations will be published shortly.

References

Van Wyk & Louw Inc. Report on a feasibility study in connection with the reopening and upgrading of approximately 22.5 km of road P89/1 between Carletonville and road 1 265. The Roads Department, Transvaal Provincial Administration. Report No. DPH 025 R-14/9/14. Pretoria, December 1978.

Louw, J.M. and A.W. Rohde. A concrete road pavement designed to bridge 50 m diameter sinkholes. Concrete/Beton, No. 27 1982 09. The Concrete Society of Southern Africa, Johannesburg.

Foundation problems on karstic limestone formations in Western Thailand – A case of Khao Laem Dam

DENNES T. BERGADO *Asian Institute of Technology, Bangkok, Thailand*
CHANIN AREEPITAK *Electric Generating Authority of Thailand, Nonthaburi*
FRIEDRICH PRINZL *University of Graz, Austria*

ABSTRACT

In Western Thailand, near the border with Burma, major difficulties were encountered during the construction of Khao Laem Dam across the Quae Noi River. The dam consists of a rockfill core with its upstream surface protected by a skin of concrete. Exploratory drilling and tunnels indicated that while the left abutment presented acceptable rock conditions, the center area of the foundation for the main body of the dam and the right abutment were composed of rocks exhibiting the faults of karstic limestone formations with cavities down to 50 m deep in the river bed. Where the investigations have shown that major cavities were present, diaphragm walls were constructed in the rocks using the overlapping piles method. In the right abutment, it was decided to excavate a series of 6 tunnels, each of 3 m diameter penetrating 3.5 km into the rock. Drilling, grouting, and cut-off wall construction were carried out between tunnels and below the lowest tunnel to ensure that the abutments were made impervious. In other areas, conventional drilling and grouting techniques were used to achieve an impervious curtain below the dam foundation. The concrete cut-off wall which includes the concrete diaphragm wall in the dam foundation and the positive cut-off wall in the right abutment appeared to be the most reliable in long term effectiveness.

INTRODUCTION

Khao Laem Dam is a multipurpose dam owned by the Electricity Generating Authority of Thailand that is being constructed across the Quae Noi River, a tributary of the famous River Quae, in Kanchanaburi Province near the border with Burma and about 230 km northwest of Bangkok. The dam is 90 m high and 920 m long made of rockfill with an upstream concrete face as the impervious layer as shown in Fig. 1. The plan of the dam showing the spillway, power station, diversion channel, and other appurtenances is given in Fig. 2. In the right abutment, a 450 m long diversion tunnel and 6 galleries with diameters of 3 m and 14 m vertical spacing were constructed with a maximum length of 3.5 km, (Fig. 3). The Quae Noi River flows in the northwest to southeast trending valley surrounded by steep mountain ranges along a regional structural feature known as the Three Pagodas Fault. At the dam site, the valley is about 800 m wide with a 50 m wide river channel. The left side steepens gradually to form a slope of 25° while the right side is a shear limestone cliff, the zone of the Three Pagodas Fault and a rugged limestone range called Khao Laem Massif.

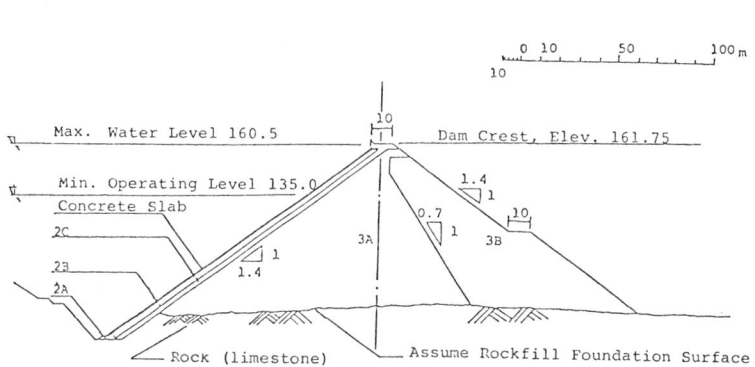

Fig. 1 - Typical Dam Section

DAM SITE GEOLOGY

The two distinct rock formations at the site are as follows (SMEC, 1979): (a) the Thung Song Group which are the rocks of the Ordovician Group consisting of argillaceous limestone, calcareous sandstone, carbonaceous rock and low-grade metamorphic rocks found mostly in the dam foundation, and (b) the Ratburi Group which are rocks of the Permian age represented by dark gray limestones on the right abutment of the dam.

The argillaceous limestone is hard, dense, bluish gray, thick to very thin beds, with general strike of 150°, moderate dip of 10° to 60° and axial plane cleavages. Differential

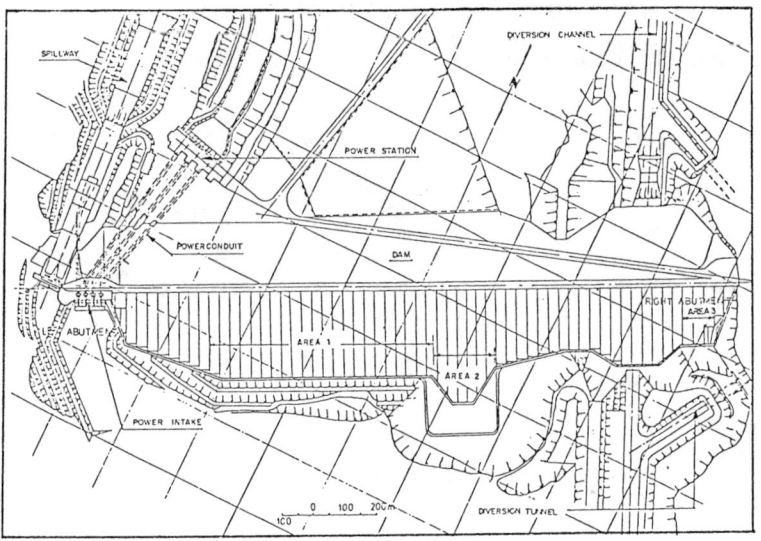

Fig. 2 - Plan of Khao Laem Dam

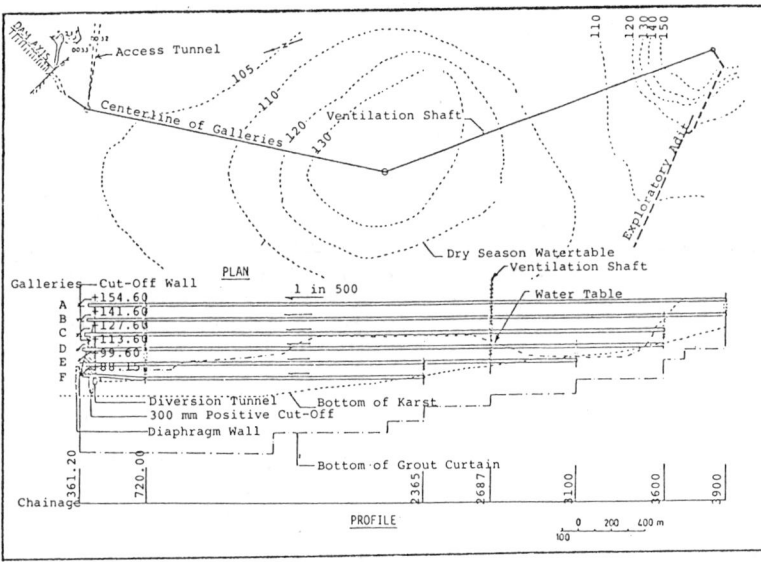

Fig. 3 - Right Abutment Cut-off

weathering features are common due to the interbedding with shale. The karst is well-developed along most of the bedding planes and axial plane cleavage in the form of solution conduits, numerous sizable cavities or isolated cavities, resulting in a "moonscape" foundation surface with localized weathering zones. On Area 1 (Fig. 2), the karst features are mostly developed down to 50 m with infilled materials of clay and sand. The direction of the karst is mostly developed down to 50 m with infilled materials of clay and sand. The direction of the karst is mostly in the north $30°$ west with vertical form. The contact between the argillaceous limestone and the calcareous sandstone is a vertical fault. The calcareous sandstone in Area 2 is mostly hard, slightly weathered with open solution joints and weathered pocket. The rock is dense, grayish and interbedded with limestones. When the calcium carbonate is leached out, the rock becomes either a soft and friable decalcified sandstone or a loose clayey sand material. The karst distribution is in the form of isolated subhorizontal conduits. The carbonaceous rock type is non-karstic which consists of graphitic shale, dark gray limestone and siltstones with gouge seams. The rock is well-laminated, medium hard with calcite or quartz intervenation and interbedded with thin to very thin limestone beds. Earthy calcite and clay coat the fracture surface along with pyrite mineralization. Block weathering is found on the rock surfaces. The contact between the carbonaceous rock and calcareous sandstone is a vertical fault.

The rocks of the Ratburi Group is summarized in Table 1. Located mostly in the right abutment, the rock contains 2-3% solution cavities. The bedding dips 35°W or is massive and unbedded. The infilled cavities are found from the cliff face to about ch2000 in the southern part of the galleries (Fig. 3). Over much of this length, high permeability is known to exist and the water table on the downstream side of the galleries never rises above 110 to 120 m even in the wet season. The Three Pagodas Fault Zone at the contact between the Ratburi Group and the Thung Song Gr. altered the Ratburi Group into microfractured and brecciated limestones. The planar fault forms a prominent overhanging cliff face on the right abutment.

METHODS OF KARSTIC ROCK TREATMENT

There are many dam sites in the world that have similar karstic foundation problems. Londe (1970) cited many successful projects. In all cases, about half of the total cost of dam construction was spent in treating karstic foundation. The foundation treatment at Khao Laem Dam ranks one of the most complex and difficult. Many karstic treatment methods were used. They are briefly described in the next sections. Two main sections were treated (Areepitak, 1983), namely: the dam foundation and the right abutment. For a concrete cut-off wall alone, there was about 77,000 m² of area in the dam foundation and about 362,000 m² on the right abutment.

Diaphragm Wall Construction

With the purpose of achieving effective, permanent and non-erodible seepage barrier, concrete diaphragm walls were constructed down to a maximum depth of 50 m in three different areas of high degree karst foundation (0.2 m to 10 m wide solution channels). As shown in Fig. 2 and 4, these areas were Area 1, Area 3 and diversion channel area. A detailed investigation was performed by 3-D geology mapping and closely spaced (6 m spacing) exploratory holes. Water pressure tests were done on the exploratory holes. These holes were than grouted with low pressure for the purpose of reducing the concrete absorption during the diaphragm wall construction. The diaphragm wall was constructed by placing 762 mm diameter piles drilled at 615 mm centers (Fig. 4) with a down-the-hole drilling machine (Ingersoll Rand "Superdrill") mounted on a link belt crane. The primary holes were first drilled and concreted by tremie pipe method. About 3 days later, the secondary holes were drilled and concreted. A 300 mm diameter mini-diaphragm wall was also constructed below the right abutment as shown in Fig. 5 for treatment of minor karst (opening width of 5 to 200 mm) below the galleries, underwater. The same overlapping pile method of construction was used.

Table 1 - Classification of Ratburi Limestones on the Right Abutment of Dam (WATAKEEKUL, S., GRATWICK, C., FREIMAINS, I., 1982)

Rock Types	Characteristics	Remarks
1. Limestone (Typical)	Dark grey, dense, homogeneous, crystalline	Karstic
2. Brecciated Limestone	Pale grey, consists of dark grey limestone fragments 3 to 10 mm size in light grey calcareous matrix, dense but friable in somepart	Non-karstic
3. Microfractured Limestone (Faulted Zone)		Unsuitable for Grouting
3.1	Blocks bounded by joints infilled to 500 mm thick with firm to stiff, red to orange brown high plastic clay. Main joints dip $10°$-$20°$ NW Blocky nature by minor orthogonal joint.	Karstic
3.2	Light to dark grey, with zone of graphitic limestone with karstic features, friable and easily erodible.	Karstic
3.3	Grey, with distinctive white calcite patches, spotted appearance friable and easily erodible, closely spaced joint (3 to 10 mm) with clay and amorphous calcite coating.	Non-karstic
4. Carbonaceous Limestone	Black, low strength, friable and easily erodible.	Non-karstic

Flushing Method

The flushing method was utilized for washing out the infilled sand and clay material in less severe karst by pressurized water before high pressure grouting commenced. Flushing was continued until the outflow water was clear. There were many places in the plinth foundation where this method was used. Area 2 in Figs. 2 and 4 is used as a case for discussion. The flushing holes were drilled around the main row of grout curtain with a diameter of about 80 mm. Each cell is composed of 6 holes with 1.0 m spacing, 3 holes in the upstream side and 3 holes in the downstream side (Fig. 4). The design depths were from 20 to 25 m and the downhole flushing and grouting were made at 5 m stages. Pressurized air and water were injected into one of the 6 holes while the other 4 holes were plugged. The pressure used was 3 kg/cm² on the first stage and 5 kg/cm² on the other stages. By successively plugging and unplugging the holes, ejection was forced through all 6 holes. The 4 rows of high pressure grouting commenced after flushing work with a pressure of 1 kg/cm² in the first stage and 5 kg/cm² in the lower stages.

Mining Methods

Mining methods in the form of open pit excavations, adits, shafts or galleries were constructed to plug the cavities and caverns with concrete in Tarbela Dam in Pakistan (Londe, 1970). In Khao Laem Dam, these methods were mostly used to treat major karst in the fault zone in the right abutment. Figure 5 shows the system of galleries in the right abutment. Exploratory access shafts were driven as pilot tunnels to intersect natural cavities (Fig. 6). Cleaning with pressurized water, all loose materials in the cavities were removed and washed out. Concrete was then

Fig. 4 - Foundation Treatment

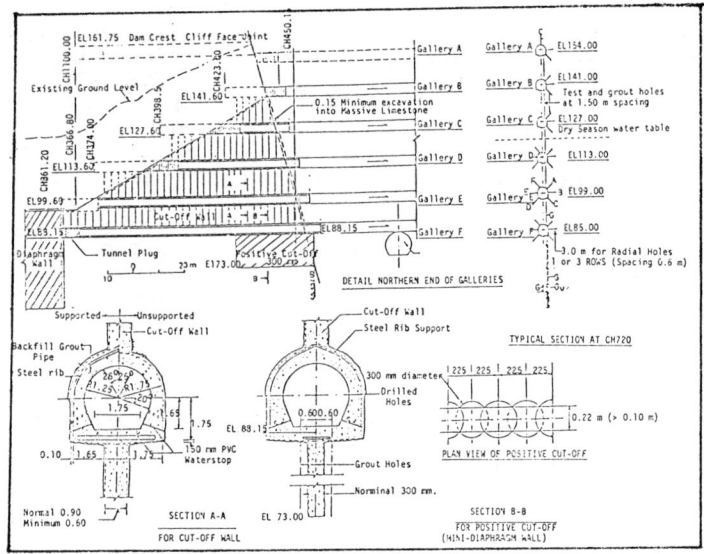

Fig. 5 - Right Abutment Cut-off

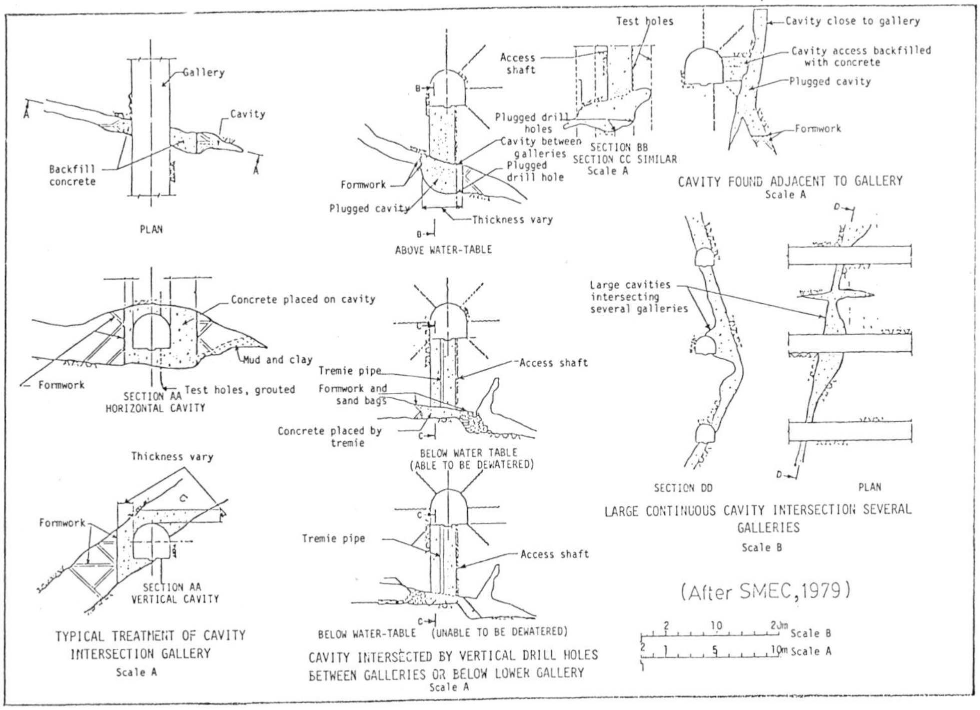

placed into the enlarged and cleaned cavities by belt conveyors, transit locomotives or concrete pumps.

Concrete-Cut-off Wall - Microfractured limestone (Table 1) was discovered in the northern part, between galleries B and F. In this deposit, the erosion and seepage potential cannot be reduced because only minimal grout takes were observed. Thus, a concrete cut-off wall was proposed between galleries (Fig. 5) from around the fully pressure-lined gallery C to the invert of gallery F.

165 mm Positive Mini-Cut-off Wall
This karstic rock treatment was applied for minor karst with infilled material for which conventional grouting was unreliable. The area for this treatment was on the southern portion of the right abutment, between the galleries, with an area of about 13,000 m². To construct this wall, 165 mm diameter vertical holes were made with spacing of 150-250 mm from the invert of the upper galleries to the crown of the lower galleries. The holes were cleaned out with high pressure water jets and examined with a TV camera. Then, the grout mix was poured into the holes by means of a 60 mm flexible pipe and worked into the crevices using needle vibrators.

Grout Curtain -
For Khao Laem Dam, the pressure grouting was applied for karstic rock treatment as primary or secondary treatment. More than 150,000 m of grout holes were

Fig. 6 - Cavity Treatment

performed on the foundation and 300,000 m on the right abutment. There were 3 to 5 rows of grouting designated as curtain grouting (A-row) in the middle, blanket grouting (B-row), both upstream and downstream, and consolidation grouting (C-row). The packer grouting method was used with a target set at 3 lugeons or grout take of 3 sacks per meter utilizing a split spacing method. The diameters of the holes were either 76 mm or 105 mm with depths ranging from 30 to 100 m. The basic grout material was portland cement and the supplemental materials were bentonite, sodium silicate, and fine sand to pebbles.

EVALUATION OF KARST TREATMENT METHODS

Although the best evaluation on the effectiveness of the karstic foundation treatment is water impounding, the following 3 methods were used to check the improvement of the karstic foundation. The groundwater table on both sides of the diaphragm wall was studied by pumping tests. The isolation between the groundwater table on both sides of the cut-off showed its integrity and its effectiveness. High lugeon areas were only found near the bottom of the cut-off due to the problems of concreting and hole collapse. Hence, grouting work was needed along the center line and at both sides of the diaphragm wall. The limitations of the diaphragm wall cut-off were the cost, workability of equipment and rate of construction. Studies of the permeability (in lugeon) of the foundation and abutment before and after treatment was also performed. From this method, a step lugeon map was obtained in which there was improvement from high to low lugeon value with the sequence of grouting. The cement absorption statistics was also studied. The effective karstic treatment method decreased the cement adsorption on each grout hole at different steps of grouting until the grouting target was reached. Observations from core drilling and from TV camera showed that the karstic features were effectively filled, all mortar contacts were well-bonded, and erodible karstic infill has been removed.

The cost and rate of construction of the various karstic rock treatment methods in Khao Laem Dam foundation and right abutment depends on the amount of concrete absorption by the foundation which is in turn influenced by the karstic condition. The workability of the equipment is also one of the factors which should be considered, such as the diaphragm wall which have been efficiently constructed by the superdrill on the dam foundation but cannot be applied in the abutment or in the steep slopes of Area 2. It was necessary to use the mini-diaphragm method, mining method and the flushing method, to remove these difficulties. The grout curtain was applied to treat rock fractures and sporadic, isolated cavities in the deeper part of the foundation. It was found uneconomical and impossible to treat this problem with other methods. Hence a combination of several methods, as primary and secondary treatment, was necessary to treat the karstic rock foundation of Khao Laem Dam.

CONCLUSIONS

From the observations and evaluations performed in this study, the following conclusions can be made:

1) The foundation rocks (Thung Song Group) contain 90% karstic rock which is argillaceous limestone, calcareous sandstone, and about 10% non-karstic rock of black shale, mudstone and siltstone. On the right abutment, the rock is limestone massif (Ratburi Group). The three Pagodas Fault, which is the contact between these two rock groups, resulted in about 100 m of fault zone and altered the rock into microfractured and brecciated limestones and sheared shale. The degree of karst in these areas are very high in the form of caves, solution channels, conduits and joints with infilled material associated with the groundwater table which is in itself erratic.
2) On the dam foundation, the karstic rock treatment included a concrete diaphragm wall in 3 locations, a grout curtain, flushing, cut-off trenches and sink shafts. On the right abutments, the karstic rock treatment has been performed from the galleries with different methods, namely: mining method, 165 mm positive cut-off wall, 300 mm mini-diaphragm wall and grout curtains.
3) The diaphragm wall and cut-off wall constructions are the most reliable and have long term effectiveness, but the cost is high. The factors considered for karstic rock treatment are the cost, workability of equipment, time schedule and karst characteristics. The combination method of karstic rock treatment is proposed to obtain an effective and economic solution.

ACKNOWLEDGEMENTS

The authors wish to thank the Electric Generating Authority of Thailand and the Snowy Mountains Engineering Consultants for supplying the necessary information. The typing efforts of Mrs. Uraiwan Singchinsuk was also appreciated.

REFERENCES

Areepitak, C., 1983, Karstic rock treatment on Khao Laem Dam foundation and abutment with the purpose of positive watertight cut-off construction, M. Eng. Thesis, Asian Institute of the Technology, Bangkok, Thailand.
Londe, P., 1970, Recent developments in the design and construction of dams and reservoirs on deep alluvium, karstic, and other unfavorable formations, Intl. Commission on Large Dams, G.R.Q. 37.
SMEC, 1979, Report of geology in Khao Laem Multipurpose Project, Vol. 1, Snowy Mountains Engineering Consultants, Australia.
Watakeekul, S., Gratwick, C. and Freimanis, I., 1982, A cut-off wall in karstic limestones within the right abutment of Khao Laem Dam, A.A. Balkema Publisher, Netherlands.

Construction on dolomite in South Africa

FRITZ VON M.WAGENER & PETER W.DAY *Jones and Wagener, Inc., Rivonia, South Africa*

ABSTRACT

Damage to structures and loss of life has been more severe on dolomite than on any other geological formation in southern Africa. The subsidence which occurs on dolomitic terrain following development or during dewatering has given dolomite a notorious reputation and engineers and geologists became reluctant to recommend development on the material.

This has led to the pioneering of founding methods for a wide variety of structures aimed at reducing the risk of severity of damage due to subsidence settlement. Structures successfully founded on dolomitic terrain include residential and industrial buildings, gold mine reduction works and shaft structures, tailings dams, water retaining structures and road and rail links.

In this paper various methods of construction, some of which were developed by the authors, are presented. It commences with a classification of a dolomite site in terms of overburden thickness followed by a discussion of the relevant construction methods. The methods include mattresses of compacted soil supported by pinnacles or "floating" in residuum, deep foundations such as caissons and the use of specialised piling techniques and soil improvement by dynamic consolidation.

INTRODUCTION

Damage to structures and loss of life has been more severe on dolomite than on any other geological formation in southern Africa. Construction problems have been experienced on dolomite since the advent of industrial development in this country. The first recorded problem was in 1910 when engineers attempted to sink a shaft through dolomite in the Far West Rand of the Transvaal province to reach the gold-bearing reefs of the famous Witwatersrand rocks. So great was the inflow of water and mud into the shaft that the project had to be abandoned. It was only in 1937 that the first shaft was successfully sunk in this area by making use of the cementation process to stabilise the residuum and control the inflow of water.

On the Far West Rand the dolomite is divided into a number of ground-water compartments by impermeable syenite dykes. Mining was being hampered by water-inflow along water-bearing fissures connecting large reservoirs in the dolomite with the Witwatersrand rock being mined. During dewatering of three of these ground-water compartments, widespread damage occurred as a result of large scale surface instability in the form of sinkholes and dolines. On the 12th December 1962, the three-storied crusher plant of the West Driefontein Mine disappeared into a steep-sided sinkhole 55m (180ft) in diameter and 30m (100ft) deep. There was no warning as the structure and 29 employees vanished without trace within a few minutes.

Similar problems, but on a much smaller scale were being experienced on non-dewatered dolomite in other parts of the Transvaal province. It was found that, soon after development started in an area, subsidence in the form of sinkholes and dolines took place. It was realised that these subsidences were triggered by water ingress and that they occurred in areas where certain conditions existed in the overburden and in the underlying dolomite.

These subsidences gave dolomite a notorious reputation and engineers and geologists became reluctant to develop on dolomitic terrain.

TYPICAL SOIL PROFILE

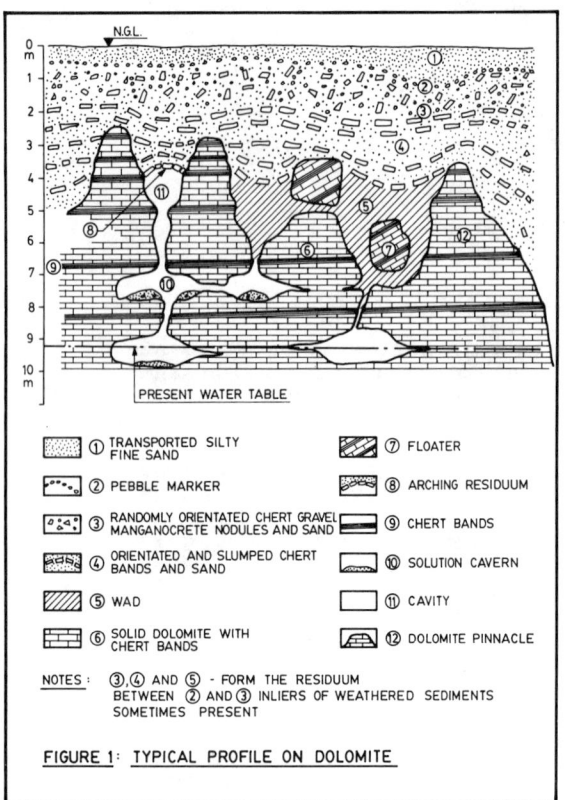

FIGURE 1: TYPICAL PROFILE ON DOLOMITE

A typical soil profile on dolomite as it occurs in the Transvaal is shown in Figure 1. The site is often covered by a blanket of transported material which is occasionally underlain by a pebble marker before the residuum is encountered. In a chert-rich dolomite the residuum usually grades from a coarse angular chert gravel to the finer insoluble components of dolomite namely wad and clay. This is followed by a highly irregular rockhead consisting of pinnacles and boulders. Cavities may be found in the residuum between the pinnacles with solution caverns in the upper layers of the "solid" dolomite. A vertical scale has been included on the side of Figure 1 but it must be realised that this scale could vary within a wide range and could probably be multiplied by a factor ranging between 0,1 and 20.

On the Far West Rand inliers of weathered sedimentary strata are sometimes found overlying the residuum.

On a dolomite site the consistency or in-situ strength of the material usually reduces with depth from the relatively competent chert gravel to the compressible wad before it increases sharply as "solid" rock is encountered. This is contrary to the situation on most geological formations where the consistency improves with depth. This represents one of the problems when developing on dolomite.

TYPES OF SETTLEMENT OR DISTRESS

The main problem associated with the development on dolomite is settlement. Three types of settlement can be distinguished as follows:

(i) NORMAL SETTLEMENT due to a combination of immediate elastic settlement and consolidation settlement. The magnitude of this settlement will depend on the load, thickness and properties of the residuum and transported soil cover. A structure founded partially on rock and partially on loose residuum may suffer from differential settlement.

(ii) SUDDEN SUBSIDENCE SETTLEMENT: SINKHOLES due to the appearance of a sinkhole caused by the collapse of an arch which spanned over a cavity in the residuum.

The size of the sinkhole formed will depend on the thickness of residuum and depth of water table. A large sinkhole cannot develop in an area with a thin soil cover or a high water table.

(iii) GRADUAL SUBSIDENCE SETTLEMENT: DOLINES. A doline is formed when the arching conditions required for the formation of a sinkhole do not develop. Large dolines can be formed when the water table is lowered in areas which contain thick layers of wad. The damage caused by a doline can be substantial but as it happens slowly it is not a threat to life as is a sinkhole.

Following the geotechnical investigation of a site on dolomite, one should be able to quantify the relative risk of occurrence and probable magnitude of the three types of settlement listed above. If possible economical and safe foundation solutions which should cater for the expected movement may then be adopted. Should there prove to be a relatively high risk of sinkholes and dolines developing, it may become uneconomical to provide a safe design and such a site may have to be abandoned.

The risk of developing on dolomite should be pointed out to the client at the outset of the investigation. He should be made aware that there will always be a risk on dolomite especially if time and cost of investigation and construction are to be kept within reasonable limits.

CLASSIFICATION OF A SITE

It has been proposed by Wagener (1981) that a dolomite site be classified according to the thickness of overburden from ground level to the tops of pinnacles and boulders. The founding methods proposed in the next section are closely related to the Class of dolomite site.

On completion of the fieldwork and evaluation of results the site can be classifed as follows:

CLASS A - Pinnacle and boulder dolomite at or near the surface: $c<3m$ (10ft).

CLASS B - Pinnacle and boulder dolomite overlain by moderately thick overburden: $3m<c<15m$.

CLASS C - Pinnacle and boulder dolomite overlain by thick overburden: $c>15m$ (50ft).

Where c is the average thickness of overburden to tops of pinnacles and boulders.

PRECAUTIONARY MEASURES

It is common knowledge that water is the triggering mechanism in the majority of cases of distress on dolomite. It is thus essential that the concentrated ingress of water into the ground be avoided at all times and this also means during construction. The precautions with respect to water are summarised below and are necessary irrespective of which foundation solution is adopted.

. Landscape the site in such a way that concentrated ingress or ponding of water is avoided. Parking areas and open areas between buildings may have to be surfaced.

. Backfill trenches for services so that the degree of compaction is at least equivalent to and the permeability less than that of the surrounding material.

. Check water-bearing services at regular intervals for leaks. Where large movements are expected these services should be flexible.

. In critical areas water-bearing services may have to be located above the ground.

. Stormwater canals should be lined over critical areas and should discharge well away from development.

. Liquid retaining structures should be provided with an impermeable membrane below the structure to prevent infiltration into the subsurface.

METHODS OF FOUNDING

The methods of founding which are commonly used in South Africa are listed below. They are closely related to the Class of site proposed earlier. A founding method has limited application where a site Class is given in brackets.

FOUNDATION DESCRIPTION	SITE CLASS
(i) CONVENTIONAL FOUNDATIONS	CLASS A, B & C
(ii) MATTRESS OF IMPROVED MATERIAL	CLASS A, B & C
(iii) FOUNDING ON PINNACLES	CLASS A
(iv) DEEP FOUNDATIONS Piling Shafts Caissons	CLASS B & (C)
(v) DYNAMIC CONSOLIDATION	CLASS B & C
(vi) REINFORCED EARTH	CLASS B & C

TABLE 1

SUMMARY OF FOUNDATION METHODS

(After Wagener 1982)

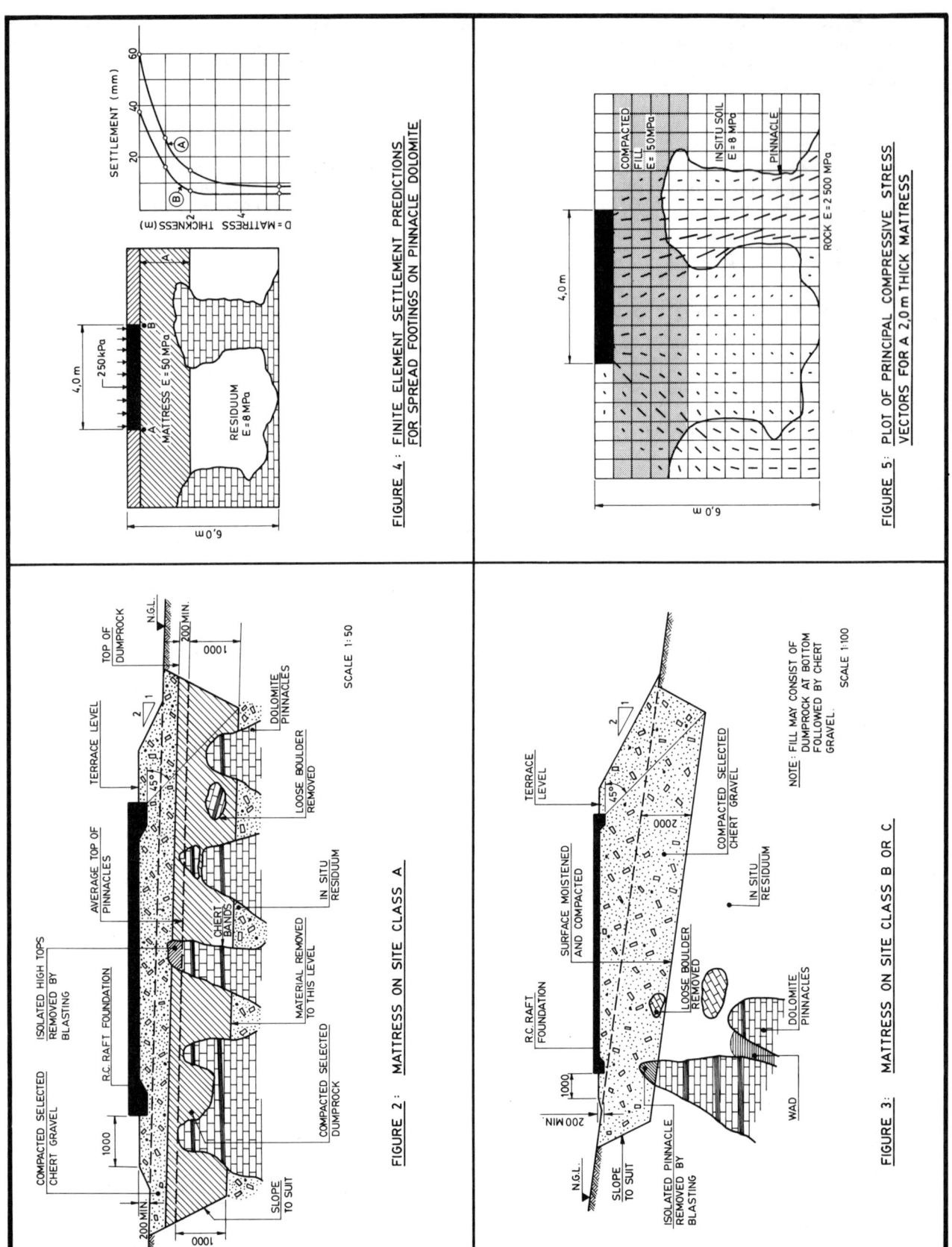

FIGURE 4: FINITE ELEMENT SETTLEMENT PREDICTIONS FOR SPREAD FOOTINGS ON PINNACLE DOLOMITE

FIGURE 5: PLOT OF PRINCIPAL COMPRESSIVE STRESS VECTORS FOR A 2,0 m THICK MATTRESS

FIGURE 2: MATTRESS ON SITE CLASS A

FIGURE 3: MATTRESS ON SITE CLASS B OR C

(i) CONVENTIONAL FOUNDATIONS - Sites Class A, B & C

Conventional foundations such as strip or spot footings are often used on dolomite for light structures such as houses and light structural steel buildings where the risk of sinkholes and dolines forming is found to be acceptable. On a Class A site, where pinnacles and boulders are close to the surface or are exposed in the foundation excavation, differential settlement is often a problem. It is good practice to use brickforce in all brick walls on such a site. In areas where differential settlements of more than 10mm are expected it is advisable to use split construction as is commonly used when building on heaving clay or collapsing sand, Jennings et al (1962). Where rock has to be removed in foundation trenches conventional foundations become costly.

(ii) MATTRESS OF IMPROVED MATERIAL - Sites Class A, B & C

Introduction

As mentioned earlier the consistency of the soil-cover on a dolomite site often deteriorates with depth in addition to being extremely variable and difficult to assess by conventional means. If a certain thickness of this soil-cover is removed and replaced under controlled conditions, a mattress of known thickness and with known strength parameters can be formed. This concept has been used with success by the authors for the founding of numerous industrial and domestic structures since 1975 on dolomite sites in South Africa.

The functions of the mattress are:

(i) To control the total and differential settlement.

(ii) To reduce the foundation stress to an acceptable level at the underside of the mattress.

(iii) To reduce the risk of the formation of small sinkholes and dolines.

An advantage of this method is that it enables an in-situ evaluation of the site to be made during construction. If conditions are found to differ from those predicted, it is possible to excavate deeper or to revert to an alternative founding method.

It has been found that this method of founding is economical and often much cheaper than alternatives such as piling provided good quality fill material is readily available. However, during execution, it is necessary to maintain close contact between geotechnical consultant, designer and contractor as design and construction modifications may become necessary as the excavation progresses and more information on the ground conditions become apparent.

The thickness of the mattress will depend on a number of factors, the most important of these being the properties and thickness of soil-cover and the sensitivity of the proposed structure to settlement. This aspect will be discussed in the next sections.

Method of construction

The method of constructing a mattress will depend on the site Class and backfill material available. Where pinnacles and boulders occur near the surface (Site Class A) the construction procedure would be as follows (Refer Figure 2):

- Remove material to a depth of say 1m below tops of pinnacles and large boulders. It may be necessary to blast off the tops of a few pinnacles which project above the general pinnacle level.

- Inspect bottom of excavation and order additional excavation where pinnacles are spaced further apart or where very loose material is present. Level bottom of excavation, moisten and compact with small vibratory rollers or by hand.

- Where available place selected dumprock into the excavation to an elevation which is just higher than the tops of pinnacles. Compact by means of a vibratory roller of static mass of at least 8 tons. Use ample water during the compaction process. Ten passes of this roller are usually sufficient for a 1m thick dumprock layer.

- The remainder of the terrace is built up to the required elevation by using selected chert gravel or other suitable fill. It is emphasised that the material be placed under controlled conditions.

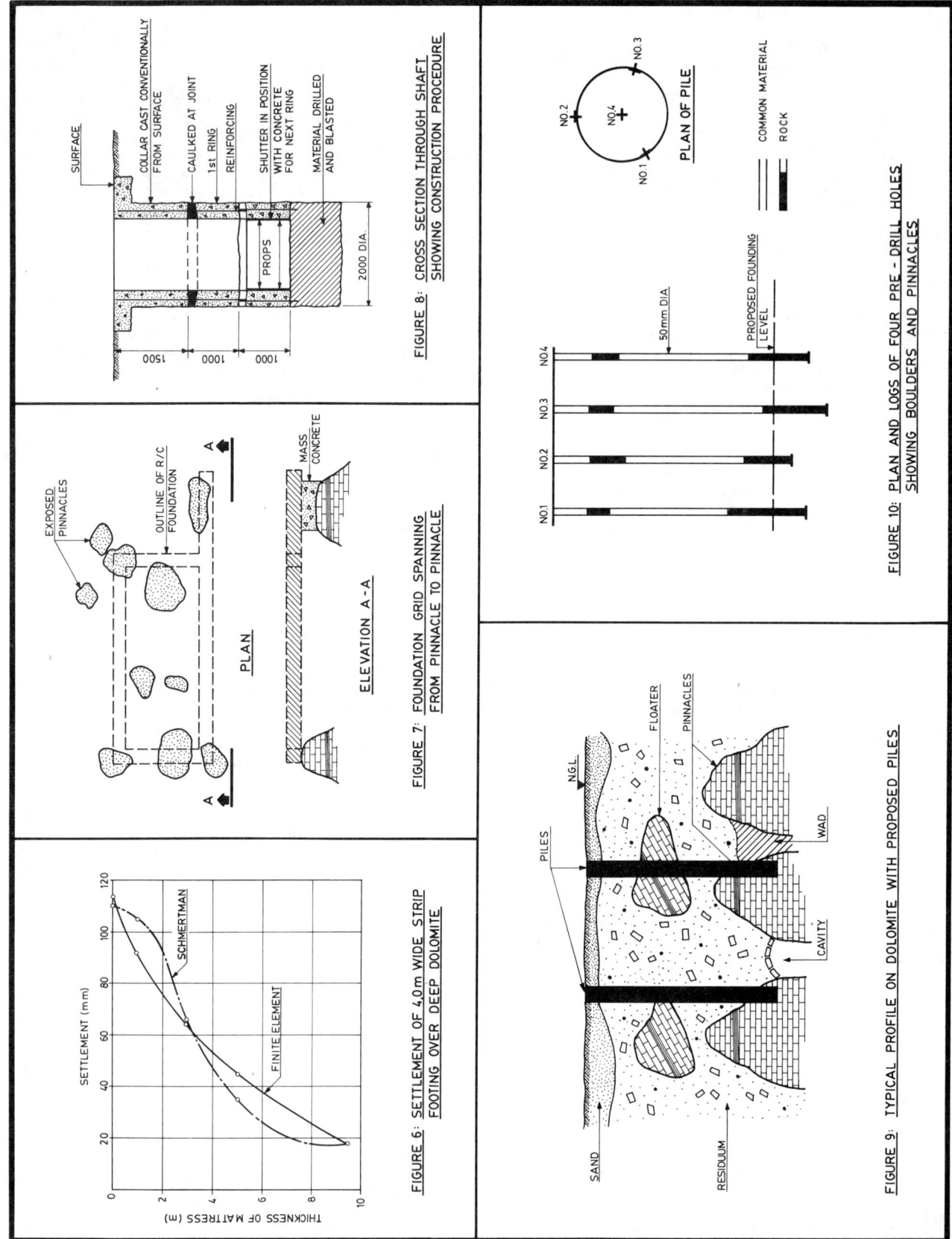

Where heavy settlement-sensitive structures, such as bridges, headgears, mine winders or compressors have to be founded on a Class A site the dumprock between the pinnacles may be replaced by mass concrete. In this case it is usually necessary to prove pinnacles by drilling and to have boulders removed. For light structures, such as houses, it may be possible to reduce the thickness of mattress considerably.

The treatment on site Class B and C is shown on Figure 3. Backfill usually consists of selected chert gravel placed under controlled conditions. Thick layers of permeable dumprock should be avoided.

The foundation of the structure which is to be founded on the mattress should be as shallow as possible as there is no point in providing a mattress and then excavating through it to found the structure. For this reason a raft-type foundation as shown in Figures 2 and 3 is often used in conjunction with a mattress.

Design of mattress

A mattress will be most effective where its thickness is greater or comparable to the size of the footings to be built on it or where the depth of cover over the top of pinnacles is small.

The results of an analysis on a Class A site of settlement versus mattress thickness is shown in Figure 4 (Day 1981). It can be seen that for the geometry of site and footing chosen the optimum mattress thickness is about 2,0m. The principal compressive stress vectors for a 2m thick mattress are plotted in Figure 5 and show how the mattress tends to distribute the load onto the pinnacles by arching.

In Figure 6 the settlement analysis for a similar footing on a Class B site, with a 10m cover over rock, is shown. In this case there is no dramatic reduction in settlement with increased mattress thickness.

WIDTH OF FOOTING	% REDUCTION OF SETTLEMENT
1m	84
2m	72
4m	57
8m	22
16m	6

TABLE 2

PERCENTAGE REDUCTION IN SETTLEMENT ACHIEVED BY USING A 4m THICK MATTRESS OVER A 16m LAYER OF RESIDUUM

It must be realised that mattresses have their limitations as is shown in Table 2. In this case a 4m thick mattress overlies a 16m layer of residuum (elastic properties as shown in Figure 4). The table summarises the effect of the presence of the mattress on settlements of footings of various widths.

The stress distribution below a mattress approximates to a "45 degree load-spread". It can readily be shown that for a square footing with sides half the mattress-thickness, only 4% of the stress reaches the base of the mattress. This value becomes 25% for a square footing with sides twice the mattress thickness. A mattress thus spreads the load very effectively to the weaker underlying layers.

Subsidence risk

When using a mattress the risk of small sinkholes and dolines occurring below the structure is reduced for the following reasons:

. The material at founding level is exposed and features such as palaeosinkholes or slumping chert-bands can be investigated.

. The mattress is usually less permeable than the original soil reducing the risk of water seeping in below the structure.

. The mattress will form a relatively strong roof to any cavity which may be forming. By reinforcing the mattress with, say metal strips, (reinforced earth) additional strength may be obtained. It should be pointed out that a rigid roof could be dangerous as distress signs in the structure, signalling the need for remedial action, are preferable to a sudden sinkhole.

(iii)   FOUNDING ON PINNACLES - Site Class A

Where abundant pinnacles occur close to the surface they can be used as "supports" for a reinforced concrete grid spanning from pinnacle to pinnacle. Brink (1979) presents a case history of a transmitter building which was founded in this way. Where a structure is light it does not really matter if it is founded on boulders rather than pinnacles. However, this may not be the case with heavier structures where it will usually be necessary to prove pinnacles before they are used as supports. This founding method is shown diagramatically in Figure 7.

A disadvantage of the method is that the design can only be completed once pinnacles have been exposed and mapped.

(iv)   DEEP FOUNDATIONS   - Sites Class B and (C)

Under this section piles, shafts and caissons are considered. These methods are usually used for the founding of major and settlement-sensitive structures such as silos, cooling towers, crushers, mine winders and bridges.

It is the authors' opinion that piles should be used on dolomite only when other methods of founding are not feasible. Large floaters of chert and dolomite, steeply sloping pinnacles which can deflect the pile shaft, water and cavities are the main problem when installing piles (Refer Figure 9). After reaching the desired depth it is also necessary to prove the competency of the rock below founding level.

Benoto and Mitsubishi piling rigs which incorporate casing oscillators, heavy grabs, chisels and a thick-walled temporary casing are usually employed. The authors often specify drilling and pre-blasting techniques to shatter the rock in advance of piling. Percussion boreholes of 50mm (2 in) diameter are drilled with a trackdrill at the position of a pile and a log is prepared showing floaters of rock and pinnacles as illustrated in Figure 10. Explosive charges are then inserted at selected depths in the boreholes to shatter the rock. Sinking may then proceed through the broken rock.

A further type of pile which has been successfully used on dolomite is a percussion borehole pile. A 300 to 600mm percussion borehole is drilled with a down-the-hole percussion hammer. A problem with this type of pile is that the drillrods can be broken against steeply sloping pinnacle sides or wedged by gravel collapsing into the hole.

The use of large-diameter bored (augered) piles, precast or displacement type piles are not recommended on dolomite due mainly to boulders and chert gravel in the overburden and the topography of the underlying bedrock.

Conventional shaft sinking techniques, as illustrated in Figure 8, can be used where heavy loads have to be carried by the foundations. The smallest practical diameter for a shaft is 2,0m which provides ample working space and a relatively safe working environment if the shaft is lined as work proceeds.

Caissons can also be sunk on dolomite terrain. A caisson normally consists of a steel cutting-edge with a short section of lining which is cast on top of the cutting edge. Material is removed from inside the caisson, which then sinks under its own weight. The shaft is extended by adding concrete lining at ground level while sinking continuously until sound founding material is reached. An advantage of caissons over shafts is that they can be used in unstable ground conditions without fear of collapse as the cutting edge can always be kept close to the bottom of the excavation.

It is reiterated that the competency of the material below founding level must be proved, for instance by proof drilling, when using these methods.

(v)   DYNAMIC CONSOLIDATION   - Sites Class B & C

This method consists of the lifting of a large "pounder" by means of a crane or tripod and dropping it onto the ground. It has been used in many parts of the world to consolidate material even where this is below water. It is thought that much of the compaction is achieved by shock waves which radiate from the "pounder" when it strikes the ground.

The method has been used with reasonable success on dolomite in South Africa (Wagener 1982) where a 12 ton pounder was dropped from a height of 18m. Reasonable compaction was achieved over a 4 to 6m layer.

A problem on a site with chert gravel residuum is that the material near the surface is often reasonably competent and tends to protect the poorer underlying material and thereby prevents meaningful improvement.

(vi)   REINFORCED EARTH   - Sites Class B & C

Should it be necessary to design a mattress to span over cavities of given dimensions it is possible to provide metal or fibre strips at the bottom of the mattress. The strips will act as the tensile member of a composite section with the soil above being the compressive member.

The method has been considered for projects in South Africa but has not been used. McKittrick (1979) describes a project in Pennsylvania where the method was used in the relocation of a highway.

CONCLUSIONS

The problems experienced on dolomite by South African geotechnical engineers and geologists over the past 25 years have lead to the development of construction methods which are presented in this paper. The methods have been used successfully for the founding of a wide variety of domestic and industrial structures. It must be stressed that the risk when developing on dolomite cannot be eliminated entirely if investigation and construction costs are to remain within reasonable limits. However, the discriminate use of the proposed methods and the control of water ingress could reduce this risk to acceptable limits.

REFERENCES

Brink, A.B.A. 1979. Engineering geology of southern Africa. Building Publications, Pretoria. South Africa.

Day, P.W. 1981. Dumprock and chert gravel mattresses. Seminar on the engineering geology of dolomite areas. Department of Geology. University of Pretoria. South Africa.

Jenning, J.E. and Evans, G.A. 1962. Practical procedures for building on expansive soil areas. The South African Builder.

McKittrick, E.V. 1979. Special uses of reinforced earth in the U.S.A. C.R. Col. Int. Reinforcement des Sols. Paris.

Wagener, F. von M. 1981. Construction methods and remedial measures. Seminar on the engineering geology of dolomite areas. Department of Geology. University of Pretoria. South Africa.

Wagener, F.von M. 1982. Engineering Construction on Dolomite. PhD thesis, University of Natal, South Africa. Distributed by Geo. Div. of SAICE, Box 61019, Marshalltown, South Africa.

High-volume grouting to control sinkhole subsidence

CHRISTOPHER R. RYAN *Geo-Con, Inc., Pittsburgh, Pennsylvania, USA*

ABSTRACT

Sink-hole subsidence caused by underground mining activities and natural solution cavities is a problem found in many areas of the United States. High-volume grouting techniques have been used to correct the problem by filling the underground voids on many sites, both as a preventative and as a remedial measure.

Key issues in designing a grouting program for sinkhole subsidence are such technical considerations as grout strength requirements, appropriate grout consistencies and drilling techniques. Grout materials have to be selected that are economical yet which will fit the technical requirements. Typically cement-flyash, cement-bentonite or flyash grouts are used.

Mixing equipment is high-capacity; pumps, lines, drill casing, etc. all have to be sized to allow large quantities of grout material to be placed very rapidly. Some projects may take 1,000 tons per day or more. Some of the same techniques can also be applied to smaller projects.

Several case examples are presented to illustrate aspects of mix design and placement technique:

- Limerock subsidence, Tampa, FL - Cement bentonite grout at low solids ratio.

- Limerock subsidence, Tampa, FL - Special casing drills to penetrate multiple voids economically.

- Mine subsidence, Pittsburgh, PA - High-ratio flyash-cement grout and a high-volume placement system.

- Mine subsidence, Birmingham, AL - Controlling subsidence at 700 ft. depth in flowing water.

High volume equipment using economical materials make this technique economically attractive on an increasing number of projects.

Introduction

Sinkhole subsidence can be found in areas throughout the United States. Causes of subsidence may be man's activities, such as mining, or natural phenomena, such as solution of limestone formations. In most cases, surficial soils leak into the underground cavities until a depression, or sinkhole, is created at the surface. We see with some regularity news accounts of cars or buildings being swallowed by sinkholes.

The basic remedial technique common to all the examples cited in this paper is grouting. Grouting is the injection of a fluid-like material underground; the material later takes on some degree of structural strength to provide support to a collapsing void. A secondary benefit of many grouting programs is that the grout blocks the flow of underground water that may be causing or exacerbating the sinkhole progression. Grouting is a fairly common engineering technique that is used to strengthen building foundations, top flow under dams, raise settled highways, etc. When used for prevention of sinkhole subsidence, however, there are usually two key differences:

- Volume. In general, the grout quantities for this type of work far exceed the amounts on typical grouting projects.

- Grout Strength. Since the grout is used primarily as a filler, the structural properties of the grout are of less importance than on a typical project.

The implications of these differences are discussed in the following paragraphs and illustrated in the case examples that follow.

Equipment

Repairs of sinkholes almost always involve relatively large quantities of materials. To lower the expense of the work, it is important to size the mixing and placement equipment to suit the job. The first step is to estimate the approximate volume of materials involved, and decide on an appropriate project schedule. Typical projects range from 1,000 c.y. per day down to 50 c.y. per day. Both higher and lower placement rates may be reasonable for specific projects.

Once the rate is decided, the viscosity of the grout must be set. A more viscous grout will tend not to travel so far nor penetrate smaller cracks and fissures. Typically, low viscosity grouts may have a slump of 5 - 10 inches as measured in the standard concrete slump test. Low viscosity grouts can have a slump too high to measure. Typically a cement-bentonite grout may have a viscosity of 15 - 30 centipoises. In the field, a simple device called a Marsh Funnel may be used to measure viscosity. A low viscosity grout may be used to fill thinner or partially filled voids and solution channels.

The rate and placement determine the size and type of the rest of the job components. The grout line must be large enough to accomodate the flow; typically, line size ranges from about two to six inches. Drill holes should accomodate grout flow equal to that of the grout supply line. Pumps, mixers, etc. all need to be selected in accordance with the project requirements.

Grout Mixture

Since the volumes are large, the cost of the grout materials play a much larger factor in job economics than for a typical project. It is important to identify local materials for the project that can be easily handled and obtained at low cost. Flyash, lime dust and crusher screenings are examples of these types of materials. Minimum amounts of cement should be used to achieve required strengths. On most void filling projects in rock, ratios of 10:1 (filler:cement) will provide strengths of 200-500 psi and will be more than adequate. Some projects have used low solids ratio cement-bentonite grouts (final strength 10-20 psi) to grout voids in soils and weak rocks. There is little rationale for placing grout whose strength far exceeds the surrounding soil or rock structure. The examples that follow illustrate the uses of various grout mixes.

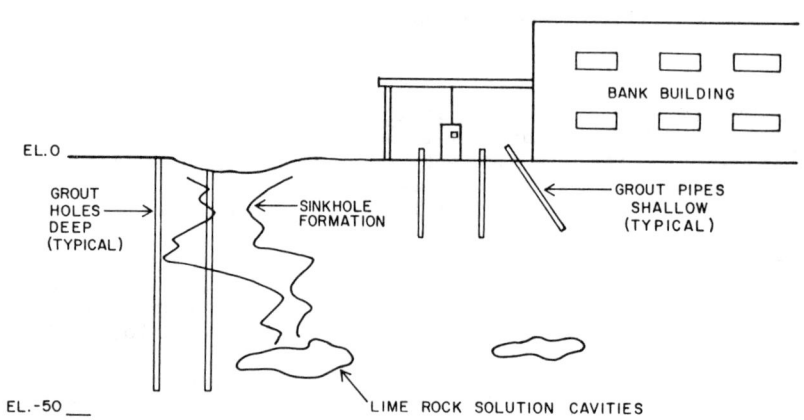

FIGURE 1
Florida Bank Subsidence - Profile

Tampa, Florida -- Lime Rock Sinkholes

A drive-in bank building was experiencing sinkholes around the building and parking lot. The project engineer used geophysical techniques to locate several underground voids and incipient sinkholes; borings further confirmed the presence of solution cavities. The eventual solution (Fig. 1) involved the drilling of deep drill holes to intersect the underground voids and driving shallow grout pipes in a shallow zone around the building (Fig. 2). A low viscosity cement-bentonite grout was used to provide for maximum penetration of the partially filled voids and flow channels. Low strength was sufficient since the soil structure itself was relatively weak. A small grout pump (100 gal/min. capacity), mixer (1 c.y. capacity) and grout lines (1 1/4 in.) were more than adequate to place the 50-100 c.y./day required for the work.

FIGURE 2
Driving Pipes for Shallow Grouting

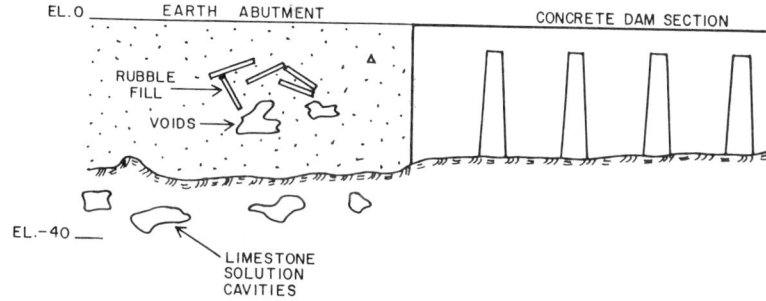

FIGURE 3
Florida Dam Sinkholes - Profile

A drive-in bank building was experiencing sinkholes around the building and parking lot. The project engineer used geophysical techniques to locate several underground voids and incipient sinkholes; borings further confirmed the presence of solution cavities. The eventual solution (Fig. 1) involved the drilling of deep drill holes to intersect the underground voids and driving shallow grout pipes in a shallow zone around the building (Fig. 2). A low viscosity cement-bentonite grout was used to provide for maximum penetration of the partially filled voids and flow channels. Low strength was sufficient since the soil structure itself was relatively weak. A small grout pump (100 gal/min. capacity), mixer (1 c.y. capacity) and grout lines (1 1/4 in.) were more than adequate to place the 50 - 100 c.y./day required for the work.

Tampa Florida -- Lime-Rock Sinkholes

A major concrete water supply dam had a potentially dangerous condition on one of its earthen abutments. Embankment material was being eroded downwards into underground solution cavities, resulting in sinkholes on the dam crest and increased flow through the abutment. The project was complicated by the presence of concrete slabs and other construction rubble in the abutment (Figure 3). The grout selection called for a thicker cement-bentonite grout with greater strength because of the importance of the structure. It was necessary to insert casings though the rubble into the rock and voids below. This is the most difficult type of drilling. The selected drill incorporated a special swing-out over-reaming bit to drill a casing in. Once the bit was extracted, the same casing served as the grout pipe. A picture of the drill and a standard double-tub grout plan is shown in Figure 4.

Pittsburgh, Pennsylvania -- Mine Subsidence

Subsidence is a relatively common problem in areas where coal is mined. In the example presented, an area of 19 residences was affected by a mine located relatively close to the surface. In a typical mine void, slabs of rock fall into the mine as the pillars crush. Spalled rock allows surficial soils to be washed into the mine, creating surface subsidence and sinkholes. In this case, a two foot deep depression had developed in the street and some of the homes were showing signs of distress.

The solution, as shown in Figure 5 was to use low slump grout to create barriers around the affected area and then to use high slump grout on the interior to completely fill the affected area. High volumes justified a major grout plant installation capable of 120 c.y/hr. (Fig. 6). The large volume of grout required large size grout holes (Fig. 7). The grout line used was 5 inch steel pipe (Fig. 8).

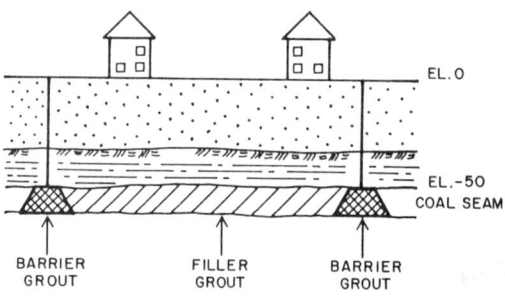

FIGURE 5
Pittsburgh Subsidence Project - Profile

FIGURE 4
Casing Drill and Double-Tub Grout Plant

FIGURE 6
Mix Plant

FIGURE 7
Drilling Grout Holes

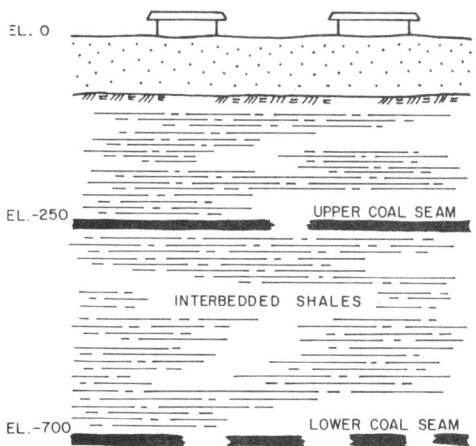

FIGURE 9
Alabama Subsidence Project - Profile

FIGURE 8
Grout Placement

FIGURE 10
Cracks in Residence Foundation

Birmingham, Alabama -- Mine Subsidence

Most subsidence and sinkhole problems have origins within about a hundred feet of the ground surface. This example is unusual because of the depth of the mine that was creating the problems at the surface. Although there were two mined seams present, it was determined that the lower seam, (Fig. 9) 700 ft. deep was the root of the problem that was causing distress to a large area of a suburban community (Fig. 10). The solution in this case was to fill portions of the mine to create additional support. The grout selected was a relatively low-slump cement-flyash-sand-stone mixture. The mixture was varied according to the requirements of each hole. A central mix plant was used to batch the materials which were then transported to the pump at the holes in readymix trucks.

CONCLUSION

The previous examples show how relatively low cost grouting materials and procedures can be used to correct existing sinkhole problems or reduce the probability that future sinkholes or subsidence will occur. The techniques, when properly applied, are flexible to adjust to underground conditions and to work around surface structures and obstructions. Grouting is an economical and technically sound approach to many of these problems.

Collapse and compaction of sinkholes by dynamic compaction

CHRISTIAN A. GUYOT *Terra Firma, Inc., Murrysville, Pennsylvania, USA*

ABSTRACT

When highways or structures of large dimensions have to be located in areas where sinkholes could develop, the major problem is to find the cavities or the loosened soils and to design a foundation system compatible with the geotechnical parameters of the site. The Dynamic Compaction technique consists of repeated applications of high energy impacts onto the ground surface. This results in the collapse and consequent improvement of the loose ground structure. The technique can also be used to precisely locate cavities in "sinkhole" areas and to provide high bearing capacities and minimum long-term settlements of the proposed foundations.

General Description of the Dynamic Compaction Technique

Dynamic Compaction is a soil improvement technique which uses the repeated application of high energy impacts to engineer the in-situ soil into a part of the foundation system. In its present form, the technique was developed in Europe in the late 1960's. Steel pounders, weighing 10 to 20 tons (occasionally more) are dropped from a height of 10 to 40 meters.

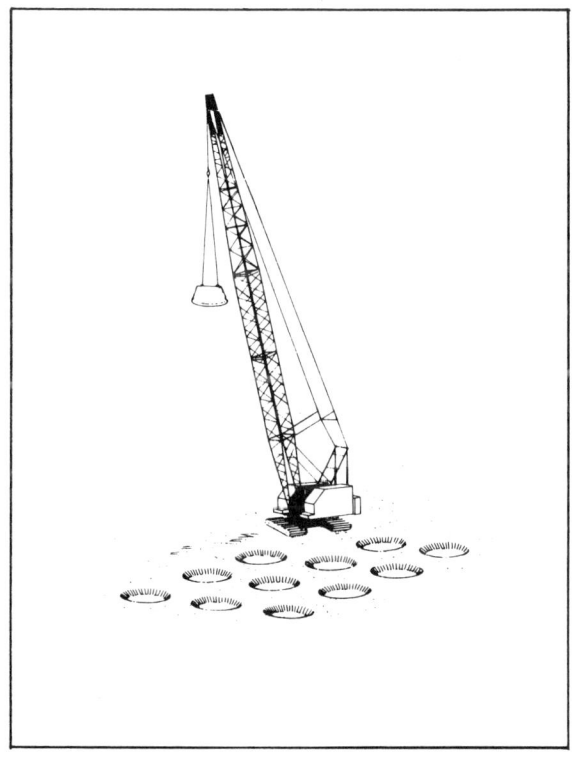

Figure 1: View of a Dynamic Compaction Rig.

In non-saturated material the shock waves, created by the impact, affect the soil skeleton directly and reorganize the soil particles into a denser arrangement.

In saturated soils, the energy is first transmitted to the groundwater. The excess pore water pressure disrupts the soil particles and rearranges them as it dissipates; thus increasing the soil density and strength. Shallow foundations and higher bearing pressures can then be used.

Soil formations as deep as 20 meters can be improved very economically. Deeper treatment can be achieved by dropping exceptionally large weights or using Dynamic Compaction in conjunction with wick drains, vibratory compaction techniques, microexplosives, grouting, etc.

Experience, engineering and comprehensive monitoring are the keys for a successful Dynamic Compaction treatment. Accurate prediction of results and future behavior of the project can be made during the design stage while on-site control, using proven state-of-the-art testing techniques assures that specifications are met.

Effective improvement can be achieved in material ranging from rockfills to silt, sanitary and other landfills and even certain clays.

Dynamic Compaction to Locate Sinkholes

When a large housing, commercial, industrial or highway project crosses areas where sinkholes are known, or suspected, to occur the primary problem is the precise, and timely, location of any voids, cavities or loose zones within the soil mass. The limited number of conventional in-situ borings carried out for the entire site during the geotechnical investigation achieves a very poor evaluation of the number, size and location of the sinkholes present. The use of geophysics more accurately enables us to determine the presence of dissolutions in the deep karstic formations. However, the degree of accuracy of these techniques is not sufficient to insure that all cavities or potential sinkholes have been located and, hence, treated.

A new approach to locating sinkholes directly from the ground surface - without the help of numerous in-situ testings - was introduced several years ago in South Africa and later on in the United States. It involves the use of Dynamic Compaction. The first parameter to be determined is the energy-per-drop (weight x height of drop) in accordance with the depth to be treated. A general rule of thumb indicates that the treated depth is approximately equal to the square root of the energy per drop, i.e. a 15 ton weight falling from 20 meters will influence 8 to 9 meters below ground surface. The treatment is usually carried out in three phases involving three different patterns.

During the first phase, the pounding occurs on a square gride. The distance between impacts can vary from 3 to 10 meters depending on the type of soil to be treated. In the second phase, fresh impacts are placed in between impacts of the first phase. The third - or ironing phase - consists of dropping the weight, in a continuous pattern, over the entire site. Additional phases may be necessary, when variable soil conditions are encountered, to ensure a uniform densification.

Surface measurements, during the compaction process, consist of systematically determining the crater depth, top and bottom diameter, volume, and any noticeable heave or lateral deformation occurring around the crater. The net volume of the crater (crater volume less heave) is a representation of the amount of strain induced at depth at that particular location. When the strain distribution over the entire site is analyzed at the end of each phase, it is possible to determine changes in the soil condition at depth. Any cavities or loose layers within the treated zone appear clearly during the treatment. As energy is eventually distributed over all the site, all anomalies of the soil conditions will be evident. Treatment of these anomalies will generally require in-situ testing followed by an additional compaction program designed on a case by case basis to suit the conditions of the particular location.

Case History at the Bayonet Point Hospital, Hudson, Florida

The most notable topographic feature at the site of the proposed hospital was the occurrence of a number of circular and elongated depressions in the center of the site. Some of these drop-outs were 1.5 to 3 meters deep and 3 to 4 meters in diameter. The test borings encountered a very erratic subsurface profile. The first layer consisted of relatively clean cohesionless loose to dense fine sands on a depth of 0.6 to 10 meters. Standard penetration resistance values recorded in the sands ranged from less than 1 to 50 blows per foot. In 13 out of 16 borings, the sands were underlain by 0.6 to 4 meters of clayey sands or sandy clays. Penetration resistance values in the clayey soils ranged from 1 to 13 blows per foot.

The overburden soils were underlain by the limestone formation. The formation was typically described as a mixture of calcareous clayey and silty sands and sandy clays with limestone fragments and interbedded layers of harder limestone. In addition to being somewhat variable in composition, the formation had undergone considerable post-depositional change due to solution-related activity. The groundwater level was approximately 1 to 2 meters deep. Three soil related problems were to be dealt with:

   a) The variable and often loose condition of the near surface sands.

   b) The potential for consolidation settlements to occur in the clayey strata.

   c) The potential for future ground subsidence due to sinkhole activity.

A Dynamic Compaction treatment was designed to achieve the following specifications:

   a) A minimum average of 75% relative density in the sands and silty sands to a depth of 7.5 meters.

   b) Locate and collapse the cavities in the upper 7.5; provided that the before-treatment standard penetration resistance values within the meter of soil above the cavity does not exceed 25 blows per foot.

   c) Locate and collapse cavities within the top 4.5 meters; with no limiting condition.

The work was performed using a 15-ton pounder falling from 20 meters. Three phases were usually applied on the site, with a distance of 9 meters between them during the first phase.

Fourth and Fifth phases were used in areas exhibiting strains after the completion of the first three. (See Table 1 below).

Grid (feet)	Location	Blows/print	Energy txm/m^2	Average Dynamic Settlement (inches)	Cumulative Settlement
30x30	Primary	8	24	4.77	
30x30	Primary	9	27	3.85	8.62
30x30	Primary	9	27	2.76	11.38
30x30	Intermediate	9	27	3.80	15.18
30x30	Intermediate	10	30	3.65	18.83
Over-lapping	Continuous	1	50	3.5	22.33
Void	At observed sink hole locations	10	--	---	
--	--	Total of 8895 blows	185		27.33

Table 1: Summary of grid spacing, energy and induced settlement

The testing and control consisted mainly of the systematic measurement of the crater dimensions and lateral heave in the vicinity of the impact. Three piezometers were installed at 3.5, 4.5, and 5.5 meters in order to monitor the increase and dissipation of excess pore-water pressure. Pressuremeter tests and standard penetration tests were performed before, during and after the treatment.

After each phase of pounding, a strain contour map (see example in figure 2) was drawn. The representation by contour made it easy to precisely locate areas where high subsidence occurred. In those areas, additional treatment was performed as described in Table 1. During the course of the work, the site subsided approximately 0.6 meters. Locally, up to 0.75 meters of settlement was observed corresponding to an induced strain of 10% of the thickness to be treated.

The statistical analysis of the standard penetration resistance values (see figure 3) was made using the Gibbs and Holtz relative criteria.

Conclusions

The Dynamic Compaction technique has successfully located, collapsed and compacted the top 9 meters of loose silty sands at the Bayonet Point Hospital. A total of 44 borings were carried out before and after treatment. This represents an unusual density of control (approximately one out of every 180 square meters of building). The net allowable bearing capacity was raised to 3 kilograms per square centimeter.

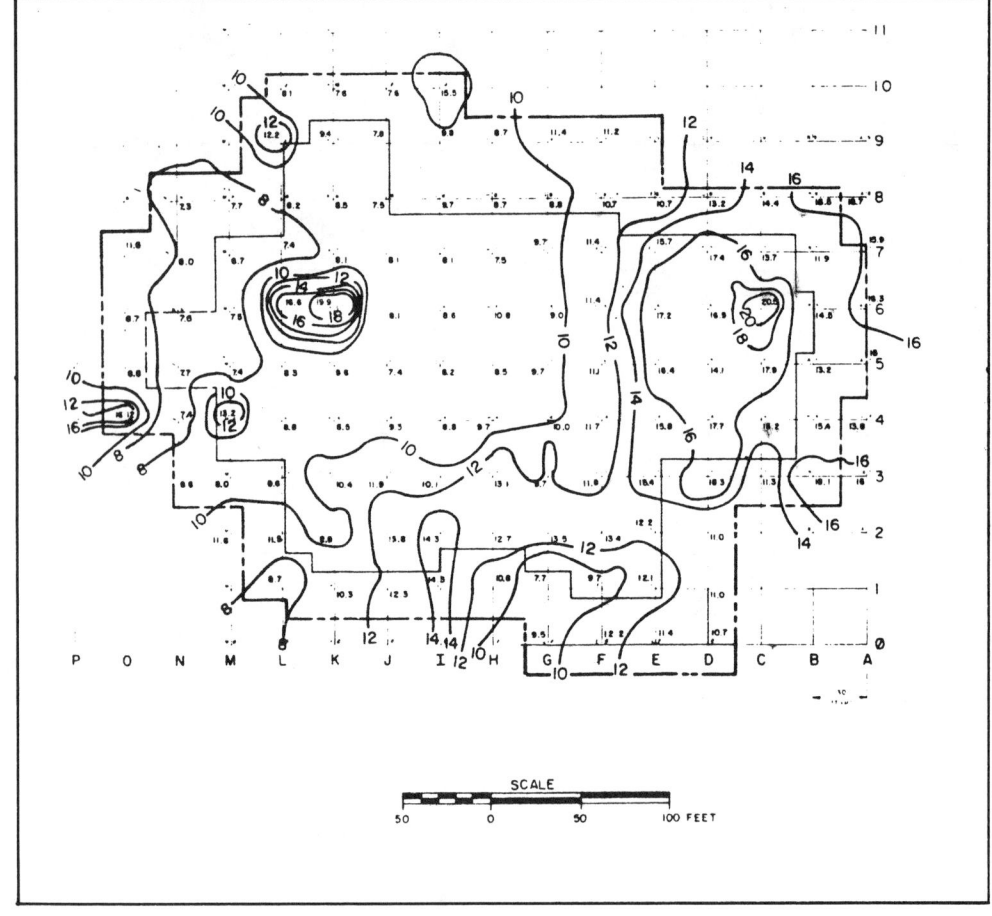

Figure 2: Strain contours to identify cavities and looser soils.

Pressuremeter results show that differential settlement will be limited to a maximum of 0.6 centimeters. After the treatment the site was graded and covered by a thick rubber liner in order to minimize the amount of water seeping through the top sandy layers which could reach the limestone and induce subsequent dissolutions.

References

Terra Firma, Inc., Dynamic Compaction brochure: P O Box 48, Murrysville, PA 15668

Menard Inc., Final Report, Bayonet Point Hospital, February 1980: 10 Duff Road, Pittsburg, PA 15235

LETCO Report of subsurface investigation, proposed hospital for Hospital Corporation of America, Hudson, Fl. Letco Job #T 3877, January 1980.

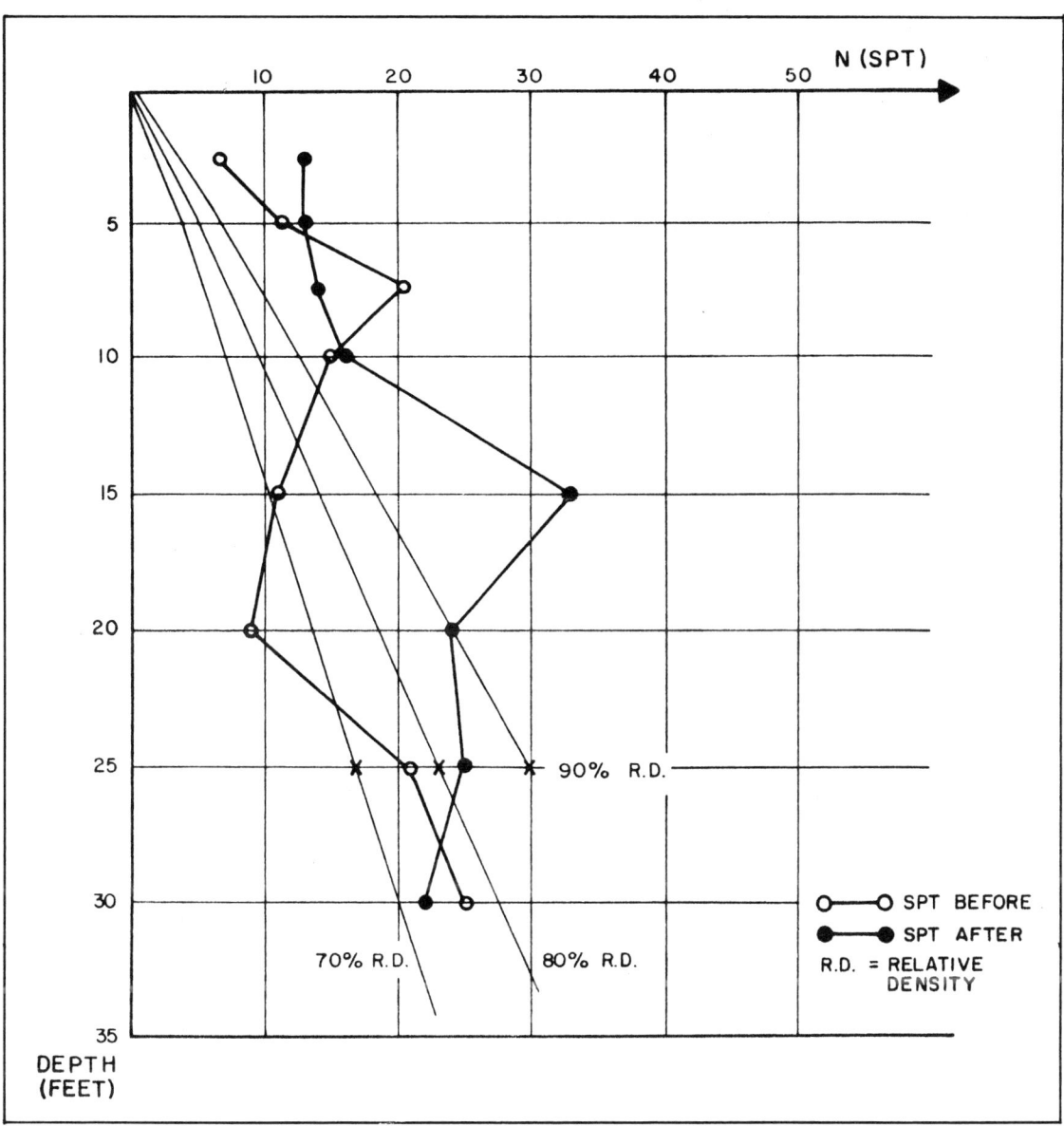

Figure 3: Comparison of standard penetration resistance values before and after dynamic compaction.

Late paper to Part 3: Sinkhole development in reclaimed smectitic spoil

MAURICE B.DUSSEAULT *Geological Engineering, Waterloo, Ontario, Canada*
J.DON SCOTT *University of Alberta, Edmonton, Canada*
STEVE MORAN *Alberta Geological Survey, Edmonton, Canada*

ABSTRACT

Sinkholes are observed to develop on reclaimed strip mine spoil in Alberta for several years after relevelling. The sinkholes are small, no more than a metre in depth and width, and are due to a delayed stoping mechanism in uncompacted smectitic spoil. The mining is carried out by draglines and spoil is cast in windrows. Relevelling consists of removing windrow tops and placement in valleys in a very poorly compacted state. Subsidence is triggered by rewetting of the smectitic over-consolidated sediments, and, where subsidence is large (1-1.5 m), sinkholes develop. The article describes the differential subsidence and presents the mechanism for the sinkhole development. Possible mitigating procedures are discussed.

INTRODUCTION

Large scale surface mining of flat-lying Cretaceous and Tertiary lignitic and sub-bituminous coals is carried out in Alberta and Saskatchewan for generation of electricity. Current reclamation practice involves returning mined lands to an agricultural productivity equal to the premining state. Research into productivity, moisture, geochemistry, erosion, and subsidence of reclaimed sites began in the late 1970's. In saline sodic spoil of the Canadian Plains and northern American states, a concern is differential subsdience in excess of one metre in conjuction with the development of sinkholes depressions. The nature and origins of the sinkholes will be discussed, and corrective measures presented.

GEOLOGY AND MATERIALS

The coal in the Plains region is in strata of Cretaceous to early Tertiary age (100-55 MYBP) that have a gentle regional dip of several metres per kilometer to the southwest (Figure 1). The strata are typical of coal measures in many parts of the world. The coal formed near a sea coast where lagoonal to swampy conditions occurred: flat esturarine plains. The overburden therefore displays a considerable lateral and vertical variability. After coal deposition ceased, sedimentation continued and burial to depths of perhaps 1000 m took place. The depth of burial was greater in the west, as regional subsidence was greatest in the Alberta Syncline. In early Tertiary times, the sedimentary wedge was elevated above sea level, and erosion has dominated to the to the present, with the exception of the deposition of a veneer of glacial deposits.

The coal overburden comprises sand, silt and clay in various proportions and mixtures. Few beds are lithified, although all strata are highly overconsolidated due to gravity load. None of the deposits display any significant tectonic deformation or structure. The majority of the strata are of clayey texture, containing more than 25% clay sizes. Even unlithified sand commonly contains 5-15% clay sizes. The clay mineralogy is dominated by smectite (swelling clays), and the other clay minerals are illite, kaolinite and chlorite, usually present in minor amounts. The origin of the smectite is from devitrification of volcanic ash brought in by winds carrying fine-grained volcanic debris from the southwest during the time of deposition. Before mining, the bulk density of the overburden is in the range of 2.0 to 2.3, as determined from bulk density geophysical logs. Only where a carbonate cement is present is a material of density greater than 2.4 found, and then only rarely.

MINING METHOD

Figure 2 is a schematic diagram of a coal strip mine in a simple stratigraphy. The overburden can be easily mined by dragline without blasting. Strata are spoiled into windrows, levelled and resoiled, all within a year or two. Mining disaggregates the dense smectitic overburden, creating a ramdom lump spoil with a volume about 20% higher than in the undisturbed state. The dragline dumping seems to compact the crests of the windrows adequately, but the flanks of the spoil piles are considerably looser. During levelling, windrow tops are dozered into valleys, completely disaggegating the material and further concentrating loose debris in the valley areas. In Canada, levelling during winter can also result in some entrapment of snow and frozen lumps.

Our observations show that differential subsidence and sinkhole formation is entirely due to the post-reclamation underground geometry of alternating undercompacted windrows (Figure 3). Sinkholes are found exclusively in the valley regions, and are correlated to areas of particularly large settlements.

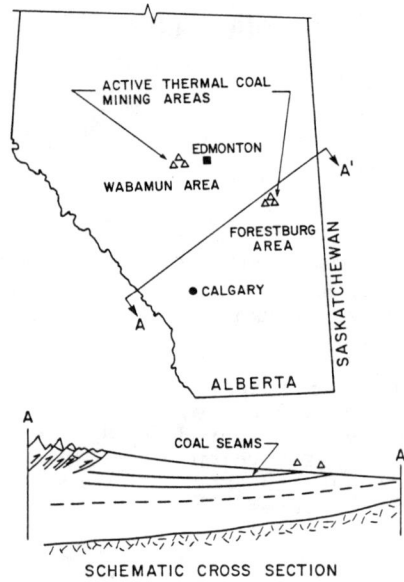

Fig. 1 ACTIVE MINE AREAS IN ALBERTA

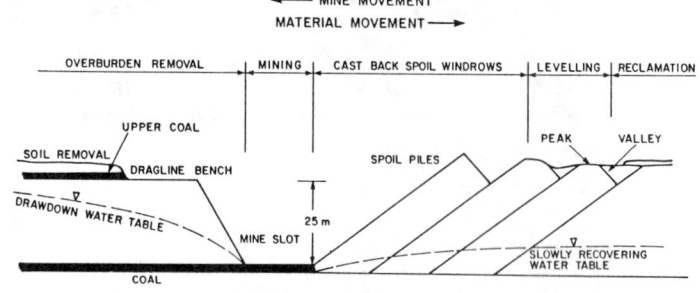

Fig. 2 TYPICAL STRIP MINE OPERATION

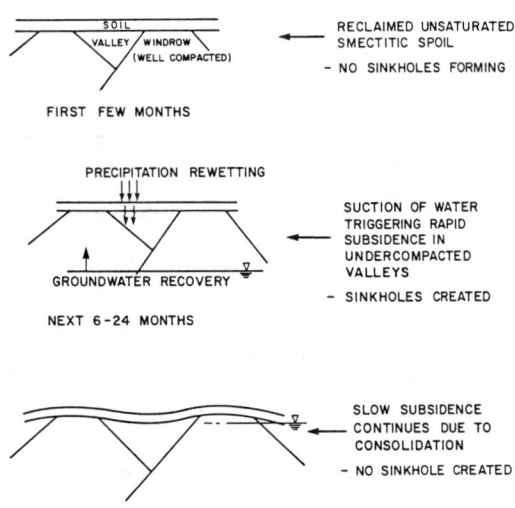

Fig. 3 MECHANISM OF DIFFERENTIAL SUBSIDENCE AND SINKHOLE FORMATION TIME

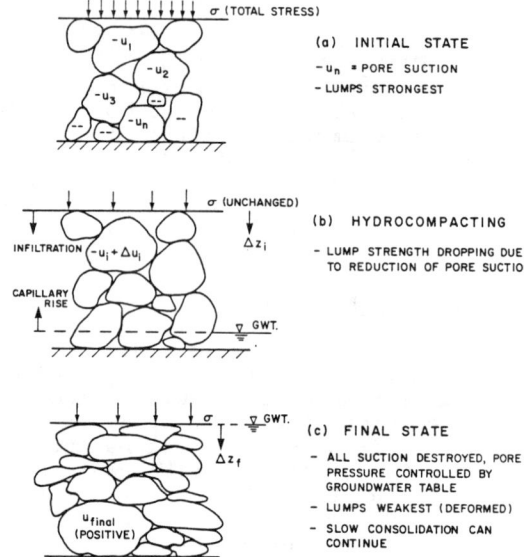

Fig. 4 DESTRUCTION OF MACRO-POROSITY UPON REWETTING

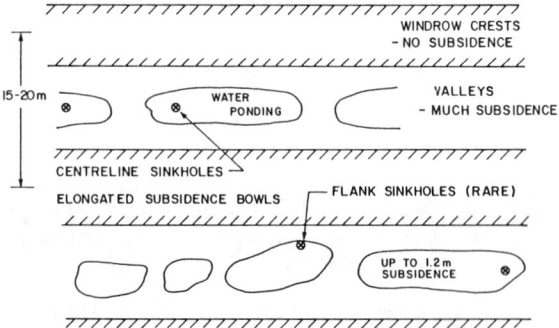

Fig. 5 TYPICAL LOCATION OF SINKHOLES (SCHEMATIC)

SUCTION, SLAKING AND DISPERSION

Differential subsidence is not immediate. Upon resoiling, carried out by scraper units, the ground surface is essentially level. Usually within a half year, differential subsidence becomes apparent, and continues for a long time. Sinkholes have been observed to develop as late as two to four years after levelling, but can be seen earlier in most areas. The subsidence is a water-triggered phenomenon. Water table recovery and rainfall or snow melt runoff make water available to the spoil, resulting in gradual settlements, particularly in the undercompacted valleys. The settlement is the result of destruction of the macroporosity by lump softening when suction is destroyed.

The smectitic materials are overconsolidated and therefore have a matrix suction potential for water solely due to the matrix structure. Also, the salinity of the pore fluids is on the order of 8 to 20 g/litre, therefore a high osmotic suction potential also exists if the available moisture is of a salinity less than the pore fluid. Under suction, the spoil lumps retain some integrity, as shear strength is proportional to the negative pore water pressure. Access to water destroys the suction, leading to slaking and dispersion, loss of lump strength, and consequent loss in macroporosity (Figure 4). Air is slowly expelled and replaced by water. Dispersion of smectite can also take place, although transport of the dispersed smectite seems not to occur because the large voids are destroyed and any dispersed materials simply blocks the throats of the smaller pores.

The time-dependency of settlement is due to several processes: delayed ingress of water, stress arch development, and consolidation if the void ratio of the material is above the equilibrium void ratio at the particular stress state. Because of the smectite, permeabilities of the spoil become very small once water is available. We have observed continuing slow settlements up to 50 mm year on a paved road over 10 year old spoil. Early movements are much more rapid, as much as 500 mm in the first 18 months.

SINKHOLE MORPHOLOGY

The sinkholes are found near the centres of the elongated depressions in the valleys (Figure 5). Typically, the open portion is 0.5-0.6 m in diameter, side slopes are 70°, and the depth to the central debris pile is 0.3-0.6 m (Figure 6). The cavity slopes away from the central portion of the sinkholes and is infilled about the rim, although sinkholes usually develop in the deepest portion of the subsidence trough. We have observed caving zones at surface with downthrown blocks which have not matured into sinkholes, although they may yet do so if not infilled by cultivation. Also, several sinkholes have been observed completely underwater in the ponds which often form in the basins. This attests to the extremely low permeability as groundwater table surveys show that the potentiometric surface in the lower spoil is five to six metres below surface.

The sinkholes are not associated with the tension cracks that form at the curved rims of the depressions (Figure 7) although both are related to the subsidence. Only a few sinkholes develop in the depressions, several dozen having been observed to date. However, in regions of thick spoil valleys, it is not unreasonable to expect that more will appear in the future, as the whole process is delayed in time.

STOPING MECHANISM

The sinkholes are not due to subsurface dispersion and water transport of fines away from a site. Neither are they due to the melting of large amounts of entrapped ice or snow as they occur in summer levelled areas with equal frequency. Rather, they are the result of a stress arching phenomenon combined with a stoping process by which settlement voids are transmitted upwards.

The mechanism is sketched in Figure 8. As the groundwater table rises and compaction is triggered, the valleys deform, and zones of tension and compression are built up. Soil replacement is usually done by scrapers and dozers operating directly on the spoil, so a surface zone that is relatively well compacted is created. This zone is stiffer and more brittle than the underlying material and enters into a compressive state as the stress arch develops. The mechanism is verified by the location of tensions cracks and by the angles of the sinkhole walls. When sufficient arching has occurred, the top layer acts as a beam, and gravitational collapse will occur when the centre portion of the beam is undermined by the stoping mechanism.

Sinkholes will only develop in those valleys which exhibit large subsidence, at least 600 mm it seems, and only if the stoping tends to concentrate the voids centrally. Also, we have not observed many sinkholes on land which has been levelled but not resoiled, suggesting that a stiffer surface layer helps the process develop fully.

ELIMINATING THE PROBLEM

Filling the sinkholes is easy, but the general problem of differential subsidence is the root cause and has deleterious effect on agriculture because of water ponding and consequent potential for soil salinization. It is desireable to eliminate the cause rather than treat only the worst symptoms.

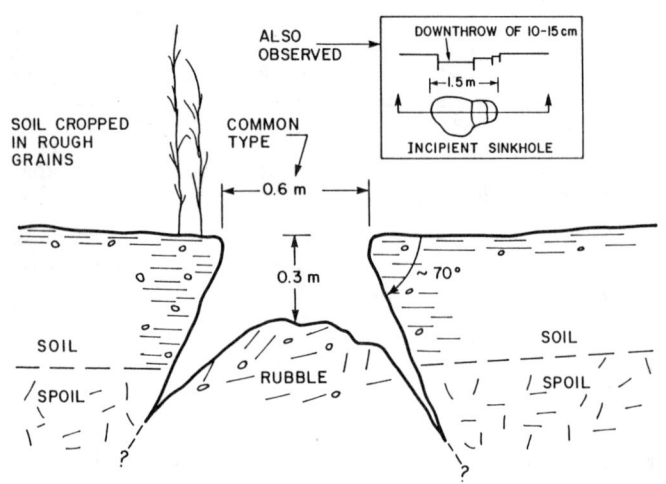

Fig. 6 SINKHOLE FEATURES

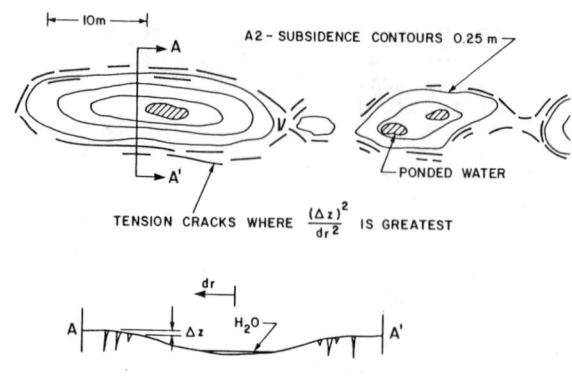

Fig. 7 AREAL LOCATION OF TENSION CRACKS

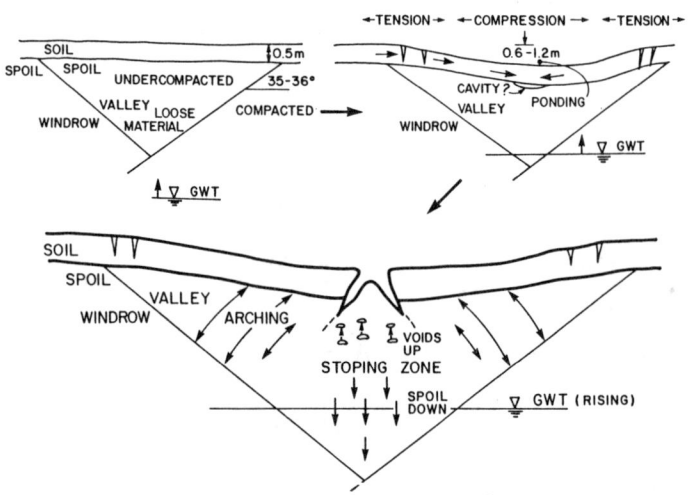

Fig. 8 STOPING MECHANISM OF SINKHOLE FORMATION

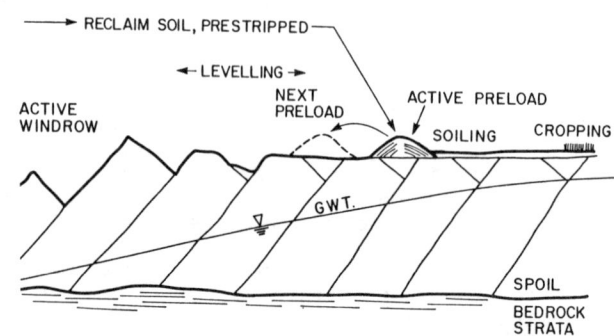

Fig. 9 TRAVELLING COMPACTION PRELOAD

In Alberta and in North Dakota, where similar subsidence has been reported, mining companies discontinued winter regrading operations some years ago to avoid burial of snow, ice and frozen spoil, which have been identified as factors in the subsidence process.

At least one company has addressed the problem by extending its reclamation sequence to allow one year between initial regrading and topsoil placement. This allowed the initial formation of subsidence depressions, which were regraded before topsoil placement. Although this practice has undoubtedly eliminated part of the settlement problems on the final reclaimed surface, it has not cured the problem. This approach has the further drawback that the settlement process is controlled not by time alone but by both time and spoil resaturation. Since spoil resaturation is a complex process that is controlled by a host of site specific factors, prediction of the rate of resaturation is difficult at the present, making it impossible to rationally design an appropriate waiting period before applying topsoil in a given mine area. Indications are that in many areas of thick, highly smectitic clayey overburden, several years are required for resaturation. Because of problems of erosion and dispersion of smectitic spoil, such a lengthy period of exposure is environmentally unacceptable. Theoretically it might be possible to accelerate and manage resaturation of the spoil and as a consequence, surface subsidence, by the application of water to the spoil surface. The absence of a water supply of adequate quantity and quality for such an undertaking at most sites in the water- deficient prairies, combined with costs that are likely prohibitive make it unlikely that such irrigation could have general applicability.

At least one company has modified its procedures for initial regrading of the spoil by using scrapers rather than dozers to transfer material from windrows to troughs. The greater compaction possible using this technique could well alleviate the settlement problem. Monitoring is continuing to evaluate the effectiveness of this approach. Groenewold and Winczewski (1977) reported that subsidence at one mine in North Dakota was less in areas where scrapers were used to grade the spoil surface than in adjacent areas where dozers were used.

Another approach to compacting the interwindrow troughs is suggested by the results of laboratory compaction tests on spoil material. We found that in dry spoil, in which the macropores are air filled, compaction and reduction of pore volume occurred very rapidly under gravity loading. This effect could be used to minimize final subsidence by stockpiles that were 5 m to 10 m high and serve as a travelling preload to compact these trough areas (Figure 9). It seems possible that total material handling costs might be reduced by such a system.

ACKNOWLEDGEMENTS

This work was conducted as part of the Plains Hydrology and Reclamation Project of the Alberta Research Council. The research has been supported by the Alberta Heritage Savings Trust Fund and administrated by the Reclamation Research Technical Advisory Committee (RRTAC) of the Land Conversation and Reclamation Council of the Alberta Government. Also, the first two authors are deeply appreciative for the support of academic research by the Natural Sciences and Engineering Research Council of Canada.

REFERENCES

Dusseault, M.B., Scott, J.D., Soderberg, H. and Moran, S., 1984. Swelling Clays and Post-Reclamation Mine Subsidence in Alberta. 5th International Conference on Expansive Soils, Adelaide, S.A., pp. 131-136.

Dusseault, M.B., Scott, J.D., Zinter, G. and Moran, S., 1984. Simulation of Spoil Pile Subsidence. Fourth Australia-New Zealand Conference on Geomechanics, Perth, WA., Volume 1, pp. 94-100.

Groenewold, G.H., and Winczewski, L.M., 1977. Probable Causes of Surface Instability in Contoured Strip-Mine Spoils-Western North Dakota. North Dakota Academy of Science, Proceedings, V. 31, p. 160-167.